Manfred T. Reetz, Zhoutong Sun, and Ge Qu

Enzyme Engineering

Manfred T. Reetz, Zhoutong Sun, and Ge Qu

Enzyme Engineering

Selective Catalysts for Applications in Biotechnology, Organic Chemistry, and Life Science

WILEY-VCH

Authors

Dr. Manfred T. Reetz
Max-Planck-Institut für Kohlenforschung
Mülheim an der Ruhr
Germany

Prof. Zhoutong Sun
Tianjin Institute of Industrial
Biotechnology
Chinese Academy of Sciences
China

Prof. Ge Qu
Tianjin Institute of Industrial
Biotechnology
Chinese Academy of Sciences
China

Cover Images: © Zhoutong Sun, Manfred T. Reetz, and Ge Qu; © theasis/Getty Images

Library of Congress Card No.: applied for

British Library Cataloguing-in-Publication Data
A catalogue record for this book is available from the British Library.

Bibliographic information published by the Deutsche Nationalbibliothek
The Deutsche Nationalbibliothek lists this publication in the Deutsche Nationalbibliografie; detailed bibliographic data are available on the Internet at <http://dnb.d-nb.de>.

Print ISBN: 978-3-527-35033-9
ePDF ISBN: 978-3-527-83687-1
ePub ISBN: 978-3-5278-3688-8
oBook ISBN: 978-3-527-83689-5

Typesetting Straive, Chennai, India
Printing and Binding CPI Group (UK) Ltd, Croydon, CR0 4YY

C118969_030123

Contents

Preface

The term *directed evolution* is relevant in two very different research areas: (i) the genetic manipulation of functional RNAs as pioneered by Sol Spiegelmann in 1967 in a Darwinian manner and continued by a number of present-day experts for potential applications, inter alia, as aptamers; and (ii) the genetic manipulation of DNA with the aim to engineer the catalytic profiles of enzymes as catalysts in organic and pharmaceutical chemistry and in biotechnology. Both approaches concern the age-old dream of imitating evolution in Nature as it has developed for billions of years, but of course not wanting to wait for such long periods! This monograph is a comprehensive treatise on the second research area.

A number of books have appeared on the directed evolution of enzymes (also called laboratory evolution), generally published by editors who asked scientists of their choice to write specific chapters on certain special topics. A different publishing approach is to write a monograph by a single author, specifically in the field of directed evolution. This would enable a more critical and comparative view of the different mutagenesis approaches and strategies practiced in this exciting research field. It was first presented in the Wiley-VCH book by Manfred T. Reetz, titled *Directed Evolution of Selective Enzymes: Catalysts for Organic Chemistry* (2016). The limitation of this seminal monograph is the fact that the alternative to directed evolution, namely rational enzyme design as a protein engineering technique, was not at all highlighted. During the last six years, the methodology development of directed evolution has continued to produce notable advances, as have the number of applications in synthetic organic chemistry, natural products synthesis, and pharmaceutical chemistry. The time for a second edition of the Reetz monograph seemed to be ripe. However, the editors of Wiley-VCH made a better and more challenging suggestion, namely to produce a unique monograph which covers both approaches to protein engineering with the evolution of selective catalysts. For this transdisciplinary task, two additional authors were willing to participate, a microbiologist with experience in applications of enzymes in chemistry and biotechnology and a bioinformatician with expertise in the computational design of proteins. Thus, readers may find the languages of chemistry and biology meet to talk about protein engineering in this book.

The present monograph with the title, *Enzyme Engineering: Selective Catalysts for Applications in Biotechnology, Organic Chemistry, and Life Science*, brings together three experts who have written a comprehensive treatise on both approaches to protein engineering, namely directed evolution and rational enzyme design. We believe that it is ideally suited for master's and doctoral students in university courses, but also for postdocs and other advanced scientists at different research institutions and those already working in industry. The first chapter introduces the basic concepts, goals, and mutagenesis methods as well as the areas of practical applications. Thereafter, several chapters follow in which the complexity of the material is gradually increased, with further details and information on effective mutagenesis strategies being offered. Importantly, the differences, limitations, and advantages of the different approaches are critically analyzed.

The monograph also contains chapters representing special topics, which are nevertheless important. These include information on how to evolve protein robustness under operating conditions as an alternative to traditional enzyme immobilization, developing promiscuous enzymes, as for example artificial metalloenzymes, and learning lessons from directed evolution, which reveals hitherto unnoticed mechanistic intricacies, thereby enabling further effective methodology developments, which include machine learning. The final chapter constitutes a concise resume, offers ideas for future research and challenging applications in organic and pharmaceutical chemistry and biotechnology, and gives tips on patent problems. The importance of ethical issues is likewise stressed.

The reader will discover in several chapters how semi-rational directed evolution and rational enzyme design have merged, specifically by focusing mutations at sites lining the binding pocket. The graphic on the front cover is a succinct reminder of this strategy. It features the key as a symbol of the substrate entering the binding pocket. The optimal shape at the active site for ensuring high stereo- and/or regioselectivity can be achieved by protein engineering.

March 2023

Manfred T. Reetz, Zhoutong Sun, and Ge Qu
Max-Planck-Institut für Kohlenforschung
Mülheim an der Ruhr
Germany

Tianjin Institute of Industrial Biotechnology
Chinese Academy of Sciences
China

About the Authors

Manfred T. Reetz (born 1943) obtained his doctoral degree in synthetic organic chemistry in 1969 under the direction of Ulrich Schöllkopf at Göttingen University in Germany followed by a postdoctoral stay with Reinhard W. Hoffmann at Marburg University. He then performed independent research in various institutions in Germany and the USA, including directorship of the Max-Planck-Institut für Kohlenforschung in Mülheim for two decades, where he pioneered the concept of directed evolution of stereo- and regio-selective enzymes as catalysts in organic and pharmaceutical chemistry. After his first retirement in 2011, he continued research at Marburg University and at the Tianjin Institute of Industrial Biotechnology, Chinese Academy of Sciences, and returned to Mülheim in 2019.

Zhoutong Sun obtained his doctoral degree in microbiology with Prof. Sheng Yang at the Shanghai Institutes for Biological Sciences, Chinese Academy of Sciences, in 2012. He then moved to Nanyang Technological University in Singapore as a research fellow in metabolic engineering. One year later, he joined the Reetz group as a postdoc at MPI für Kohlenforschung and Marburg University in directed evolution and biocatalysis. In 2016, he became a full professor at the Tianjin Institute of Industrial Biotechnology, Chinese Academy of Sciences. His research interests are in enzyme engineering, metabolic engineering, and synthetic biology.

Ge Qu received his PhD in bioinformatics from Adam Mickiewicz University in Poland in 2016. Currently, he is working as an associate professor at the Tianjin Institute of Industrial Biotechnology, Chinese Academy of Sciences, in the group of Professor Zhoutong Sun. His major research interests are the discovery and computational engineering of enzymes that have potential as industrial biocatalysts.

1 Introduction to Directed Evolution and Rational Design as Protein Engineering Techniques

1.1 Methods and Aims of Directed Enzyme Evolution

For decades organic chemists have viewed enzymes as possible catalysts in their toolbox of synthetic methods with a great deal of skepticism, traditional textbooks hardly mentioning them. Indeed, even modern monographs on organic chemistry provide little information on the importance of these biocatalysts [1]. On the other hand, textbooks on enzymes in organic chemistry and biotechnology have interested mainly specialists [2], not organic chemists. Apart from psychological reasons, the actual limitations of enzymes were known to everyone, which are as follows:

- Insufficient robustness under operating conditions.
- Insufficient activity.
- Narrow substrate scope.
- Insufficient or wrong stereoselectivity.
- Insufficient or wrong regioselectivity.
- Sometimes product inhibition.

In addition, organic chemists were not trained to handle enzymes and therefore preferred to be content with the use and development of their own synthetic homogeneous and heterogeneous catalysts, knowing that enzymes cannot catalyze many or most of the important reaction types that dominate modern synthesis. *Today all of these problems can be addressed and generally solved by applying directed evolution*, as summarized in a 2016 monograph [3] authored by Manfred T. Reetz, and in more recent reviews [4]. It mimics Darwinian evolution as it occurs in nature, but it does not constitute real natural evolution. The process consists of several steps (Scheme 1.1), beginning with mutagenesis of the gene encoding the enzyme of interest. The library of mutated genes is then inserted into a bacterial or yeast host, such as *Escherichia coli* or *Pichia pastoris*, respectively, which is plated out on agar plates. After a growth period, single colonies appear, each originating from a single cell, which now begins to express the respective protein variants. This ensures the required linkage between phenotype and genotype. Multiple copies of transformants as well as wild type (WT)

Enzyme Engineering: Selective Catalysts for Applications in Biotechnology, Organic Chemistry, and Life Science, First Edition. Manfred T. Reetz, Zhoutong Sun, and Ge Qu.

appear, which unfortunately decreases the quality of libraries and increases the screening effort. Colony harvesting must be performed carefully because cross contamination leads to the formation of inseparable mixtures of mutants with concomitant misinterpretations. The colonies are picked by a robotic colony picker (or manually using toothpicks) and placed individually in the wells of 96- or 384-format microtiter plates, which contain nutrient broth. Portions of each well content are then placed in the respective wells of another microtiter plate where the screening (Chapter 2) for a given catalytic property ensues. In some (fortunate) cases, a sufficiently improved variant (hit) is identified in such an initial library. If this does not happen, which proves to be the case most often, then the gene of the best variant is extracted and used as a template in the next cycle of mutagenesis/expression/screening (Scheme 1.1). This exerts "evolutionary pressure," the underlying characteristic of directed evolution.

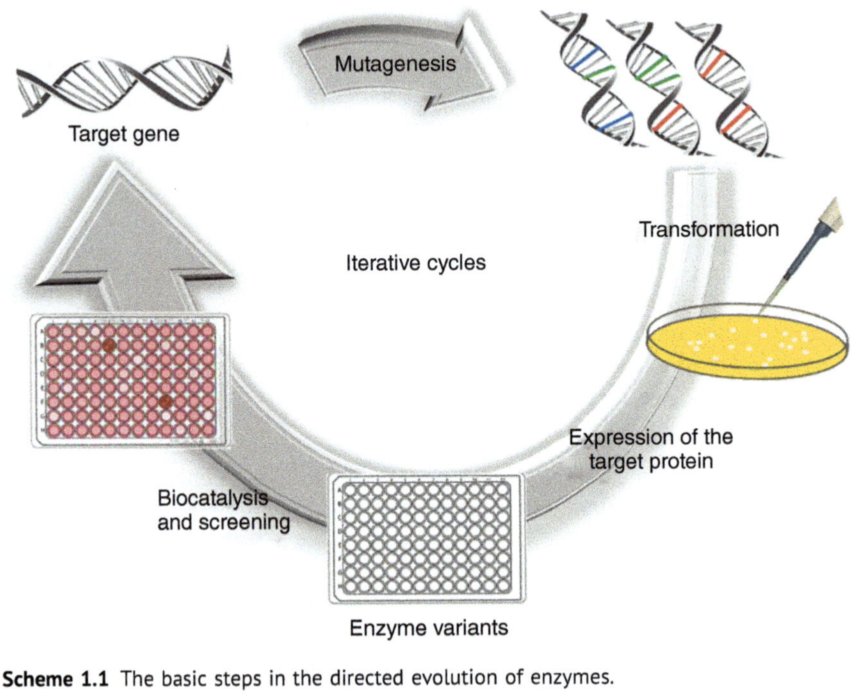

Scheme 1.1 The basic steps in the directed evolution of enzymes.

If a library in a recursive mode fails to harbor an improved mutant/variant, the Darwinian process ends abruptly at a local minimum on the fitness landscape. Fortunately, researchers have developed ways to escape from such local minima ("dead ends") (Chapter 4).

Directed evolution is thus an alternative to so-called "rational design" in which the researcher utilizes structural, mechanistic, and sequence information, possibly flanked by computational aids, to perform site-directed mutagenesis at a given position in a protein (Section 1.3). Needed for this alternative is the molecular biological technique of site-specific mutagenesis with exchange of an amino acid at a specific position in a protein by one of the other 19 canonical amino acids, as established by Michael Smith in the late 1970s [5a], which led to the Nobel Prize [5b]. The method is based on designed synthetic oligonucleotides and has been used extensively by Fersht [6] and numerous other groups in the study of enzyme mechanisms. Application of the Smith technique in rational enzyme design has been shown to be successful mainly when aiming to increase protein robustness (Section 1.3 and Chapter 6). *However, when aiming for enhanced or reversed enantioselectivity, diastereoselectivity, and/or regioselectivity, rational design is much more difficult, in which case directed evolution is generally the preferred strategy* [3, 4].

Directed evolution of enzymes is not as straightforward as it may appear to be. The challenge in putting Scheme 1.1 into practice has to do with the vastness of protein sequence space. High structural diversity is easily achieved in random mutagenesis, but the experimenter is quickly confronted by the so-called "numbers problem," which in turn relates to the screening effort (bottleneck). When mutagenizing a given protein, the theoretical number of variants N is described by Eq. (1.1), which is based on the use of all 20 canonical amino acids as building blocks [3, 4]:

$$N = 19^M X!/[(X - M)!M!] \tag{1.1}$$

where M denotes the total number of amino acid substitutions per enzyme molecule and X is the total number of residues (size of protein in terms of amino acids). For example, when considering an enzyme composed of 300 amino acids, 5700 different mutants are possible if only one amino acid is exchanged randomly, 16 million if two substitutions occur simultaneously, and about 30 billion if three amino acids are substituted simultaneously.

Calculations of this kind pinpoint a dilemma that accompanies directed evolution to this day, namely how to probe the astronomically large protein sequence space efficiently. One strategy is to limit diversity to a point at which screening can be handled within a reasonable time, but excessive reduction of diversity reduction should be avoided because then the frequency of hits in a library begins to diminish dramatically. Finding the optimal compromise constitutes the primary issue of this monograph. A very different strategy is to develop selection systems rather than experimental platforms that require screening (Chapter 2). In a selection system, the host organism thrives and survives because it expresses a variant having the catalytic characteristics that the researcher wants to evolve. A third approach is based on the use of various types of display systems, which are sometimes called "selection systems," although they are more related to screening. These issues are delineated in Chapter 2, which serves as a guide for choosing the appropriate system. Since it is extremely difficult to develop genuine selection systems or display platforms for directed evolution of stereo- and regioselective enzymes, researchers had to devise medium- and high-throughput screening systems (Chapter 2).

1.2
History of Directed Enzyme Evolution

Readers of this monograph may not be interested in historical aspects, but certainly doctoral students and postdocs who want to learn directed evolution are advised to read this section carefully. Much can be learned by seeing in some detail how a given field developed, some contributions being interesting but perhaps lacking essential conclusions for further work, others actually opening a new research field by posting seminal publications. At the end of this section, the most important developments are summarized in a "timeline" scheme.

Scientists have strived for a long time to "reproduce" or mimic natural evolution in the laboratory. In 1965–1967 Spiegelman et al., performed a "Darwinian experiment with a self-duplicating nucleic acid molecule" (RNA) outside a living cell [7]. It was believed that this mimics an early precellular evolutionary event. Later investigations showed that Spiegelman's RNA molecules were not truly self-duplicating, but his contributions marked the beginning of a productive new research area on RNA evolution, fueled seminally by Szostak and Joyce [8]. Directed evolution at RNA level is a very different field of research with totally different goals, focusing on selection of RNA aptamers, selection of catalytic RNA molecules, or evolution of RNA polymerase ribozyme and ribozymes by continuous serial transfer. Directed evolution in this area has been reviewed [8b]. The term "directed evolution" in the area of protein engineering was used as early as 1972 by Francis and Hansche, describing an *in vivo* system involving an acid phosphatase in *Saccharomyces cerevisiae* [9]. In a population of 10^9 cells, spontaneous mutations in a defined environment were continuously screened over 1000 generations for their influence on efficiency and activity of the enzyme at pH 6. A single mutational event (M1) induced a 30% increase in the efficiency of orthophosphate metabolism. The second mutational event (M2 in the region of the structural gene) led to an adaptive shift in the pH optimum and in the enhancement of phosphatase activity by 60%. Finally, the third event (M3) induced cell clumping with no effect on orthophosphate metabolism [9].

In the 1970s, further contributions likewise describing *in vivo* directed evolution processes appeared sporadically. The contribution of Hall utilizing the classical microbiological technique of genetic complementation constitutes a prominent example [10]. In one of the earliest directed evolution projects in the Hall group, new functions for the *ebgA* (*ebg* = evolved ß-galactosidase) were explored (Scheme 1.2) [10a]. Growth of different carbohydrates as the energy source was the underlying evolutionary principle. Wildtype (WT) *ebgAo* is an enzyme showing very little or no activity toward certain carbohydrates such as the natural sugar lactose. It was shown, inter alia, that for an *E. coli* strain with *lac2* deletion to obtain the ability to utilize lactobionate as the carbon source, a series of mutations must be introduced in a particular order in the *ebg* genes. It was also found experimentally when growing cells on different carbon sources, that in some cases old enzyme functions remain unaffected or are actually improved.

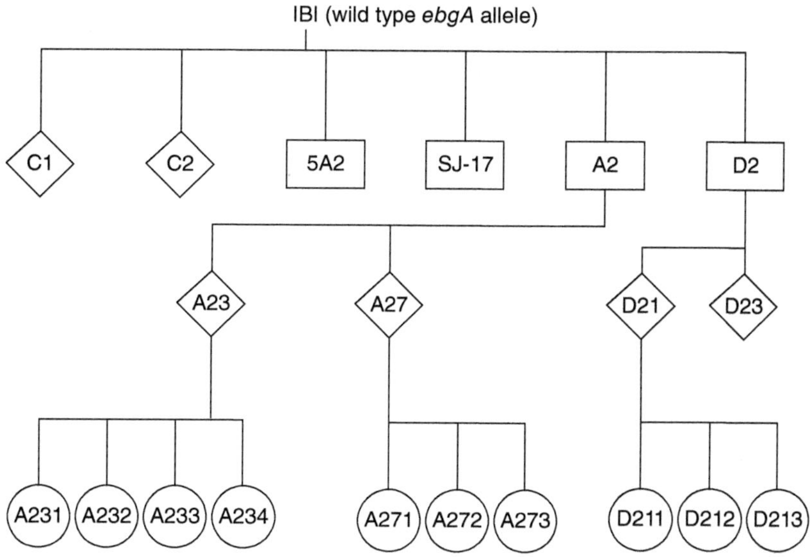

Scheme 1.2 Pedigree of *ebgA* alleles in evolved strains [10a]. Strain 1B1 carries the wild-type allele, *ebgAO*. Strains on line one have a single mutation in the *ebgA* gene; those in line two have two mutations in *ebgA;* those in line three have three mutations in *ebgA*. All strains are *ebgR*-. Strains enclosed in rectangles were selected for growth on lactose; those enclosed in diamonds were selected for growth on lactulose; those in circles were selected for growth on lactobionate. This pedigree shows only the descent of the *ebgA* gene; i.e. strains SJ-17, A2, 5A2, and D2 were not derived directly from IBI, but their *ebgA* alleles were derived directly from the *ebgA* allele carried in IBI.

Two decades later the technique was extended by Kim et al. [11a]. It may have inspired other groups to study and develop new evolution experiments, e.g. by Lenski et al., who investigated parallel changes in gene expression after 20 000 generations of evolution in bacteria [11b], and more recently by Liu et al., who implemented a novel technique for continuous evolution [11c], including a phage-assisted embodiment [11d].

Although originally not specifically related to directed evolution, developments such as the Kunkel method of mutational specificity based on depurination [12a] deserves mention because it was used two decades later in mutant library design based on error-prone rolling circle amplification (epRCA) [12b]. These and many other early developments inspired scientists to speculate about potential applications of directed evolution in biotechnology. As can be seen, these and other developments were suggesting that directed evolution could be generalized with the emergence of a general protein engineering method. Indeed, in 1984 Eigen and Gardiner summarized these intriguing perspectives by emphasizing the necessity of self-replication in molecular *in vitro* evolution [13]. At the time the best self-replication system for the laboratory utilized the replication of single-stranded RNA by the replication enzyme of the coliphage Qf3. The logic of laboratory Darwinian evolution involving recursive cycles of gene mutagenesis, amplification, and selection was formulated. However, the generation of bacterial colonies on agar plates for ensuring the required genotype–phenotype relation, as employed later by essentially all directed evolution researchers, was not considered. It should be remembered that in the early 1980s the polymerase chain reaction (PCR) for high-fidelity DNA amplification had not yet been developed. Following its announcement in the 1980s by Mullis [14], which won him the Nobel Prize, completely new perspectives emerged in many fields, including directed evolution.

Parallel to these developments, researchers began to experiment with different types of mutagenesis methods to generate mutant libraries which were subsequently screened or selected for an enzyme property, generally protein thermostability. Sometimes such methods were introduced without any real applications at the time of publication. Only a few early representative developments are highlighted here. In 1985 Matsumura and Aiba subjected kanamycin nucleotidyltransferase (cloned into a single-stranded bacteriophage M13) to hydroxylamine-induced chemical mutagenesis [15a]. It had been known for a long time that certain chemicals can induce mutations in genes. Following recloning of the mutagenized gene of the enzyme into the vector plasmid pTB922, the recombinant plasmid was employed to transform *Bacillus stearothermophilus* so that more stable variants could be identified by screening. About 12 out of 8000 transformants were suspected to harbor thermostabilized variants, the best one out of the 12 then being characterized as having a single-point mutation and a stabilization of 6 °C. Unfortunately, a second cycle for further improvement was not reported. Several other early studies of T4 lysozyme using chemical mutagenesis appeared, summarized by Matthews in a 2010 review [15b]. Today it is accepted that the formation of improved enzymes in an initial mutant library does not (yet) constitute an evolutionary process and that at least one additional cycle of mutagenesis/expression/screening as shown in Scheme 1.1 is required before the term directed evolution applies.

In light of this, the report by Hageman and coworkers in 1986, in which for the first time two mutagenesis/screening cycles were described, is indeed a hallmark event. In their efforts to enhance the thermostability of kanamycin nucleotidyltransferase by a true evolutionary process, they employed a mutator strain [16]. Mutator strains such as *E. coli* XL1-red induce randomly mutagenized plasmid libraries because they contain defects in one or more of their DNA repair genes [17]. Basically, Hageman's seminal study consisted of cloning the gene that encodes the enzyme from a mesophilic organism, introducing the gene into an appropriate thermophile, and selecting for activity at the higher growth temperatures of the host organism (in this case *B. stearothermophilus*). The host organism is resistant to the antibiotic at 47 °C, but not at temperatures above 55 °C. Upon passing a shuttle plasmid through the *E. coli mutD5* mutator strain and introduction into *B. stearothermophilus*, a point mutation that led to resistance to kanamycin at 63 °C was identified, namely Asp80Tyr. Using this as a template, the second round was performed under higher selection pressure at

70 °C, leading to the accumulation of mutation Thr130Lys, the respective double mutant Asp80Tyr/Thr130Lys showing even higher thermostability (Scheme 1.3) [16]. *The Darwinian character of this approach to thermostabilization of proteins is self-evident, and opened the door to a new research area.* Importantly, in a follow-up study, it was demonstrated that the best mutant also shows higher resistance to hostile solvents such as aqueous dimethylformamide (DMF) solutions and even urea, relative to WT [18].

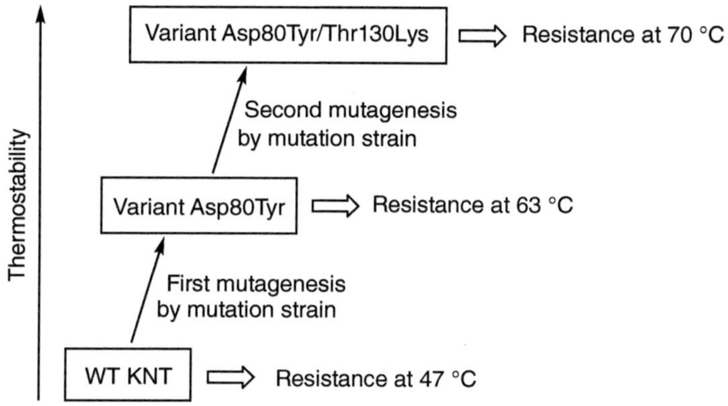

Scheme 1.3 First example of directed evolution of thermostability of an enzyme. Kanamycin nucleotidyltransferase (KNT) served as the enzyme and a mutator strain as the random mutagenesis technique in an iterative manner. Source: Adapted from Liao et al. [16].

As noted above, the original site-specific mutagenesis of Smith allows the specific exchange of any amino acid in a protein by any one of the other 19 canonical amino acids [5], but if all 19 mutants are needed, then 19 separate experiments are required. This could be expensive and time-consuming. The generation of *random* mutations all at once in a given experiment at a single residue or at a defined multiresidue randomization site was not developed until later. Early on (during the period 1983–1988), several variations of cassette mutagenesis based on the use of "doped" synthetic oligodeoxynucleotides, which allows the combinatorial introduction of all 19 canonical amino acids at a given position, were developed [19]. These and similar studies were performed for different reasons, not all having to do with enzyme catalysis. The early study by Wells et al., based on saturation mutagenesis is highlighted here [20]. Saturation mutagenesis means focused randomization at a chosen site in the protein (Chapter 3). The Wells study constitutes a clever combination of rational design and directed evolution for the purpose of increasing the robustness of the serine protease subtilisin so that it shows enhanced resistance to chemical oxidation by H_2O_2 relative to WT [20]. At the time it was known that residue Met222 constitutes a site at which undesired oxidation occurs. Therefore, saturation mutagenesis was performed at this position, which led to several improved variants showing resistance to 1 M H_2O_2 as measured by the reaction of *N*-succinyl-L-Ala-L-Ala-L-Pro-L-Phe-*p*-nitroanilide, including mutants Met222Ser, Met222Ala, and Met222Leu [20].

Cassette mutagenesis as a form of saturation mutagenesis was developed early on. As pointed out by Ner, Goodin, and Smith in 1988, a disadvantage of cassette mutagenesis is the fact that the synthetic oligodeoxynucleotides in form of a cassette have to be introduced between two restriction sites, one on either side of the to be randomized sequence [21a]. Since the restriction sites had to be generated by standard oligodeoxynucleotide mutagenesis, additional steps were necessary prior to the actual randomization procedure. Therefore, an improved version was developed using a combination of the known primer extension procedure [21a] and Kunkel's method of strand selection [12a]. The technique uses a mixed pool of oligodeoxynucleotides prepared by contaminating the monomeric nucleotides with low levels of the other three nucleotides so that the full-length oligonucleotide contains on average one to two changes/molecule [21b]. It was employed in priming *in vitro* synthesis of the complementary strand of cloned DNA fragments in M13 or pEMBL vectors, the latter having been passed through the *E. coli* host. The method allows random point mutations as well as codon replacements. Scheme 1.4 illustrates the case of the *MAT*a1 gene from *Saccharomyces cerevisae* [21b].

Despite these advancements, further improvements with respect to generality, efficiency, and ease of performance were necessary. Indeed, these began to appear in the late 1980s. They included the generation of mutant libraries using spiked oligodeoxyribonucleotide primers according to Hermes et al. [22]. A particular important study was published by Pease and coworkers at the Mayo Clinic, which is based on the use of overlap extension polymerase chain reaction (OE-PCR) for site-specific mutagenesis and saturation mutagenesis [23a], see Chapter 3 for details. It has greatly influenced directed evolution [3, 4]. OE-PCR can also be used in insertion/deletion mutagenesis [24].

In the 1980s, further interesting contributions appeared, including a publication by Loeb et al. [25]. Accordingly, ß-lactamase mutants that render *E. coli* resistant to the antibiotic carbenicillin by replacing the DNA sequence corresponding to the active site with random nucleotide sequences can be evolved without exchanging the codon encoding catalytically active Ser70. The inserted oligonucleotide $Phe^{66}XXXSer^{70}XXLys^{73}$ contains 15 base pairs of chemically synthesized random sequences that code for 2.5 million amino acid exchanges. It should be noted that ß-lactamase is an ideal enzyme with which randomization-based protein engineering can be performed because a simple and efficient *selection* system is available (Chapter 2).

Major advancements followed in the 1990s, these allowing saturation mutagenesis-based simultaneous randomization at more than one residue site. Based on some of these developments, the so-called QuikChange™ protocol for saturation mutagenesis emerged in 2002 [26], see Chapter 3 for details. Another important version of saturation mutagenesis is the "megaprimer" method of site-specific mutagenesis introduced by Kamman et al. [27] and improved

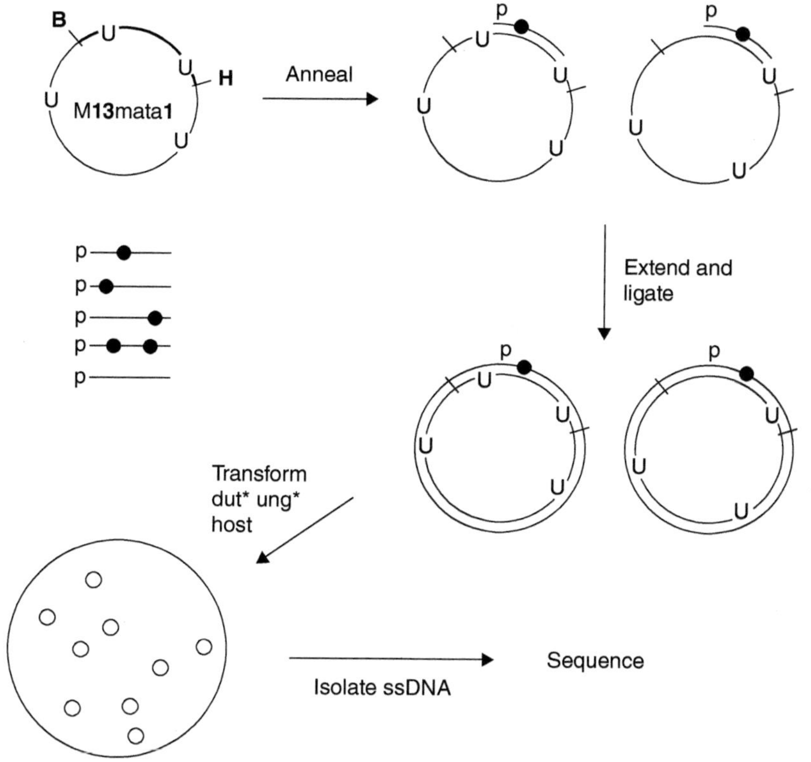

Scheme 1.4 Mixed oligonucleotide mutagenesis of the gene MATa1 from *Saccharomyces cerevisae*. Source: Adapted from Zoller and Smith [21b].

by Sarkar and Sommer [28]. The overall procedure is fairly straightforward and easy to perform, but it also has limitations as discussed in Chapter 3. Early developments of site-directed mutagenesis, which can also be used for randomization, were summarized by Tao in 1992 [29].

In 1989 a *landmark study* was published by Leung, Chen, and Goeddel describing error-prone polymerase chain reaction (epPCR) [30a], but it was not applied to enzymes until a few years later (see below). It relies on *Taq* polymerase or similar DNA polymerases, which lack proofreading ability (no removal of mismatched bases). To control the mutational rate, the reaction conditions need to be optimized by varying such parameters as the $MgCl_2$ or $MnCl_2$ concentrations and/or employing unbalanced nucleotide concentrations (see details in Section 3.3). The mutation rate can be controlled empirically by varying such parameters as the $MgCl_2$ or $MnCl_2$ concentrations and/or employing unbalanced amounts of nucleotides [30b].

The first applications of epPCR are due to Hawkins and Winter in 1992 [31], who reported *in vitro* selection and affinity maturation of antibodies from combinatorial libraries. The creation of large combinatorial libraries of antibodies was a new area of science at the time, as shown by Lerner and coworkers using different techniques, including the use of OE-PCR [32]. It should be noted that epPCR suffers from various limitations, which are discussed in Chapter 3. To this day the technique continues to be employed, especially when X-ray structural data of the protein are not available. A different but seldom used molecular biological random mutagenesis method was developed and applied in 1992/1993 by Zhang et al., to increase the thermostability of aspartase as a catalyst in the industrially important addition reaction of ammonia to fumarate with formation of L-aspartic acid [33]. Unbalanced nucleotide amounts were used in a special way, but diversity is lower than in the case of epPCR [33b].

In 1993, Chen and Arnold published a paper describing the use of random mutagenesis in the quest to increase the robustness of the protease subtilisin E in aqueous medium containing a hostile organic solvent (DMF) [34]. This study is described here in detail, because at the Nobel Prize announcement on 3rd October 2018, it was highlighted as constituting the seminal turning point in directed evolution (see Nobel Prize Lecture by Frances H. Arnold [4b] and also her contribution in developing artificial metalloenzymes as highlighted in Chapter 7). The only difference in the 1993 study was the use of epPCR as the random mutagenesis method (although the term epPCR is not mentioned in the paper), and the fact that not thermostability was aimed for as Hageman et al. had already described [16], but resistance to DMF in an otherwise aqueous environment [34]. This paper does not involve any increase in the activity of the WT in the absence of a cosolvent, as Arnold et al. actually pointed out [34]. *Thus, the aim was not to increase the activity of an enzyme* [4b], which is sometimes misunderstood. The traditional aim in directed evolution to enhance enzyme activity is a different challenge, which was addressed much later and is still relevant today [3].

In the Arnold paper, the mutations of three variants obtained earlier by rational design were first combined with formation of the respective triple mutant Asp60Asn/Gln103Arg/Asn218Ser to which was added a fourth point mutation Asp97Gly, leading to variant Asp60Asn/Gln103Arg/Asn218Ser/Asp97Gly ("4M variant") [34]. The *Hind*III/*Bam*HI DNA fragment of 4M subtilisin E from residue 49 to the C-terminus was then employed as the template for PCR-based random mutagenesis (epPCR). Thus, this diverges a little from epPCR as originally developed by Leung, Chen, and Goeddel [30], which addresses the whole gene. The PCR conditions were modified so that the mutational frequency increased (including the use of $MnCl_2$). An easy-to-perform prescreen for activity was developed using agar plates containing 1% casein, which upon hydrolysis forms a halo. The roughly identified active mutants were then sequenced and used as catalysts in the hydrolysis of *N*-succinyl-L-Ala-L-Ala-L-Pro-L-Met-*p*-nitroanilide and *N*-succinyl-L-Ala-L-Ala-L-Pro-L-Phe-*p*-nitroanilide. Upon going through three cycles of random mutagenesis, the final best hit (PC3) was identified as having a total of 10 point mutations. The catalytic efficiency of variant PC3 relative to WT subtilisin E in aqueous medium containing different amounts of DMF is shown in Scheme 1.5 [34]. This study constitutes the second example of more than one cycle of mutagenesis/expression/screening in directed evolution [16].

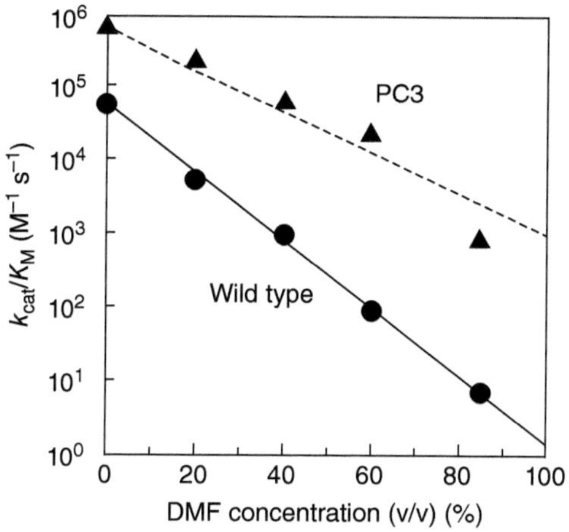

Scheme 1.5 Catalytic efficiency of WT subtilisin E and variant PC3 as catalysts in the hydrolytic cleavage of *N*-succinyl-L-Ala-L-Ala-L-Pro-L-Met-*p*-nitroanilide. Source: Chen and Arnold [34].

In addition to epPCR and OE-PCR, other important mutagenesis methods were developed in the late 1980s. *A key study* by Pease, Horton, and coworkers is a prime example [23b] (Scheme 1.6). It is an extension of their earlier work on OE-PCR [23a]. Fragments from two genes that are to be recombined are first produced by separate PCR, the primers being designed so that the ends of the products feature complementary sequences (Scheme 1.6). Upon mixing, denaturing, and reannealing the PCR products, those strands that have matching sequences at their 3′ ends overlap and function as primers for each other. Extension of the overlap by a DNA polymerase leads to products in which the original sequences are spliced together. This recombinant technique for producing chimeric genes was called splicing by overlap extension (SOE), which also allows the introduction of random errors (mutations). The technique was illustrated using two different mouse class-I major histocompatible genes. However, at the time it was not exploited by the biotechnology community active in directed evolution.

In 1994, a novel gene mutagenesis method called DNA shuffling was reported by Stemmer [35a]. It is a process that simulates sexual evolution as it occurs in nature. In the original study, ß-lactamase served as the enzyme, the selection system is based on the increased resistance to an antibiotic. DNA shuffling is illustrated here when starting with mutants of a given enzyme previously produced by some other mutagenesis technique (Scheme 1.7). Family shuffling, also introduced in 1994, is a variation that in many cases constitutes the superior

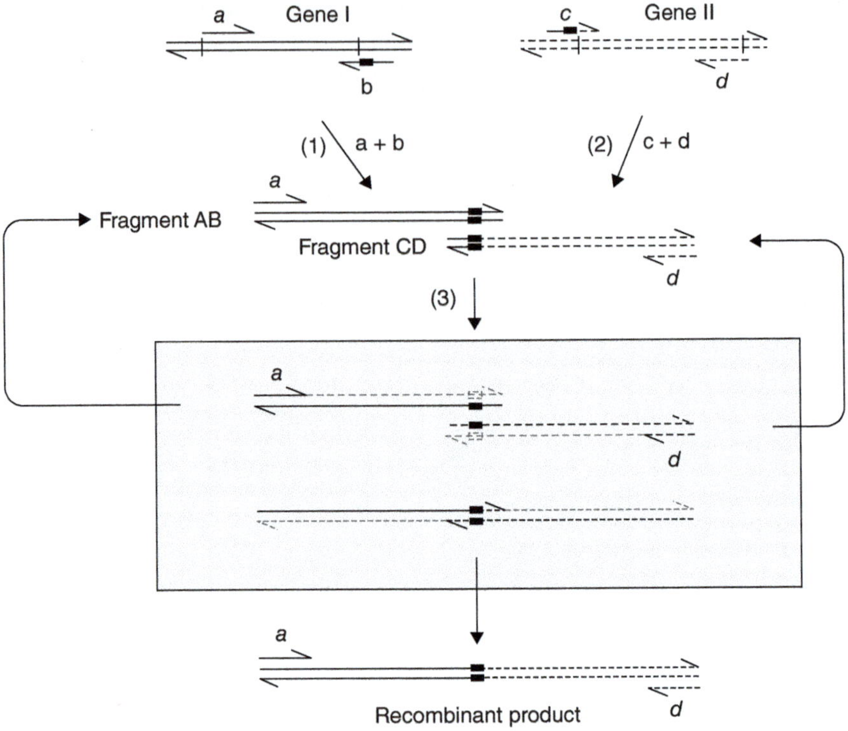

Scheme 1.6 Steps in the recombinant technique of splicing by overlap extension (SOE), illustrated here using two different genes. Source: Horton et al. [23b]/With permission of Elsevier.

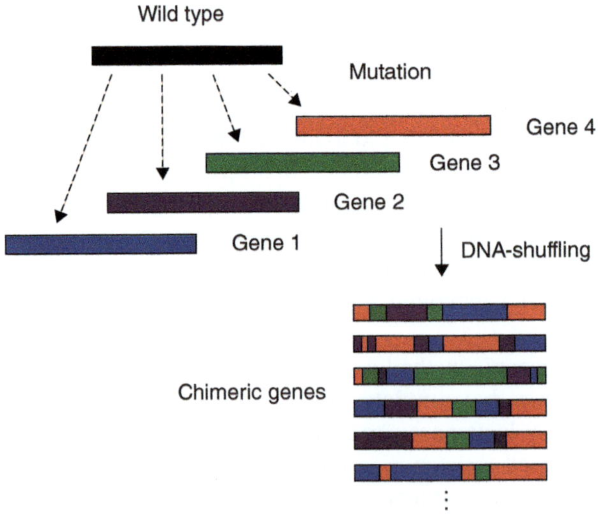

Scheme 1.7 DNA shuffling starting from a single gene encoding a given enzyme.

approach [35b]. See Chapter 3 for a description of this seminal technique and other recombinant methods.

These landmark papers sparked a great deal of further research in the area of directed evolution in the 1990s. In many of the studies, recombinant and/or non-recombinant methods were applied to shed light on the mechanism of enzymes, but usually, only initial mutant libraries were considered. To this day directed evolution is sometimes employed in the quest to study enzyme mechanisms rather than for the purpose of evolving altered enzymes for practical purposes. Contributions by Hilvert [4d, 36] and Benkovic et al. [37] are prominent examples, as are the studies of Hecht et al. concerning binary patterning [38]. In an informative overview by Lutz and Benkovic which appeared in 2002, many of these and other early developments in directed evolution including applications were assessed [39].

Phage display, invented by George P. Smith in 1985 [40], was originally not intended for directed evolution of enzymes, but it inspired Gregory Winter to apply it with the aim of developing antibodies as medicines [41]. *These seminal developments led to the Chemistry Nobel Prize 2018* [40c, 41b] for Smith and Winter, which they shared with Arnold [4b]. Despite the evolutive nature of such systems, application of phage display in directed evolution of stereo- and/or regioselective enzymes is problematic and has limitations [42]. Although flow cytometry had been invented decades ago, it was not combined with fluorescence-activated cell sorter technology (FACS) for application in directed evolution until much later, especially relevant in vaccine development, as demonstrated by the early contributions of Georgiou and coworkers [43]. *The water-in-oil emulsion technology, elegantly developed by Griffiths and Tawfik, is likewise a step forward* [44]. All of these selection platforms, which are really screening techniques (Chapter 2), are useful in many protein engineering applications, but to this day their utilization in the laboratory evolution of stereo- and/or regioselective enzymes remains marginal (see Chapters 2 and 3).

The generation of selective catalytic monoclonal antibodies can also be considered to be based on evolutionary principles, but despite impressive contributions [45], these biocatalysts have not entered the stage of practical applications in stereoselective organic chemistry or biotechnology. This appears to be due to the fact that the immune system functions on the basis of binding, and not on catalytic turnover [45a].

Turning to a different subject in directed evolution, *an important contribution by Patrick and Firth describing algorithms for designing mutant libraries based on statistical analyses influences the field to this day* [46]. Ostermeier developed a similar metric [47], and Pelletier et al., extended these statistical models [48]. Reetz employed the Patrick/Firth mathematics in the establishment of a particularly easy-to-use statistical system, especially useful when evolving selectivity, activity, or stability [49]. It is available free of charge on the Reetz website (http://www.kofo.mpg.de/en/research/biocatalysis).

Up to this point, evolving protein robustness by the different approaches has been the focus of this section. However, at the time organic chemists were not impressed by what was going on in the field of directed evolution, as already noted in Section 1.1. In contrast, the Reetz group, at the time composed only of organic chemists, realized that stereo- and regioselectivity stood at the heart of modern organic chemistry and that this could be an exciting, albeit difficult goal in directed evolution. Recently, four former postdocs published a historical account that delineates how Manfred T. Reetz, originally with little interest in enzymes, switched fields [50]. In the 1990s, it was known that enzyme immobilization sometimes influences enantioselectivity [51] but in a completely unreliable manner. Moreover, rational enzyme design (Section 1.3) had not been successful in engineering stereoselectivity. *Therefore, Reetz set out to develop a fundamentally new approach to asymmetric catalysis: The directed evolution of stereoselective enzymes as catalysts in organic chemistry and biotechnology, the initial positive results being published in 1997* [49]. The underlying concept is very different from the traditional development of chiral synthetic transition metal catalysts or organocatalysts because the stepwise increase in stereoselectivity can be expected to emerge as a consequence of the evolutionary pressure exerted in each cycle (Scheme 1.8) [49]. However, at the time this was an extremely slim hope because there was no precedence from which one could be guided. Evolving stereoselectivity seemed to be much more challenging than stability engineering. Enhancing and even reversing enantioselectivity were dream goals in those days, and thus a highly risky endeavor.

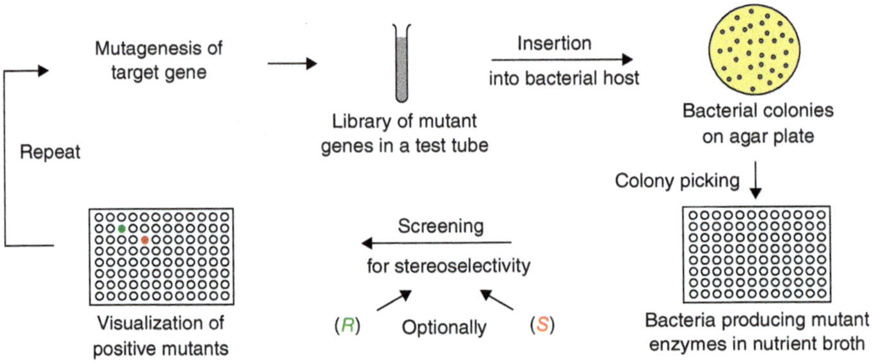

Scheme 1.8 Concept of directed evolution of stereoselective enzymes with (*R*)- or (*S*)-selective mutants being accessible on an optional basis. Source: Adapted from Refs. [49a, b].

In a proof-of-principle study in 1997, the lipase from *Pseudomonas aeruginosa* (PAL) was used as the enzyme in the hydrolytic kinetic resolution of ester **1** (Scheme 1.9) [49a, b]. WT PAL is a poor catalyst in this reaction because the selectivity factor reflecting the relative rate of reaction of (*R*)- and (*S*)-**1** amounts to only $E = 1.1$ with slight preference for (*R*)-**2**. Four cycles of epPCR at low mutation rate led to variant A showing enhanced enantioselectivity ($E = 11$), which caused great excitement in the Reetz group. Today such a value would hardly be publishable! The mutant is characterized by four point mutations S149G/S155L/V476/F259L which accumulated in a stepwise manner (Scheme 1.10) [49a, b]. At the time, high- or medium-throughput *ee*-assays were not available. An on plate pretest, as well as a UV/Vis-based screening system for identifying enantioselective lipase mutants (300–600 transformants/day), had to be developed first, based on the gradual release of *p*-nitrophenolate **3** (see Chapter 2 for more details as well as other screening systems subsequently developed by various groups).

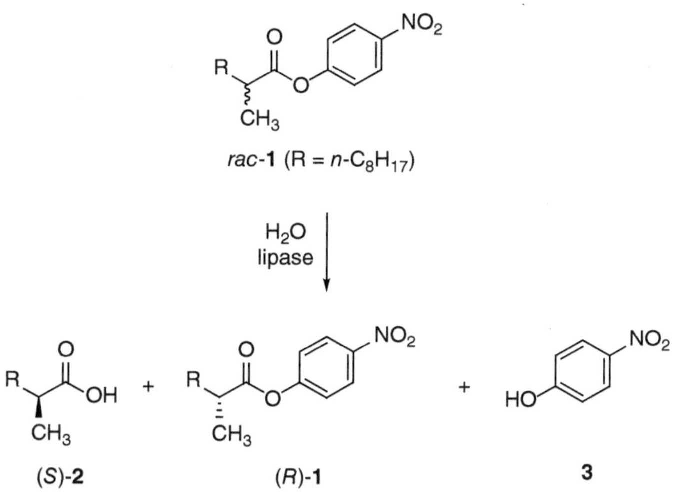

Scheme 1.9 Hydrolytic kinetic resolution of *rac*-**1** catalyzed by the lipase from *Pseudomonas aeruginosa* (PAL). Source: Adapted from Refs. [49a, b].

Although a selectivity factor of $E = 11$ does not suffice for practical applications, *this study set the stage for the rapid development of directed evolution of stereoselective enzymes, which ensued in many laboratories worldwide.* Progress up to 2004 covering several different enzyme types was summarized in 2004 reviews [52], which also emphasized the importance of methodology development. Indeed, this aspect of directed evolution continues to be important [3, 4a].

Since subsequent epPCR experiments proved to be disappointing ($E = 13$), the doctoral student Steffi Wilensek, working in the Reetz group, suggested, inter alia, focused saturation mutagenesis at sites lining the binding pocket using so-called NNK codon degeneracy (20 canonical amino acids as building blocks). It was the first time that this technique was applied to enzyme stereoselectivity [53]. Scheme 1.11 shows the first saturation mutagenesis experiment at a site next to the binding pocket for enhancing stereoselectivity. Today, some researchers wonder why it took so long to realize that this approach is the way to go. It worked indeed ($E = 30$), but at the time the combination of saturation mutagenesis, epPCR, and DNA shuffling was even better ($E = 51$) [53].

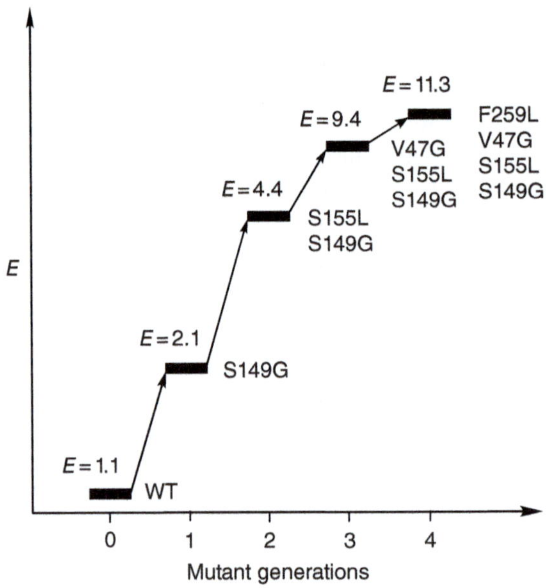

Scheme 1.10 First example of directed evolution of a stereoselective enzyme. The model reaction involved the hydrolytic kinetic resolution of *rac*-**1** catalyzed by the lipase PAL, four rounds of epPCR being used as the gene mutagenesis method. Source: Adapted from Refs. [49a, b].

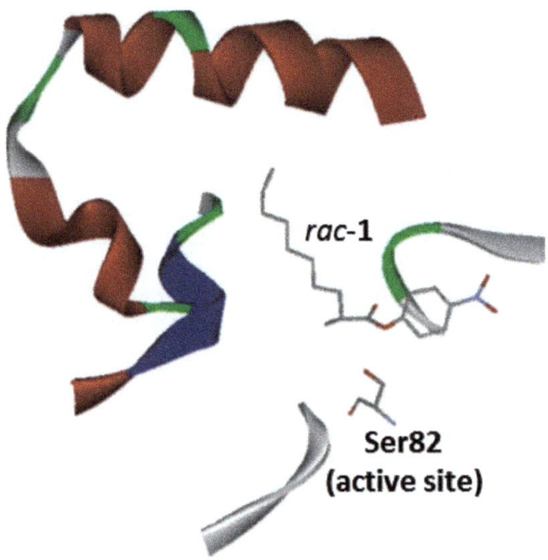

Scheme 1.11 First example of focused saturation mutagenesis at sites next to a substrate in an enzyme's binding pocket, in this case, compound 1 bound in *P. aeruginosa* (PAL) in the study aimed at enhancing stereoselectivity. Source: Adapted from Refs. [53, 54].

In 2005, the first step toward systematization of the idea apparent in Scheme 1.11 was proposed with the emergence of the combinatorial active-site saturation test (CAST), which was a true breakthrough [55]. The initial model example, although not optimized using later improved CAST versions [3, 4], concerned the enhancement of activity of PAL as the catalyst in the reaction of substrates not accepted by the lipase. Shortly thereafter, iterative saturation mutagenesis (ISM) was developed, intended for use when an initial library does not contain a mutant with a fully acceptable catalytic profile [56]. CAST and ISM are illustrated in Scheme 1.12 [56]. ISM can also be applied with the aim of increasing protein robustness, according to which saturation mutagenesis is focused on remote sites displaying high B-factors which indicate flexibility. In this way, B-FIT and the metric B-FITTER were developed [57]. CAST/ISM and B-FTTER can be downloaded free of charge from the Reetz website (http://www.kofo.mpg.de/en/research/biocatalysis).

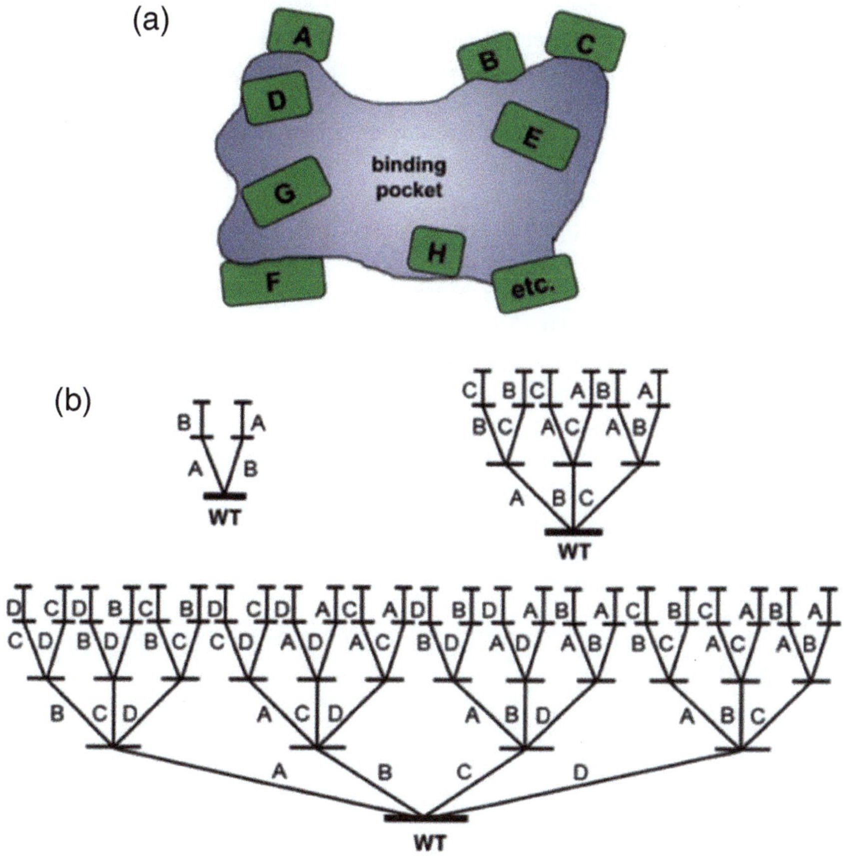

Scheme 1.12 Illustration of ISM. (a) Sites chosen for ISM; (b) 2-, 3-, and 4-site ISM systems. Source: Adapted from Reetz et al. [56].

Further improvements of CAST/ISM and other innovations from different groups between 2006 and 2021 include:

- First example of directed evolution of an enantioselective artificial metalloenzyme [58].
- Genetic selection system for evolving enantioselectivity [59].
- Reduced amino acid alphabets in directed evolution [60].
- Efficient saturation method for difficult-to-amplify templates [61].
- Regio- and stereoselective P450-catalyzed steroid hydroxylation [62].
- How to escape from local minima in fitness landscapes [63].
- Reducing codon redundancy in saturation mutagenesis libraries [64].
- Assembly of designed oligonucleotides (ADO) as a tool in synthetic biology [65].
- Triple code saturation mutagenesis [66].
- Directed evolution of regioselectivity [67].
- Beating bias in protein engineering by error-free on-chip solid-phase synthesis of mutant libraries [68].
- Achieving reversal of regioselectivity of Baeyer–Villiger monooxygenases [69].
- Machine learning in directed evolution of stereoselectivity and other parameters [70].
- Focused rational iterative saturation mutagenesis (FRISM) [71].
- Unveiling long-range conformational and cooperative mutational effects on multiple catalytic traits [72].
- Engineering of $P450_{BM3}$ to catalyze carbene and nitrene transfer reactions [4b, 73].
- Improving the hydrolysis activity of a lipase by exchanging the catalytic triad from Ser-His-Asp to Cys-His-Asp [74].

The most important single advances and events in directed evolution of enzymes from the early phase to the present day are summarized in Scheme 1.13.

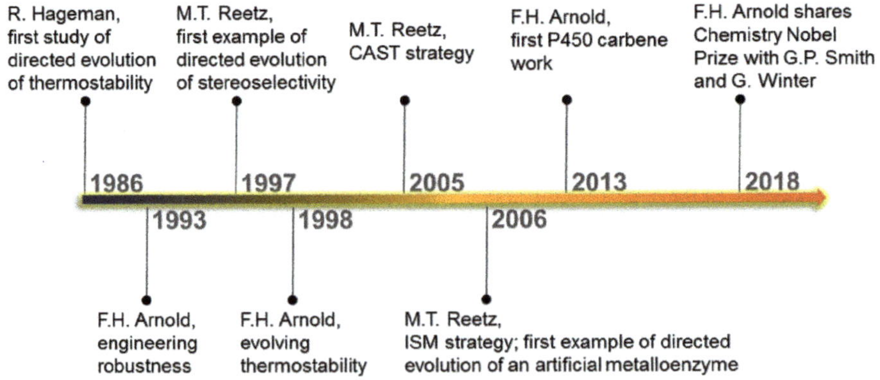

Scheme 1.13 Timeline of enzyme-directed evolution.

1.3
Methods and Aims of Rational Design of Enzymes

The aims of rational enzyme design, namely obtaining useful and robust biocatalysts for organic chemistry and biotechnology, are basically similar to those of directed evolution, but the methods are very different. With the advent of site-specific mutagenesis as pioneered by Michael Smith [5] (Section 1.1), protein engineers have at their disposal a reliable technique to replace any amino acid in the protein with one of the other 19 canonical amino acids. The challenge is to identify a given residue as a potential hot spot where such an exchange is likely to lead to an improvement of the respective catalytic performance. Maximal efficiency would come about if only a single prediction would be necessary, i.e. not to perform all 19 site-specific mutagenesis experiments. The problem is actually even more difficult because it is unlikely that modification at only one residue suffices. Nevertheless, the advantage over directed evolution would be that rational design circumvents the time-consuming screening of large mutant libraries. Several review articles have appeared [75].

Originally, rational design was applied to enhance the robustness of enzymes [76]. As already pointed out, Wells improved the oxidative stability by combining rational design and directed evolution [20]. *A general approach to rationally engineering enzyme thermostability was developed shortly thereafter by Eijsink and others* [77a]. Accordingly, especially H-bonds and salt bridges on the protein surface need to be introduced by the properly designed amino acid exchanges. It was demonstrated that well-designed salt-bridge networks may tolerate the disorder of increased thermal motion to improve thermostability [77b, c]. Today, a number of criteria and aids are being used:

- H-bonds and salt bridges on protein surfaces.

During natural evolution, thermophilic proteins obtained charged residues on the surface [77d], and thus could form more salt bridges, in contrast to their mesophilic homologs [77e, f]. Harnessing this observation, protein thermal stability can be improved by introducing salt bridges. For example, six designed salt bridges on a mesophilic β-glucosidase can independently increase the melting temperature range from 0.7 to 8.8 °C, respectively. Interestingly, out of the six-point thermostable variants, the combination of a quadruple mutant generated a melting temperature increase of 15.7 °C, suggesting that epistatic effect also exists when engineering thermostability. In the view of protein thermodynamics, the reason salt bridges enhance protein thermostability is the reduction of heat capacity of unfolding [77g]. Apart from thermostability, stabilized salt bridges on the surface of *Bacillus subtilis* Lipase A increases organic solvent resistance [77h].

- Conformational dynamics.

As dynamical entities, the motion of residues in proteins, especially the ones located in loops nearby the active site has been recognized to influence the catalytic properties of enzymes [78a, b]. Several studies have shown that conformational dynamics guided enzyme engineering is successful in manipulating substrate recognition [78c], altering stereoselectivity [78d, e], relieving product inhibition [78f], increasing activity [78e, g, h], and elevating thermostability [78i], among other properties.

- Steric hindrance in the substrate binding pocket.

By reshaping the substrate-binding cavity, introducing or eliminating steric clashes is a useful means to control catalytic selectivity. In the engineering of enantioselectivity of an epoxide hydrolase, Janssen and coworkers introduced steric hindrance to prevent unwanted substrate binding modes by replacing the small amino acid residues with bulky ones (Trp, Tyr, and Phe) at target positions [79a]. Likewise, Sun and coworkers implemented the same strategy to control the regioselective glucosylation of a glycosyltransferase [79b]. Moreover, in the P450 world, replacing the bulky natural phenylalanine residue F87 (numbering according to $P450_{BM3}$) with alanine leads to the variant F87A that can be used as a promising starting mutant for further mutations by directed evolution. In this way, the substrate binding pocket can be reshaped to accommodate bulky substrates or even alter stereo- and regioselectivity [62, 79c]. Using F87A and further rational design alone is generally more difficult.

- Rosetta computational package for enzyme redesign.

In de novo design of enzymes [80a], the highly acclaimed Rosetta package [80b] can be utilized to obtain promiscuous enzymes. A recent example by Wu and coworkers concerns a computational redesign toward an aspartase, a highly specific ammonia lyase, to yield C–N lyases with cross-compatibility of noncanonical amino acids (ncAAs) with high conversion (up to 99%), regioselectivity >99% and enantioselectivity >99% [80c], and the Rosetta field has been reviewed [4h].

- Machine learning for designing biosensors.

Unger et al. developed machine learning approaches, including random forest modeling and generalized linear modeling, for rapid binding-pocket redesign leading to a selective and sensitive serotonin sensor [81].

In addition to the abovementioned approaches, further methodologies have recently emerged. As outlined in Section 3.4, the so-called focused rational iterative site-specific mutagenesis (FRISM) is of particular significance [4a, 71, 82]. As a promising fusion of rational design and directed evolution, it requires only a few predicted mutants, which means that the screening effort is limited to very small collections of mutants. Mutant libraries as in typically directed evolution are not involved [71b]. FRISM was first proposed in protein engineering of *Candida antarctica* lipase B (CalB) to access all possible stereoisomers of chiral esters bearing multiple stereocenters in a fully stereodivergent manner [71a]. In another case study, FRISM was adopted in manipulating the regioselectivity of the glycosyltransferase UGT74AC2 with no trade-off in activity [79b].

References

1. (a) Streitwieser, A., Heathcock, C., and Kosomer, E.M. (1998). *Introduction to Organic Chemistry*, 4e. Pearson College Div. (b) Vogel, P. and Houk, K.N. (2019). *Organic Chemistry: Theory, Reactivity and Mechanisms in Modern Synthesis*. Weinheim: Wiley-VCH. (c) Caron, S. (ed.) (2020). *Practical Synthetic Organic Chemistry: Reactions, Principles, and Techniques*. Wiley. (d) Fürstner, A. (2014). Catalysis for total synthesis. *Angew. Chem. Int. Ed.* **53**: 8587–8598. (e) Ley, S. (2018). The engineering of chemical synthesis: humans and machines working in harmony. *Angew. Chem. Int. Ed.* **57**: 5185–5203. (f) Kitson, P.J., Marie, G., Francoia, J.-P. et al. (2018). Digitization of multistep organic synthesis in reactionware for on-demand pharmaceuticals. *Science* **359** (6373): 314–319. (g) Blakemore, D.C., Castro, L., Churcher, I. et al. (2018). Organic synthesis provides opportunities to transform drug discovery. *Nat. Chem.* **10**: 383–394.
2. (a) Drauz, K., Gröger, H., and May, O. (ed.) (2012). *Enzyme Catalysis in Organic Synthesis*, 3e. Weinheim: Wiley-VCH. (b) Faber, K. (2011). *Biotransformations in Organic Chemistry*, 6e. Heidelberg: Springer. (c) Liese, A., Seelbach, K., and Wandrey, C. (2006). *Industrial Biotransformations*, 2e. Weinheim: Wiley-VCH.
3. Reetz, M.T. (2016). *Directed Evolution of Selective Enzymes: Catalysts for Organic Chemistry and Biotechnology*. Weinheim: Wiley-VCH.
4. Recent reviews of directed evolution of enzymes: (a) Qu, G., Li, A., Acevedo-Rocha, C.G. et al. (2020). The crucial role of methodology development in directed evolution of selective enzymes. *Angew. Chem. Int. Ed.* **59** (32): 13204–13231. (b) Arnold, F.H. (2019). Innovation by evolution: bringing new chemistry to life (nobel lecture). *Angew. Chem. Int. Ed.* **58** (41): 14420–14426. (c) Gargiulo, S. and Soumillion, P. (2021). Directed evolution for enzyme development in biocatalysis. *Curr. Opin. Chem. Biol.* **61**: 107–113. (d) Zeymer, C. and Hilvert, D. (2018). Directed evolution of protein catalysts. *Annu. Rev. Biochem.* **87** (1): 131–157. (e) Chen, K. and Arnold, F.H. (2020). Engineering new catalytic activities in enzymes. *Nat. Catal.* **3**: 203–213. (f) Bunzel, H.A., Anderson, J.L.R., and Mulholland, A.J. (2021). Designing better enzymes: insights from directed evolution. *Curr. Opin. Struct. Biol.* **67**: 212–218. (g) Reetz, M.T. (2022). Witnessing the birth of directed evolution of stereoselective enzymes as catalysts in organic chemistry. *Adv. Synth. Catal.* https://doi.org/10.1002/adsc.2022004466. (h) Lovelock, S.L., Crawshaw, R., Basler, S. et al. (2022). The road to fully programmable protein catalysis. *Nature* **606** (7912): 49–58.
5. (a) Smith, M. (1985). In vitro mutagenesis. *Annu. Rev. Genet.* **19**: 423–462. (b) Smith, M. (1994). Synthetic DNA and biology (nobel lecture). *Angew. Chem. Int. Ed.* **33** (12): 1214–1221.
6. Fersht, A. (1999). *Structure and Mechanism in Protein Science*, 3e. New York: W. H. Freeman and Company.
7. (a) Mills, D.R., Peterson, R.L., and Spiegelman, S. (1967). An extracellular Darwinian experiment with a self-duplicating nucleic acid molecule. *Proc. Natl. Acad. Sci. U.S.A.* **58** (1): 217–224. (b) Spiegelman, S. (1971). An approach to the experimental analysis of precellular evolution. *Quart. Rev. Biophys.* **4** (2&3): 213–253.
8. (a) Adamala, K., Engelhart, A.E., and Szostak, J.W. (2015). Generation of functional RNAs from inactive oligonucleotide complexes by non-enzymatic primer extension. *J. Am. Chem. Soc.* **137** (1): 483–489. (b) Joyce, G.F. (2007). Forty years of in vitro evolution. *Angew. Chem. Int. Ed.* **46** (34): 6420–6436. (c) Blain, J.C. and Szostak, J.W. (2014). Progress toward synthetic cells. *Annu. Rev. Biochem.* **83**: 615–640. (d) Sun, H. and Zu, Y. (2015). Aptamers and their applications in nanomedicine. *Small* **11** (20): 2352–2364. (e) Mayer, G., Ahmed, M.S., Dolf, A. et al. (2010). Fluorescence-activated cell sorting for

aptamer SELEX with cell mixtures. *Nat. Protoc.* **5** (12): 1993–2004.

9. Francis, J.C. and Hansche, P.E. (1972). Directed evolution of metabolic pathways in microbial populations. I. Modification of acid-phosphatase pH optimum in *S. cerevisiae*. *Genetics* **70** (1): 59–73.
10. (a) Hall, B.G. (1977). Number of mutations required to evolve a new lactase function in *Escherichia coli*. *J. Bacteriol.* **129** (1): 540–543. (b) Hall, B.G. (1978). Experimental evolution of a new enzymatic function. II. Evolution of multiple functions for *EBG* enzyme in *E. coli*. *Genetics* **89** (3): 453–465. (c) Hall, B.G. (1981). Changes in the substrate specificities of an enzyme during directed evolution of new functions. *Biochemistry* **20** (14): 4042–4049.
11. (a) Hwang, B.Y., Oh, J.M., Kim, J., and Kim, B.G. (2006). Pro-antibiotic substrates for the identification of enantioselective hydrolases. *Biotechnol. Lett.* **28** (15): 1181–1185. (b) Cooper, T.F., Rozen, D.E., and Lenski, R.E. (2003). Parallel changes in gene expression after 20,000 generations of evolution in *Escherichia coli*. *Proc. Natl. Acad. Sci. U.S.A.* **100** (3): 1072–1077. (c) Esvelt, K.M., Carlson, J.C., and Liu, D.R. (2011). A system for the continuous directed evolution of biomolecules. *Nature* **472** (7344): 499–503. (d) Leconte, A.M., Dickinson, B.C., Yang, D.D. et al. (2013). A population-based experimental model for protein evolution: effects of mutation rate and selection stringency on evolutionary outcomes. *Biochemistry* **52** (8): 1490–1499.
12. (a) Kunkel, T.A. (1985). Rapid and efficient site-specific mutagenesis without phenotypic selection. *Proc. Natl. Acad. Sci. U.S.A.* **82** (2): 488–492. (b) Fujii, R., Kitaoka, M., and Hayashi, K. (2006). Error-prone rolling circle amplification: the simplest random mutagenesis protocol. *Nat. Protoc.* **1** (5): 2493–2497.
13. Eigen, M. and Gardiner, W. (1984). Evolutionary molecular engineering based on RNA replication. *Pure Appl. Chem.* **56** (8): 967–978.
14. (a) Mullis, K.B. (1994). The polymerase chain-reaction (nobel lecture). *Angew. Chem. Int. Ed.* **33** (12): 1209–1213. (b) Glick, B.R., Pasternak, J.J., and Patten, C.L. (2010). *Molecular Biotechnology: Principles and Applications of Recombinant DNA*. Washington, DC: ASM Press.
15. (a) Matsumura, M. and Aiba, S. (1985). Screening for thermostable mutant of kanamycin nucleotidyltransferase by the use of a transformation system for a thermophile, *Bacillus stearothermophilus*. *J. Biol. Chem.* **260** (28): 15298–15303. (b) Baase, W.A., Liu, L., Tronrud, D.E., and Matthews, B.W. (2010). Lessons from the lysozyme of phage T4. *Protein Sci.* **19** (4): 631–641.
16. Liao, H., Mckenzie, T., and Hageman, R. (1986). Isolation of a thermostable enzyme variant by cloning and selection in a thermophile. *Proc. Natl. Acad. Sci. U.S.A.* **83** (3): 576–580.
17. Muteeb, G. and Sen, R. (2010). Random mutagenesis using a mutator strain. *Methods Mol. Biol.* **634**: 411–419.
18. Liao, H.H. (1993). Thermostable mutants of kanamycin nucleotidyltransferase are also more stable in proteinase K, urea, detergents, and water-miscible organic solvents. *Enzyme Microb. Technol.* **15** (4): 286–292.
19. (a) Matteuchi, M.D. and Heyneker, H.L. (1983). Targeted random mutagenesis: the use of ambiguously synthesised oligonucleotides to mutagenize sequences immediately 5′ of an ATG initiation codon. *Nucleic Acids Res.* **11** (10): 3113–3121. (b) Hui, A., Hayflick, J., Dinkelspiel, K., and de Boer, H.A. (1984). Mutagenesis of the three bases preceding the start codon of the ß-galactosidase mRNA and its effect on translation in *Escherichia coli*. *EMBO J.* **3** (3): 623–629. (c) Dreher, T.W., Bujarski, J.J., and Hall, T.C. (1984). Mutant viral RNAs synthesized in vitro show altered aminoacylation and replicase template activities. *Nature* **311** (5982): 171–175. (d) Seeburg, P.H., Colby, W.W., Capon, D.J. et al. (1984). Biological properties of human c-Ha-ras1 genes mutated at codon 12. *Nature* **312** (5989): 71–75. (e) Schultz, S.C. and Richards, J.H. (1986). Site-saturation studies of

beta-lactamase: production and characterization of mutant ß-lactamases with all possible amino acid substitutions at residue 71. *Proc. Natl. Acad. Sci. U.S.A.* **83** (6): 1588–1592. (f) Derbyshire, K.M., Salvo, J.J., and Grindley, N.D. (1986). A simple and efficient procedure for saturation mutagenesis using mixed oligodeoxynucleotides. *Gene* **46** (2–3): 145–152. (g) Reidhaar-Olson, J.F. and Sauer, R.T. (1988). Combinatorial cassette mutagenesis as a probe of the informational content of protein sequences. *Science* **241** (4861): 53–57. (h) Oliphant, A.R., Nussbaum, A.L., and Struhl, K. (1986). Cloning of random-sequence oligodeoxynucleotides. *Gene* **44** (2–3): 177–183.

20. Estell, D.A., Graycar, T.P., and Wells, J.A. (1985). Engineering an enzyme by site-directed mutagenesis to be resistant to chemical oxidation. *J. Biol. Chem.* **260** (11): 6518–6521.

21. (a) Ner, S.S., Goodin, D.B., and Smith, M. (1988). A simple and efficient procedure for generating random point mutations and for codon replacements using mixed oligodeoxynucleotides. *DNA* **7** (2): 127–134. (b) Zoller, M.J. and Smith, M. (1982). Oligonucleotide-directed mutagenesis using M13-derived vectors: an efficient and general procedure for the production of point mutations in any fragment of DNA. *Nucleic Acids Res.* **10** (20): 6487–6500.

22. Hermes, J.D., Parekh, S.M., Blacklow, S.C. et al. (1989). A reliable method for random mutagenesis – the generation of mutant libraries using spiked oligodeoxyribonucleotide primers. *Gene* **84** (1): 143–151.

23. (a) Ho, S.N., Hunt, H.D., Horton, R.M. et al. (1989). Site-directed mutagenesis by overlap extension using the polymerase chain-reaction. *Gene* **77** (1): 51–59. (b) Horton, R.M., Hunt, H.D., Ho, S.N. et al. (1989). Engineering hybrid genes without the use of restriction enzymes – gene-splicing by overlap extension. *Gene* **77** (1): 61–68.

24. Lee, J., Shin, M.K., Ryu, D.K. et al. (2010). Insertion and deletion mutagenesis by overlap extension PCR. *Methods Mol. Biol.* **634**: 137–146.

25. Dube, D.K. and Loeb, L.A. (1989). Mutants generated by the insertion of random oligonucleotides into the active-site of the ß-lactamase gene. *Biochemistry* **28** (14): 5703–5707.

26. Hogrefe, H.H., Cline, J., Youngblood, G.L., and Allen, R.M. (2002). Creating randomized amino acid libraries with the QuikChange Multi Site-Directed Mutagenesis Kit. *Biotechniques* **33** (5): 1158–1160.

27. Kammann, M., Laufs, J., Schell, J., and Gronenborn, B. (1989). Rapid insertional mutagenesis of DNA by polymerase chain-reaction (PCR). *Nucleic Acids Res.* **17** (13): 5404–5404.

28. Sarkar, G. and Sommer, S.S. (1990). The megaprimer method of site-directed mutagenesis. *Biotechniques* **8** (4): 404–407.

29. Reikofski, J. and Tao, B.Y. (1992). Polymerase chain reaction (PCR) techniques for site-directed mutagenesis. *Biotechnol. Adv.* **10** (4): 535–547.

30. (a) Leung, D.W., Chen, E., and Goeddel, D.V. (1989). A method for random mutagenesis of a defined DNA segment using a modified polymerase chain reaction. *Technique* **1**: 11–15; (b) Cadwell, R.C. and Joyce, G.F. (1994). Mutagenic PCR. *PCR Methods Appl.* **3** (6): S136–S140.

31. Hawkins, R.E., Russell, S.J., and Winter, G. (1992). Selection of phage antibodies by binding affinity. Mimicking affinity maturation. *J. Mol. Biol.* **226** (3): 889–896.

32. Barbas, C.F., Bain, J.D., Hoekstra, D.M., and Lerner, R.A. (1992). Semisynthetic combinatorial antibody libraries: a chemical solution to the diversity problem. *Proc. Natl. Acad. Sci. U.S.A.* **89** (10): 4457–4461.

33. (a) Zhang, J., Li, Z.-Q., and Zhang, H.-Y. (1992). An enzymatic method for random- (site-specific) mutagenesis of Ginseng gene in vitro. *Chin. Biochem. J.* **8** (1): 115–120. (b) Zhang, H.Y., Zhang, J., Lin, L. et al. (1993). Enhancement of the stability and activity of aspartase by random and site-directed mutagenesis. *Biochem. Biophys. Res. Commun* **192** (1): 15–21.

34. Chen, K.Q. and Arnold, F.H. (1993). Tuning the activity of an enzyme for unusual environments – sequential random mutagenesis of subtilisin-E for catalysis in dimethylformamide. *Proc. Natl. Acad. Sci. U.S.A.* **90** (12): 5618–5622.

35. (a) Stemmer, W.P.C. (1994). Rapid evolution of a protein in-vitro by DNA shuffling. *Nature* **370** (6488): 389–391. (b) Stemmer, W.P. (1994). DNA shuffling by random fragmentation and reassembly: in vitro recombination for molecular evolution. *Proc. Natl. Acad. Sci. U.S.A.* **91** (22): 10747–10751.

36. Otten, R., Pádua, R.A.P., Bunzel, H.A. et al. (2020). How directed evolution reshapes the energy landscape in an enzyme to boost catalysis. *Science* **370** (6523): 1442–1446.

37. (a) Posner, B.A., Li, L.Y., Bethell, R. et al. (1996). Engineering specificity for folate into dihydrofolate reductase from *Escherichia coli*. *Biochemistry* **35** (5): 1653–1663. (b) Warren, M.S., Marolewski, A.E., and Benkovic, S.J. (1996). A rapid screen of active site mutants in glycinamide ribonucleotide transformylase. *Biochemistry* **35** (27): 8855–8862.

38. Kamtekar, S., Schiffer, J.M., Xiong, H.Y. et al. (1993). Protein design by binary patterning of polar and nonpolar amino acids. *Science* **262** (5140): 1680–1685.

39. Lutz, S. and Benkovic, S. (2002). Engineering protein evolution. In: *Directed Molecular Evolution of Proteins* (ed. S. Brakmann and K. Johnsson), 177–213. Weinheim: Wiley-VCH.

40. (a) Smith, G. (1985). Filamentous fusion phage: novel expression vectors that display cloned antigens on the virion surface. *Science* **228** (4705): 1315–1317. (b) Smith, G.P. and Petrenko, V.A. (1997). Phage display. *Chem. Rev.* **97** (2): 391–410. (c) Smith, G.P. (2019). Phage display: simple evolution in a petri dish (nobel lecture). *Angew. Chem. Int. Ed.* **58**: 14428–14437.

41. (a) Clackson, T., Hoogenboom, H.R., Griffiths, A.D., and Winter, G. (1991). Making antibody fragments using phage display libraries. *Nature* **352** (6336): 624–628. (b) Winter, G. (2019). Harnessing evolution to make medicines (nobel lecture). *Angew. Chem. Int. Ed.* **58**: 14438–14445.

42. Dröge, M.J., Boersma, Y.L., van Pouderoyen, G. et al. (2006). Directed evolution of *Bacillus subtilis* lipase A by use of enantiomeric phosphonate inhibitors: crystal structures and phage display selection. *ChemBioChem* **7** (1): 149–157.

43. (a) Georgiou, G., Stathopoulos, C., Daugherty, P.S. et al. (1997). Display of heterologous proteins on the surface of microorganisms: from the screening of combinatorial libraries to live recombinant vaccines. *Nat. Biotechnol.* **15** (1): 29–34. (b) Daugherty, P.S., Iverson, B.L., and Georgiou, G. (2000). Flow cytometric screening of cell-based libraries. *J. Immunol. Methods* **243** (1–2): 211–227.

44. Griffiths, A.D. and Tawfik, D.S. (2000). Man-made enzymes – from design to in vitro compartmentalisation. *Curr. Opin. Biotechnol.* **11** (4): 338–353.

45. (a) Hilvert, D. (2000). Critical analysis of antibody catalysis. *Annu. Rev. Biochem.* **69**: 751–793. (b) Schultz, P.G. and Lerner, R.A. (1993). Antibody catalysis of difficult chemical transformations. *Acc. Chem. Res.* **26** (8): 391–395.

46. (a) Firth, A.E. and Patrick, W.M. (2005). Statistics of protein library construction. *Bioinformatics* **21** (15): 3314–3315. (b) Firth, A.E. and Patrick, W.M. (2008). GLUE-IT and PEDEL-AA: new programmes for analyzing protein diversity in randomized libraries. *Nucleic Acids Res.* **36**: W281–W285.

47. Bosley, A.D. and Ostermeier, M. (2005). Mathematical expressions useful in the construction, description and evaluation of protein libraries. *Biomol. Eng.* **22** (1–3): 57–61.

48. Denault, M. and Pelletier, J.N. (2007). Protein library design and screening: working out the probabilities. In: *Protein Engineering Protocols* (ed. K.M. Arndt and K.M. Müller), 127–154. Totowa, NJ: Humana Press.

49. (a) Reetz, M.T., Zonta, A., Schimossek, K. et al. (1997). Creation of enantioselective biocatalysts for organic chemistry by in vitro evolution. *Angew. Chem. Int. Ed.* **36** (24): 2830–2832. (b) Reetz, M.T. (1999). Strategies for the development of enantioselective catalysts. *Pure Appl. Chem.* **71** (8): 1503–1509. (c) Zha, D., Wilensek, S., Hermes, M. et al. (2001). Complete reversal of enantioselectivity of an enzyme-catalyzed reaction by directed evolution. *Chem. Commun.* 2664–2665.

50. Acevedo-Rocha, C.G., Hollmann, F., Sanchis, J., and Sun, Z. (2020). A pioneering career in catalysis: Manfred T. Reetz. *ACS Catal.* **10**: 15123–15139.
51. Reviews and works to immobilizations: (a)Guisan, J.M. (ed.) (2006). *Immobilization of Enzymes and Cells*, 2e. Humana Press. (b) Sheldon, R.A. and van Pelt, S. (2013). Enzyme immobilisation in biocatalysis: why, what and how. *Chem. Soc. Rev.* **42**: 6223–6235. (c) Gray, C.J., Weissenborn, M.J., Eyers, C.E., and Flitsch, S.L. (2013). Enzymatic reactions on immobilised substrates. *Chem. Soc. Rev.* **42**: 6378–6405. (d) Tielmann, P., Kierkels, H., Zonta, A. et al. (2014). Increasing the activity and enantioselectivity of lipases by sol–gel immobilization: further advancements of practical interest. *Nanoscale* **6**: 6220–6228. (e) Engelmann, C., Ekambaram, N., Johannsen, J. et al. (2020). Enzyme immobilization on synthesized nanoporous silica particles and their application in a bi-enzymatic reaction. *ChemCatChem* **12**: 2245–2252.
52. (a) Reetz, M.T. (2004). Controlling the enantioselectivity of enzymes by directed evolution: practical and theoretical ramifications. *Proc. Natl. Acad. Sci. U.S.A.* **101** (16): 5716–5722. (b) Lutz, S. and Patrick, W.M. (2004). Novel methods for directed evolution of enzymes: quality, not quantity. *Curr. Opin. Biotech.* **15** (4): 291–297.
53. Reetz, M.T., Wilensek, S., Zha, D., and Jaeger, K.-E. (2001). Directed evolution of an enantioselective enzyme through combinatorial multiple-cassette mutagenesis. *Angew. Chem. Int. Ed.* **40** (19): 3589–3591.
54. Reetz, M.T. (2006). Directed evolution of enantioselective enzymes as catalysts for organic synthesis. In: *Advances in Catalysis*, vol. **49** (ed. B.C. Gates and K. Knözinger), 1–69. San Diego: Elsevier.
55. Reetz, M.T., Bocola, M., Carballeira, J.D. et al. (2005). Expanding the range of substrate acceptance of enzymes: combinatorial active-site saturation test. *Angew. Chem. Int. Ed.* **44** (27): 4192–4196.
56. Reetz, M.T., Wang, L.W., and Bocola, M. (2006). Directed evolution of enantioselective enzymes: iterative cycles of CASTing for probing protein-sequence space. *Angew. Chem. Int. Ed.* **45** (8): 1236–1241.
57. (a) Reetz, M.T., Carballeira, J.D., and Vogel, A. (2006). Iterative saturation mutagenesis on the basis of B factors as a strategy for increasing protein thermostability. *Angew. Chem. Int. Ed.* **45** (46): 7745–7751. (b) Reetz, M.T. and Carballeira, J.D. (2007). Iterative saturation mutagenesis (ISM) for rapid directed evolution of functional enzymes. *Nat. Protoc.* **2** (4): 891–903.
58. Reetz, M.T., Peyralans, J.J.P., Maichele, A. et al. (2006). Directed evolution of hybrid enzymes: evolving enantioselectivity of an achiral Rh-complex anchored to a protein. *Chem. Commun.* 4318–4320.
59. Reetz, M.T., Höbenreich, H., Soni, P., and Fernández, L. (2008). A genetic selection system for evolving enantioselectivity of enzymes. *Chem. Commun.* 5502–5504.
60. Reetz, M.T. and Wu, S. (2008). Greatly reduced amino acid alphabets in directed evolution: making the right choice for saturation mutagenesis at homologous enzyme positions. *Chem. Commun.* 5499–5501.
61. Sanchis, J., Fernández, L., Carballeira, J.D. et al. (2008). Improved PCR method for the creation of saturation mutagenesis libraries in directed evolution: application to difficult-to-amplify templates. *Appl. Microbiol. Biotechnol.* **81** (2): 387–397.
62. Kille, S., Zilly, F.E., Acevedo, J.P., and Reetz, M.T. (2011). Regio- and stereoselectivity of P450-catalysed hydroxylation of steroids controlled by laboratory evolution. *Nat. Chem.* **3** (9): 738–743.

63. Gumulya, Y., Sanchis, J., and Reetz, M.T. (2012). Many pathways in laboratory evolution can lead to improved enzymes: how to escape from local minima. *Chem. Bio. Chem.* **13** (7): 1060–1066.
64. Kille, S., Acevedo-Rocha, C.G., Parra, L.P. et al. (2013). Reducing codon redundancy and screening effort of combinatorial protein libraries created by saturation mutagenesis. *ACS Synth. Biol.* **2** (2): 83–92.
65. Acevedo-Rocha, C.G. and Reetz, M.T. (2014). Assembly of designed oligonucleotides: a useful tool in synthetic biology for creating high quality combinatorial DNA libraries. *Methods Mol. Biol.* **1179**: 189–206.
66. (a) Sun, Z., Lonsdale, R., Wu, L. et al. (2016). Structure-guided triple code saturation mutagenesis: efficient tuning of the stereoselectivity of an epoxide hydrolase. *ACS Catal.* **6** (3): 1590–1597. (b) Sun, Z., Lonsdale, R., Ilie, A. et al. (2016). Catalytic asymmetric reduction of difficult-to-reduce ketones: triple code saturation mutagenesis of an alcohol dehydrogenase. *ACS Catal.* **6** (3): 1598–1605.
67. Wang, J., Li, G., and Reetz, M.T. (2017). Enzymatic site-selectivity enabled by structure-guided directed evolution. *Chem. Comm.* 3916–3928.
68. Li, A., Acevedo-Rocha, C.G., Sun, Z. et al. (2018). Beating bias in directed evolution of proteins: combining high-fidelity on-chip solid-phase gene synthesis with efficient gene assembly for combinatorial library construction. *ChemBioChem.* **19** (3): 221–228.
69. Li, G., Garcia-Borras, M., Fürst, M. et al. (2018). Overriding traditional electronic effects in biocatalytic Baeyer–Villiger reactions by directed evolution. *J. Am. Chem. Soc.* **140** (33): 10464–10472.
70. Selected papers focusing on machine learning in directed evolution: (a) Feng, X., Sanchis, J., Reetz, M.T., and Rabitz, H. (2012). Enhancing the efficiency of directed evolution in focused enzyme libraries by the adaptive substituent reordering algorithm. *Chem. Eur. J.* **18**: 5646–5654. (b) Cadet, F., Fontaine, N., Li, G. et al. (2018). A machine learning approach for reliable prediction of amino acid interactions and its application in the directed evolution of enantioselective enzymes. *Sci. Reports* **8** (1): 16757. (c) Li, G., Dong, Y., and Reetz, M.T. (2019). Can machine learning revolutionize directed evolution of selective enzymes? *Adv. Synth. Catal.* **361** (11): 2377–2386. (d) Wu, Z., Kann, S.J., Lewis, R.D. et al. (2019). Machine learning-assisted directed protein evolution with combinatorial libraries. *Proc. Natl. Acad. Sci. U.S.A.* **116**: 8852–8858. (e) Mazurenko, S., Prokop, Z., and Damborsky, J. (2020). Machine learning in enzyme engineering. *ACS Catal.* **10**: 1210–1223. (f) Wittmann, B.J., Johnston, K.E., Wu, Z., and Arnold, F.H. (2021). Advances in machine learning for directed evolution. *Curr. Opin. Struct. Biol.* **69**: 11–18. (g) Biswas, S., Khimulya, G., Alley, E.C. et al. (2021). Low-N protein engineering with data-efficient deep learning. *Nat. Methods* **18**: 389–396. (h) Li, G., Qin, Y., Fontaine, N.T. et al. (2021). Machine learning enables selection of epistatic enzyme mutants for stability against unfolding and detrimental aggregation. *ChemBioChem* **22** (5): 904–914.
71. (a) Xu, J., Cen, Y., Singh, W. et al. (2019). Stereodivergent protein engineering of a lipase to access all possible stereoisomers bearing multiple stereocenters. *J. Am. Chem. Soc.* **141** (19): 7934–7945. (b) Li, D., Wu, Q., and Reetz, M.T. (2020). Focused rational iterative site-specific mutagenesis (FRISM). *Methods Enzymol.* **643**: 225–242.
72. Acevedo-Rocha, C.G., Li, A., D'Amore, L. et al. (2021). Pervasive cooperative mutational effects on multiple catalytic enzyme traits emerge via long-range conformational dynamics. *Nat. Commun.* **12** (1): 1621.
73. (a) Coelho, P.S., Brustad, E.M., Kannan, A., and Arnold, F.H. (2013). Olefin cyclopropanation via carbene transfer catalyzed by engineered cytochrome P450 enzymes. *Science* **339** (6117): 307–310. (b) Coelho, P.S., Wang, Z.J., Ener, M.E. et al. (2013). A serine-substituted P450 catalyzes highly

efficient carbene transfer to olefins *in vivo*. *Nat. Chem. Biol.* **9** (8): 485–487. (c) Hyster, T.K., Farwell, C.C., Buller, A.R. et al. (2014). Enzyme-controlled nitrogen-atom transfer enables regiodivergent C–H amination. *J. Am. Chem. Soc.* **136** (44): 15505–15508. (d) Hammera, S.C., Knight, A.M., and Arnold, F.H. (2017). Design and evolution of enzymes for non-natural chemistry. *Curr. Opin. Green Sustainable Chem.* **7**: 23–30.

74. Cen, Y., Singh, W., Arkin, M. et al. (2019). Artificial cysteine-lipases with high activity and altered catalytic mechanism created by laboratory evolution. *Nat. Commun.* **10** (1): 3198.

75. (a) Widmanna, M., Pleiss, J., and Samland, A.K. (2012). Computational tools for rational protein engineering of aldolases. *Comput. Struct. Biotechnol. J.* **2** (3): e201209016. (b) Planas-Iglesias, J., Marques, S.M., Pinto, G.P. et al. (2021). Computational design of enzymes for biotechnological applications. *Biotech. Adv.* **47**: 107696. (c) Huang, P.-S., Boyken, S.E., and Baker, D. (2016). The coming of age of *de novo* protein design. *Nature* **537**: 320–327. (d) Leman, J.K., Weitzner, B.D., Lewis, S.M. et al. (2020). Macromolecular modeling and design in Rosetta: recent methods and frameworks. *Nat. Methods* **17**: 665–680.

76. (a) Brannigan, J.A. and Wilkinson, A.J. (2002). Protein engineering 20 years on. *Nat. Rev. Mol. Cell Biol.* **3**: 964–970. (b) Cedrone, F., Ménez, A., and Quéméneur, E. (2000). Tailoring new enzyme functions by rational redesign. *Curr. Opin. Struct. Biol.* **10**: 405–410.

77. (a) Eijsink, V.G.H., Bjørk, A., Gåseidnes, S. et al. (2004). Rational engineering of enzyme stability. *J. Biotechnol.* **113** (1–3): 105–120. (b) Missimer, J.H., Steinmetz, M.O., Baron, R. et al. (2007). Configurational entropy elucidates the role of salt-bridge networks in protein thermostability. *Protein Sci.* **16** (7): 1349–1359. (c) Lee, C.W., Wang, H.J., Hwang, J.K., and Tseng, C.P. (2014). Protein thermal stability enhancement by designing salt bridges: a combined computational and experimental study. *PLoS One* **9**: e112751. (d) Fukuchi, S. and Nishikawa, K. (2001). Protein surface amino acid compositions distinctively differ between thermophilic and mesophilic bacteria. *J. Mol. Biol.* **309**: 835–843. (e) Kumar, S., Tsai, C.J., and Nussinov, R. (2000). Factors enhancing protein thermostability. *Protein Eng.* **13**: 179–191. (f) Szilagyi, A. and Zavodszky, P. (2000). Structural differences between mesophilic, moderately thermophilic and extremely thermophilic protein subunits: results of a comprehensive survey. *Structure* **8**: 493–504. (g) Chan, C.-H., Yu, T.-H., and Wong, K.-B. (2011). Stabilizing salt-bridge enhances protein thermostability by reducing the heat capacity change of unfolding. *PLoS One* **6** (6): e21624. (h) Cui, H., Eltoukhy, L., Zhang, L. et al. (2021). Less unfavorable salt bridges on the enzyme surface result in more organic cosolvent resistance. *Angew. Chem. Int. Ed.* **60** (20): 11448–11456.

78. (a) Kreß, N., Halder, J.M., Rapp, L.R., and Hauer, B. (2018). Unlocked potential of dynamic elements in protein structures: channels and loops. *Curr. Opin. Chem. Biol.* **47**: 109–116. (b) Petrovic, D., Risso, V.A., Kamerlin, S.C.L., and Sanchez-Ruiz, J.M. (2018). Conformational dynamics and enzyme evolution. *J. R. Soc. Interface* **15** (144): 20180330. (c) Ouedraogo, D., Souffrant, M., Vasquez, S. et al. (2017). Importance of Loop L1 dynamics for substrate capture and catalysis in *Pseudomonas aeruginosa* d-arginine dehydrogenase. *Biochemistry* **56**: 2477–2487. (d) Liu, B., Qu, G., Li, J. et al. (2019). Conformational dynamics-guided loop engineering of an alcohol dehydrogenase: capture, turnover and enantioselective transformation of difficult-to-reduce ketones. *Adv. Synth. Catal.* **361**: 3182–3190. (e) Qu, G., Bi, Y., Liu, B. et al. (2022). Unlocking the stereoselectivity and substrate acceptance of enzymes: proline induced loop engineering test. *Angew. Chem. Int. Ed.* **61**: e202110793. (f) Han, S.-S., Kyeong, H.-H., Choi, J.M. et al. (2016). Engineering of the conformational dynamics of an enzyme for relieving the product inhibition. *ACS*

Catal. **6**: 8440–8445. (g) Kaczmarski, J.A., Mahawaththa, M.C., Feintuch, A. et al. (2020). Altered conformational sampling along an evolutionary trajectory changes the catalytic activity of an enzyme. *Nat. Commun.* **11**: 5945. (h) Karamitros, C.S., Murray, K., Winemiller, B. et al. (2022). Leveraging intrinsic flexibility to engineer enhanced enzyme catalytic activity. *Proc. Natl. Acad. Sci. U.S.A.* **119** (23): e2118979119. (i) Parra-Cruz, R., Jager, C.M., Lau, P.L. et al. (2018). Rational design of thermostable carbonic anhydrase mutants using molecular dynamics simulations. *J. Phys. Chem. B* **122** (36): 8526–8536.

79. (a) Wijma, H.J., Floor, R.J., Bjelic, S. et al. (2015). Enantioselective enzymes by computational design and *in silico* screening. *Angew. Chem. Int. Ed.* **54**: 3726–3730. (b) Li, J., Qu, G., Shang, N. et al. (2021). Near-perfect control of the regioselective glucosylation enabled by rational design of glycosyltransferases. *Green Synth. Catal.* **2**: 45–53. (c) Ortiz de Montellano, P.R. (2010). Hydrocarbon hydroxylation by cytochrome P450 enzymes. *Chem. Rev.* **110**: 932–948.

80. (a) Röthlisberger, D., Khersonsky, O., Wollacott, A.M. et al. (2008). Kemp elimination catalysts by computational enzyme design. *Nature* **453** (7192): 190–195. (b) Zanghellini, A., Jiang, L., Wollacott, A.M. et al. (2006). New algorithms and an in silico benchmark for computational enzyme design. *Protein Sci.* **15** (12): 2785–2794. (c) Cui, Y., Wang, Y., Tian, W. et al. (2021). Development of a versatile and efficient C–N lyase platform for asymmetric hydroamination via computational enzyme redesign. *Nat. Catal.* **4**: 364–373.

81. Unger, E.K., Keller, J.P., Altermatt, M. et al. (2020). Directed evolution of a selective and sensitive serotonin sensor via machine learning. *Cell* **183** (7): 1986–2002.

82. Reetz, M.T. (2022). Making enzymes suitable for organic chemistry by rational protein design. *ChemBioChem* e202200049.

2
Screening and Selection Techniques

2.1
Introductory Remarks

Directed evolution of activity, stereo- and regio-selectivity, as well as thermostability requires efficient screening or selection methods [1]. The terms "screening" and "selection" are sometimes used interchangeably, which may cause some confusion. In fact, the two terms are distinctly different [2]. Screening utilizes analytical tools such as automated gas chromatography (GC) or high-performance liquid chromatography (HPLC), but also robotic instruments based on UV/vis, fluorescence, multiplex mass spectrometry, or circular dichroism. In sharp contrast, selection [3] involves an experimental platform in which the host organism, such as *Escherichia coli*, has a growth and survival advantage because it harbors an enzyme or mutants thereof with a desired (evolved) catalytic profile (Scheme 2.1). Ideally, only those colonies can be expected to appear on agar plates that harbor the desired improved variants. In principle, this is a highly attractive feature. Unfortunately, the fact that many examples of "selection" have been reported does not mean that this method can be used in a general way to identify improved mutants in directed evolution, especially when aiming for stereoselectivity [1–3]. As will be seen, whatever approach is chosen, screening or selection, the linkage of genotype to phenotype must be maintained.

2.2
Screening Methods

A comprehensive overview of screening systems used in the directed evolution of activity, substrate scope, and stereoselectivity as well as thermostability can be found in a 2006 monograph edited by J.-L. Reymond [1a], which was later augmented by review articles [1b–i, 2, 3]. We begin by focusing on some basic principles. The advent of directed evolution of stereoselective enzymes in 1997 included the first medium-throughput ee-assay [4a], which sparked research on further and more efficient medium- and high-throughput ee-screening systems [4b, c, 5]. In the original study and subsequent reports [4b, c, 5], involving the hydrolytic kinetic resolution of a racemic *p*-nitrophenyl ester catalyzed by the

Enzyme Engineering: Selective Catalysts for Applications in Biotechnology, Organic Chemistry, and Life Science, First Edition. Manfred T. Reetz, Zhoutong Sun, and Ge Qu.

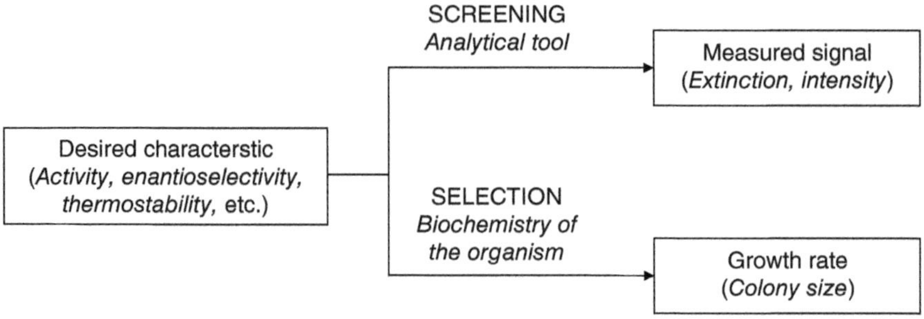

Scheme 2.1 Screening versus selection in directed evolution. Source: Acevedo-Rocha et al. [2]/With permission of Elsevier.

lipase PAL (see Schemes 1.9 and 1.10) as a function of time (typically during the first eight minutes). The released *p*-nitrophenolate has a yellow color, which can be monitored by a UV/vis plate reader. It is highlighted here because the underlying principle forms the basis of many other assays needed for different enzyme types. Measuring the time-dependent formation of *p*-nitrophenolate released from the racemic ester provides information only about the approximate overall rate, not of the enantioselectivity. Therefore, a trick had to be devised [4]. The (*R*)- and (*S*)-substrates were prepared separately in enantiomerically pure form by the synthetic Evans method that was then studied pairwise on 96-well microtiter plates, allowing 48 mutants to be assayed for enantioselectivity on one plate. Whenever the slopes of the absorption/time curves differ considerably, a hit is identified, which is subsequently used as the catalyst in the kinetics of the real hydrolytic kinetic resolution of the racemic *p*-nitrophenyl ester. This allows the exact determination of the selectivity factor E (relative rate of reaction of one enantiomer compared to that of its mirror image). Whenever the slopes are very similar or even overlap, stereoselectivity is unacceptable. Figure 2.1 shows the respective curves of the wildtype (WT) lipase, indicating extremely low selectivity (experimentally, $E = 1.4$), and a typical variant displaying enhanced enantioselectivity [4a].

The measurement of about 500–800 plots of this kind was possible per day. However, this does not have to be performed for all of the transformants because an on-plate pretest for activity can be carried out (Figure 2.2). Only the active clones are subsequently studied by the UV/vis assay or directly by kinetics. The agar plate contains tributyrin, which on hydrolysis forms a halo easily identified by the human eye. This is a classical test used to assess lipase or esterase activity, which was simply adapted for high-throughput analysis [4a, c, 5]. The so-called casein test for identifying active proteases is also based on the appearance of halos on casein-spiked agar plates [6].

The idea that the initial rates of reaction of (*R*)- and (*S*)-enantiomers are crudely assessed separately in the wells of microtiter plates forms the basis of many other ee-assays [1, 4, 5]. The so-called umbelliferone-based fluorescence assay is an example that depends on the use of surrogate substrates; it can

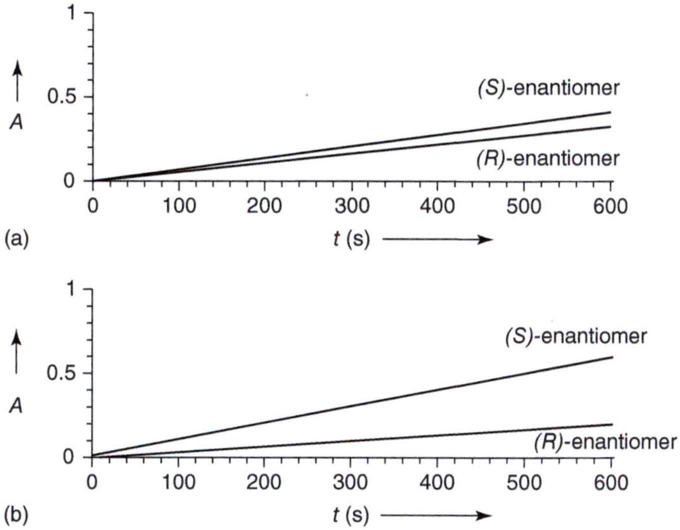

Figure 2.1 (a) Time course of the lipase-catalyzed hydrolysis of two enantiomeric *p*-nitrophenyl esters, (*S*)- and (*R*) (Schemes 1.9 and 1.10) separately using WT enzyme with poor stereoselectivity. (b) Time course of the lipase-catalyzed hydrolysis of the two enantiomeric *p*-nitrophenyl esters using an enzyme variant with enhanced (*S*)-enantioselectivity. Source: Refs. [4a, c, 5]/ John Wiley & Sons.

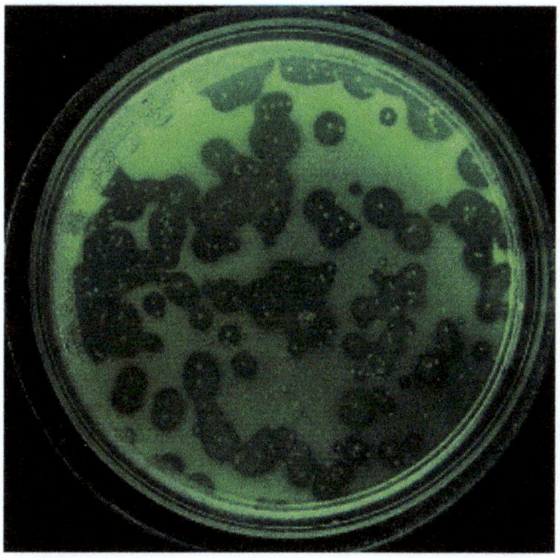

Figure 2.2 On-plate pretest for lipase activity based on halos that form upon hydrolysis of tributyrin. White dots represent bacterial colonies harboring active lipases; those having no (clear) black background contain inactive mutants. Source: Reetz et al. [4c]/with permission of Elsevier.

also be used for activity determination alone [1a]. In one embodiment, a periodate-coupled fluorogenic assay was used specifically for hydrolases such as lipases, esterases, and epoxide hydrolases (Scheme 2.2) [1a, 7].

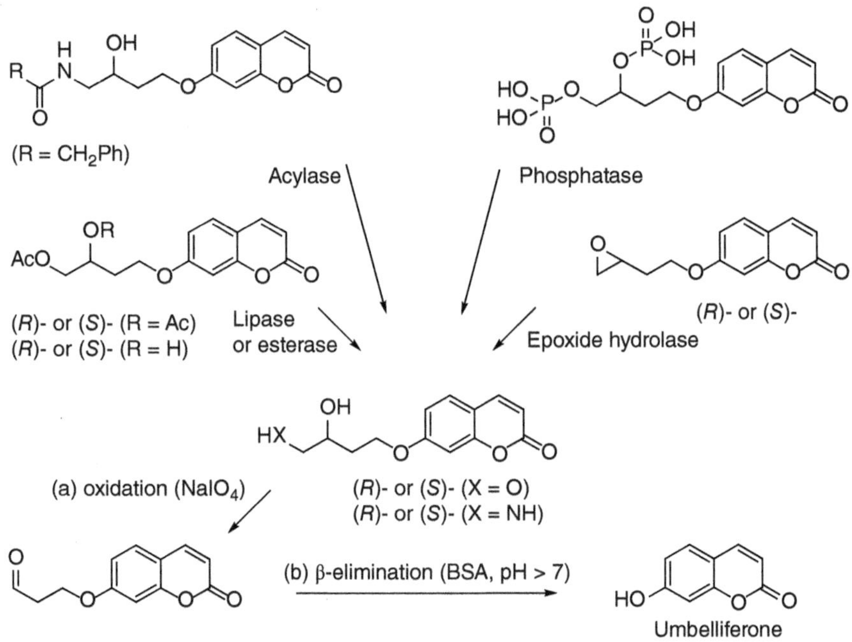

Scheme 2.2 Periodate-coupled fluorogenic assay designed for assessing the enantioselectivity of hydrolases. Source: Adapted from Refs. [1a, 7].

Whenever enantioselectivity is not involved, i.e. when only activity assessment is aimed for, these and similar surrogate-dependent UV/vis- or fluorescence-based screens can be applied. But again, they suffer from the disadvantage that surrogate substrates will not be used in practical (industrial) applications. The mutants evolved for these substrates are likely to show different selectivity and activity when the corresponding "real" substrates are employed for industrial applications. Therefore, alternative assays are generally preferred. The report of an enzyme-coupled UV/vis-based assay for assessing the activity of lipases or esterases is an example, a measure of enantioselectivity also being possible by using "real" (*R*)- and (*S*)-substrates separately (Scheme 2.3) [8a]. Lipase- or esterase-catalyzed hydrolysis forms acetic acid, which is converted by acetyl-CoA synthetase (ACS) to acetyl-CoA in the presence of ATP and coenzyme A (CoA). Citrate synthase (CS) is used to catalyze the reaction between acetyl-CoA and oxaloacetate, leading to citrate. The required oxaloacetate is generated from L-malate and NAD^+ in the presence of L-malate dehydrogenase (L-MDH). Initial rates of acetic acid formation are accessible by monitoring the increase in absorption at 340 nm due to the increase in NADH concentration.

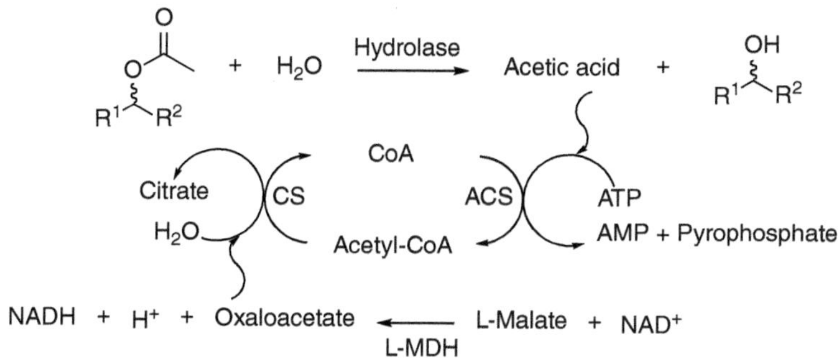

Scheme 2.3 An enzyme-coupled assay for assessing the activity of lipases or esterases, with measurement of the apparent enantioselectivity also being possible when using (*R*)- and (*S*)-substrates separately. Source: Baumann et al. [8a]/John Wiley & Sons.

An alternative high-throughput ee-screen also utilizing enzyme-coupled reactions can be applied to any enzyme-catalyzed stereoselective formation of chiral alcohols (e.g. ADH reduction of prochiral ketones, P450-catalyzed oxidative hydroxylation of achiral compounds, and lipase- or esterase-catalyzed hydrolysis of esters) [8b]. Further screens, which also avoid the use of surrogates, have been developed [1], e.g. those that measure the pH change in a lipase- or esterase-catalyzed reaction by means of appropriate dye indicators [9].

Yet another approach utilizes isotopically labeled pseudo-enantiomers in hydrolytic or oxidative kinetic resolution, or using (pseudo) meso substrates in hydrolytic or oxidative desymmetrization (Scheme 2.4) [10–12]. For labeling, deuterium, ^{13}C, or ^{15}N isotopes are generally used, with quantitative high-throughput detection of enantioselectivity being possible by applying multiplexing mass spectrometry (MS) [10a–d]. This unique concept was applied in the directed evolution of an enantioselective lipase and an epoxide hydrolase, with up to 10 000 ee-determinations being possible per day in optimized systems [10b]. In an industrial example, directed evolution of a nitrilase as the catalyst in the desymmetrization of 1,3-dicyano-3-hydroxypropane needed in the chemoenzymatic synthesis of the cholesterol-lowering therapeutic drug Lipitor®, one of the cyano-groups was ^{15}C-labeled; enantioselectivity was then measured by multiplexing MS, which allowed thousands of samples to be analyzed in one day [10c]. IR spectroscopy can be used as an attractive alternative to the expensive multiplexing MS because it is very effective and requires only a relatively cheap instrument [11]. Even fast NMR spectroscopy can be employed [12]. It must be noted that the concept shown in Scheme 2.4 does not apply to all asymmetric processes that can be detected by these otherwise efficient screens, e.g. enantioselective reduction of prochiral ketones. In contrast to the IR-based system, the multiplexing MS instrument is currently too expensive for most academic laboratories.

Screening methods with high throughput have also been developed for oxidoreductases, the majority of which require the cofactors $NAD(P)^+$ and

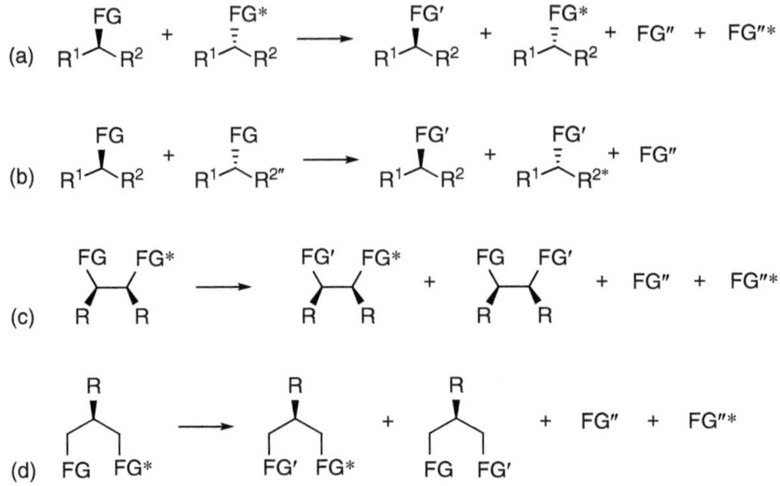

Scheme 2.4 Utilization of isotopically labeled pseudo-enantiomers in high-throughput screening of mutant libraries generated by directed evolution [10]. (a) Asymmetric transformation of a mixture of pseudo-enantiomers involving cleavage of the functional groups FG and labeled functional groups FG*. (b) Asymmetric transformation of a mixture of pseudo-enantiomers involving either cleavage or bond formation at the functional group FG; isotopic labeling at R^2 is indicated by the asterisk. (c) Asymmetric transformation of a pseudo-meso substrate involving cleavage of the functional groups FG and labeled functional groups FG*. (d) Asymmetric transformation of a pseudo-prochiral substrate involving cleavage of the functional group FG and labeled functional group FG*. Source: Reetz et al. [10a]/John Wiley & Sons.

NAD(P)H [13]. However, due to background signals originating from cell lysates, the absorbance of NAD(P)H at 340 nm cannot generally be used to monitor its formation or depletion as a function of time in a fully reliable manner. Consequently, the redox reactions need to be coupled with a dye-forming reaction, of which various types have been developed [13]. In some cases, focusing on the oxidized form $NAD(P)^+$ is an alternative, with absorption at 580 nm being monitored by a UV/vis-plate reader. In this case, an agar plate pretest based on the identification of white spots on a purple background is useful for spotting active clones. In special applications, other tests are called for, as in the case of P450-catalyzed terminal oxidative hydroxylation of linear alkanes, in which *p*-nitrophenylethers are used as surrogate substrates that generate the yellow *p*-nitrophenolate for straightforward monitoring by a plate reader [13b]. Among other related developments, several convenient colorimetric screening assays have been developed for application in the directed evolution of fungal laccases as catalysts in the conversion of plant biomass [13c]. Other examples include a high-throughput color-based screening system for assaying the activity of terpene-synthetases in cyclization reactions [13d], and a microplate assay for real-time screening of aldolases [13e]. Apart from the dependence of $NAD(P)^+$ or NAD(P)H, a newly reported HTS approach can monitor the activity of P450 fatty acid decarboxylases based on the colorimetric detection of H_2O_2 consumption [13f]. In addition, new trends toward high-throughput screening in terms of carbonyl groups have recently been reported [14]. Carbonyl groups in ketones or aldehydes can easily form chromogenic products with particular molecular probes. Taking this advantage, high-throughput color-based screening systems were developed for assaying the activity of carboxylic acid reductases using amino benzamidoxime (ABAO) as a probe [14a], and for assaying the activity of ketone reductases utilizing 2,4-dinitrophenylhydrazine (DNPH) [14b] or *para*-methoxy-2-amino benzamidoxime (PMA) [14c].

Traditionally, chiral GC and HPLC were considered to be slow processes. However, especially with the advent of directed evolution of stereoselective enzymes [4a], the need for high- or at least medium-throughput arose. As has been stressed so often, developing efficient screening systems goes hand in hand with the construction of small but smart mutant libraries (Chapter 1). During the past 15 years, methodology development of smart mutant libraries has continued at a rapid pace [15], which means that today fancy assays are usually no longer needed, since automated GC or HPLC can be used, specifically when connected robotically to the respective microtiter plates. However, in the early directed evolution days, such automated GC or HPLC systems did not exist. The first attempt to implement a platform for medium-throughput ee-determination in the directed evolution of an enantioselective enzyme utilized two GC instruments in one unit (Scheme 2.5), rather than two columns in one unit [16]. This saves space, but in many labs, this may not be a current

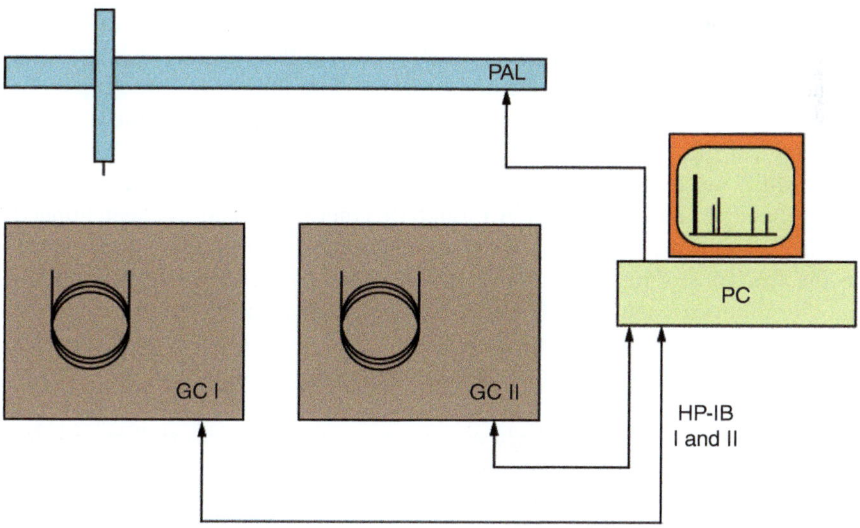

Scheme 2.5 First example of a medium-throughput GC unit used in the directed evolution of an enantioselective enzyme (lipase from *Pseudomonas aeruginosa*, PAL) [16]. It contained two GC instruments and one PC, with this part of the platform being robotically connected to the respective 96-well microtiter plates. Source: Reetz et al. [16]/With permission of Elsevier.

issue. Today, automated GC instruments for handling samples taken directly from microtiter plates, including washing steps, are commercially available at a cost of less than $50 000.

Numerous more recent examples of such GC (HPLC) platforms have been reported. An example is the directed evolution of the haloalkane dehalogenase DhaH-31 [16]. Depending upon the nature of the product or starting material, about 300–700 samples can be handled per day [17, 18]. Automated HPLC has been used in a similar medium-throughput manner [17]. In most cases, the respective throughputs are not as high as in GC-based systems, but since the generation of small and smart libraries is continually improving [15], these screening systems suffice. Other recent high-throughput screening examples include a prazole sulfide monooxygenase [19] and a malonyl-CoA synthetase [20].

The concept of pooling mutant libraries offers an interesting way to reduce the overall screening effort to a notable extent [21a]. An early improved system, the directed evolution of the enoate reductase YqjM as a catalyst in the enantioselective conjugate reduction of 3-substituted cyclohexenone derivatives, was studied [21b]. Analysis of the pooling and screening results demonstrated that the screening effort can be reduced by a factor of at least 2 [21b].

In addition to pooling mutant libraries, screening pooled racemic compounds in asymmetric catalysis is also a fantastic task. However, how to simultaneously optimize multiple, diverse substrates with high enantioselectivity is a challenge in asymmetric catalysis [22a]. The conventional analytical methods for ee determination include HPLC or supercritical fluid chromatography (SFC) with chiral stationary phases (CSPs) combined with isocratic elution and UV–vis detection. Due to the dependence of baseline separation and the removal of interfering signals before analysis, these methods are therefore limited in throughput. To overcome this obstacle, the Jacobsen group investigated a new pipeline coupling supercritical fluid chromatography and mass spectrometry (SFC-MS)that was successfully used to complete diverse ee measurements simultaneously in Pictet–Spengler reactions [22a]. Although this method has not yet been applied in directed enzyme evolution, it can be envisioned as a promising technique in the scenarios of high-throughput screening of enantioselective variants. In another innovative study, a MS-based screening system using chiral man-made catalysts by monitoring backreactions of quasi-enantiomeric products was developed, but its application in protein engineering has not been reported to date [22b].

Allosteric transcription factors can be utilized as a powerful means for high-throughput *in vivo* screening in biocatalytic systems. However, most transcription factors suffer from a narrow concentration range of the corresponding target-ligands [23a]. For example, the transcription factor NadR from *Bacillus subtilis* (BsNadR), which in response to niacin concentration, is oversensitive to metabolically synthesized niacin. Even trace amounts of niacin lead to extremely strong repressive activity [23b]. To make it functional over a wide range of niacin concentration, the Zhou group used epPCR mutagenesis on the natural BsNadR, and obtained the desensitized BsNadR mutants with distinct niacin operating ranges [23b]. As such, they developed a niacin biosensor (NAsensor) using a green fluorescent protein (GFP) as a reporter, responding to a wide range of niacin concentrations (0–50 mM) [23b]. Subsequently, NAsensor was successfully applied to the high-throughput *in vivo* screening of three different enzymes, nitrilase, amidase, and nitrile hydratase (Scheme 2.6) [23b].

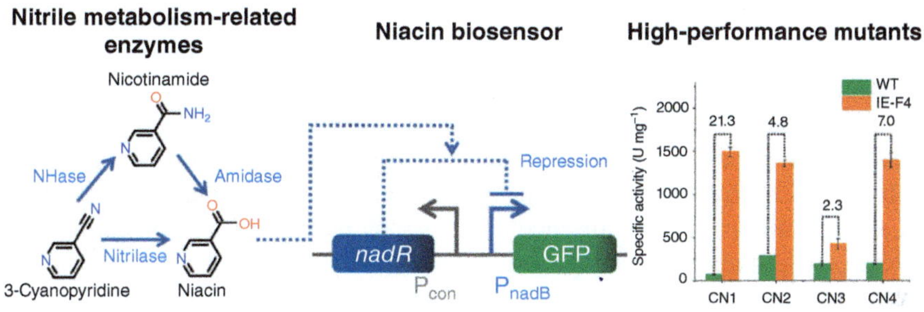

Scheme 2.6 Developing a desensitized niacin repressive biosensor and its application in the directed evolution of nitrile metabolism-related enzymes. Source: Han et al. [23b]/American Chemical Society.

Fluorescence-based screening is being used widely for many applications. An example is the design of an aptamer-based biosensor for high-throughput ee-determination [24], but its application in directed evolution of selective enzymes still needs to be demonstrated. Two seminal reviews of high-throughput screening in protein engineering cover the most recent developments, including a list of FACS-based screening platforms and microfluidic assays [1g, h]. Microfluidic screening as part of lab-on-a-chip is indeed rapidly emerging for different areas of potential industrial application, including directed evolution [25–29].

In an interesting development, the rapid exploration of the substrate scope of newly discovered enzymes or mutants generated by directed evolution has been reported. To achieve this goal, cluster-screening was developed, as in the case of new cytochrome P450 monooxygenases (CYPs) [30]. A library of 51 compounds to be tested with two different enzymes of this type (CYP 154E1) and (CYP 154A8) was organized into nine groups according to their structural properties. Following CYP-catalyzed oxidation, automated GC/MS was applied, with the generation of practical data.

In parallel, *in silico* high-throughput screening techniques have also been developed recently. Similar to directed evolution, *de novo* computational protein design programs (e.g. Rosetta Design) can generate a vast set of designed variants, which can be evaluated by *in silico* screening methods [31]. For example, high-throughput molecular dynamics simulations combined with the geometric criteria for near attack conformations are capable of screening enantioselective enzymes [32]. It was demonstrated that invoking *in silico* screening into protein engineering can reduce the wet-lab work and efficiently obtain enhanced biocatalysts.

2.3
Selection Methods

Tracing back to the first DNA shuffling study, antibiotic cefotaxime was increased stepwise as selection pressure to evolve the activity of TEM-1 ß-lactamase in *E. coli* as the catalyst [33]. In another prominent example, an aspartate aminotransferase was engineered with extended substrate acceptance in which the selection system used the auxotrophy of an *E. coli* strain deficient in the branched-chain amino acid transferase gene *ilvE*, which cannot grow on a minimal plate in the absence of the supplements valine, isoleucine, and leucine [34]. Later on, employing auxotrophic strains that grow only when a given (mutant) enzyme is generated or evolved, has been implemented in other studies as well. However, in such cases, even low enzyme activity may meet the needs of cell growth, which hampers the development of an efficient selection system. A possible way to tackle this problem is to tune transcription with an enzyme-degradation tag which reduces intracellular protein concentrations to relatively low levels, a concept thereupon was applied successfully to chorismate mutase [35]. Another example of genetic complementation is discovering genes encoding latent glucokinase activity in an overexpression library using a recombinant *E. coli* strain that was unable to accept glucose as a carbon source [36]. *In vivo* chemical complementation prevents some of the drawbacks of conventional genetic complementation (each enzyme needs a new assay), mainly based on the use of reporters such as ß-galactosidase or amino acid selectable markers of yeast, which are linked to the substrate of interest [37].

Aside from utilizing auxotrophy hosts, a distinct selection method was devised for increasing the activity of an *N*-acyl amino acid racemase (NAAAR) by directed evolution, with the aim of establishing a system enabling dynamic kinetic resolution (DKR) of amino acids on an industrial scale [38]. Interestingly, rapid enantio-differentiation from a rapidly racemizing mixture of chiral *N*-acyl amino acids was ensured by a D-acylase known to be highly stereoselective. Selection pressure with the appearance of improved NAAAR variants was enforced by linking the racemization rate to the viability of the *E. coli* host. It was guaranteed by disabling its natural L-methionine biosynthetic pathway, as well as eliminating a D-amino acid racemization pathway. The chemistry involved in this system is shown in Scheme 2.7 [38]. The improved NAAAR combined with the stereoselective D-acylase ensured efficient DKR of different amino acids at an industrially practical level. In more details, the initial mutagenesis was performed using a mutator strain (XL1-Red) (>10^7 variants), which delivered an improved variant G291E showing higher activity than WT. This hot spot was then subjected to saturation mutagenesis (100 variants), which led to a better variant G291D. It was consequently subjected to epPCR (>10^5 variants), generating the final double

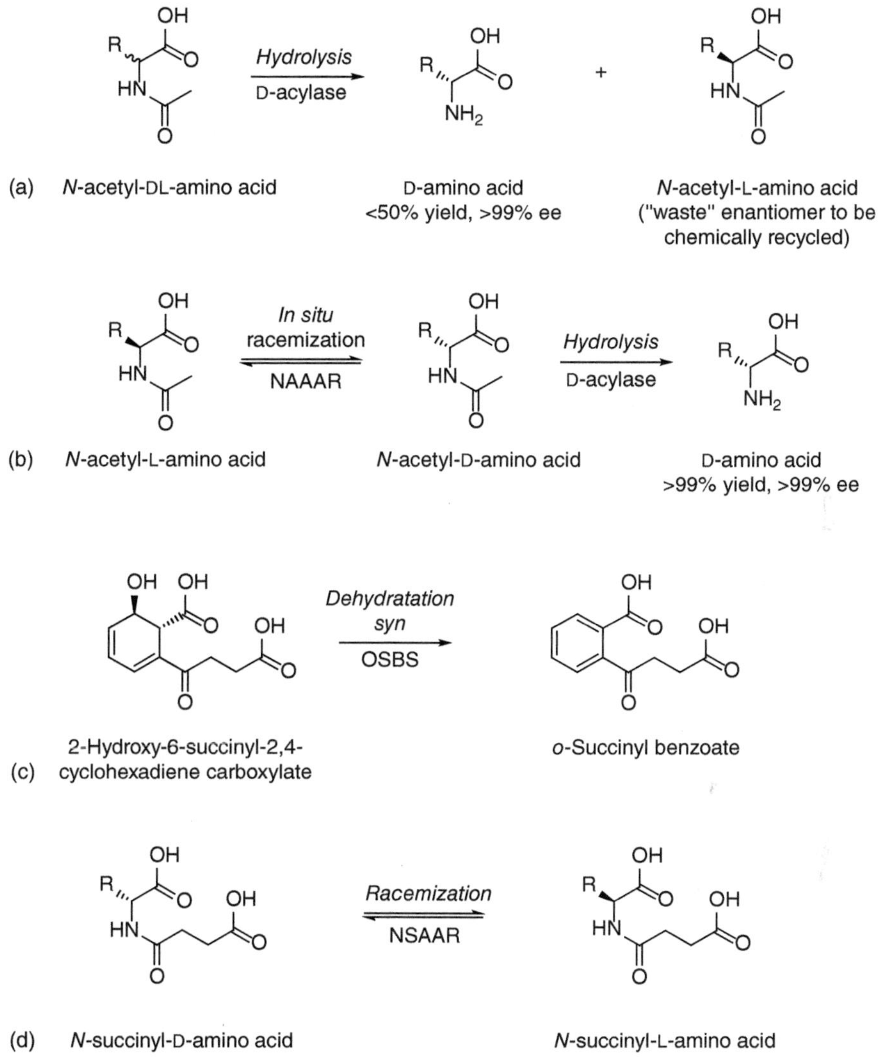

Scheme 2.7 (a) Kinetic resolution of NAcetyl-DL-amino acid using D-Acylase as a catalyst; (b) NAAAR paired with acylase in dynamic kinetic resolution; (c) NAAAR functions as an *o*-succinyl benzoate synthase (OSBS) in the Menaquinone Pathway; (d) *N*-succinyl-amino acid racemase (NSAAR) participated in detoxification of D-amino acid. Source: Baxter et al. [38]/American Chemical Society.

mutant G291D/F323Y with a sixfold increase in activity relative to WT NAAAR. Notably, this elegant system involves directed evolution of activity, not enantioselectivity.

Likewise, Fahrig-Kamarauskaite et al. developed an *in vivo* selection system for improving the activity of a chorismate mutase from *Mycobacterium tuberculosis* (MtCM) [39]. This selection system was constructed based on the defect strain *E. coli* KA12 that has a deletion of the chorismate mutase-encoding genes. As such, it requires the introduction of a functional chorismate mutase gene to survive on minimal medium that lacks phenylalanine and tyrosine [39]. Consequently, four iterative cycles of mutagenesis and selection were carried out with the aim of improving the activity of MtCM. The best MtCM variant showed a k_{cat}/K_m value with a 270-fold improvement over that of the starting enzyme [39].

Another case study of designed growth-selection pressure in directed evolution and pathway engineering utilizes a related method with the aim of increasing the efficiency of an NADPH-dependent homophenylalanine dehydrogenase [40]. In fact, several growth-based, high-throughput selection systems using cofactor recycling strategies have been developed for both NADPH- and NADH-dependent oxygenases. Such studies include converting the cofactor-dependence of cyclohexanone monooxygenase from NADPH to NADH [41], and improving the activity of carboxylic acid reductase [42]. They do not involve selection-based engineering of stereoselectivity.

In an earlier work, a simple and efficient on-plate selection system for identifying active epoxide hydrolases in an *E. coli* strain had been devised [43]. It is known that epoxides are toxic to many organisms, while hydrolysis with the formation of the respective diol generally causes detoxification. Thus, an epoxide hydrolase under defined conditions with more reactivity will lead to better cell growth and survival. As such, prior to the normal directed evolution procedure, agar plates containing *E. coli* were first treated with various amounts of chiral epoxide. After culturing for several days, visual inspection of the plates is able to identify positive (active) hits in a large mutant library [43]. The resultant mutants can then be isolated, characterized, and tested as catalysts in the hydrolytic kinetic resolution of racemic epoxides as substrates for possible enhanced enantioselectivity. With the aim of verifying the viability of this crude but useful preselection system, an agar plate harboring *E. coli* and an epoxide was charged with 92 inactive and 4 active epoxide hydrolase mutants at defined positions. After incubation, the active variants were successfully identified via visual inspection [43]. As such, this system can be automated for high-throughput identification of active epoxide hydrolases from large libraries of mutants. In more details, it may require the simple transfer of fresh transformants manually or automatically by a robot (e.g. QPix or Genetix) to the epoxide-containing agar plates harboring *E. coli* in microtiter plate format (96- or 384-well). Hundreds of such plates are possible to be produced per day, while an extension to stereoselectivity selection still needs to be developed.

The aforementioned contributions and various other examples of selection-based directed evolution constitute impressive achievements, yet genetic selection is still not general [2]. A major reason is that it is difficult to construct such experimental platforms for many enzyme types. It is even more challenging to develop such selection systems when engineering stereoselectivity, and indeed, very few examples have been reported [2]. The fundamental question is: why should an organism have a growth advantage just because it harbors an enantioselective enzyme mutant? If this could be properly addressed, then only colonies harboring variants with enhanced enantioselectivity for a defined asymmetric reaction would appear on agar plates, which would be of tremendous advantage (Scheme 2.8).

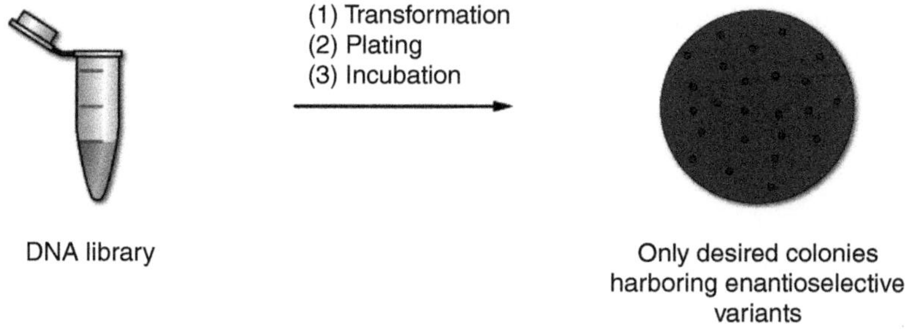

Scheme 2.8 Genetic selection in the directed evolution of enantioselective enzymes. Source: Adapted from Acevedo-Rocha et al. [2].

With the aim of identifying enantioselective lipases, a system utilizing pro-antibiotic substrates has been designed [44]. Scheme 2.9 shows the general strategy, which involves monitoring the growth of *E. coli* or *Exixiguobacterium acetylicum* cells during hydrolysis of (*R*)- and (*S*)-esters. To evaluate the viability of this interesting concept, several lipases were tested without the employment of stereoselective hydrolytic kinetic resolution. A possible problem that may be encountered in such a scenario is the fact that surrogates are required as substrates, which would not be used in directed evolution aimed at real

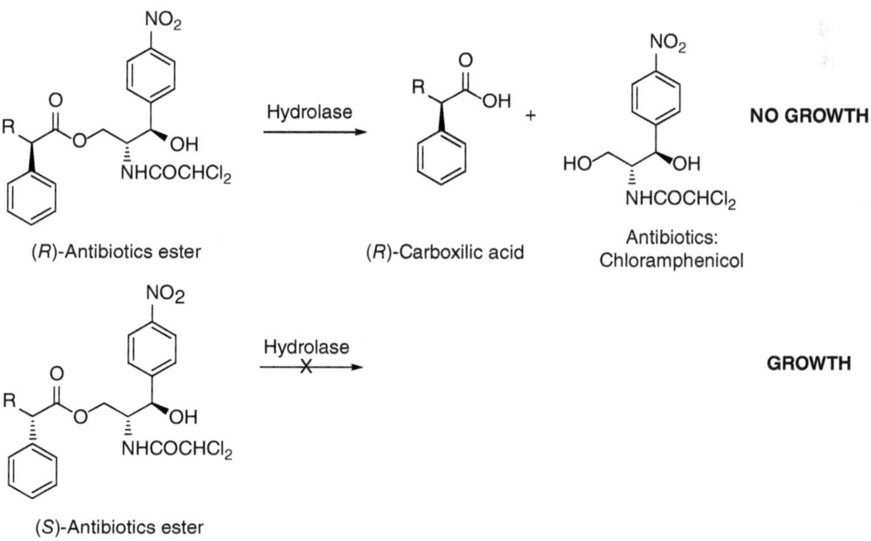

Scheme 2.9 A growth-based selection method based on pro-antibiotic substrates. Source: Hwang et al. [44]/Springer Nature.

(industrial) applications (see discussion concerning surrogates as substrates in Section 2.2).

A conceptionally distinct approach relies on pseudo-enantiomers in which one of the pseudo-enantiomeric pairs has an isosteric yet toxic component, as shown in Scheme 2.10 [45]. Via a bond breaking reaction, the appropriately designed substrate generates a product in one enantiomeric form (e.g. (*R*)-product) and an energy source for cell growth, while the mirror-image substrate constitutes an isosteric pseudo-enantiomer which upon bond breakage releases a poison (inhibition of cell growth or cell death) (Scheme 2.10).

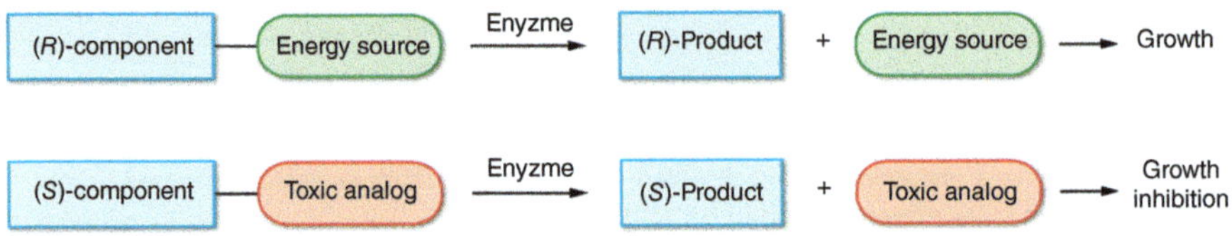

Scheme 2.10 A genetic selection system for directed evolution of enantioselective enzymes in kinetic resolution. Source: Adapted from Reetz et al. [45a].

Due to the necessity of a stereo relevant bond-breaking reaction, the concept is not general. Nonetheless, a proof-of-principle study was constructed by using the lipase-catalyzed hydrolytic kinetic resolution of an acetate as the model reaction. The lipase from *Candida antarctica* B (CALB) was chosen as the enzyme, and *Pichia pastoris* as the host organism [45]. Both the (*R*)- and (*S*)-substrate can be hydrolyzed to generate acetate, which serves as an energy source for the host organism. Therefore, the host organism can survive either hydrolysis of the enantiomers. To add selection pressure favoring one of the enantiomers, a pseudo-racemate was designed as outlined in Scheme 2.10. Initially, the acetate and the sterically similar fluoro-acetate of isopropylidene glycerol (IPG; **2**) were prepared as (*S*)-**1** and (*R*)-**4**, respectively (Scheme 2.11). After hydrolysis, (*S*)-**1** would provide acetic acid (**3**) as an energy source, while in the case of (*R*)-**4**, it would be expected to generate fluoroacetic acid (**5**) as a poison. The latter inhibits the aconitase step of the essential citric acid cycle.

The WT enzyme in the kinetic resolution of *rac*-**1** slightly favors the formation of (*S*)-**2** ($E = 1.9$). The experiment was therefore designed to reverse the enantioselectivity with enhanced (*R*)-selectivity. Preexperiments demonstrated that the two starting compounds themselves are non-toxic to the host organism, and the pseudo-racemates need not consist exactly of a 1 : 1 mixture. However, excessive formation of fluoroacetic acid (**5**) can cause undesired immediate cell death. As such, the amount of (*R*)-**4** had to be decreased to an optimal level [45]. In the following optimization experiments in liquid cultures and solid plates, selection plates with 0.3% (17 mM) of the acetate (*S*)-**1** and 0.003% of the fluoro-acetate (*R*)-**4** were found to be optimal. Thereafter, NNK codon degeneracy-based saturation mutagenesis at a 2-residue site next to the CALB binding pocket at Leu278/Ala281 was carried out based on the optimized

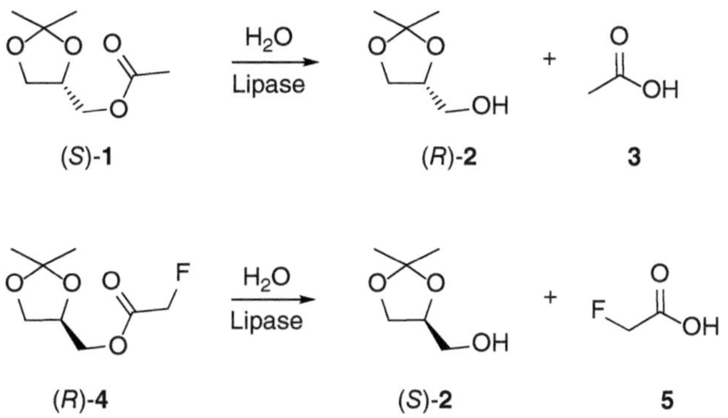

Scheme 2.11 A selection system utilizing a pseudo-racemate (*S*)-**1**/(*R*)-**4** in the CALB-catalyzed hydrolytic kinetic resolution [45]. (Note that the designation of absolute configuration upon going from (*S*)-**1** to (*R*)-**2** or from (*R*)-**4** to (*S*)-**2** switches according to the priority rules of the Cahn – Ingold – Prelog convention). Source: Reetz et al. [45a]/Royal Society of Chemistry.

system. After incubation, the fairly small library was spread out on an agarose plate, and approximately 70–80 colonies were observed. Out of them, 10 largest ones were selected and characterized by sequence determination, which were then tested as catalysts in the hydrolytic kinetic resolution of the real racemic acetates (1 : 1 mixture of (*S*)-**1** and (*R*)-**1**). Eight of the ten variants inversed the enantioselectivity and led to the expected preferential reaction of (*S*)-**1** [45]. The percentage of false positives is only 20%, reflecting that the selection system is viable.

As a proof of principle study, it still suffers from several drawbacks that impede better results. First of all, *P. pastoris* is probably not the most commonly used as a host organism. Next, the library size was too small to ensure enough sequential diversity. Simultaneous saturation mutagenesis at, for example, a 10-residue site would most likely lead to variants showing considerably higher (*S*)-selectivity. Finally, it would be interesting to see how the analogous system performs in which the acetate and fluoro acetate are interchanged, which could harvest CALB variants with opposite enantioselectivity.

At the same time, a different genetic selection system for enhancing or inverting enantioselectivity of a lipase was reported, which is based on an alternative concept [46]. In this study, a lipase from *B. subtilis* (LipA) was used as the catalyst to catalyze the hydrolytic kinetic resolution of *rac*-**6** (Scheme 2.12). Similar to the acetate *rac*-**1** used in the above study, this is also a "difficult" substrate. First, WT LipA was shown to be slightly (*R*)-selective ($E = 1.8$). On this basis, the authors aimed for a reversal of enantioselectivity with the evolution of an (*S*)-selective variant [46].

To this end, a dual selection system was developed by using aspartate esters **7** and **8**, as well as LipA inhibitors **9** and **10** as substrates, respectively (Scheme 2.13) [46]. Initially, a mutant library was generated by saturation mutagenesis on residues 132–136, which was then transformed into the aspartate

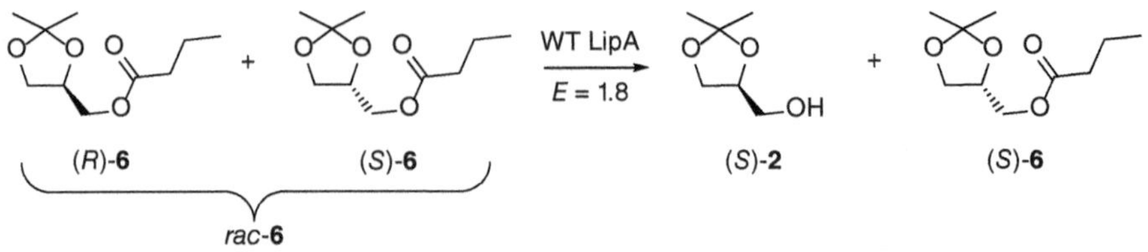

Scheme 2.12 Hydrolytic kinetic resolution of *rac*-**6** directed by LipA. Source: Adapted from Boersma et al. [46].

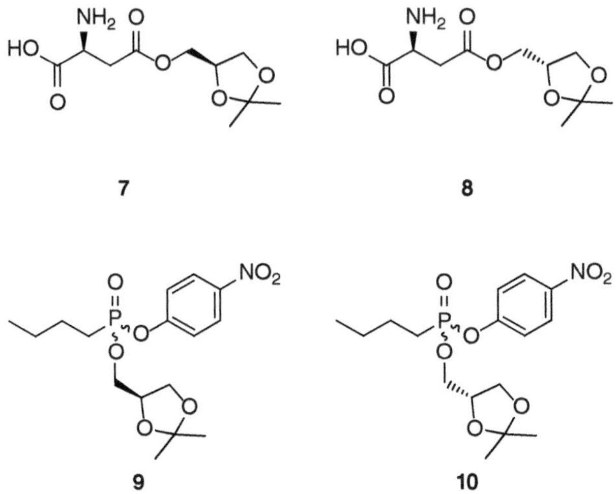

Scheme 2.13 Compounds used in devising a selection system for the evolution of (*S*)-selective LipA variants as catalysts in the hydrolytic kinetic resolution of *rac*-**9**. Source: Adapted from Boersma et al. [46].

auxotrophic strain *E. coli* K-12 PA340/T6. Secondly, this strain was cultured with a selective minimal medium that contains aspartate ester **7**. Only those LipA mutants that hydrolyze this ester to release aspartate allow cell growth. Moreover, the phosphonate inhibitor **10** can be added in this system to minimize the growth of bacteria that express less enantioselective variants [46].

In the first round of incubation, approximately 1000 bacterial colonies appeared on the plates within two days. Subsequently, an increased amount of phosphonate inhibitor was added in the second cycle, which led to 750 colonies left. After assaying the top-ranked 50 largest colonies, the most enantioselective variant was identified as Asp133Gly/Met134Leu/Ile135Asn, with $E = 12$, favoring (*S*)-**2** in the hydrolytic kinetic resolution toward *rac*-**6** [46]. Intriguingly, seven other mutants preferring (*S*)-**2** were also identified, although with lower *E*-values. As commented in the previous study [45], more endeavor is needed to explore how far enantioselectivity can be boosted. Using surrogates as substrates may hinder its employment in practical applications.

As aforementioned, there are multiple meanings of the term "selection" [2]. Apart from the strict sense of the word that relies on cell growth and cell survival for selection (a terminology which the authors of this monograph prefer), the term is also used when referring to various systems in which proteins are displayed on the surface of an organism, e.g. phage display in combination with fluorescence-activated cell sorting (FACS) being a commonly used detection technique [1a, 47]. When detecting proteins with specific binding properties (e.g. antibodies), the display systems are particularly well suited in this scenario. However, such systems are generally less reliable in cases involving catalytic binding and turnover. Still, the merits of display systems are obvious, one being the enormous size of libraries ($\approx 10^{10}$) that can be rapidly screened [1a, 47, 48].

To date, display systems have been devised in directed evolution with the aim of enhancing or inverting enantioselectivity [47]. An early study employed phage display with the aim of evolving an enantioselective mutant of *B. subtilis* lipase (LipA) as the catalyst in the hydrolytic kinetic resolution of *rac*-**6** [49]. By using the similar strategy that was utilized in the dual selection approach featured in Scheme 2.12, the same types of inhibitors were prepared and attached covalently to a highly porous silicate carrier (SIRAN) in the phage-display system (Scheme 2.14). Thereafter, saturation mutagenesis was carried out at large randomization sites. In the first round, c. 8.3×10^{10} phages were incubated with the phosphonate inhibitor immobilized on the solid carrier SIRAN and allowed to react for 15 minutes. After a few rounds of selection illustrated in Scheme 2.14, a LipA mutant showing inverted enantioselectivity (33% ee at 32% conversion) was isolated [49b]. Although inversion of enantioselectivity was achieved according to the plan, the degree of stereoselectivity remains meager. A few reasons for this somewhat disappointing result have been postulated. The most likely one is that phage display, which is based on (selective) complexation rather than catalysis [49b].

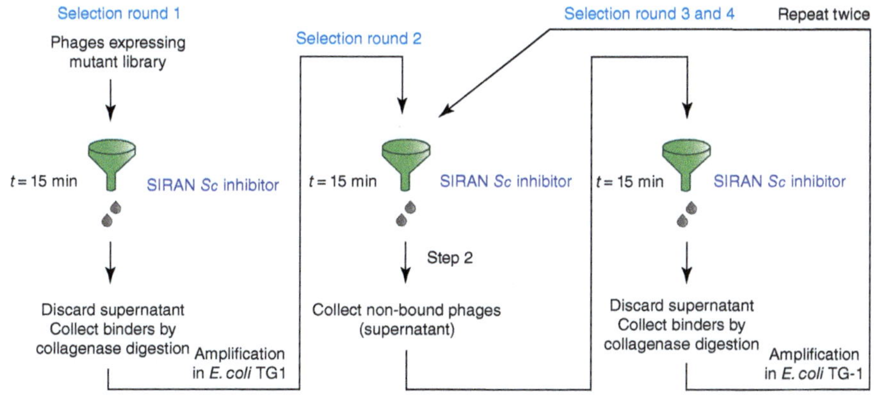

Scheme 2.14 Dual selection system for evolving enantioselectivity of a *Bacillus subtilis* lipase. Source: Droge et al. [49b]/John Wiley & Sons.

In a more successful case study, a yeast surface display system employing *Saccharomyces cerevisiae* was used for the purpose of evolving horseradisch peroxidase (HRP) as a catalyst in the enantioselective oxidative reaction of L- and D-tyrosinol [50]. In this work, covalent attachment of products via enzyme display was adopted to assess activity. The absolute configuration of the substrate was labeled by using fluorescent dye (Alexa 488) via radical chemistry, as shown in Scheme 2.15. Two 2×10^6-sized libraries were constructed by extensive epPCR and simultaneous saturation mutagenesis at sites lining the binding pocket, respectively. And then the two libraries were subjected to FACS analysis. In the latter case, the selection harvested mutants with a four- to eightfold improved enantioselectivity for the D- and the L-substrate, respectively. However, no stereoselective mutants were found in the previous library [50]. As the first case of directed evolution of enzyme stereoselectivity using a eukaryotic organism for surface displaying, it was suggested that the scope of enzyme yeast surface display could be extended by employing "any nontoxic substrate that can be conjugated to a standard linker" [50]. Nevertheless, more research is needed to generalize this interesting concept.

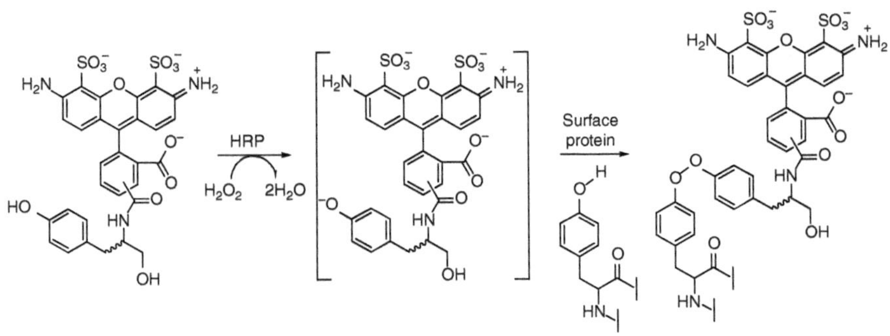

Scheme 2.15 Radical polymerization of L- or D-tyrosine and Alexa-488 derivatives using horseradish peroxidase (HRP) as the catalyst. Source: Lipovsek et al. [50]/Elsevier.

As demonstrated in a proof-of-concepts study, single-cell FACS-based high throughput was capable of identifying enantioselective enzymes [51]. An esterase from *Pseudomonas aeruginosa* (EstA) was chosen as the catalyst toward the hydrolytic kinetic resolution of the chiral ester derived from 2-methyldecanoic acid, (*R*)- and (*S*)-**11** (Scheme 2.16). The transformation is the same as the one that was used in the first example of directed evolution of an enantioselective enzyme (lipase) (Chapter 1) [4a].

A WT EstA with a poor selectivity in the hydrolytic kinetic resolution of *rac*-**11** ($E = 1.2$, favoring (*S*)-**12**), was employed as an ideal starting point. The two enantiomers of substrate *rac*-**11** were labeled with green and red fluorescent dyes, respectively. During the analytical procedure, the tyramide ester

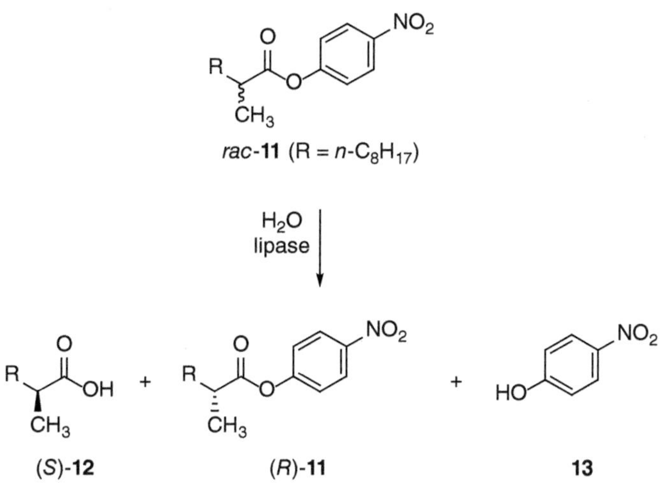

Scheme 2.16 EstA-catalyzed hydrolytic kinetic resolution and used in FACS-based assessment of enantioselectivity. Source: Adapted from Becker et al. [51].

substrates undergo hydrolysis and subsequent peroxidase-mediated formation of short-lived radicals, which ensures the immediate covalent attachment of reaction products to the surface of the esterase-proficient bacterial cell (Scheme 2.17) [51]. As the attached products are fluorescence labeled, cells with reactive mutants can be identified by FACS. This system allowed 10^8 cells to be assayed individually within only a few hours.

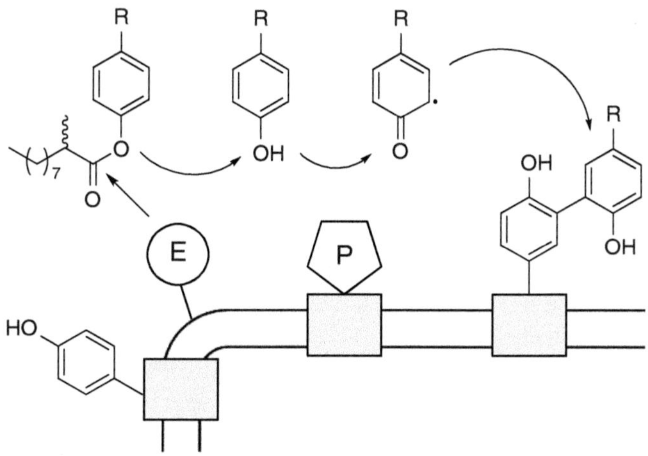

Scheme 2.17 Coupling reactions ensuring covalent attachment of tyramide species on the surface of *E. coli* cells [51], E, and P indicate esterase and peroxidase, respectively. Source: Becker et al. [51]/John Wiley & Sons.

In the attempt to evolve enantioselectivity, two distinct pseudo-enantiomeric substrates (*S*)-**14** and (*R*)-**15** were used, labeled with indicator groups 2,4-dinitrophenyl (DNP) and biotin, respectively (Scheme 2.18). As a consequence, the hydrolysis triggers green or red fluorescence signaling (*R*)- and (*S*)-selectivity, respectively, via the designed experimental setup.

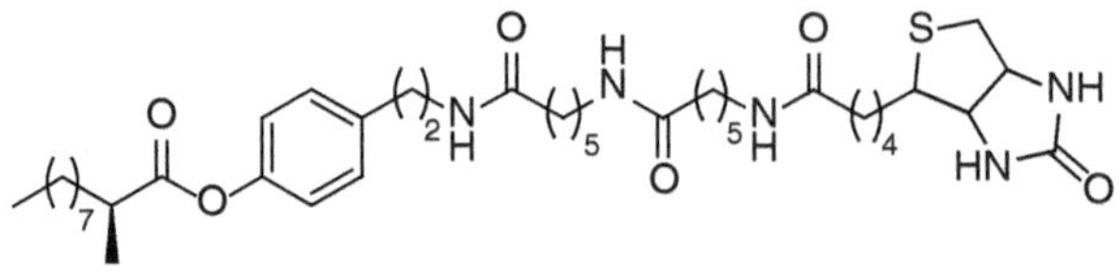

(*S*)-**14**

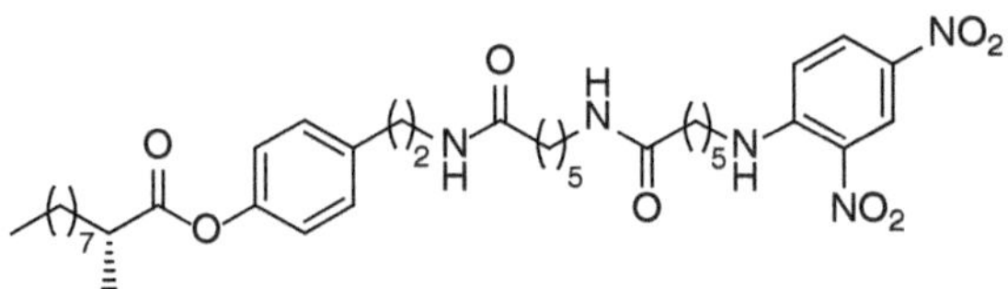

(*R*)-**15**

Scheme 2.18 Two distinct labeled enantiomers of 2-MDA tyramide ester substrates (*S*)-**14** and (*R*)-**15**. Source: Becker et al. [51]/John Wiley & Sons.

The designed cell labeling was programmed individually by conjugating horseradish peroxidase (P) to the cell surface. In the case of DNP tyramide labeling, an Alexa Fluor 488-labeled antibody was used to detect green fluorescent, while the red fluorescence of biotin tyramide deposition was detected by a streptavidin (*R*)-phycoerythrin conjugate. Flow cytometry can distinguish between the individual labeled cells; the typical resultant FACS histograms are displayed in Figure 2.3 [51].

To test the viability, more than 10^7 clones generated by epPCR were rapidly analyzed by FACS. The identification of (*R*)-selective mutants in such a large library was initiated by incubation with a 1 : 1 mixture of (*S*)-**14** and (*R*)-**15** after applying the peroxidase. As a result, only 86 active clones were obtained from the 6.8×10^7 FACS-analyzed cells. The enantioselectivity was then determined for 35 randomly chosen mutants by kinetics under standard conditions using (*S*)- and (*R*)-**11** (Scheme 2.16), respectively. The best variant is a triple mutant W185R/G224D/G263S with $E = 16.3$ in favor of the (*R*)-**12** as determined in separate experiments. In addition, a few variants were identified representing very slight enhancement of (*S*)-selectivity [51].

The authors envisioned that the somewhat moderate stereoselectivity could be caused by the sub-optimal choice of the mutagenesis method (epPCR). In fact, a previous study using the same substrate (*rac*-**11**) but a lipase instead of an esterase, multiple cycles of epPCR also delivered a mutant showing only

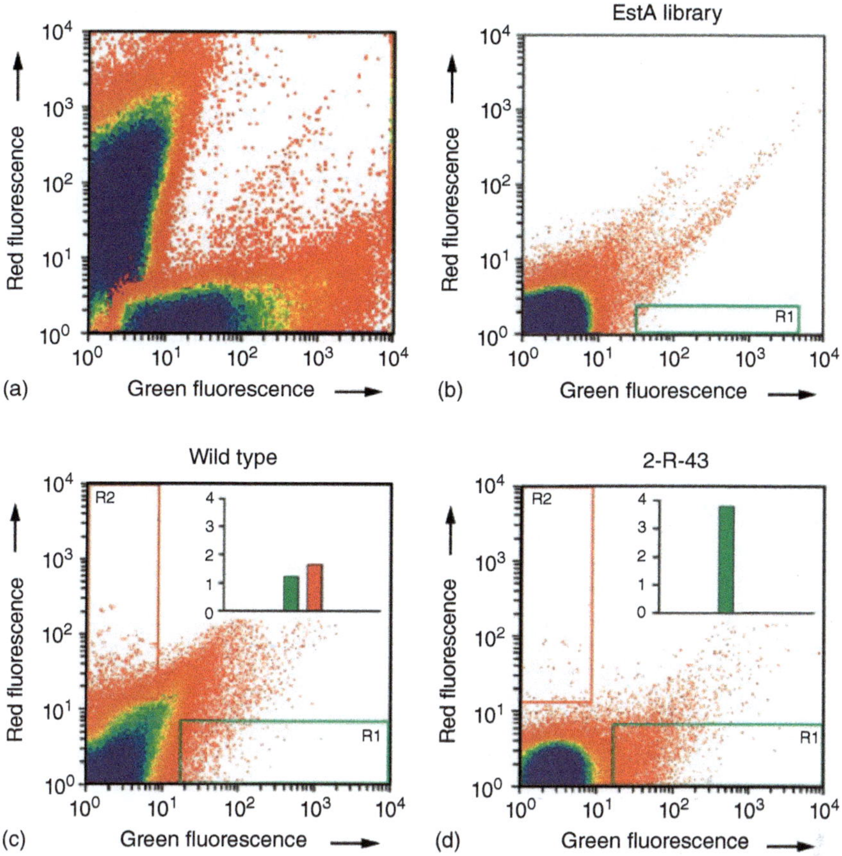

Figure 2.3 FACS-based high-throughput screening for evolving stereoselectivity [51]. (a) Overlay of flow-cytometry analyses of esterase-displaying cells that were incubated for one hour with either (*S*)- or (*R*)-enantiomer of tyramide ester. (b) EstA library sort. The green window indicates the sorting gate. (c, d) FACS histogram of WT EstA (c) and clone 2-R-43 (d) after five minutes incubation with a 1 : 1 mixture of both enantiomeric substrates and fluorescence staining. The inlet shows the percentage of cells within the respective green and red gate. Source: Becker et al. [51]/John Wiley & Sons.

moderate enantioselectivity ($E = 11$) [4a]. In contrast, later experiments using saturation mutagenesis at CASTing sites proved to be much more efficient ($E = 594$ in favor of (*S*)-**12**) [4b]. Simultaneous saturation mutagenesis using NNK-degenerated codons at the 10 amino acid positions adjacent to the binding pocket could be even more effective if combined with FACS-based flow-cytometry, which would cover an almost astronomically large protein sequence space ($\approx 10^{15}$ variants) that cannot be completely screened by any other current method. The optimal (*R*)- and (*S*)-selective mutants would then be expected to be harvested in such a large library.

Another work [52] reported the combination of *in vivo* selection along with the earlier system in which one enantiomer generates a cell-death product [45]. The model substrate comprises ester **16** with an (*R*)-configuration in the acid part and ester **18** with the opposite (*S*)-configuration (Scheme 2.19) [52]. After hydrolysis, the product glycerol (**17**) can be afforded as an energy source for the *E. coli* host, whilst hydrolysis of the pseudo-enantiomer **18** generates the dibromide **19**, which is toxic for the host. It evolved from low to high enantioselectivity. The best mutant, which had been evolved in an earlier study, showed an *E*-value of 100 (*R*). The viability of this concept was demonstrated by the growth behavior of *E. coli* in medium supplemented with mixtures of the two substrates in the presence of esterases BS2, PestE, or CL1. In addition, this concept was further confirmed by using a different esterase, PFE [52].

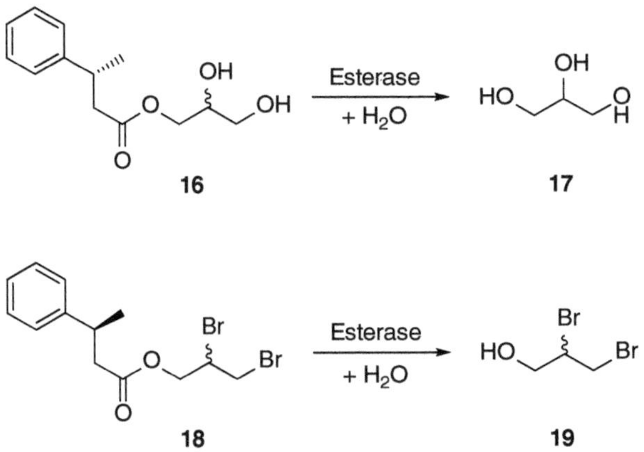

Scheme 2.19 The esterase PFE-directed hydrolytic kinetic resolution toward the pseudo-racemate (*R*)-**16**/(*S*)-**18**. Source: Fernandez-Alvaro et al. [52]/John Wiley & Sons.

In parallel, a series of *in vitro* methodologies have also been developed [1, 2], including lysate assays and ribosome display [53], and especially systems based on *in vitro* compartmentalization (IVC) [3c, 48d]. In the latter system [54], an *in vitro* transcription/translation reaction mixture was added to stirred mineral oil harboring appropriate surfactants such as Span80 and Tween80. The resultant emulsion with particles has a mean droplet diameter of about 2.6 μm, which is comparable in size to *E. coli* cells (Scheme 2.20). Such droplets proved to be successful in enabling the translation of several proteins, including dihydrofolate reductase and the DNA methyltransferase *Hae*III.

After this initial report [54], the IVC-concept has been expanded to include directed evolution of proteins and RNAs for catalytic and binding properties as well as regulatory activities [3c, 48d]. The general use of the technology appeared to be somewhat limited because the polydisperse emulsions make quantitative assessments impossible and the fact that the addition of new reagents to preformed droplets is not trivial. However, these problems were

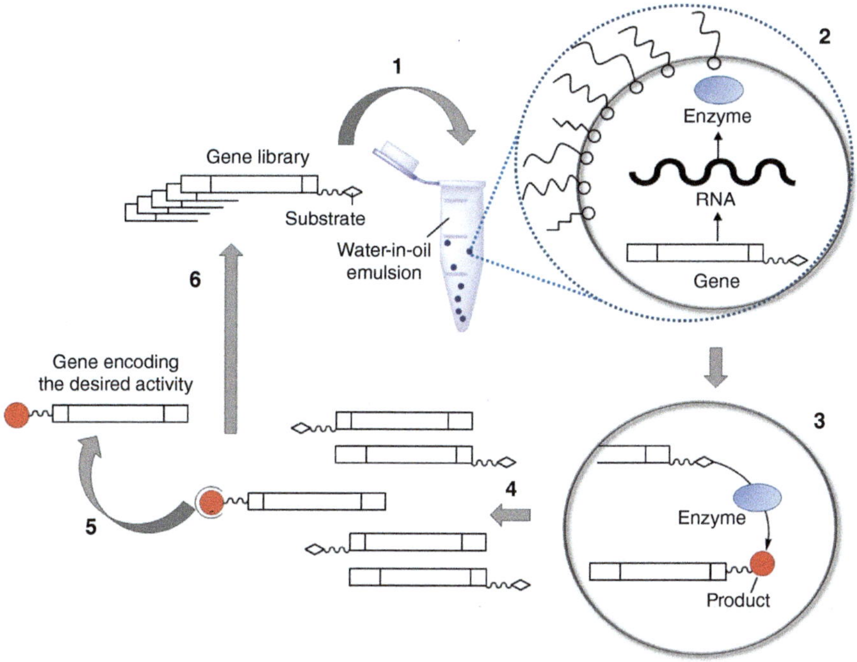

Scheme 2.20 Schematic representation of gene selection in compartmentalized oil-in-droplet emulsions. Source: Tawfik and Griffiths [54]/Springer Nature.

partially solved by the further development of droplet-based microfluidic devices [25]. Some excellent examples include a high-throughput cellulase screening system for activity [26a] and a highly parallel microfluidic droplet device which enables single-molecule counting for digital enzyme detection [26b], among other developments [26c–g]. By combining droplet microfluidic screening with next-generation DNA sequencing, an ultrahigh-throughput technique for probing enzyme sequence-function relationships has been successfully devised [26d]. This method was employed to determine glycosidase variants displaying higher activity and stability. In aqueous microdroplets, all members of a library resulting from deep mutational scanning [27] were assayed and sorted. Both of the sorted and unsorted variant pools were subsequently characterized by high-throughput DNA sequencing (Figure 2.4) [26g]. This system depends on an in-house constructed microfluidic droplet sorter that requires specialized instrumentation.

These microfluidic devices have been profitably applied in many cases, including physiologically relevant cell-based assays (e.g. proliferation of human cells) [28], genetic analysis, and retroviral display using droplet-based microfluidics [29]. It remains to be seen how well droplet-based technologies perform in the broader context of directed evolution of stereo- and regio-selective enzymes as catalysts in organic chemistry and biotechnology.

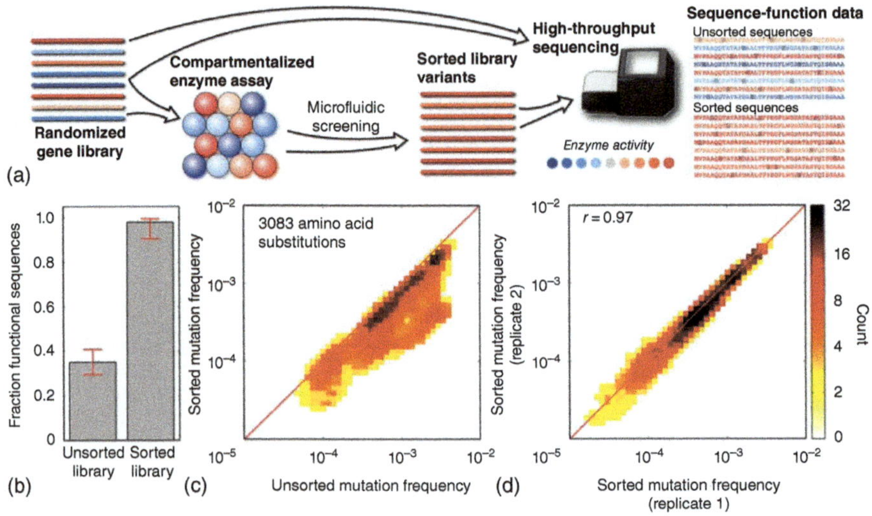

Figure 2.4 Deep mutational scanning-assisted ultrahigh-throughput sequence-function mapping [26g]. (a) Overview of the mapping procedure; (b) Droplet-based microfluidic screening leading to the recovery of functional sequences from the initial random mutagenesis library; (c) Frequency of 3083 amino acid exchanges in the sorted and unsorted glycosidase libraries, and (d) Reproducibility of the sequence-function mapping protocol with two independent replicas showing good consistent with amino acid frequencies. Source: Romero et al. [26g] with permission of National Academy of Sciences, U.S.A.

Conclusively, a plethora of seminal studies have appeared with the focus on selection systems (including display systems) for directed evolution of enantioselective enzymes, while in other cases, the respective techniques have not yet been tested for this purpose. Although the various approaches and concepts are clear, the majority of success has so far been limited to the directed evolution of enantio-, diastereo-, and regio-selective enzymes. Nevertheless, further efforts and ideas in this intriguing research area may lead to new assessments.

2.4 Conclusions and Perspectives

The decision of whether screening or selection should be used depends upon the particular goal of a directed evolution project. On-plate pretests for activity, whenever available, are always recommended. Nowadays, numerous enzymes, from either genome mining or directed evolution, are emerging at unprecedented rates. However, the necessary validation of their catalytic properties is time-consuming and labor-intensive, which often involves screening or selection at a considerable throughput. On the one hand, growth-coupled, UV/vis/fluorescence-based or molecular probe-dependent screening systems can be chosen specifically based on the signals of enzyme-catalyzed (see Section 2.2). On the other hand, exerting growth selection pressures on the host strains, including antibiotic resistance, auxotrophy, or detoxification, is also capable of evolving desired variants (see Section 2.3).

When attempting to induce improved or reversed stereo- and/or regio-selectivity, screening assays as opposed to selection platforms are currently the only practical options. In these cases, screening is the bottleneck of the whole process. For this reason, great efforts have gone into developing molecular biological methods for generating higher-quality mutant libraries [15] (see also Chapters 3 and 4). With these advanced genetic techniques and strategies, the analytical effort has been reduced generally to the screening of several thousand transformants or less. A state-of-the-art automated GC or HPLC should be used. In the future, more efforts to merge advanced mutagenesis methods and strategies (Chapter 4) with improved analytical techniques can be expected in the protein evolution of selective enzymes [31, 55].

References

1. Reviews of screening and selection systems: (a)Reymond, J.-L. (2006). *Enzyme Assays—High-Throughput Screening, Genetic Selection and Fingerprinting*. Weinheim: Wiley-VCH. (b) Reymond, J.-L., Fluxa, V.S., and Maillard, N. (2008). Enzyme assays. *Chem. Commun.* **1**: 34–46. (c) McLachlan, M.J., Sullivan, R.P., and Zhao, H. (2009). Directed enzyme evolution and high-throughput screening. In: *Biocatalysis for the Pharmaceutical Industry: Discovery, Development and Manufacturing* (ed. J. Tao, G.-Q. Lin and A. Liese), 45–64. New York: Wiley. (d) Boersma, Y.L., Dröge, M.J., and Quax, W.J. (2007). Selection strategies for improved biocatalysts. *FEBS J.* **274** (9): 2181–2195. (e) Martínez, R. and Schwaneberg, U. (2013). A roadmap to directed enzyme evolution and screening systems for biotechnological applications. *Biol. Res.* **46** (4): 395–405. (f) Wojcik, M., Telzerow, A., Quax, W.J., and Boersma, Y.L. (2015). High-throughput screening in protein engineering: recent advances and future perspectives. *Int. J. Mol. Sci.* **16** (10): 24918–24945. (g) Bhattacharya, S., Margheritis, E.G., Takahashi, K. et al. (2022). NMR-guided directed evolution. *Nature* **610**: 389–393. (h) Zeng, W., Guo, L., Xu, S. et al. (2020). High-throughput screening technology in industrial biotechnology. *Trends Biotechnol.* **38** (8): 888–906. (i) Vanella, R., Kovacevic, G., Doffini, V. et al. (2022). High-throughput screening, next generation sequencing and machine learning: advanced methods in enzyme engineering. *Chem. Commun.* **58**: 2455–2467.
2. Acevedo-Rocha, C.G., Agudo, R., and Reetz, M.T. (2014). Directed evolution of stereoselective enzymes based on genetic selection as opposed to screening systems. *J. Biotechnol.* **191** (10): 3–10.
3. Reviews focusing mainly on selection: (a)Taylor, S.V., Kast, P., and Hilvert, D. (2001). Investigating and engineering enzymes by genetic selection. *Angew. Chem. Int. Ed.* **40** (18): 3310–3335. (b) Lin, H. and Cornish, V.W. (2002). Screening and selection methods for large-scale analysis of protein function. *Angew. Chem. Int. Ed.* **41** (23): 4402–4425. (c) Aharoni, A., Griffiths, A.D., and Tawfik, D.S. (2005). High-throughput screens and selections of enzyme-encoding genes. *Curr. Opin. Chem. Biol.* **9** (2): 210–216. (d) van Sint Fiet, S., van Beilen, J.B., and Witholt, B. (2006). Selection of

biocatalysts for chemical synthesis. *Proc. Natl. Acad. Sci. U.S.A.* **103** (6): 1693–1698.

4. (a) Reetz, M.T., Zonta, A., Schimossek, K. et al. (1997). Creation of enantioselective biocatalysts for organic chemistry by in vitro evolution. *Angew. Chem. Int. Ed. Engl.* **36** (24): 2830–2832. (b) Reetz, M.T., Prasad, S., Carballeira, J.D. et al. (2010). Iterative saturation mutagenesis accelerates laboratory evolution of enzyme stereoselectivity: rigorous comparison with traditional methods. *J. Am. Chem. Soc.* **132** (26): 9144–9152. (c) Reetz, M.T. (2006). Directed evolution of enantioselective enzymes as catalysts for organic synthesis. In: *Advances in Catalysis*, vol. **49** (ed. B.C. Gates and K. Knözinger), 1–69. San Diego: Elsevier.
5. Reetz, M.T. (2006). High-throughput screening systems for assaying the enantioselectivity of enzymes. In: *Enzyme Assays – High-Throughput Screening, Genetic Selection and Fingerprinting* (ed. J.-L. Reymond), 41–76. Weinheim: Wiley-VCH.
6. Chen, K.Q. and Arnold, F.H. (1993). Tuning the activity of an enzyme for unusual environments – sequential random mutagenesis of subtilisin-E for catalysis in dimethylformamide. *Proc. Natl. Acad. Sci. U.S.A.* **90** (12): 5618–5622.
7. Badalassi, F., Wahler, D., Klein, G. et al. (2000). A versatile periodate-coupled fluorogenic assay for hydrolytic enzymes. *Angew. Chem. Int. Ed.* **39** (22): 4067–4070.
8. (a) Baumann, M., Stürmer, R., and Bornscheuer, U.T. (2001). A high-throughput-screening method for the identification of active and enantioselective hydrolases. *Angew. Chem. Int. Ed.* **40** (22): 4201–4204. (b) Li, Z., Bütikofer, L., and Witholt, B. (2004). High-throughput measurement of the enantiomeric excess of chiral alcohols by using two enzymes. *Angew. Chem. Int. Ed.* **43** (13): 1698–1702.
9. (a) Janes, L.E., Löwendahl, A.C., and Kazlauskas, R.J. (1998). Quantitative screening of hydrolase libraries using pH indicators: identifying active and enantioselective hydrolases. *Chem. Eur. J.* **4** (11): 2324–2331. (b) Liu, A.M.F., Somers, N.A., Kazlauskas, R.J. et al. (2001). Mapping the substrate selectivity of new hydrolases using colorimetric screening: lipases from *Bacillus thermocatenulatus* and *Ophiostoma piliferum*, esterases from *Pseudomonas fluorescens* and *Streptomyces diastatochromogenes*. *Tetrahedron: Asymmetry* **12** (4): 545–556.
10. (a) Reetz, M.T., Becker, M.H., Klein, H.-W., and Stöckigt, D. (1999). A method for high-throughput screening of enantioselective catalysts. *Angew. Chem. Int. Ed.* **38** (12): 1758–1761. (b) Schrader, W., Eipper, A., Pugh, D.J., and Reetz, M.T. (2002). Second-generation MS-based high-throughput screening system for enantioselective catalysts and biocatalysts. *Can. J. Chem.* **80** (6): 626–632. (c) DeSantis, G., Wong, K., Farwell, B. et al. (2003). Creation of a productive, highly enantioselective nitrilase through gene site saturation mutagenesis (GSSM). *J. Am. Chem. Soc.* **125** (38): 11476–11477.
11. Tielmann, P., Boese, M., Luft, M., and Reetz, M.T. (2003). A practical high-throughput screening system for enantioselectivity by using FTIR spectroscopy. *Chem. Eur. J.* **9** (16): 3882–3887.
12. Reetz, M.T., Eipper, A., Tielmann, P., and Mynott, R. (2002). A practical NMR-based high-throughput assay for screening enantioselective catalysts and biocatalysts. *Adv. Synth. Catal.* **344** (9): 1008–1016.
13. (a) Johannes, T.W., Woodyer, R.D., and Zhao, H. (2006). High-throughput screening methods developed for oxidoreductases. In: *Enzyme Assays: High-Throughput Screening, Genetic Selection and Fingerprinting* (ed. J.-L. Reymond), 77–93. Weinheim: Wiley-VCH. (b) Schwaneberg, U., Schmidt-Dannert, C., Schmitt, J., and Schmid, R.D. (1999). A continuous spectrophotometric assay for P450 BM-3, a fatty acid hydroxylating enzyme, and its mutant F87A. *Anal. Biochem.* **269** (2): 359–366. (c) Pardo, I., Chanaga, X.,

Vicente, A.I. et al. (2013). New colorimetric screening assays for the directed evolution of fungal laccases to improve the conversion of plant biomass. *BMC Biotech.* **13** (1): 90. (d) Lauchli, R., Rabe, K.S., Kalbarczyk, K.Z. et al. (2013). High-throughput screening for terpene-synthase-cyclization activity and directed evolution of a terpene synthase. *Angew. Chem. Int. Ed.* **52** (21): 5571–5574. (e) Ma, H., Enugala, T.R., and Widersten, M. (2015). A microplate format assay for real-time screening for new aldolases that accept aryl-substituted acceptor substrates. *ChemBioChem* **16** (18): 2595–2598. (f) Xu, H., Liang, W., Ning, L. et al. (2020). Directed Evolution of P450 Fatty Acid Decarboxylases via High-Throughput Screening towards Improved Catalytic Activity. *ChemCatChem* **12** (1): 80–84.

14. (a) Horvat, M., Larch, T.-S., Rudroff, F., and Winkler, M. (2020). Amino benzamidoxime (ABAO)-based assay to identify efficient aldehyde-producing *Pichia pastoris* clones. *Adv. Synth. Catal.* **362** (21): 4673–4679. (b) Zhou, J., Xu, G., Han, R. et al. (2016). Carbonyl group-dependent high-throughput screening and enzymatic characterization of diaromatic ketone reductase. *Catal. Sci. Technol.* **6** (16): 6320–6327. (c) Mei, Z., Zhang, K., Qu, G. et al. (2020). High-throughput fluorescence assay for ketone detection and its applications in enzyme mining and protein engineering. *ACS Omega* **5** (23): 13588–13594.
15. Qu, G., Li, A., Acevedo-Rocha, C.G. et al. (2020). The crucial role of methodology development in directed evolution of selective enzymes. *Angew. Chem. Int. Ed.* **59** (32): 13204–13231.
16. Reetz, M.T., Kühling, K.M., Wilensek, S. et al. (2001). A GC-based method for high-throughput screening of enantioselective catalysts. *Catal. Today* **67** (4): 389–396.
17. van Leeuwen, J.G.E., Wijma, H.J., Floor, R.J. et al. (2012). Directed evolution strategies for enantiocomplementary haloalkane dehalogenases: from chemical waste to enantiopure building blocks. *ChemBioChem* **13** (1): 137–148.
18. (a) Reetz, M.T., Daligault, F., Brunner, B. et al. (2004). Directed evolution of cyclohexanone monooxygenases: enantioselective biocatalysts for the oxidation of prochiral thioethers. *Angew. Chem. Int. Ed.* **43** (31): 4078–4081. (b) Kille, S., Zilly, F.E., Acevedo, J.P., and Reetz, M.T. (2011). Regio- and stereoselectivity of P450-catalysed hydroxylation of steroids controlled by laboratory evolution. *Nat. Chem.* **3** (9): 738–743.
19. Liu, F., Geng, Q., Zhao, C. et al. (2022). Colorimetric high-throughput screening method for directed evolution of prazole sulfide monooxygenase. *ChemBioChem* e202200228.
20. Yang, T., Ye, Z., and Lynch, M.D. (2022). "Multiagent" screening improves directed enzyme evolution by identifying epistatic mutations. *ACS Synth. Biol.* **11** (5): 1971–1983.
21. (a) Polizzi, K.M., Parikh, M., Spencer, C.U. et al. (2006). Pooling for improved screening of combinatorial libraries for directed evolution. *Biotechnol. Progr.* **22** (4): 961–967. (b) Bougioukou, D.J., Kille, S., Taglieber, A., and Reetz, M.T. (2009). Directed evolution of an enantioselective enoate-reductase: testing the utility of iterative saturation mutagenesis. *Adv. Synth. Catal.* **351** (18): 3287–3305.
22. (a) Wagen, C.C., McMinn, S.E., Kwan, E.E., and Jacobsen, E.N. (2022). Screening for generality in asymmetric catalysis. http://chemrxiv.org. (b) Müller, C.A. and Pfaltz, A. (2008). Mass spectrometric screening of chiral catalysts by monitoring the back reaction of quasienantiomeric products: palladium-catalyzed allylic substitution. *Angew. Chem. Int. Ed.* **47** (18): 3363–3366.
23. (a) Snoek, T., Chaberski, E.K., Ambri, F. et al. (2020). Evolution-guided engineering of small-molecule biosensors. *Nucleic Acids Res.* **48** (1): e3. (b) Han, L., Liu, X., Cheng, Z. et al. (2022). Construction and application of a high-throughput in vivo screening platform for the evolution of nitrile

metabolism-related enzymes based on a desensitized repressive biosensor. *ACS Synth. Biol.* **11** (4): 1577–1587.

24. Feagin, T.A., Olsen, D.P., Headman, Z.C., and Heemstra, J.M. (2015). High-throughput enantiopurity analysis using enantiomeric DNA-based sensors. *J. Am. Chem. Soc.* **137** (12): 4198–4206.
25. Fallah-Araghi, A., Baret, J.C., Ryckelynck, M., and Griffiths, A.D. (2012). A completely *in vitro* ultrahigh-throughput droplet-based microfluidic screening system for protein engineering and directed evolution. *Lab Chip* **12** (5): 882–891.
26. (a) Ostafe, R., Prodanovic, R., Lloyd Ung, W. et al. (2014). A high-throughput cellulase screening system based on droplet microfluidics. *Biomicrofluidics* **8** (4): 041102. (b) Guan, Z., Zou, Y., Zhang, M. et al. (2014). A highly parallel microfluidic droplet method enabling single-molecule counting for digital enzyme detection. *Biomicrofluidics* **8** (1): 014110. (c) Baret, J.C., Miller, O.J., Taly, V. et al. (2009). Fluorescence-activated droplet sorting (FADS): efficient microfluidic cell sorting based on enzymatic activity. *Lab Chip* **9** (13): 1850–1858. (d) Agresti, J.J., Antipov, E., Abate, A.R. et al. (2010). Ultrahigh-throughput screening in drop-based microfluidics for directed evolution. *Proc. Natl. Acad. Sci. U.S.A.* **107** (9): 4004–4009. (e) Kintses, B., Hein, C., Mohamed, M.F. et al. (2012). Picoliter cell lysate assays in microfluidic droplet compartments for directed enzyme evolution. *Chem. Biol.* **19** (8): 1001–1009. (f) Abate, A.R., Hung, T., Mary, P. et al. (2010). High-throughput injection with microfluidics using picoinjectors. *Proc. Natl. Acad. Sci. U.S.A.* **107** (45): 19163–19166. (g) Romero, P.A., Tran, T.M., and Abate, A.R. (2015). Dissecting enzyme function with microfluidic-based deep mutational scanning. *Proc. Natl. Acad. Sci. U.S.A.* **112** (23): 7159–7164.
27. (a) Hietpas, R.T., Jensen, J.D., and Bolon, D.N. (2011). Experimental illumination of a fitness landscape. *Proc. Natl. Acad. Sci. U.S.A.* **108** (19): 7896–7901. (b) Jacquier, H., Birgy, A., Le Nagard, H. et al. (2013). Capturing the mutational landscape of the beta-lactamase TEM-1. *Proc. Natl. Acad. Sci. U.S.A.* **110** (32): 13067–13072.
28. Clausell-Tormos, J., Lieber, D., Baret, J.C. et al. (2008). Droplet-based microfluidic platforms for the encapsulation and screening of mammalian cells and multicellular organisms. *Chem. Biol.* **15** (5): 427–437.
29. Granieri, L., Baret, J.C., Griffiths, A.D., and Merten, C.A. (2010). High-throughput screening of enzymes by retroviral display using droplet-based microfluidics. *Chem. Biol.* **17** (3): 229–235.
30. von Buhler, C., Le-Huu, P., and Urlacher, V.B. (2013). Cluster screening: an effective approach for probing the substrate space of uncharacterized cytochrome P450s. *ChemBioChem* **14** (16): 2189–2198.
31. Lovelock, S.L., Crawshaw, R., Basler, S. et al. (2022). The road to fully programmable protein catalysis. *Nature* **606** (7912): 49–58.
32. (a) Wijma, H.J., Floor, R.J., Bjelic, S. et al. (2015). Enantioselective enzymes by computational design and in silico screening. *Angew. Chem. Int. Ed.* **54** (12): 3726–3730. (b) Arabnejad, H., Bombino, E., Colpa, D.I. et al. (2020). Computational design of enantiocomplementary epoxide hydrolases for asymmetric synthesis of aliphatic and aromatic diols. *ChemBioChem* **21** (13): 1893–1904. (c) Li, R., Wijma, H.J., Song, L. et al. (2018). Computational redesign of enzymes for regio- and enantioselective hydroamination. *Nat. Chem. Biol.* **14** (7): 664–670.
33. Stemmer, W.P.C. (1994). Rapid evolution of a protein in vitro by DNA shuffling. *Nature* **370** (6488): 389–391.
34. Yano, T., Oue, S., and Kagamiyama, H. (1998). Directed evolution of an aspartate aminotransferase with new substrate specificities. *Proc. Natl. Acad. Sci. U.S.A.* **95** (10): 5511–5515.
35. Neuenschwander, M., Butz, M., Heintz, C. et al. (2007). A simple selection strategy for evolving highly efficient

enzymes. *Nat. Biotechnol.* **25** (10): 1145–1147.

36. Miller, B.G. and Raines, R.T. (2004). Identifying latent enzyme activities: substrate ambiguity within modern bacterial sugar kinases. *Biochemistry* **43** (21): 6387–6392.
37. Lin, H., Tao, H., and Cornish, V.W. (2004). Directed evolution of a glycosynthase via chemical complementation. *J. Am. Chem. Soc.* **126** (46): 15051–15059.
38. Baxter, S., Royer, S., Grogan, G. et al. (2012). An improved racemase/acylase biotransformation for the preparation of enantiomerically pure amino acids. *J. Am. Chem. Soc.* **134** (47): 19310–19313.
39. Fahrig-Kamarauskaite, J., Würth-Roderer, K., Thorbjørnsrud, H.V. et al. (2020). Evolving the naturally compromised chorismate mutase from *Mycobacterium tuberculosis* to top performance. *J. Biol. Chem.* **295** (51): 17514–17534.
40. Li, H. and Liao, J.C. (2014). Development of an NADPH-dependent homophenylalanine dehydrogenase by protein engineering. *ACS Synth. Biol.* **3** (1): 13–20.
41. Maxel, S., Saleh, S., King, E. et al. (2021). Growth-based, high-throughput selection for NADH preference in an oxygen-dependent biocatalyst. *ACS Synth. Biol.* **10** (9): 2359–2370.
42. Kramer, L., Le, X., Rodriguez, M. et al. (2020). Engineering carboxylic acid reductase (CAR) through a whole-cell growth-coupled NADPH recycling strategy. *ACS Synth. Biol.* **9** (7): 1632–1637.
43. Reetz, M.T. and Wang, L.-W. (2006). High-throughput selection system for assessing the activity of epoxide hyrolases. *Comb. Chem. High Throughput Screening* **9**: 295–299.
44. Hwang, B.Y., Oh, J.M., Kim, J., and Kim, B.G. (2006). Pro-antibiotic substrates for the identification of enantioselective hydrolases. *Biotechnol. Lett.* **28** (15): 1181–1185.
45. (a) Reetz, M.T., Höbenreich, H., Soni, P., and Fernandez, L. (2008). A genetic selection system for evolving enantioselectivity of enzymes. *Chem. Commun.* **43**: 5502–5504. (b) Reetz, M.T. and Rüggeberg, C.J. (2002). A screening system for enantioselective enzymes based on differential cell growth. *Chem. Commun.* **13**: 1428–1429.
46. Boersma, Y.L., Dröge, M.J., van der Sloot, A.M. et al. (2008). A novel genetic selection system for improved enantioselectivity of *Bacillus subtilis* lipase A. *ChemBioChem* **9** (7): 1110–1115.
47. (a) Sidhu, S.S. (2005). *Phage Display in Biotechnology and Drug Discovery*. Boca Raton: CRC Press. (b) Yang, G. and Withers, S.G. (2009). Ultrahigh-throughput FACS-based screening for directed enzyme evolution. *ChemBioChem* **10** (17): 2704–2715.
48. (a) Olsen, M.J., Stephens, D., Griffiths, D. et al. (2000). Function-based isolation of novel enzymes from a large library. *Nat. Biotechnol.* **18** (10): 1071–1074. (b) Santoro, S.W. and Schultz, P.G. (2002). Directed evolution of the site specificity of Cre recombinase. *Proc. Natl. Acad. Sci. U.S.A.* **99** (7): 4185–4190. (c) Griswold, K.E., Kawarasaki, Y., Ghoneim, N. et al. (2005). Evolution of highly active enzymes by homology-independent recombination. *Proc. Natl. Acad. Sci. U.S.A.* **102** (29): 10082–10087. (d) Mastrobattista, E., Taly, V., Chanudet, E. et al. (2005). High-throughput screening of enzyme libraries: in vitro evolution of a beta-galactosidase by fluorescence-activated sorting of double emulsions. *Chem. Biol.* **12** (12): 1291–1300.
49. (a) Reetz, M.T., Rüggeberg, C.J., Dröge, M.J., and Quax, W.J. (2002). Immobilization of chiral enzyme inhibitors on solid supports by amide-forming coupling and olefin metathesis. *Tetrahedron* **58**: 8465–8473. (b) Droge, M.J., Boersma, Y.L., van Pouderoyen, G. et al. (2006). Directed evolution of *Bacillus subtilis* lipase A by use of enantiomeric phosphonate inhibitors: crystal structures and phage display selection. *ChemBioChem* **7** (1): 149–157.
50. Lipovsek, D., Antipov, E., Armstrong, K.A. et al. (2007). Selection of horseradish peroxidase variants with enhanced enantioselectivity by yeast

surface display. *Chem. Biol.* **14** (10): 1176–1185.

51. Becker, S., Höbenreich, H., Vogel, A. et al. (2008). Single-cell high-throughput screening to identify enantioselective hydrolytic enzymes. *Angew. Chem. Int. Ed.* **47** (27): 5085–5088.

52. Fernandez-Alvaro, E., Snajdrova, R., Jochens, H. et al. (2011). A combination of in vivo selection and cell sorting for the identification of enantioselective biocatalysts. *Angew. Chem. Int. Ed.* **50** (37): 8584–8587.

53. Hanes, J. and Plückthun, A. (1997). *In vitro* selection and evolution of functional proteins by using ribosome display. *Proc. Natl. Acad. Sci. U.S.A.* **94** (10): 4937–4942.

54. Tawfik, D.S. and Griffiths, A.D. (1998). Man-made cell-like compartments for molecular evolution. *Nat. Biotechnol.* **16** (7): 652–656.

55. Reetz, M.T. (2022). Making enzymes suitable for organic chemistry by rational design. *ChemBioChem* e202200049.

3
Gene Mutagenesis Methods in Directed Evolution and Rational Enzyme Design

3.1
Introductory Remarks

In Chapter 1, the basic characteristics of the most important gene mutagenesis techniques have already been introduced, including error-prone polymerase chain reaction (epPCR), saturation mutagenesis, and DNA shuffling. In the present chapter, we go into more detail by analyzing the respective advantages and disadvantages, in addition to listing other options and new developments. Any of these or other gene mutagenesis methods can be expected to result in the creation of improved enzyme variants [1, 2], but the degree of improvement and the invested amount of lab work may be very different, which is sometimes ignored. When applying directed evolution for mechanistic purposes, efficiency plays a minor role. However, when the purpose is to evolve active, stereoselective, and robust enzymes as catalysts in organic and pharmaceutical chemistry and/or biotechnology, then efficient, reliable, and fast techniques are required. Since the bottleneck of directed evolution continues to be the screening step (Chapter 2), advanced directed evolution techniques are featured here which aim for "smart" mutant libraries exhibiting a high density of notably improved variants (Section 3.2). Where appropriate, statistical analyses regarding the degree of oversampling are included, which are essential for optimal library design. This is followed by a description of current approaches to rational enzyme design (Section 3.3). It will also be shown that the two options for protein engineering no longer develop on separate tracks, but have merged in recent years (Section 3.4) [1a, 2f, 3].

3.2
Directed Evolution Approaches

3.2.1
Mutator Strains

Traditionally, gene mutagenesis was performed for different reasons using chemicals, radiation, or mutator strains. Bacterial mutator strains are based

Enzyme Engineering: Selective Catalysts for Applications in Biotechnology, Organic Chemistry, and Life Science, First Edition. Manfred T. Reetz, Zhoutong Sun, and Ge Qu.

on artificially enhanced mutation during replication, i.e. the efficiency of the DNA repair machinery has been reduced [4]. They are relatively easy to apply, requiring little or no knowledge of recombinant DNA techniques, as in the case of the *Escherichia coli* strain XL1-Red, which was engineered to have a 5000-fold higher mutation rate than wildtype (WT) [4e]. The natural mutation rate is low, amounting to about 0.0025/1000 base pairs in 30 generations due to defects in repair mechanisms, in contrast to the commercially available strain XL1-Red, which causes 0.5/1000 base pair mutations. Pitfalls when using this strain are outlined in the Agilent instruction manual [4e]. In directed evolution, mutator strains have been rarely used [4c, d], a notable example being the industrial production of chiral amines catalyzed by monoamine oxidases [4c]. Another case study of evolving antibiotic resistance utilizes mutator strain MP6 that produces an average of 29 substitutions per genome per generation, which is a fourfold higher mutation rate than that of XL1-Red [4f].

3.2.2
Error-Prone Polymerase Chain Reaction (epPCR)

For a long time, epPCR was the most popular gene mutagenesis method in directed evolution [5] (see Section 1.1). Whenever structural information about an enzyme is lacking, epPCR is certainly the preferred approach, although today the method is sometimes used even when X-ray structural information is available [6]. It introduces mutations more or less randomly over the whole gene/protein, as opposed to saturation mutagenesis, in which random mutations are restricted to defined sites in the protein (Section 3.2.4). The full-gene epPCR mutagenesis method is illustrated in Scheme 3.1 [7].

The details of epPCR involve the following steps: In the first step, the DNA of a given gene is denaturized into single strands, followed by the second step in which annealing of a primer to the DNA single strands occurs, the primers consisting of two oligonucleotides having lengths of 15–30 base pairs complementary to the ends of the amplified region. In the third step, primer extension is ensured by an appropriate DNA polymerase, which lacks proof-reading ability. Based on the original sequence, nucleotides complementary to the single-strand template are added until normal DNA double strands are restored. The *Taq* polymerase from the thermophilic bacterium *Thermus aquaticus* is used most often [5]. It incorporates "wrong" nucleotides at a fairly low frequency of $0.1–2 \times 10^{-4}$ [8]. In epPCR experiments, the error rate can be continuously increased to $1–20 \times 10^{-3}$. This is achieved by performing the polymerase chain reaction (PCR) "sloppily" so that mistakes in DNA base pairings are introduced, which encode point mutations in the protein. The mutation rate can be controlled empirically by varying such parameters as the $MgCl_2$ or $MnCl_2$ concentrations, employing unbalanced amounts of nucleotides, and utilizing higher concentrations of *Taq* polymerase. The error-rate can also be influenced by the incorporation of synthetic mutagenic dNTPs such as 8-oxo-dGTP, which is subsequently eliminated in PCR employing natural dNTPs [9]. On average, one, two, three, or more amino acid exchange events at the protein level can be

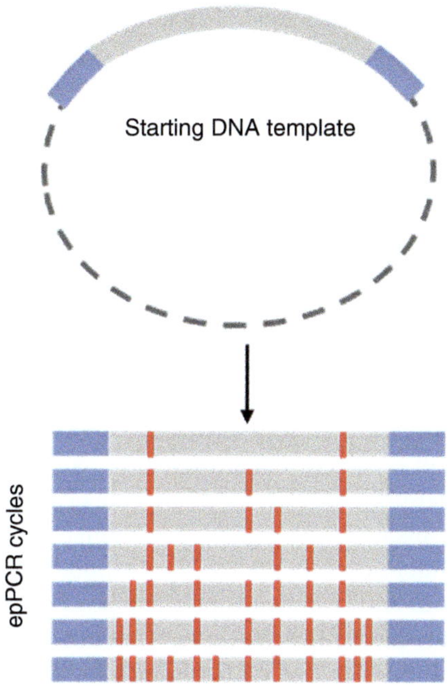

Scheme 3.1 Illustration of epPCR. Source: Adapted from Ref. [7].

induced by epPCR. Following amplification, the DNA products (new sequences) are cloned into an appropriate vector, which can be the "rate-limiting" step in terms of lab work [10]. Recently, a new method called "casting epPCR" was developed to increase the substitution rate by dividing the target sequence into a few fragments, but the benefits were not fully clear [11].

An informative review summarizing the virtues of different techniques for cloning into expression vectors, thereby enabling protein synthesis in a host, has appeared [12]. A number of epPCR kits are commercially available, but they are not identical in terms of performance. Assessing the relative efficiency by performing systematic comparative studies of the kits using different enzymes has not been done to date, although many companies offer helpful information. For example, the GeneMorph II Random Mutagenesis Kits (Stratagene) have been developed to a point where a more uniform mutational epPCR spectrum results as a consequence of partially reduced amino acid bias. It utilizes Mutazyme II DNA polymerase, an epPCR enzyme blend leading to equivalent mutation rates at A's and T's versus G's and C's. It is interesting to note that the epPCR strategy has been applied to the generation of influenza vaccine candidates [13].

The size of epPCR libraries varies according to the amount of lab work the researcher is willing to invest, with 10^3–10^5 clones being typical if screening (not selection) is involved. When a selection or fluorescence-activated cell sorter (FACS)-based screening system is available, epPCR libraries comprising 10^6–10^9 clones can be handled (Chapter 2). A detailed protocol which considers some of the limitations of epPCR was published in 2003 [14] and further developments continue to this day (see below).

Less experienced users of epPCR should be aware of all the drawbacks of this random mutagenesis technique, which is often considered to cover more or less the whole gene/protein in a "shotgun" manner [15]. Ideally, the full-gene mutagenesis method would ensure the equal occurrence of all four transitions T_s (AT → GC and GC → AT) and eight transversions T_v (AT → TA, AT → CG, GC → CG, and GC → TA), which means a probability of 16.67% for each nucleotide substitution pair as well as a T_s:T_v ratio of 0.5 [12]. Also, in an ideal technique, deletions and insertions should not occur. Bias occurs in epPCR due to the redundancy of the generic ode, among other factors. For example, AT → GC transitions and AT → TA transversions are favored. Due to statistical reasons, only single bases are replaced within the triplet codon, which restricts diversity considerably. The event of two or even three base-pair exchanges per codon is extremely unlikely. Therefore, at best one nucleotide of a given codon is exchanged, leading to just nine (instead of 64 possible) different codons encoding four to seven (instead of 20) different amino acids. The number of "designed" amino acid substitution events in reality depends on the type of original codon. For example, silent mutations are more likely for some types of codons, e.g. CGA coding for arginine, than for other types, e.g. AAC coding for asparagine. These and other sources of bias have been discussed in an experimental study flanked by a statistical analysis [16]. By analyzing every single codon of the lipase from *Bacillus subtilis*, composed of 181 residues, a model calculation was performed, which reveals the real number of enzyme variants obtained by epPCR with one mutation per gene. Employing the conventional algorithm in Eq. (3.1),

$$N = 19^M X!/[(X - M)!M!] \tag{3.1}$$

where N denotes the number of variants at maximal diversity, M the total number of amino acid substitutions, and X the number of amino acids, the theoretical number of variants assuming the absence of bias or other problems can be calculated (Table 3.1) [16]. The mutational bias for this particular model system was then estimated (Scheme 3.2).

Table 3.1 Theoretical number of variants in a library obtained for a protein consisting of 181 amino acids (lipase A from *B. subtilis*) with one to five amino acid exchanges per enzyme molecule.

Number of amino acid exchanges (*M*)	Number of variants (*N*)
1	3 439
2	5 880 690
3	6 666 742 230
4	5 636 730 555 465
5	3 791 264 971 605 760

Source: Ref. [16]/Taylor & Francis.

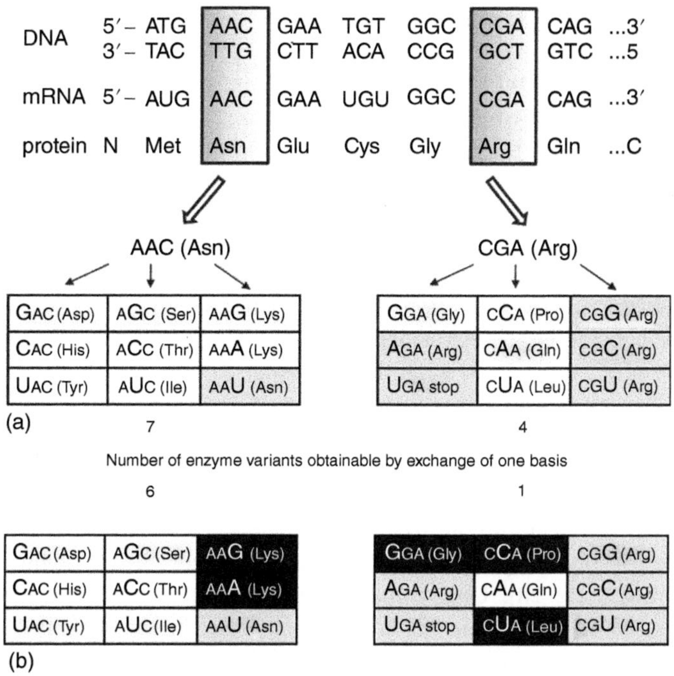

Scheme 3.2 Mutational bias of epPCR in the case of the lipase from *Bacillus subtilis*. The substitution of one nucleotide per codon results in nine new triplets which may encode four to seven different amino acids depending on the type of codon. (a) The example shows that the mutation of the codon AAC coding for asparagine can yield a maximum of seven different amino acids, whereas the mutation of the codon CGA coding for arginine can yield a maximum of four different amino acids. (b) Low frequencies of transversions G → T, C → A, G → C, and C → G result in a further decrease of diversity: for codon AAC, six different amino acid exchanges may occur, and for the GC-rich codon CGA just a single new amino acid exchange is expected. Background color coding: white shows codons that encode new amino acids, gray indicates silent mutations, or the formation of stop codons, and black shows codons that would require the formation of an unfavored basepair exchange (G → T, C → A, G → C, or C → G). The bold letters indicate nucleotides exchanged by epPCR. Source: Ref. [16]/Taylor & Francis.

It can be seen from Scheme 3.2 that serious amino acid bias occurs. In further analysis, it was concluded that the calculated library sizes represent only about 20% of the theoretical diversity in this system [16]. Potential users of epPCR should be aware of this phenomenon [15]. Consequently, here and in other systems, Eq. (3.1) cannot be used to estimate the real diversity when using epPCR. It is also important to point out that due to statistical reasons, mutations remote from the active center are more likely to occur.

Another source of undesired bias results from the exponential nature of PCR amplification [17]. Several other studies which model epPCR by statistical means and in part by experimental data have appeared in the literature. Although different approaches were taken, these analyses likewise illuminate the scope and limitations of this mutagenesis method [18]. Variations and/or improvements have appeared, e.g. by alcohol-mediated epPCR [19] or epPCR influenced by heavy water (D_2O) [20], both showing a clear shift in bias. Other DNA polymerases have been used, showing different bias [21], and still others have been reengineered for the purpose of using them in random mutagenesis [22]. In order to reduce bias, two DNA polymerases have been used in one experimental setup: *Taq* and Mutazyme combined with a recombinant mutagenesis method [21].

In a different study, a mutagenesis method called "hypermutagenic PCR" was reported, which involves all four transitions and a large proportion of transversions [23]. It can be considered to be an extension of epPCR, but thus far it has not been used very often in directed evolution. In order to enhance the proficiency of epPCR by alternative means, "megaprimer PCR of whole plasmid" (MEGAWHOP) was introduced, which is a successful PCR amplification procedure used in many other cloning applications as well [24]. It has also been applied in directed evolution not directly based on epPCR [25], e.g. when applying saturation mutagenesis in special cases [26, 27].

Guidelines and carefully worked out experimental protocols for applying epPCR are available [7, 14, 28]. Unfortunately, in some published directed evolution studies based on epPCR, details are missing, which means that it is not clear which version was actually used. Other critical remarks on the epPCR variations have been reviewed [29].

It should be noted that the abovementioned disadvantages of epPCR do not all mean that this method is fully disqualified. The user should simply know what is being generated in terms of diversity and what is not formed. Indeed, this message applies to all gene mutagenesis methods, because none are "perfect." How much screening is performed depends upon the amount of effort the researcher is willing to invest. Algorithms for estimating diversity and library (in)completeness in epPCR and other gene mutagenesis methods have been developed [28a, 30].

Once the optimal choice of an epPCR version has been made, the question of mutation rate arises. Should the epPCR conditions be chosen so that on average a single amino acid substitution occurs per enzyme, or should they be adjusted to lead to two, three, or even more exchange events? In the latter case, protein sequence space increases astronomically. A general answer cannot be given, but trends are emerging. If an assay system based on a type of "selection" with or without FACS-analysis (Chapter 2) is available, then a high mutational rate can be used, which is the preferred option. If screening based on gas chromatography (GC) or high performance liquid chromatography (HPLC) is the only possibility, as in most directed evolution studies aimed at improving activity, stereo-, and or regio-selectivity, the answer has remained unclear for some time. In the original publication reporting the directed evolution of an enantioselective enzyme (lipase from *Pseudomonas aeruginosa*, PAL) as the catalyst in a hydrolytic kinetic resolution (Schemes 1.9 and 1.10) [31], a low error rate averaging one amino acid substitution was applied. This was the general guideline in several previous papers on thermostability because it was thought at the time that as the number of mutations rises, so does the probability of deleterious unfolding [32]. In the case of the stereoselectivity study, low error-rate epPCR in all four cycles resulted in moderate catalyst improvement, as shown in Schemes 1.9 and 1.11 [31]. In contrast, *upon repeating the epPCR experiments at a higher mutational rate, averaging about three amino acid substitutions per enzyme molecule, considerably better results featuring higher degrees of enantioselectivity in the same hydrolytic kinetic resolution were observed* [33]. However, no improvements in further epPCR cycles were observed, leading to the conclusion that recursive epPCR is not the most efficient mutagenesis method for enhancing [33] or reversing enantioselectivity [34]. In the long run, it proved to be a turning point in the directed evolution of stereoselective enzymes since focused randomization at semi rationally chosen residues by means of saturation mutagenesis was considered for the first time and shown to be superior to epPCR [1a, 2f].

The above epPCR results at a high mutation rate originated from a single study cannot be generalized as such. Nevertheless, the trend suggests that creating theoretically higher diversity at a given screening effort in the expanded protein sequence space is more rewarding than opting for low diversity at the same screening effort [33]. This assessment contradicts previous studies [32]. Relevant is an earlier conclusion stating that no "optimal mutational load for protein engineering" is possible [9]. Just a few years later, several studies (not focusing on stereoselectivity) suggested that high mutational rates are in fact preferred [35]. Indeed, original recommendations concerning low mutational rate [32a] were later reversed by coming to the conclusion that "high error-rate random mutagenesis libraries are enriched in functional and improved proteins" [36].

In addition to the above improvements of epPCR leading to reduced bias, several alternative whole-gene random mutagenesis techniques have been proposed, although thus far they are rarely applied. One of several approaches combines the generality of epPCR with Kunkel-mutagenesis (Section 1.2), in which it was shown that competent cells engineered for expressing restriction endonuclease *in vivo* by epPCR are efficient in the elimination of partial and parental clones in digestion and subsequent retransformation [37]. However, error-prone rolling circle amplification (epRCA) [38] and sequence saturation mutagenesis (SeSaM) [39] are further examples, which are rarely used. epRCA is a "sloppy" form of traditional rolling circle amplification (RCA) used for diagnostic and biosensing assays [40]. Mutations are introduced by varying the $MnCl_2$ concentration. The method has been described as the "simplest random mutagenesis method" [38]. A single RCA step is necessary, followed by direct transformation of the host strain, leading to mutants characterized by three to four mutations per kilobase. Advantages include the fact that no restriction enzymes, ligases, specific primers, or special equipment such as a thermocycler are needed. However, some bias remains. The use of ϕ29DNA polymerase or mutant thereof favors $C \rightarrow T$ and $G \rightarrow A$ mutations (66%), which differs considerably from *Taq* polymerase. RCA has also been used to construct large phage display antibody libraries [41], and it has been exploited in primer extension mutagenesis [42]. In another approach, the QuikChange™ protocol for site-directed mutagenesis or saturation mutagenesis (see Section 3.2.4) was performed with added $MnCl_2$, which resulted in the construction of randomly mutagenized libraries with the claimed advantage that a ligation step becomes superfluous [43]. It remains to be seen whether these methods are in fact simple and efficient enough to be widely used in directed evolution.

The same question pertains to the 4-step SeSaM technique, which likewise reduces bias, but requires a formidable laboratory effort [39]: (i) Generation of a pool of random length DNA fragments. (ii) Utilization of a universal/degenerate base(s) at the 3′-termini of the DNA fragment pool. (iii) Application of PCR for elongation of the DNA fragment pool to full-length mutant genes. (iv) Replacement of universal/degenerate bases by the use of standard nucleotides. The second generation SeSaM-Tv protocol is an improvement, but again, it is still somewhat labor-intensive [39b].

In summary, when researchers consider the use of whole-gene random mutagenesis methods, then epPCR should be chosen. Relative to the original versions [5a], various improvements and alternatives have been reported (see discussions above). Several standard protocols are available [7, 14, 28]. To date, none of the alternative whole-gene mutagenesis techniques has replaced epPCR. A recent example of using solely epPCR without the aid of DNA shuffling or saturation mutagenesis is highlighted here. It concerns the directed evolution of prodigiosin ligase PigC as the enzyme in the condensation reaction of the bifurcated prodigiosin biosynthesis (Figure 3.1) [44]. It is noteworthy that the epPCR campaign always relies on high-throughput screening. However, in most cases, the semi-rational approach to directed evolution based on saturation mutagenesis, which ensures mutations at or near the binding pocket, is the better way to proceed [1a, 45], as noted as early as 2001 [33] and agreed on by various groups [46] (see Section 3.2.4 for details).

3.2.3 Whole Gene Insertion/Deletion Mutagenesis

In natural evolution, the types of mutations include insertions and deletions [47]. In protein engineering, undesired insertions and deletions are generally a nuisance when applying such mutagenesis techniques as epPCR, saturation mutagenesis, or DNA shuffling, but they can indeed occur. For example, the application of a high error rate when using epPCR may induce insertions and deletions. Ways to eliminate such unplanned events and other undesired mutations, specifically when applying focused saturation mutagenesis, have been developed [48] (see Section 3.2.7). However, whole-gene mutagenesis techniques enabling random insertion and deletion can in fact be beneficial. In one of the early reports, a technique for specific random insertion and substitution was described [49], to be followed by "random insertion and deletion" (RID), which allows the deletion

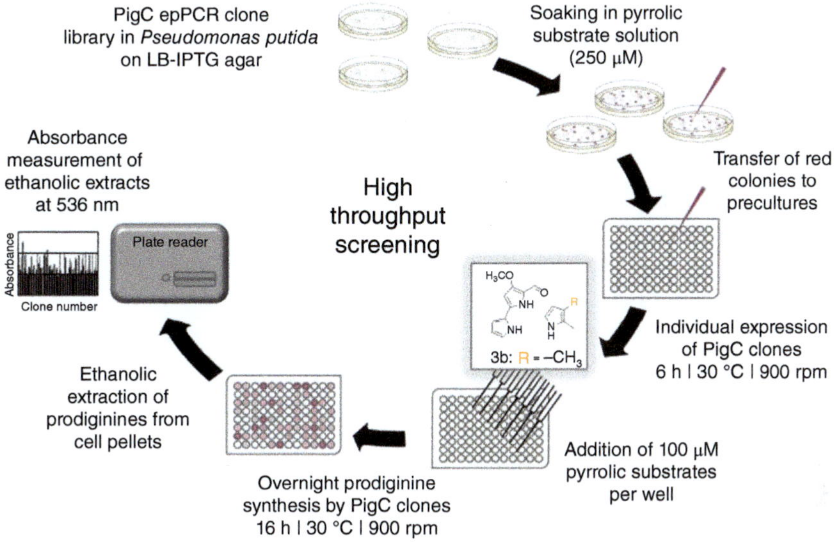

Figure 3.1 epPCR-directed evolution of prodigiosin ligase PigC with the aim to improve the activity. Source: Ref. [44]/Royal Society of Chemistry.

of an arbitrary number of consecutive bases (up to 16) at random positions and enables the insertion of a specific sequence or random sequences of an arbitrary number [50]. The 8-step protocol is summarized in Scheme 3.3.

Variations and improvements to the RID approach have been reported [51]. The RID procedure was initially applied to the directed evolution of the green fluorescent protein GFP_{UV} by replacing three randomly chosen consecutive bases with a mixture of 20 codons [50]. In *E. coli* as host, six variants were selected, several of which show quite different fluorescence properties. A yellow fluorescent protein and an enhanced green fluorescent mutant were identified, neither of which was accessible by conventional epPCR. It should be mentioned that the earliest directed evolution studies of GFPs utilized other mutagenesis methods such as epPCR [52] or saturation mutagenesis at a 20-residue region [53]. Unnatural amino acids using expanded genetic codes [54] can also be introduced by this technique [52a, 55], but undesired secondary mutations in the PCR process may also occur. Studies reporting other techniques for deletion mutagenesis have appeared, claiming simpler protocols [56, 57]. For example, phenylacetone monooxygenase (PAMO), which has a narrow substrate range in Baeyer–Villiger reactions, was converted into a cyclohexanone monooxygenase (CHMO) by *rationally* designing deletions at several residues, guided by the consensus techniques and X-ray data (Figure 3.2) [57]. Deletions were performed on residues S441, A442, and L443, and combinations thereof. Deletion of A442 led to the best results, e.g. high activity in the Baeyer–Villiger monooxygenase (BVMO)-catalyzed reaction of 2-phenylcyclohexanone, which is not accepted by WT PAMO. Nevertheless, a completely different protein

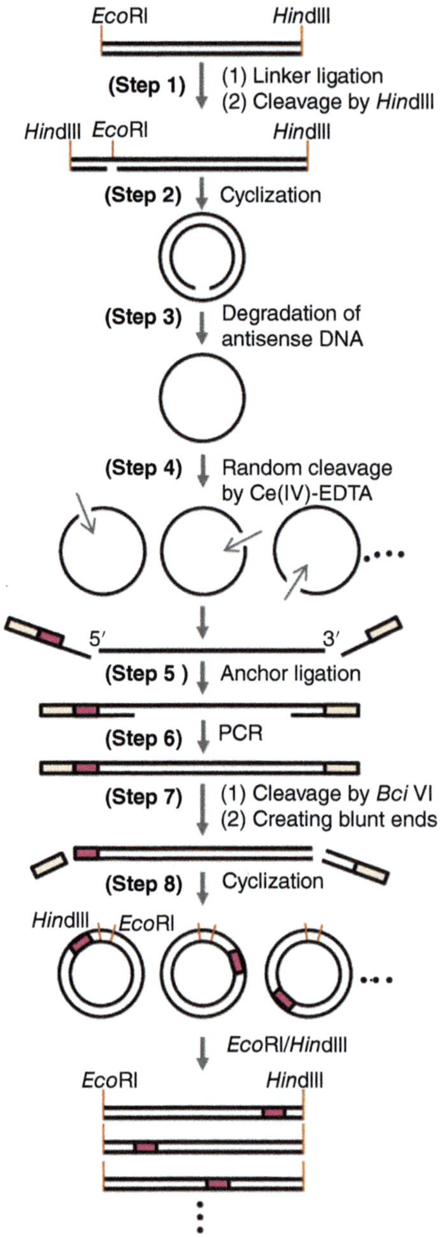
EcoRI
HindIII
(Step 1)
(1) Linker ligation
(2) Cleavage by HindIII
HindIII EcoRI
HindIII
(Step 2)
Cyclization
(Step 3)
Degradation of antisense DNA
(Step 4)
Random cleavage by Ce(IV)-EDTA
5′
3′
(Step 5)
Anchor ligation
(Step 6)
PCR
(Step 7)
(1) Cleavage by Bci VI
(2) Creating blunt ends
(Step 8)
Cyclization
HindIII
EcoRI
EcoRI/HindIII
EcoRI
HindIII

Scheme 3.3 Illustration of random insertion/deletion (RID) mutagenesis for the construction of a library of mutant genes. *Step 1*: (1) The fragment obtained by digesting the original gene with *Eco*RI and *Hin*dIII is ligated to a linker; and (2) the product is then digested with *Hin*dIII to make a linear dsDNA with a nick in the antisense chain. *Step 2*: The gene fragment is cyclized with T4 DNA ligase to make a circular dsDNA with a nick in the antisense chain. *Step 3*: The circular dsDNA is treated with T4 DNA polymerase to produce a circular ssDNA. *Step 4*: The circular ssDNA is randomly cleaved at single positions by treating with Ce(IV) – EDTA complex. *Step 5*: The linear ssDNAs, which have unknown sequences at both ends, are ligated to the 5′-anchor and the 3′-anchor, respectively. *Step 6*: The DNAs that are linked to the two anchors at both ends are amplified by PCR. *Step 7*: The PCR products are treated with *Bci*VI, leaving several bases from the 5′-anchor, at the 5′-end. The *Bci*VI treatment also deletes a specific number of bases at the 3′-end. *Step 8*: The digested products are treated with the Klenow fragment to make blunt ends and cyclized again with T4 DNA ligase. The products are treated with *Eco*RI and *Hin*dIII, and the fragments are cloned into an *Eco*RI – *Hin*dIII site of modified pUC18 (pUM). Source: Ref. [50]/Springer Nature.

engineering strategy based on Combinatorial Active-Site Saturation Test (CAST) and soliciting the PAMO crystal structure and the consensus technique in which eight homologous BVMOs were aligned, provided notably better results, as highlighted in Section 4.3.3.3 [58]. Combining the two approaches has not been attempted to date. A different deletion/insertion technique dubbed INSULT avoids subcloning and obviates the necessity of special "ultra-competent cells" in other systems [59].

While the general concept of inserting and deleting residues has not been applied very often in directed evolution, a new start was recently made by assessing unexplored regions of fitness landscapes in directed evolution via insertion/deletion mutagenesis [60]. The authors introduced a transposition-based random insertion and deletion mutagenesis (TRIAD) approach for generating libraries of random variants with short-frame insertions and deletions, the aim being to turn a phosphotriesterase into an arylesterase. The general procedure of constructing deletion libraries includes five main steps: (1) generating a TransDel insertion library by transpositioning the transposon TransDel into the target sequence; (2) digestion of a single break per variant by removing TransDel along with an extended 3 bp from the target sequence using the type IIS restriction enzyme Mlyl; and (3a) the reformation of the target sequence minus 3 bp by self-ligation, thereby generating a library of single variants; or (3b) inserting DNA cassettes (e.g. Del2 and Del3) within the break in the target sequence, followed by (4b) digestion and self-ligation and (5b) to generate double and triple variants, respectively. Likewise, generation of insertion libraries also requires five steps: (1) transpositioning the transposon TransIns into the target sequence; (2) digestion by NotI and MlyI to remove TransIns; (3) insertion of DNA cassettes Ins1, Ins3 and Ins3 (with 1, 2, and 3 randomized NNN triplets, respectively) within the target sequence to obtain the corresponding insertion libraries; (4) removing the cassettes via AcuI digestion and 3′-end digestion; and (5) self-ligation of the target sequence to elongate 3, 6, and 9 random bp,

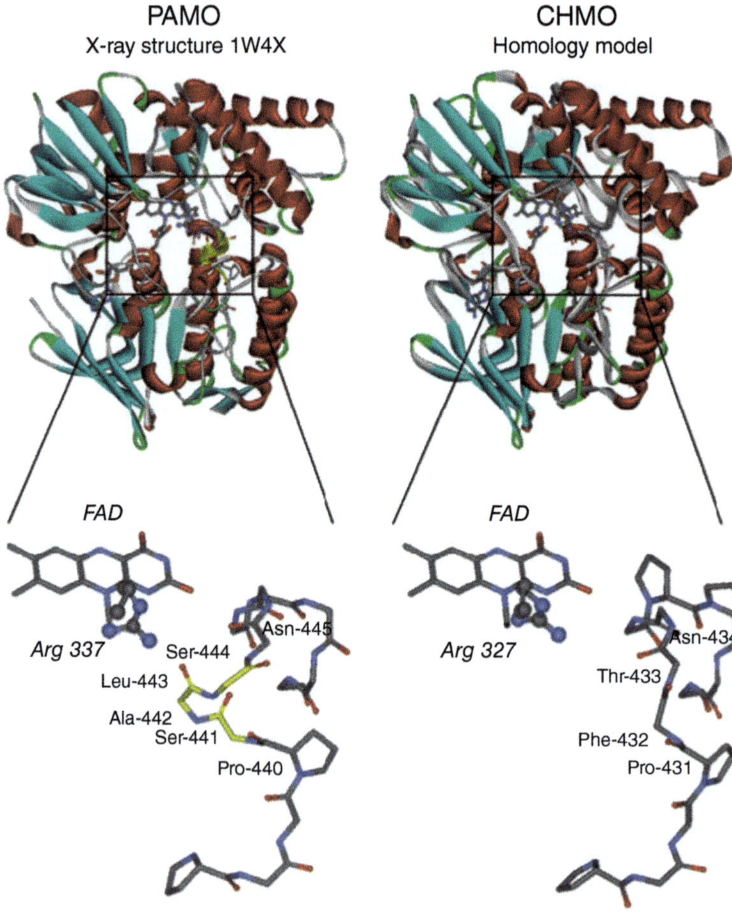

Figure 3.2 Comparison of the crystal structure of PAMO (1W4X) (left) and the homology model of CHMO (right) with 40.3% sequence identity [57]. The upper parts show the similar folds, and the lower parts are close-ups of the respective altered active sites showing the FAD cofactor as solid sticks and the catalytic arginine in a ball-and-stick manner. The yellow color highlights the presence of two additional residues in the arginine-stabilizing loop of PAMO compared with the CHMO backbone representation. Source: Ref. [57]/John Wiley & Sons.

and generate libraries of single variants with an insertion of 1, 2, and 3 triplets, respectively (Figure 3.3) [60].

AncHLD-RLuc is a bifunctional enzyme that is able to catalyze both haloalkane dehalogenase (HLD) and luciferase (LUC) reactions [61]. By utilizing the TRIAD approach, random insertion–deletion (InDel) libraries were constructed using AncHLD-RLuc as the starting enzyme (Figure 3.4) [62]. After screening of the libraries, 25 hits with improved LUC or HLD activities were selected. The selected beneficial variants were then subjected to protein

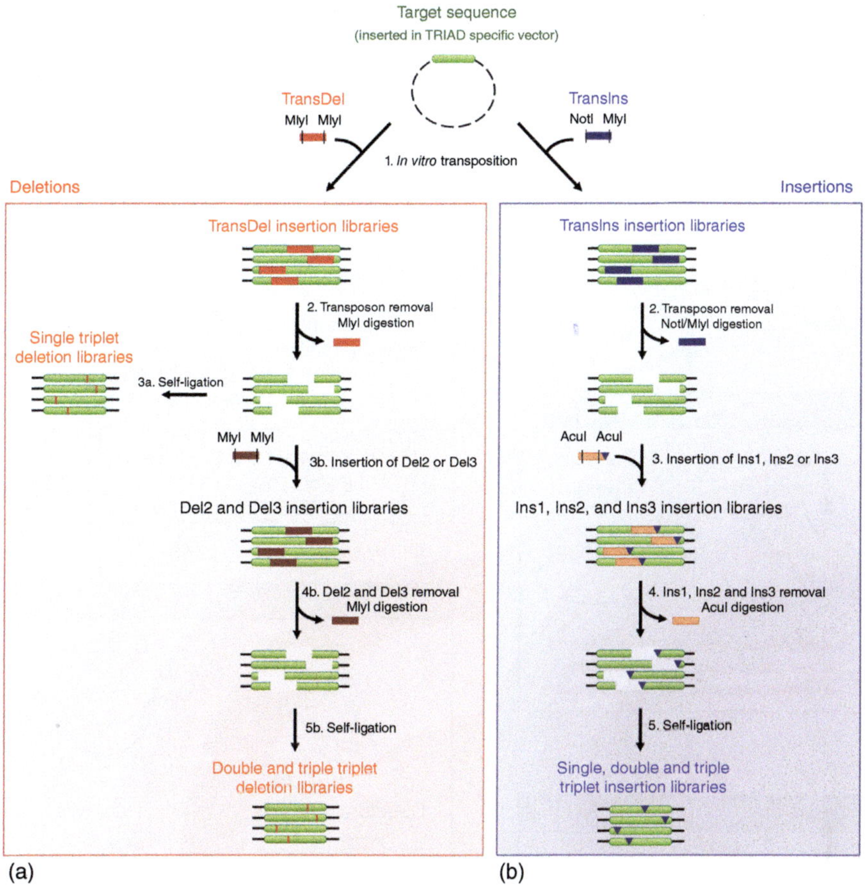

Figure 3.3 The general procedure of transposon-based mutagenesis approach (TRIAD). Deletions and insertions are shown by red vertical lines and purple triangles, respectively. Source: Ref. [60]/Springer Nature/CC BY 4.0.

dynamics and statistical analysis. Thereafter, the resultant dynamic elements that may be essential for efficient catalysis were transplanted and assembled to form a new variant, AncFT. It showed a 7000-fold increase in catalytic efficiency relative to AncHLD-RLuc and a 100-fold longer glow-type bioluminescence than the wild-type *Renilla* luciferase RLuc8 [62]. As a result of the application of TRIAD-guided library construction, it demonstrates that insertions or deletions are effective means in protein engineering, not only point substitutions.

While deletion/insertion mutagenesis in directed evolution has not been applied very often and is difficult to design by prediction, such mutations as well as point mutations are currently being considered in a completely different research area: Utilization of CRISPR genome editing in human induced pluripotent stem cells (iPSCs) [63].

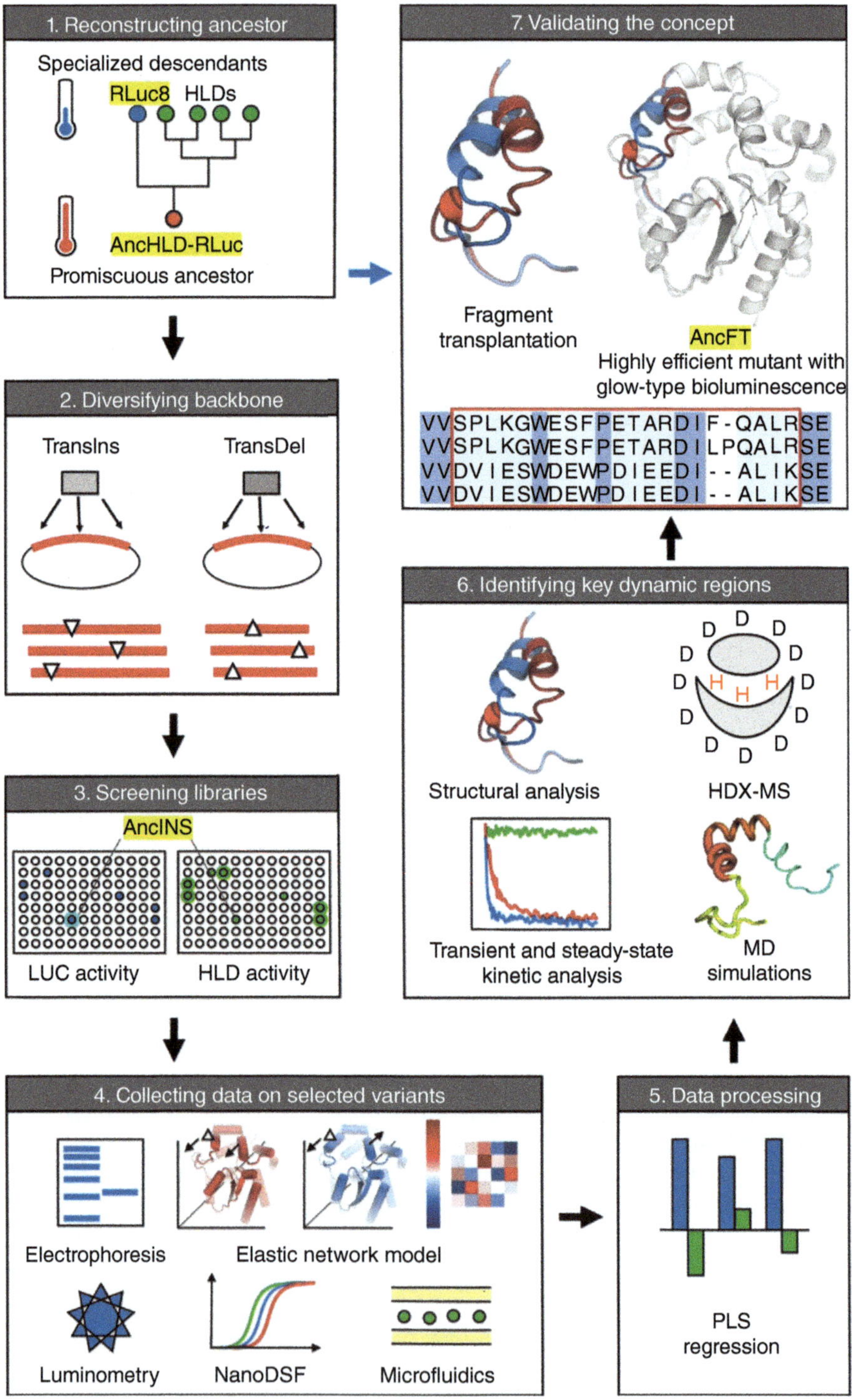

Figure 3.4 TRIAD-guided directed evolution of a bifunctional enzyme. Source: Ref. [62]/Springer Nature/CC BY 4.0.

3.2.4 Saturation Mutagenesis as a Privileged Method: Away from Blind Directed Evolution

The basic characteristics of saturation mutagenesis were already outlined in the introductory Chapter 1. It involves random amino acid exchange events at a defined single amino acid position, or at a site composed of more than one amino acid position in a protein of interest, with creation of focused mutant libraries. Various terms have been proposed for different ways to apply saturation mutagenesis [1a]. When a randomization site comprises a single amino acid position, it is often called Site Saturation Mutagenesis (SSM). When several residues are grouped into a larger randomization site, the designation "combinatorial saturation mutagenesis" (CSM) is sometimes used [1a]. Generally, the term "saturation mutagenesis" suffices.

Maximum genetic diversity in saturation mutagenesis is introduced when all 20 canonical amino acids are used as building blocks by way of NNK or NNS codon degeneracy, but reduced amino acid alphabets can be employed in clever ways (see below). Restricting randomization to a single amino acid position or a site composed of several such positions should be guided by various types of information, including X-ray structural data (if available), homology models, and/or consensus sequence information. The decision where randomization should be focused is crucial, which depends upon the purpose of a mutagenesis project [2f]:

- At hot spots identified by epPCR for manipulating theoretically any catalytic property of interest such as enhanced thermostability [64] or increased stereoselectivity [65] (but this strategy does not always work [66]).
- At positions selected by rational design for enhancing oxidative stability [67].
- At positions predicted by a consensus approach based on multiple sequence alignment (MSA) and guided by structural information for enhancing stability [68].
- At sites lining the binding pocket for manipulating activity, stereo-, and regio-selectivity and/or substrate scope [1a, 2f].
- At remote sites which can be expected to induce allosteric effects for manipulating activity, stereo-, and regio-selectivity and/or substrate scope [69].
- At sites displaying high B-factors for manipulating thermostability and/or robustness in the presence of hostile organic solvents [70].

Experience has shown that some choices are more productive than others [1a, 15]. If a completely wrong decision is made, then the probability of generating variants displaying improved catalytic profiles is low, e.g. performing saturation mutagenesis at sites lining the binding pocket in order to enhance thermostability would not be a reasonable choice, although in exceptional cases, such a mutation may in fact lead to enhanced stability (Chapter 6). In the discussion that follows, different molecular biological techniques for saturation mutagenesis are described before addressing strategic questions about how to apply them optimally.

In the introductory Chapter 1, a number of older methods of saturation mutagenesis are featured, some of which are no longer in use today due to the emergence of superior protocols. As a most general description of this focused mutagenesis method, appropriate primers that carry the genetic information encoding the desired mutational changes are designed, prepared in the lab or acquired commercially, and applied in different procedures. A bewildering number of studies have appeared and continue to be published which feature variations and improvements in saturation mutagenesis, some only reporting improved cloning procedures in specific cases (which is essential for acceptable performance). Systematic comparative studies are rare, and since they cover only a limited number of approaches [1, 2, 26b, 71–75], general conclusions are not possible. Fortunately, during the last few years trends, have emerged [1a, 15]. The following three options appear to be most popular:

- QuikChange protocol [71].
- Megaprimer approach [24, 72, 76].
- Overlap extension polymerase chain reaction (OE-PCR) [73].

The most commonly used procedure for saturation mutagenesis is the so-called QuikChange protocol of Stratagene/Agilent [71], which is based on previous studies (Chapter 1). Originally, it was developed for efficient site-specific mutagenesis, but it can be used to include saturation mutagenesis at a single specified residue or at a site composed of up to five amino acid positions. It consists of several steps: (i) plasmid amplification with a pair of complementary primers on both the sense and anti-sense strands containing designed mutations; (ii) DpnI digestion to selectively removing the methylated parental strands; and (iii) transformation into an appropriate host (e.g. *E. coli* strain) (Scheme 3.4). In the first step, a DNA polymerase such as Pfu or KOD-DNA having no nick-translation activity needs to be used. As already pointed out,

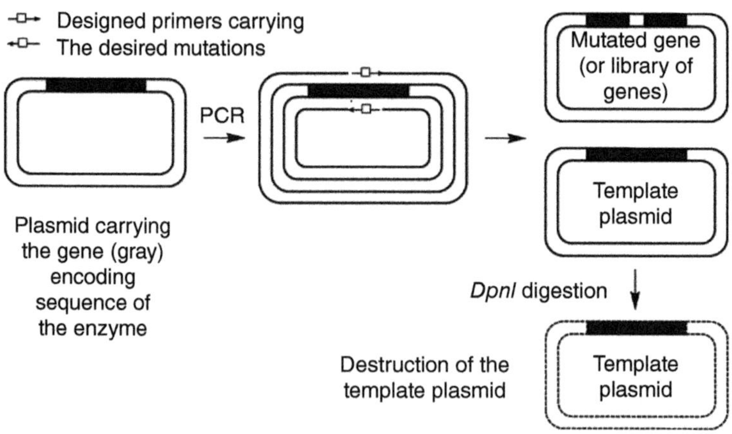

Scheme 3.4 Illustration of saturation mutagenesis based on the QuikChange™ (Stratagene/Agilent) protocol. Source: Adapted from Ref. [71].

algorithms for estimating diversity and degree of library completeness when using saturation mutagenesis have been developed [30].

A reasonable recommendation for potential users is to start with QuikChange, and in the case of poor or negative results, to continue empirically by testing the megaprimer approach or overlap extension-based OE-PCR (or combinations thereof). QuikChange may fail due to problems associated with primer length and design, especially when more than five amino acid positions are targeted simultaneously. The formal representation of the megaprimer approach is shown in Scheme 3.5 [24, 72, 74]. Two cycles of PCR are required, utilizing two flanking primers and one internal mutagenic primer carrying the designed base substitution(s). The first PCR round is carried out using the internal mutagenic primer and the first flanking primer. This PCR product, the "megaprimer," is purified and then used together with the second flanking primer for the second PCR. This ensures that the final PCR product harbors the desired mutation(s) in a specific DNA sequence.

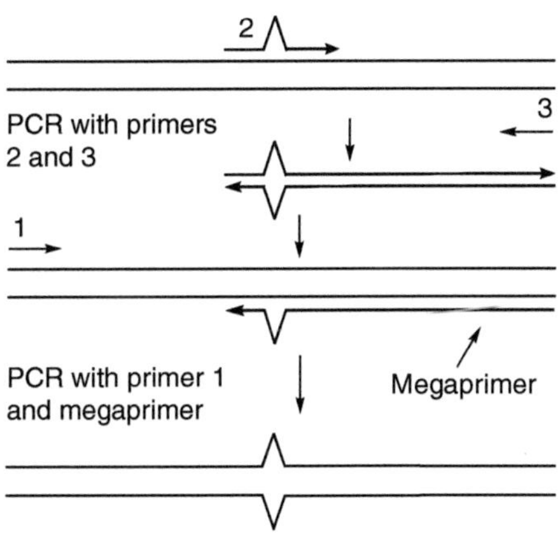

Scheme 3.5 General illustration of megaprimer PCR. Source: Adapted from Ref. [74].

As already pointed out in Chapter 1, OE-PCR is a particularly important advancement when generating focused libraries. It involves several steps, as outlined in Scheme 3.6 [73].

Even with these improvements, problems may still arise [1a], especially in the case of recalcitrant targets such as large plasmids. Unfortunately, some researchers may not be aware of this scenario, which results in poor quality libraries, unless quality checks are administered. An example for which a solution has been offered is P450-BM3 from *Bacillus megaterium*, a case of difficult-to-amplify templates [75]. Based on the concept of using non-overlapping oligonucleotides [77], a notably improved two-stage technique for creating saturation mutagenesis libraries was developed (Scheme 3.7). In the initial stage, the mutagenic primer and the anti-primer (not complementary) anneal to the template. In the second stage, the amplified sequence is utilized

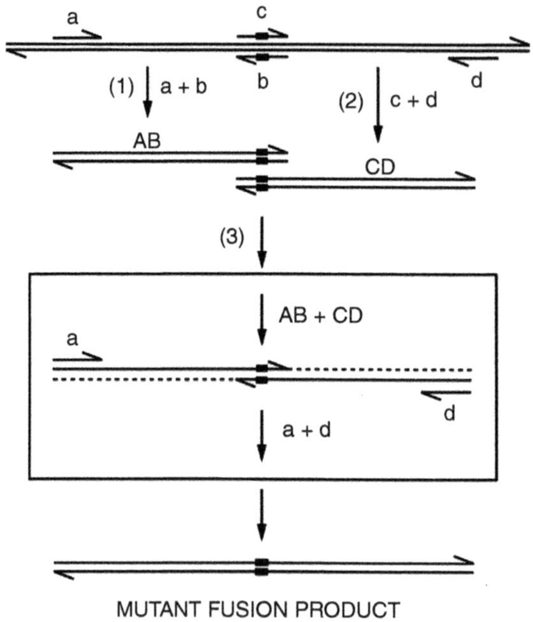

Scheme 3.6 Steps in site-directed mutagenesis by overlap extension PCR which can also be used for randomization at single residues or sites composed of more than one amino acid position. Lines with arrows represent the ds DNA and synthetic oligonucleotides with the arrows indicating the 5′ to 3′ orientation. Small black rectangles denote the site of mutagenesis. Lower-case letters refer to oligos while the PCR products are indicated by pairs of upper-case letters corresponding to the oligo primers which are employed to generate the product. The box represents the intermediate steps at which the denatured fragments anneal at the overlap and are extended b 3′ by the DNA polymerase (dotted line). Further PCR amplification occurs by additional primers "a + d." Source: Ref. [73]/With permission of Elsevier.

as a megaprimer. Importantly, sites comprising more than one residue can be randomized efficiently in a single PCR, irrespective of their location in the gene sequence. In a systematic comparative study, the virtue of this method was carefully compared with the performance of traditional QuikChange and related techniques using four different enzymes: P450-BM3 from *B. megaterium*; PAL; lipase B from *Candida antarctica*; and epoxide hydrolase from *Aspergillus niger* [75]. In all cases it proved to be superior in terms of library quality.

Several other strategies for higher-quality saturation mutagenesis libraries which lead to less amino acid bias, have been reported: (i) Synthesis of redundancy-free mutagenic primers using special mono-, di-, or tri-nucleotide phosphoramidite solutions [78]. (ii) The MAX strategy is based on the synthesis of a template and 20 selection oligonucleotides followed by hybridization [79], and its extension to contiguous codons via the ProxiMAX method [80].

Recently, a significant improvement in SSM was reported, which reduces amino acid bias and makes the appearance of all desired 19 canonical amino

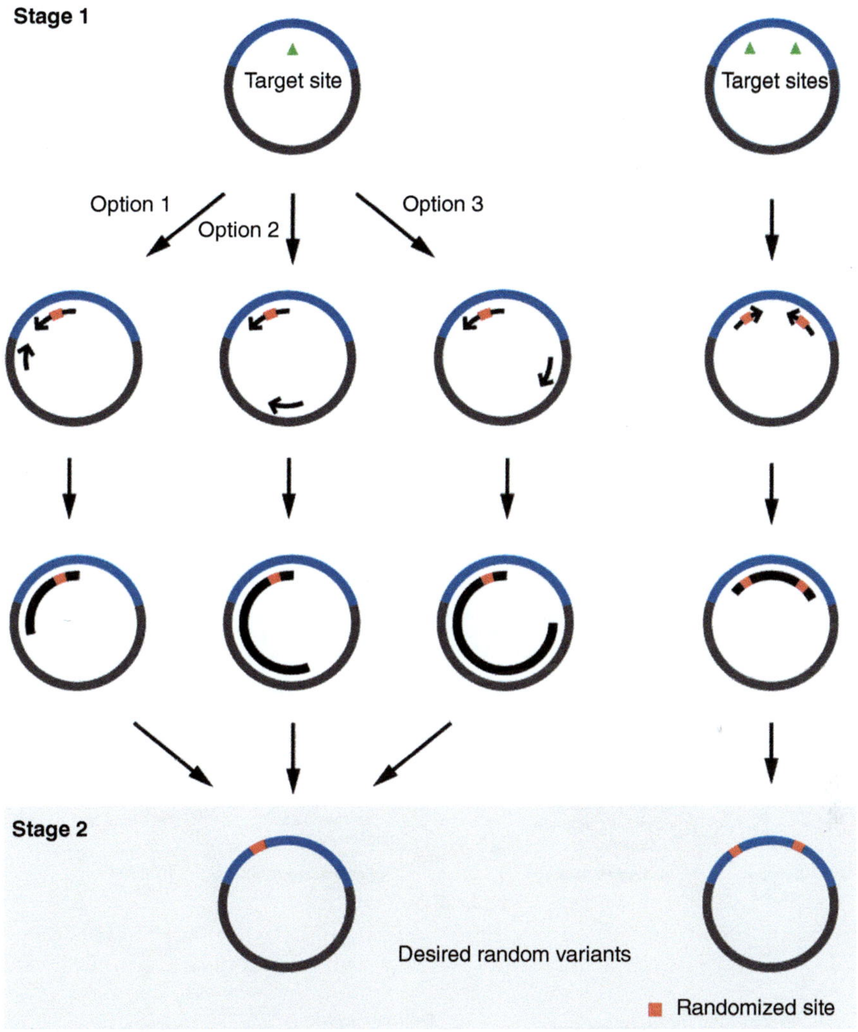

Scheme 3.7 Efficient method for saturation mutagenesis useful for cases of difficult-to-amplify templates, the scheme showing variation of the antiprimer position. The gene is represented in blue, the vector backbone in gray, and the formed megaprimer in black. In the first stage of the PCR, both the mutagenic primer (positions randomized represented by a red square) and the antiprimer (or another mutagenic primer, shown to the right) anneal to the template and the amplified sequence is used as a megaprimer in the second stage. Finally, the template plasmids are digested using *DpnI*, and the resulting library is transformed in bacteria. The scheme to the left illustrates the three possible options in the choice of the megaprimer size for a single site randomization experiment. The scheme to the right represents an experiment with two sites being simultaneously randomized. Source: Ref. [75]/Springer Nature.

acid substitutions at a defined residue highly probable [81]. Whereas most other approaches involve a one-step PCR procedure, a two-step PCR-based technique was developed in which a mutagenic primer and a non-mutagenic (silent) primer are first employed to generate a short DNA fragment. Following recovery, it is then used as a megaprimer in order to amplify the whole plasmid (Scheme 3.8). In order to illustrate the utility of this procedure, the difficult-to-amplify gene of cytochrome P450-BM3 was used. Of course, an alternative is to perform 19 site-specific mutagenesis experiments, but this can be costly and time-consuming. Compared to the two-step PCR procedure, the one-step PCR may lead to sub-optimal library quality when using large plasmids as templates [1a].

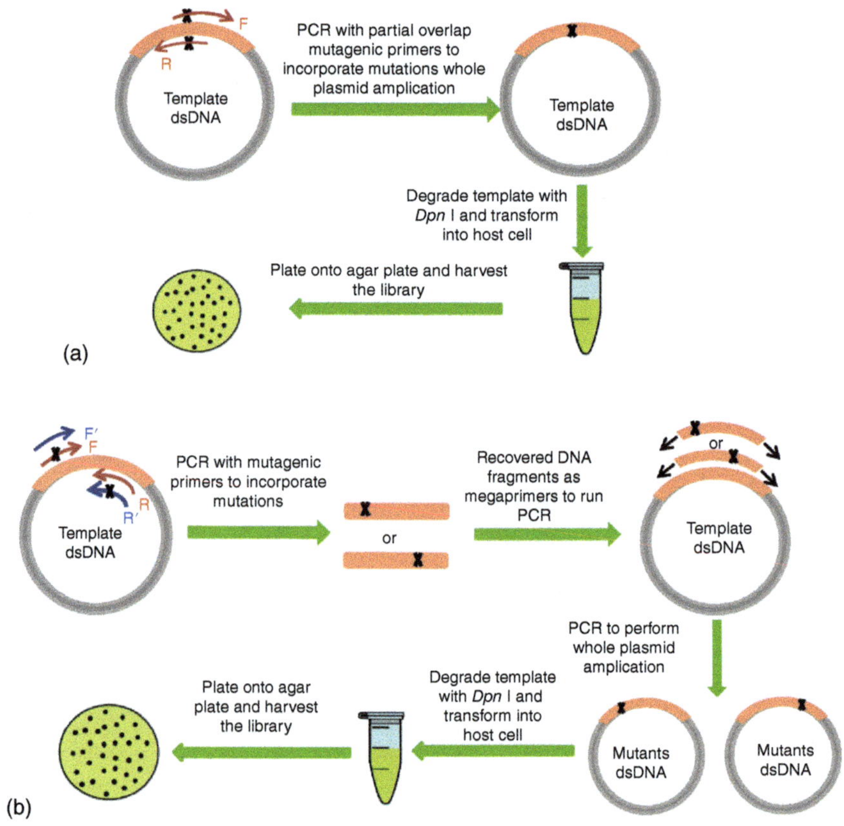

Scheme 3.8 (a) Illustration of canonical SSM library construction using the one-step PCR approach with partially overlapping primers. (b) Improved SSM library construction method using the two-step PCR approach with a mutagenic primer and a non-mutagenic (silent) primer. Source: Ref. [81]/Springer Nature/CC BY 4.0.

Other variations such as OmniChange [82a] and OSCARR [82b, c], have not been applied very often. In an interesting publication, the so-called Golden

Mutagenesis was developed, in which multi-site saturation mutagenesis was performed with automated primer design [82d]. In this approach, one to five amino acid positions can be altered simultaneously. The close relationship to CAST/Iterative Saturation Mutagenesis (ISM) was noted.

While many of these "tricks" promise higher-quality libraries, choosing the correct randomization site(s) is just as important [1a, 15]. When activity, substrate scope, stereoselectivity, and/or regioselectivity are the parameters of interest, rational design exploiting site-specific mutagenesis at the active site may be an alternative strategy. Few studies have demonstrated efficacy [83, 84]. Indeed, purely rational design is not as general as directed evolution based on focused combinatorial randomization at sites lining the binding pocket [2f]. The latter directed evolution approach was first employed in the successful attempt to enhance the enantioselectivity of lipase A from *P. aeruginosa* as the catalyst in the hydrolytic kinetic resolution of a chiral ester (Schemes 1.9 and 1.10), saturation mutagenesis being focused on a 4-residue site lining the binding pocket [33] (see also Chapter 4 for an analysis of this early experiment). In view of Emil Fischer's lock-and-key hypothesis [85] and Linus Pauling's concept regarding transition state stabilization by the protein environment [86], any attempt to reshape an enzyme's binding pocket by saturation mutagenesis at such sites for manipulating stereo- and/or regio-selectivity, or substrate acceptance appears perfectly logical [46b,c] (see also Chapter 8). It is all the more surprising that this strategy [33] was not systematized until later with the emergence of the CAST [87], a convenient acronym for this embodiment of saturation mutagenesis. Residues surrounding the binding pocket are first identified on the basis of X-ray data or homology models, designated as A, B, C, etc. (Scheme 3.9a). A limited selection of CAST sites (red-marked residues in Scheme 3.9a) is often sufficient. Each site may comprise one, two, or more amino acid positions. The problem of how to group single-residue sites into multiple-residue sites (if at all), constitutes a strategic question to be treated in Chapter 4. The libraries A, B, C, etc. may harbor improved hits, but the degree of improvement may not be acceptable. This problem can be solved by applying ISM [88a], first reported in 2005 using an epoxide hydrolase as the catalyst in the hydrolytic kinetic resolution of a racemic epoxide [88b]. Accordingly, the gene of a hit in a given library is used as a template for saturation mutagenesis at another site, and the process is continued until all sites have been visited once in an upward pathway. The schemes for 2-, 3-, and 4-site ISM systems are illustrated in Scheme 3.9b, involving 2, 6, and 24 upward pathways, respectively [2f, 88, 89]. Questions of how to identify the optimal upward pathway and how to escape from local minima are also treated in Chapter 4.

When applying saturation mutagenesis, the issue of oversampling in the screening process needs to be considered [2f, 89, 90], which is essential when choosing the optimal evolutionary strategy (Chapter 4). Two types of statistical analyses have been proposed for estimating the required degree of screening (necessary number of transformants) assuming the absence of bias: the Patrick/Firth algorithm [30a, b] or similar metrics [30c, d] which focus on the

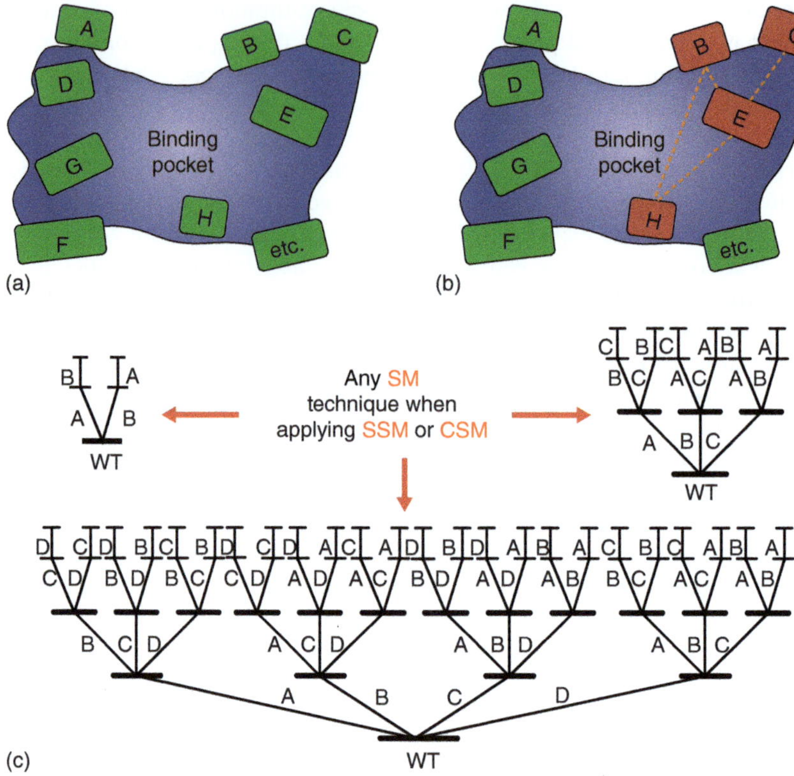

Scheme 3.9 Illustration of the CAST/ISM strategy. (a) Rationally chosen sites A, B, C, etc. for SM-guided randomization, each site being composed of 1, 2, 3 or more amino acid residues; (b) Grouping four sites (red boxes) for a simultaneous SM; (c) ISM scheme for 2-, 3-, and 4-site systems involving 2, 6, and 24 upward pathways, respectively. The red arrows indicate that an improved mutant can be obtained by SSM or CSM by using any gene-mutagenesis technique. Source: Adapted from Refs. [87, 88].

%-coverage of a library, and the recently introduced Nov metric, which refers to the first, second, third, or *n*th best mutant in a given library as a function of the number of screened transformants [30e, 91]. The calculated minimum number of screened transformants for, e.g. 95% library coverage, or for finding the best mutant, serves as a useful guide when deciding on how much effort to invest in the screening step. The Patrick/Firth approach, as featured in the computational tool GLUE-IT [30b], has been incorporated into the user-friendly CASTER computer aid for designing saturation mutagenesis libraries [70b], available free of charge on the Reetz website (http://www.kofo.mpg.de/en/research/biocatalysis). These and other computational tools are summarized in Section 3.3.

When designing saturation mutagenesis libraries, especially for controlling enzyme stereoselectivity, it is highly recommended to consider reduced amino acid alphabets with codon degeneracies other than NNK or NNS. As will be seen, this has significant advantages [1a]. Constructing proteins with fewer than the normal 20 canonical amino acids has been reported earlier for different reasons, which are related to the following questions:

- Did nature use less than 20 canonical amino acids for constructing primordial proteins before evolving the optimal full number [92a–d]?
- Do proteins constructed with notably reduced amino acid alphabets enable proper folds [92b, c]?
- Do proteins constructed with notably reduced amino acid alphabets enable proper folds as well as activity [92d]?

All three questions have been answered affirmatively. For example, a chorismate mutase was constructed using a set of nine amino acids, leading to a proper fold as well as enzyme activity [92d]. *Therefore, the generally observed success when choosing an appropriate reduced amino acid alphabet in saturation mutagenesis at appropriate sites is not so surprising, but it took some years before it was applied for the first time with the aim of engineering stereoselective enzymes in directed evolution* [90b].

As the number of amino acid positions in a randomization site increases, so does the screening effort at all degrees of library coverage. This effect is illustrated in Table 3.2, which features the case of 95% library coverage when NNK codon degeneracy (N: adenine/cytosine/guanine/thymine; K: guanine/thymine) encoding all 20 canonical amino acids is used versus NDT (D: adenine/guanine/thymine; T: thymine) encoding 12 amino acids which comprise a fairly balanced mixture of polar/non-polar, charged/non-charged, and hydrophobic/hydrophilic building blocks (Phe, Leu, Ile, Val, Tyr, His, Asn, Asp, Cys, Arg, Ser, Gly) [2f, 89, 90]. It can be concluded that saturation mutagenesis at a single residue is best performed with NNK codon degeneracy because screening only a single 96-well plate suffices for 95% library coverage. Practical differences between NNK and NDT arise when randomizing larger sites. If in a given project an efficient selection or FACS-based system is available, which in the case of stereoselectivity is rare (Chapter 2), then there is little reason to choose NDT codon degeneracy because it generates lower diversity. On the other hand, if in a given case, such assay systems are unavailable and screening is the only option, which is the rule, then NDT or even smaller amino acid alphabets should definitely be considered.

A fundamental question arises from such calculations. If a researcher has chosen a randomization site, e.g. a 3-residue site, and for practical reasons has defined an upper limit regarding the number of transformants to be screened, e.g. 5000, then two options are possible:

- Application of NNK codon degeneracy, which covers only a very small portion of the respective library, e.g. 15%.
- Use of NDT codon degeneracy, which ensures 95% library coverage.

Table 3.2 Oversampling necessary for 95% library coverage as a function of NNK versus NDT codon degeneracy and the number of amino acid positions in a randomization site.

Number of amino acid positions at one site	NNK		NDT	
	Codons	Transformants needed for 95% library coverage	Codons	Transformants needed for 95% library coverage
1	32	94	12	34
2	1 028	3 066	144	430
3	32 768	98 163	1 728	5 175
4	$>1.0\times10^{6}$	$>3.1\times10^{6}$	$>2.0\times10^{4}$	$>6.2\times10^{4}$
5	$>3.3\times10^{7}$	$>1.0\times10^{8}$	$>2.5\times10^{5}$	$>5.5\times10^{5}$
6	$>1.0\times10^{9}$	$>3.2\times10^{9}$	$>2.9\times10^{6}$	$>8.9\times10^{6}$
7	$>3.4\times10^{10}$	$>1.0\times10^{11}$	$>3.5\times10^{7}$	$>1.1\times10^{8}$
8	$>1.0\times10^{12}$	$>3.3\times10^{12}$	$>4.2\times10^{8}$	$>1.3\times10^{9}$
9	$>3.5\times10^{13}$	$>1.0\times10^{14}$	$>5.1\times10^{9}$	$>1.5\times10^{10}$
10	$>1.1\times10^{15}$	$>3.4\times10^{15}$	$>6.1\times10^{10}$	$>1.9\times10^{11}$

Source: Adapted from Ref. [89].

The question as to which option is optimal and should be chosen was addressed empirically some time ago [90a]. An NNK- and an NDT-library were constructed, and 5000 transformants were screened in each case, using an epoxide hydrolase as the catalyst in the hydrolytic kinetic resolution of a racemic substrate. This means 15% versus 95% library coverage, respectively. The two equally sized libraries were checked for quality, meaning the frequency of improved hits and the degree of improved enantioselectivity. It turned out that the NDT-library had a distinctly higher quality [90a]. In a recent more detailed study, a similar conclusion was reached, i.e. it is generally better to strive for higher library coverage at lower diversity, in contrast to considerably lower coverage at maximum diversity [93]. Nevertheless, when cleverly choosing reduced amino acid alphabets, less than 95% library coverage may well suffice [94].

A wide variety of reduced amino acid alphabets have been utilized with great success, ranging from 12 amino acids (NDT or DNT codon degeneracy) to only one amino acid as a building block (Chapter 4). In conclusion, the use of reduced amino acid alphabets when using saturation mutagenesis of the CAST-type and ISM represents a significant advancement in the directed evolution of enzymes with enhanced or reversed stereo- and regio-selectivity, activity, and substrate scope [2f, 26, 89, 90, 95]. CAST/ISM has also been applied routinely by the Arnold group and others in the directed evolution of artificial metalloenzymes [1b, c, 2d, 96]. Not surprisingly, the concept of reduced amino acid alphabets can even be applied to evolving thermostability, because in this case it is based on the rational B-FITTER algorithm [70b] (Chapter 6). The positive results of applying advanced versions of CAST/ISM by many research groups speak for themselves. Moreover, the concept of deconvolution of multi-mutational variants has helped to explain these successes [97]. Cooperative (more than traditional additive) mutational effects were routinely uncovered when applying

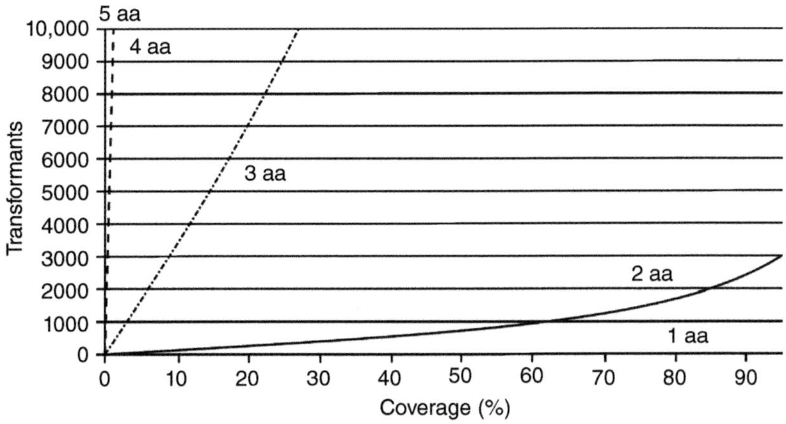

Figure 3.5 Library coverage calculated for NNK codon degeneracy at sites consisting of 1, 2, 3, 4, and 5 amino acid positions (aa = amino acids). Source: Ref. [90a]/John Wiley & Sons.

this unique technique. One of an increasing number of examples concerns the deconvolution of multi-mutational variants of an acycltransferase obtained by ISM and used in the production of valuable pharmaceutical products [98]. A deeper discussion of the deconvolution technique can be found in Chapter 8.

It is clear that ensuring 95% library coverage (or even more) is not absolutely necessary in most practical applications, but knowledge of such numbers is useful when designing saturation mutagenesis experiments in a meaningful manner. For this reason, the numbers computed for the whole range of library coverage from 0% to 95% serve as an imminently useful guide. Figures 3.5 and 3.6 depict the situation for NNK versus NDT codon degeneracy, respectively, with sites containing one, two, three, four, and five amino acid positions being considered [90a].

Based on the Patrick/Firth algorithm [30a, b], another index of interest to the experimenter was derived, namely the oversampling factor O_f starting from Eq. (3.2), where T is the number of transformants actually screened, P denotes the probability that a particular sequence occurs in the library, and F_i is the frequency [2f, 90a]:

$$T = -\ln(1 - P_i)/F_i \tag{3.2}$$

Upon substituting for F_i, the relationship reduces to Eq. (3.3), where V is the number of mutants on gene level in a given library:

$$T = -V\ln(1 - P_i) \tag{3.3}$$

thereby defining the correlation between the number of mutants V of a given library and the number of transformants T that need to be screened for a defined degree of completeness.

This leads to the oversampling factor O_f, which defines the degree of oversampling necessary for achieving a certain completeness under the assumption

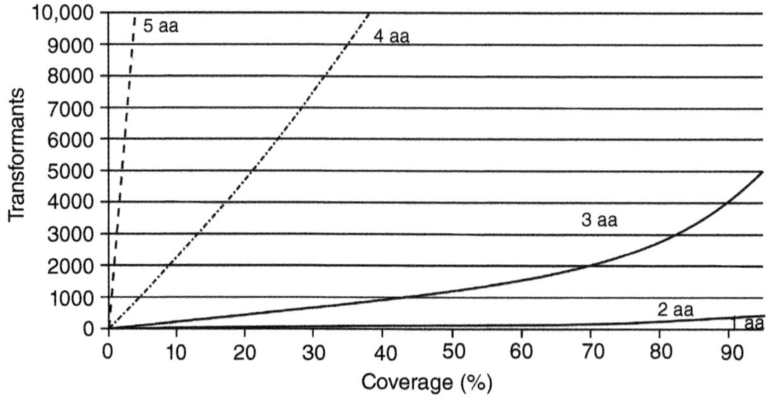

Figure 3.6 Library coverage calculated for NDT degeneracy at sites consisting of 1, 2, 3, 4, or 5 amino acid positions. Source: Ref. [90a]/John Wiley & Sons.

of no amino acid bias (Eq. (3.4)) [90a]:

$$O_f = T/V = -\ln(1 - P_i) \tag{3.4}$$

When computing the oversampling factor O_f as a function of %-coverage, the curve featured in Figure 3.7 results, spanning the whole range. For example, when aiming for 95% library coverage, the oversampling factor O_f amounts to about 3, which means that a threefold excess of transformants relative to the theoretical number of possible mutants on DNA level needs to be screened.

Researchers preferring the Nov approach, which also assumes the absence of bias, can use the respective computer aid, TopLib [91c]. The results of a Nov analysis for sites comprising one, two, three, or four amino acid positions are shown in Figure 3.8 [91a].

Obviously, there is a correlation between the Patrick/Firth and the Nov metrics, i.e. between a given %-coverage and the nth best variant, respectively. In order

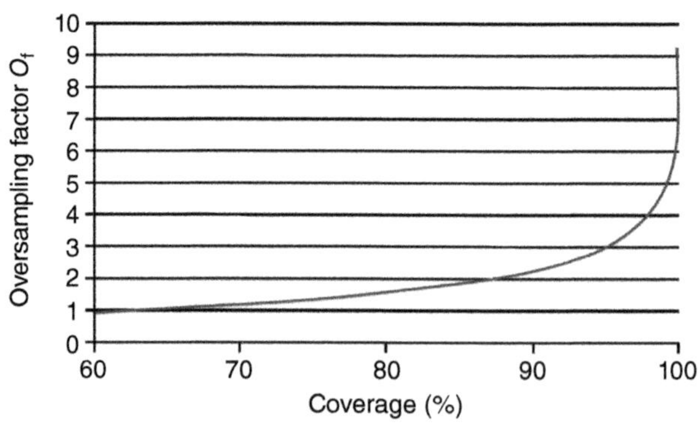

Figure 3.7 Correlation between oversampling factor O_f and percent library coverage. Source: Ref. [90a]/John Wiley & Sons.

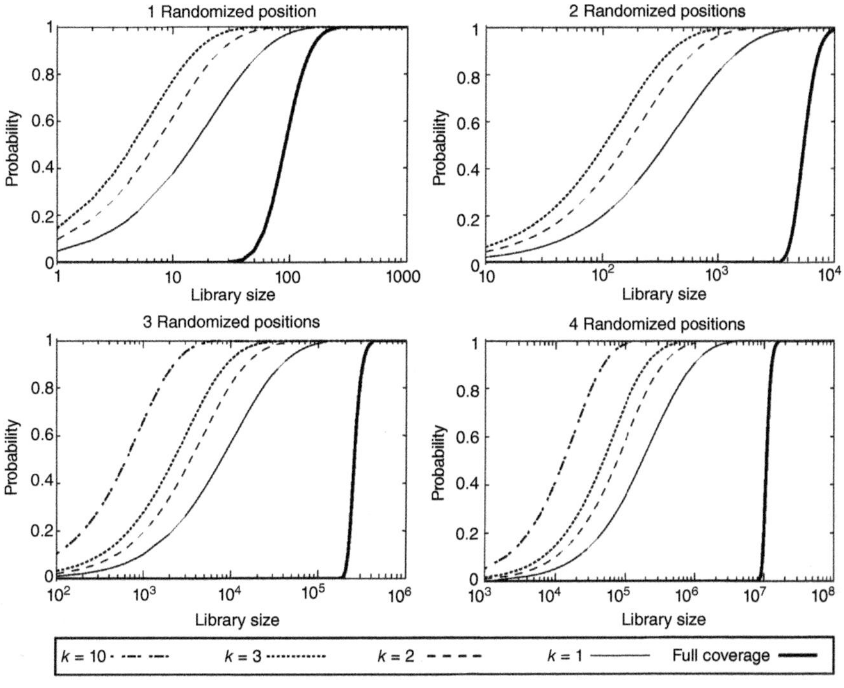

Figure 3.8 Probabilities of "full coverage" and of discovering at least one of the top k protein variants in variant space as a function of the library size when randomizing sites comprising one, two, three, and four amino acid positions in the case of NNK codon degeneracy. Source: Adapted from Ref. [91a].

to illustrate this for potential users of these metrics who may prefer one or the other statistical approach, the graph in Figure 3.9 has been computed [94a].

Whichever saturation mutagenesis method is used, *it is absolutely advisable to routinely apply the Quick Quality Control (QQC) for checking the expected diversity of a mutant library at DNA level* [99a], because screening too many plates for something that does not exist makes no sense. QQC can be performed quickly by pooling a limited number of transformants, which are then sequenced. Since the cost of sequences has gone down drastically in recent years, this small investment may well reduce wasted time and resources. The Stewart method which generates so-called *Q*-values [99b], is not as fast as QQC, but has the advantage of quantitative quality assessment.

In addition to the above advancements, several other molecular biological modifications have been reported for increasing the quality of saturation mutagenesis libraries, the goal again being the reduction of amino acid bias and the elimination of stop codons. Two recent studies deserve particular attention when randomization with the introduction of all 20 canonical amino acids as building blocks is aimed for. Traditionally, NNS or NNK codon degeneracy has been used, but as already seen, library quality may be poor. The Tang-procedure

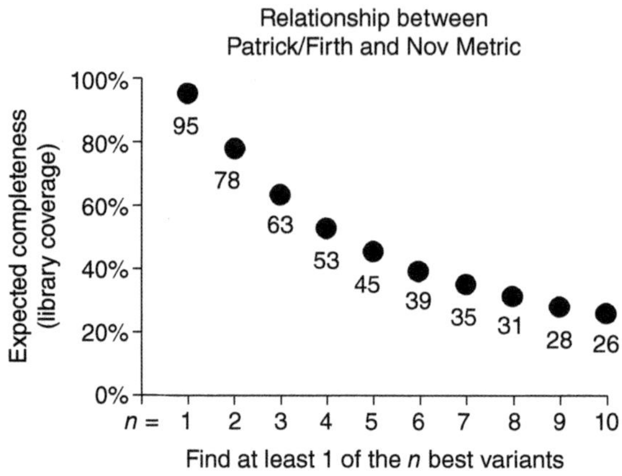

Figure 3.9 Patrick–Firth versus Nov statistical metrics. There is a relationship between the expected completeness (i.e. library coverage) algorithm by Patrick and Firth as computed by GLUE-IT and the concept of finding at least one of the *n*th-best variants (with a 95% probability) by Nov as computed by TopLib. This mathematical relationship is independent of the number of positions randomized and of the randomization scheme. Source: Ref. [94a]/American Chemical Society.

dedicated to the construction of "small-intelligent" focused libraries [100a] and the related "22c-trick" [100b] are well suited to solve these fundamental problems. The highly effective 22c-trick, generally in conjunction with CAST/ISM, has been applied routinely, inter alia, in the directed evolution of artificial metalloenzymes [1a, c]. In the Tang-procedure, a primer mixture is designed so that exactly one codon occurs per amino acid, assisted by a software tool called the DC-analyzer. Stop codons and eight rare *E. coli* are also eliminated. The Tang mixture is composed of:

- NDT codons encoding 12 amino acids (N, S, I, H, R, L, Y, C, F, D, G, and V).
- VMA codons encoding 6 amino acids (E, A, Q, P, K, and T).
- ATG codon encoding 1 amino acid (M).
- TGG codon encoding 1 amino acid (W).

When applying the "22c-trick," 22 unique codons encoding the 20 canonical amino acids are used:

- NDT codons encoding 12 amino acids (N, S, I, H, R, L, Y, C, F, D, G, and V)
- VHG codons encoding 9 amino acids.
- TGG codon encoding 1 amino acid (W).

It can be seen that in the Tang approach, four primers with two degeneracies (NDT, VMA) and two coding sequences (ATG, TGG) are involved, together targeting all 20 amino acids while theoretically eliminating all bias. A codon to amino acid ratio of 20 : 20 is ideal. The 22c trick requires two more codons. Nevertheless, the difference between the two techniques is not great, and both approaches have advantages and disadvantages [100b]. From a cost perspective, which has been analyzed using statistical methods [101], the total number of primers should be minimized, because this increases with *n* number of residues. In the case of the Tang approach, one site requires four equimolar mixed sense or antisense primers, but this can double to eight depending on the technique. Along such lines, a notably higher number of primers is required when considering two or three residues, namely 16 or 64 in the case of sense or antisense and 32 or 128 sense and antisense separately synthesized primers, respectively [100b].

In the study describing the "22c-trick", the screening efforts for NNN, NNK(S), and 22c were compared when using 2- or 3-residue sites (Figure 3.10). The benefits of the "22c-trick" are evident.

Using a BVMO CHMO as the model enzyme, the QCC [99a] was applied to an NNK library and the respective 22c-library (Figure 3.11). Here again, the superior library quality of "22c trick" is obvious [100b].

As will be further documented in Chapter 4 with the help of case studies, *the use of reduced amino acid alphabets in CAST/ISM is the preferred technique for manipulating stereo- and regio-selectivity as well as activity* [15].

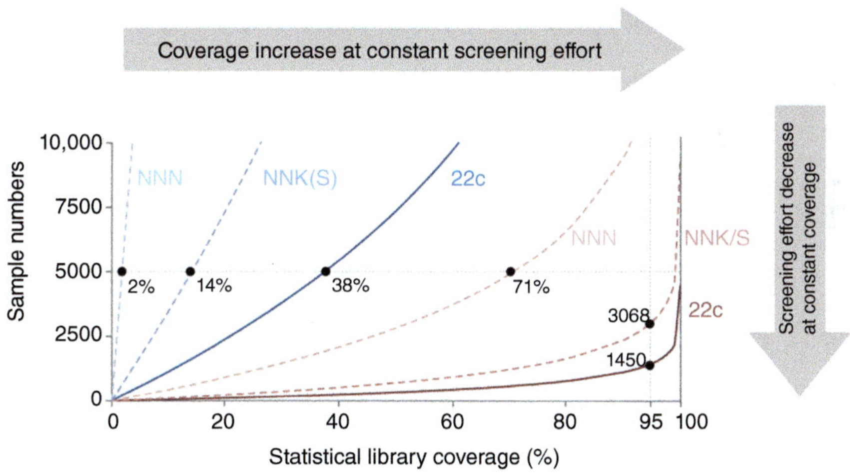

Figure 3.10 Screening effort required for different randomization schemes regarding sites composed of two or three amino acid residues. The choice of codon degeneracy dictates the sampling size for a desired statistical coverage of the library. For a 95% library coverage targeting two amino acid residues (red lines), 3068 samples have to be screened in the case of NNK/S, whereas only 1450 are necessary when applying the 22c-trick (53% lower screening effort). However, if the assumed capacity of medium-throughput systems is limited to 5000 samples, the library coverage drops to 71% when using NNN degeneracy. Similarly, when targeting three amino acid residues (blue lines) and limiting the sample size to 5000 colonies or transformants, the library coverage changes drastically to 38%, 14%, and 2% in the case of the 22c-trick, NNK/S, and NNN, respectively. Source: Ref. [100b]/American Chemical Society.

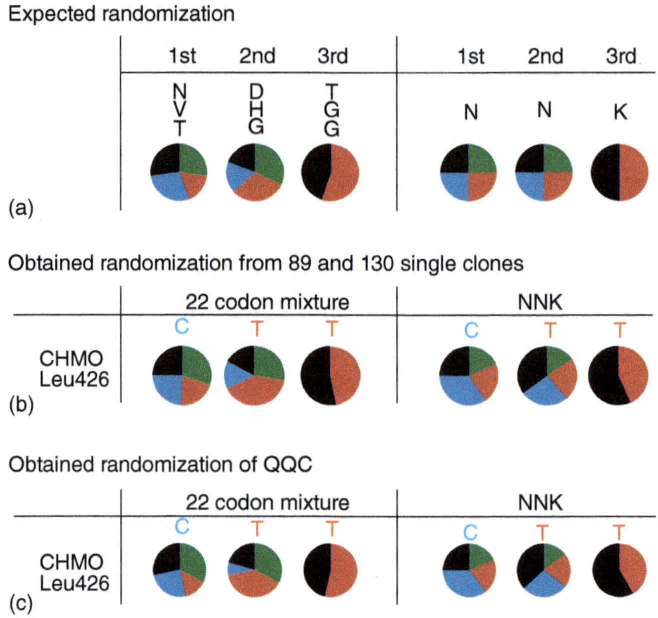

Figure 3.11 Distribution of nucleotide bases in the randomized residue Leu426 of CHMO. The percentual distribution of nucleotides is shown in pie diagrams for each of the three randomized bases using the 22c-trick (left) and NNK (right) degeneracies. (a) Theoretical expected distribution. (b) Experimental distribution calculated from the sequencing of 89 and 130 individual clones from the 22c-trick and NNK libraries, respectively. (c) Experimental Quick Quality Control from colony pooling. The nucleotide base guanidine (G) is depicted in black, adenosine (A) in green, threonine (T) in red, and cytosine (C) in blue. Source: Ref. [100b]/American Chemical Society.

Rather than utilizing reduced amino acid alphabets, expanded genetic codes encoding noncanonical amino acids can also be applied in directed evolution [102]. In doing so, different strategies are possible:

- Utilize more than 20 amino acids as building blocks in a randomization procedure.
- Introduce one or more non-natural amino acids at a strategic site and then use such a mutant as a template for directed evolution employing a designed reduced amino acid alphabet for saturation mutagenesis.

Expanded genetic codes have already been applied in the generation of artificial metalloenzymes [102c] and in manipulating the spectral properties of green fluorescent proteins [102d]. From a practical point of view, notable advantages still need to be demonstrated, but this may well be a matter of time. Introducing site-specifically an amino acid having a synthetic side-chain characterized by a diphosphane or dipyridine moiety, which strongly binds transition metals, *en route* to artificial metalloenzymes, offers exciting perspectives. Chapter 7 focuses on different ways to construct artificial metalloenzymes.

3.2.5 DNA Shuffling and Related Recombinant Gene Mutagenesis Methods

Methods which cause the breaking and rejoining of DNA in new combinations are called recombinant techniques [1a]. Numerous embodiments of this general approach to gene mutagenesis have been developed, the most prominent being DNA shuffling [103]. Three general versions are possible: (i) single-gene shuffling; (ii) shuffling of a set of mutants generated from a single gene, as illustrated in Scheme 1.8, or (iii) family shuffling using two or more homologous enzymes. In summary, genes are digested with DNase to generate random double-stranded oligonucleotide fragments of 10–50 base pairs, which are then PCR-amplified and purified from agarose gel. Reassembly by a DNA polymerase is achieved in a primerless PCR, followed by amplification in a second PCR employing a pair of flanking primers, leading to full-length mutant genes. In this way, the desired crossovers with the formation of mutations are achieved. An experimental comparison has shown that family shuffling ensures dramatically higher diversity than single-gene DNA shuffling (Figure 3.12) [103b]. In fact, single-gene shuffling provides libraries with >97% identical enzymes.

DNA shuffling is a multi-step process, which is fairly easy to perform, but such issues as optimization, limitations, and potential pitfalls should be considered. For best performance, several variables need to be optimized, including fragment size for controlling the frequency of crossovers [103], temperature cycle during reassembly, amount of assembly, and the number of cycles in amplification [17, 103]. *Thus, some experience is necessary for optimal performance.* In the case of family shuffling, homology should amount to >70% for acceptable library quality; self-hybridization with the formation of WT enzymes should also be minimized [103]. Point mutations may occur during the PCR step, which are difficult to distinguish from the "normal" crossover shuffling mutations. In principle, it should be possible to increase "additional" point mutations by using high concentrations of $MnCl_2$, if so desired, but usually the opposite is strived

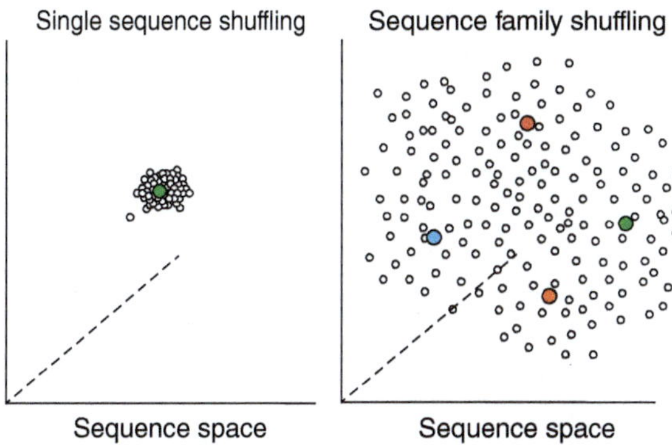

Figure 3.12 Searching sequence space by single-gene shuffling versus family shuffling. Source: Ref. [103b]/Springer Nature.

for. *Whichever form of DNA shuffling is chosen, researchers should be aware that amino acid bias probably exists in the formation of mutant libraries, caused by the necessity to apply several PCR steps*. Oddly enough, this has not been a concern in the scientific community.

In an interesting variation of DNA shuffling, designed synthetic nucleotides in a type of saturation mutagenesis are added during the reassembly process, which Stemmer dubbed *C*ombinatorial *M*ultiple-*C*assette *M*utagenesis (CMCM) [103c]. It was successfully applied in the directed evolution of stereoselective mutants of the PAL (for details see Schemes 1.9 and 1.10, and Chapter 4) [33]. In order to assess crossover efficiency and to improve DNA shuffling, probe hybridization in macroarray format was developed, which allows the analysis of chimeric DNA libraries [30a]. This procedure also reveals any bias that may occur in shuffling experiments. Algorithms for estimating diversity and library completeness in DNA shuffling have also been developed [104], although most researchers choosing DNA shuffling ignore this possibility.

Studies focused on improving DNA shuffling protocols have been summarized in several reviews [17] and detailed experimental procedures are available [104]. An improved version of family shuffling using single-stranded DNA (ssDNA) has also been reported in which the gene is cleaved by restriction enzymes instead of DNase I [105]. Accordingly, single-stranded DNA templates in place of double-stranded templates are employed in the DNase I fragmentation. ssDNA templates reduce the probability of homo-duplex formation, thereby increasing the quality of DNA shuffling libraries as demonstrated in a comparative study using dsDNA versus ssDNA templates [105c]. In a more comprehensive comparative study, which includes practical experimental protocols, "restriction enzyme-mediated" family shuffling was shown to be superior [106]. A recent example of using only DNA shuffling without the support of saturation mutagenesis or epPCR concerns the enhancement of activity of a laccase [107]. In another rare example, glutathione transferases (GST) from plants *Phaseolus vulgaris* and *Glycine max* were engineered by employing directed evolution through DNA shuffling without the further assistance of epPCR or saturation mutagenesis (Figure 3.13). The generated chimeral GST enzyme (PvGmGSTUG) displayed enhanced glutathione hydroperoxidase activity and displayed unusual cooperative kinetics toward 1-chloro-2,4-dinitrochlorobenzene, which is not its natural substrate [108].

The potential use of DNA shuffling can be guided by typical experimental protocols available in the literature [105107]. Alternatively, several bio-companies offer commercial DNA kits, e.g. PickMutant DNA Shuffling Kit (Canvax/Spain) or JBS DNA-Shuffling Kit (Jena Bioscience/Germany).

In yet another shuffling approach, the staggered extension process (StEP) was developed based on cross hybridization of growing gene fragments as the DNA polymerase-catalyzed primer extension process occurs [109]. Following denaturation, the primers anneal and extend under conditions that limit extension, which ensures that the primers reanneal to different parent sequences throughout the multiple cycles randomly. Then the recombinant full-length

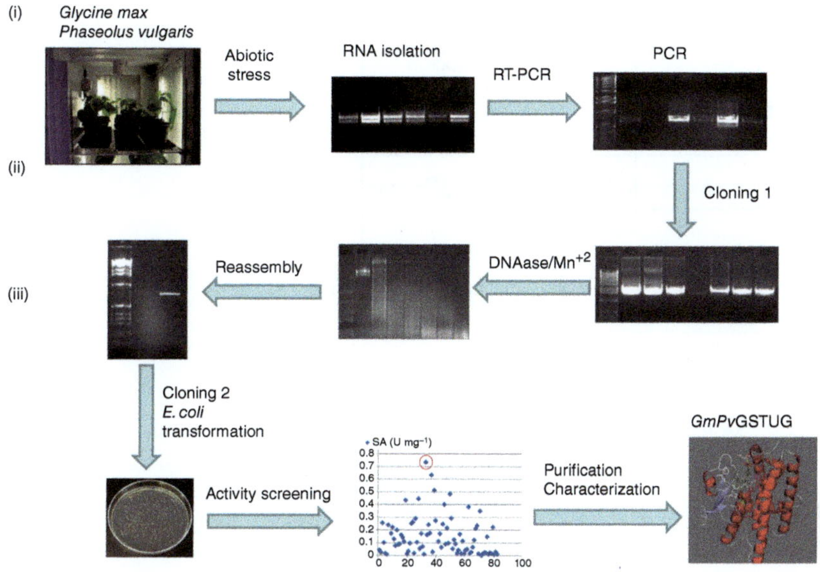

Figure 3.13 Workflow of the directed evolution of the enzyme PvGmGSTUG via DNA shuffling. Source: Ref. [108]/from Frontiers Media S.A., CC BY 4.0.

gene products are PCR amplified. This approach has not been used by many groups.

In addition to these variations, a number of improved recombinant procedures were reported early on that focused on various issues such as the problem of low homology. These include random chimeragenesis on transient templates (RACHITT) [110], degenerate oligonucleotide gene shuffling (DOGS) [111] in comparison with random drift mutagenesis (RNDM) [112], degenerate incremental truncation for the creation of hybrid enzymes (ITCHY) [113], Thio-ITCHY [114], SCRATCHY (a combination of ITCHY and DNA shuffling) [115], sequence homology-independent protein recombination (SHIPREC) [116], sequence-independent site-directed chimeragenesis (SISDC) [117], recombined extension on truncated templates (RETT) [118], recombination-dependent exponential amplification polymerase chain reaction (RDA-PCR) [119], and structure-based combinatorial protein engineering (SCOPE) [120]. It should be mentioned that most of these techniques have not been used very often following their initial publication.

Another recombinant method is "biased mutation-assembly," according to which a library is created by overlap extension PCR with DNA fragments from a WT protein and phenotypically advantageous mutant genes [121]. The number of assembled mutations in the WT gene is controlled stochastically by the mixing ratio of the WT fragments to the mutant DNA fragments (Scheme 3.10).

Adding designed synthetic oligonucleotides to a mixture of gene fragments prior to reassembly as a form of semisynthetic DNA shuffling is a more or

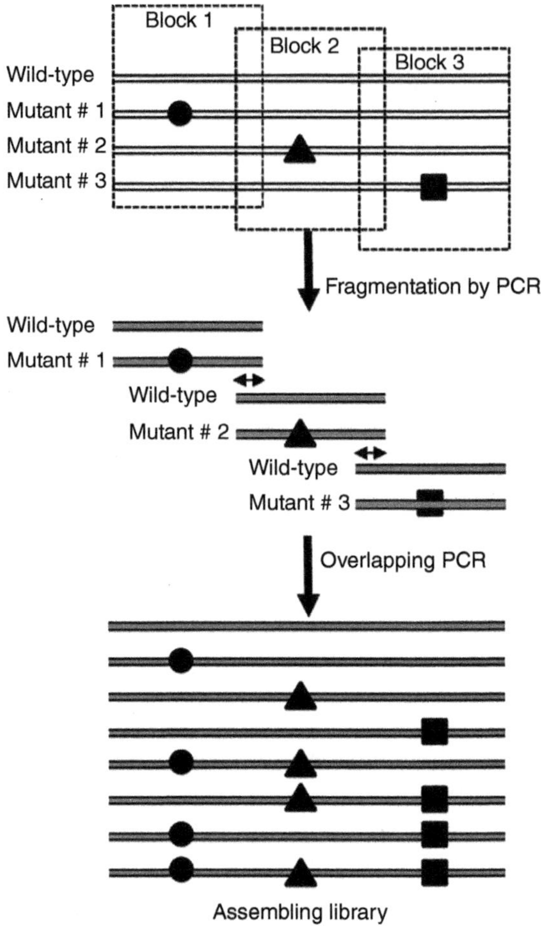

Scheme 3.10 A schematic example of biased mutation-assembling, assuming a basis set of three mutations. The circle, triangle, and square each represent one mutation. A block represents a portion of the gene containing one mutation and represents a recombination unit. The double-headed arrows represent overlapping sequences between adjacent blocks and these overlapping sequences hybridize during PCR recombination. Source: Ref. [121]/Oxford University Press.

less routine technique, as for example, in the engineering of the biosynthetic pathway of the natural product Avermectin for pharmaceutical production [122]. It is reminiscent of the Stemmer-method of CMCM [103c], as discussed above. This interesting method was later extended and dubbed "incorporating synthetic oligonucleotides via gene reassembly" (ISOR) [123], in which insertions and deletions may occur (see Section 3.2.3).

Hundreds of DNA shuffling patents were applied for or issued early on, which made collaborative efforts between academic and industrial groups difficult. An interesting alternative to DNA gene shuffling is "artificial shuffling," a process, which starts not by fragmenting genes but by assembling DNA fragments previously designed and generated by other means. Sequence information is often used as a guide. Three independent studies initiated this new approach, which tolerates fairly low homology and minimizes undesired self-hybridization of parental genes (low WT-appearance in the mutant libraries) [124–126]. In one efficient embodiment called "Assembly of Designed Oligonucleotides" (ADO), two strategies for linking fragments are possible (Scheme 3.11) [126, 127]. A review of various assembly methods, including the Gibson technique [128], appeared in 2015 [129].

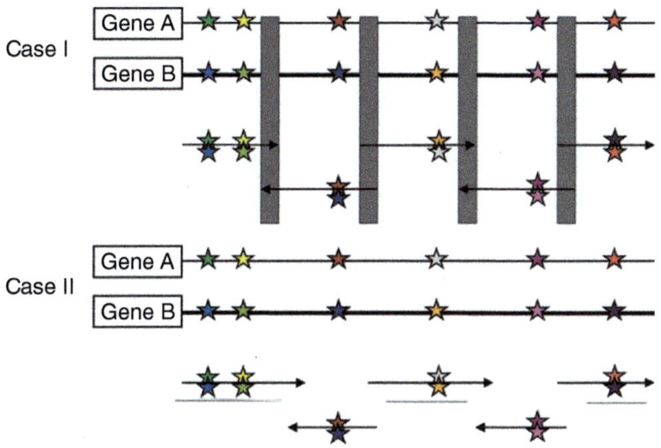

Scheme 3.11 General concept of ADO with two strategies for the linking of fragments being possible (cases I and II). In case I, the two genes A and B to be virtually shuffled are aligned; the different colored stars refer to information that encodes different amino acids, while oligonucleotide fragments with both colored stars in the same position of the parent gene denote the synthetic oligonucleotide fragment with degenerate nucleotides. The gray blocks denote conserved regions of sequence that can be used as the linking parts with homologous recombination. Case II shows no homology between flanking oligos, which can be assembled by ligation between ssDNA and an unknown terminal sequence. Source: Ref. [126]/John Wiley & Sons.

Thus far, optimized ADO has not been applied very often in directed evolution projects [130], but with the rapid development of optimized saturation mutagenesis [1a], a combination of both methods could offer new perspectives. Moreover, the gene assembly step in ADO offers many possibilities as a tool in synthetic biology for creating high-quality DNA libraries [126]. The benefits and limitations of ADO as a gene assembly method can be summarized as follows [127]:

- Fast library construction within hours by changing the primer composition.
- Useful for introducing diversity not just in enzyme-coding genes, but especially when manipulating metabolic pathways and even genomes.
- Rational engineering is possible by controlling the mutagenesis sites.

- Initial costs may be formidable, but it pays off if many libraries are planned.
- Contiguous codons may increase costs if there are no degenerate codons.
- Depending upon the particular system, repetitive sequences may be difficult to assemble without errors.

Finally, DNA shuffling has been applied successfully in areas beyond enzymes as active and selective biocatalysts. Examples are evolving adeno-associated viruses as synthetic delivery vectors [131], generating a bacteriocin gallocin D with activity against vancomycin [132], detecting the relationships between synonymous mutations and RNA toxicity [133], investigating the key residues responsible for gene transplantation [134], and others.

3.2.6 Circular Mutation and Other Domain Swapping Techniques

The mutagenesis methods discussed thus far all involve random or focused amino acid substitutions. A different possibility is protein engineering by circular permutation, meaning the creation of a different order in the protein sequence in the absence of point mutations [135, 136]. This idea was inspired by the structure and function of circularly permutated proteins known to occur in nature, of which more than 2000 have been identified. Circular permutation was discussed as early as 1979 [136]. Two steps are required: (i) covalent linkage of the native N- and C-termini by means of a short and flexible peptide linker, followed by (ii) cleavage of an existing peptide bond, the end result being a shift of the termini elsewhere. Thus, the primary sequence has been rearranged by means of termini relocation.

As a protein engineering technique for evolving improved catalytic profiles, it entered the stage much later [135, 137]. In a model system, the effect of relocating the termini of the lipase from *Candida antarctica* B (CALB) was investigated [135, 138]. The new locations of the N- and C-termini were designed to occur at positions 283 and 282 of WT CALB, respectively, in the hope of influencing local backbone flexibility and possibly active site accessibility. In transesterification reactions using model compounds and vegetable oil, one variant (cp283) showed improved catalyst performance in terms of activity relative to WT [138b] Using the best variants, the effect of circular permutation on enantioselectivity was also investigated using several chiral substrates [138b]. Stereoselectivity was not compromised, and in some cases, slight improvements were actually observed. Thus, it seems that circular permutation is not readily suited as a directed evolution method for enhancing or inverting stereoselectivity. Combining the method with loop truncation near the active site may lead to a different picture. This multi-step process was tested with CALB (Scheme 3.12), leading to a change in the enzyme quaternary structure, and shifting from a monomeric form to a dimer with domain-swapped N-terminal segment. This structurally induced enhanced hydrolytic lipase activity, but the effect on enantioselectivity was not reported [138e].

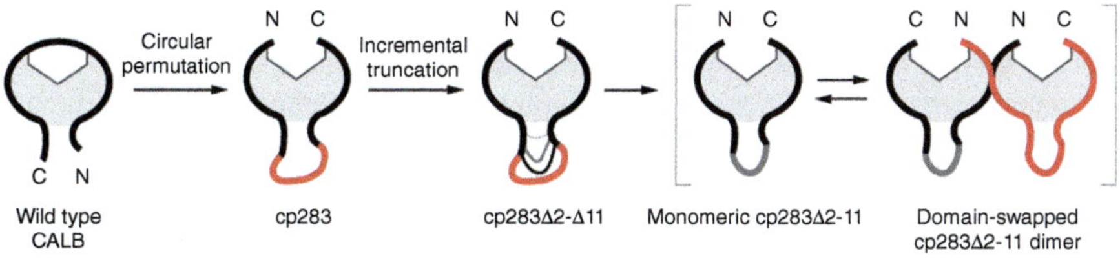

Scheme 3.12 Schematic overview of CALB engineering by circular permutation and subsequent incremental truncation of the newly created surface loop in cp283, the most active variant among the lipase permutants. Source: Adapted from Ref. [138e].

In a comprehensive study appearing in 2015, the following drawbacks of conventional circular permutation as a gene mutagenesis method for creating libraries of variants were noted [139]: (i) undesired sequence modification at the new termini of circular permutants; (ii) either relatively inefficient blunt-end ligation during library construction or redesign of transposons for optimal expression are required. Consequently, an engineered transposon for efficient generation of random circular permutation libraries was developed [139]. The method enables the use of sticky-end ligation and possible external tunability for the expression of circular permutants. These improvements are significant and go beyond earlier modifications [138], as delineated by the researchers of this study [140].

In summary, traditional circular permutation has been used effectively for more than two decades, mostly in the study of the structure and function of proteins, but its adaptation as a gene mutagenesis method in directed evolution was not realized until much later. Thus far, its role in laboratory evolution in biotechnology has been limited, the reliable control of such parameters as activity, stereoselectivity, regioselectivity, and thermostability being difficult relative to the use of other gene mutagenesis methods, especially saturation mutagenesis (Section 3.2.4). With the potential emergence of increased predictive power when applying circular permutation in order to change a given catalytic parameter, the method may play a greater role in future directed evolution studies.

Domain swapping (including 3D domain swapping) is the more general approach, as evidenced by a number of studies which traditionally focused mainly on questions regarding natural evolution as well as structural and mechanistic aspects of protein science [141]. The technique has also been used to generate hybrid enzymes characterized by novel catalytic profiles. Progress until 2000 was summarized in a review article [142]. Incremental truncation has been included [114]. Since then, further studies have appeared concerning fundamental questions in evolutionary biology but also aimed at metabolic engineering, where the method is sometimes combined with other mutagenesis techniques such as epPCR or saturation mutagenesis [143a–i]. Structure-inspired subdomain exchanges have also been applied to the directed evolution of enantioselective BVMOs [143j]. Nevertheless, it is not clear whether both enhancement and reversal of enantioselectivity in BVMOs can be evolved reliably using such domain swapping. Notwithstanding these important contributions, it remains to be seen whether domain swapping can be developed into a general and efficient method for creating useful enzymes as catalysts in organic chemistry and biotechnology. Recently, a domain-swapping-based trimerization strategy was utilized to enhance the thermal stability and bestow metal binding ability [143k].

3.2.7 Solid-Phase Combinatorial Gene Synthesis as a PCR-Independent Mutagenesis Method for Mutant Library Creation

3.2.7.1 Introductory Remarks

A number of biotech companies offer the commercial production of chemically synthesized genes and even synthetic saturation mutagenesis libraries, although the latter are still somewhat expensive. What are the potential advantages of such an expenditure relative to applying conventional saturation mutagenesis in one's own laboratory? As already emphasized, despite the general success of the CAST/ISM approach, it is characterized by some degree of amino acid bias (as other PCR-based mutagenesis methods such as epPCR) [1a]. This means that it is not at all certain whether all of the designed mutants will actually appear in the respective saturation mutagenesis libraries. Moreover, many colonies harboring WT will also be present, which likewise reduces library quality. Finally, undesired insertions and deletions are also likely to occur. Amino acid bias leads to a reduced quality of mutant libraries and thus "wasted" screening effort has several origins [48], including:

- The degeneracy of the genetic code causes favored and disfavored amino acid exchange events, depending upon the particular amino acid.
- The PCR steps in a typical saturation mutagenesis experiment involve amplification, and these processes are imperfect and thus prevent some designed amino acid exchanges from occurring, in addition to causing unplanned insertions and deletions.
- Imperfect annealing temperatures may cause similar problems.
- Imperfect quality of primers used in library construction also lead to bias.

A seminal development in directed evolution is highlighted below, which promises to change the whole research area, namely the optimal utilization of solid-phase combinatorial gene synthesis. Two major types of chemical gene syntheses have been developed earlier for various purposes: polymerase cycling assembly (PCA)-dependent and PCA-independent approaches. Both methods involve recursive cycles of traditional phosphoramidite chemistry comprising base deprotection, coupling, capping, and oxidation, but PCA uses PCR [144]. These gene synthesis methods play a central role in synthetic biology and biotechnology in general, especially in the manipulation of proteins, metabolic pathways, and entire genomes [145]. PCA-dependent methods involve, inter alia, PCR-extension, which has some disadvantages as summarized in a review on gene synthesis [146]. PCA-independent techniques utilize the synthesis of oligonucleotides on solid supports followed by assembly catalyzed by enzymes involved in DNA repair and/or ligation rather than polymerases [146]. In this respect, DNA microarrays have been developed, allowing the high-throughput synthesis of defined sequences [146, 147]. Most of these advancements utilize Agilent technology [148]. Unfortunately, these microarray techniques limit the size of fragments to about 200 bases. Two other early approaches likewise deserve mention: the so-called Blue Heron solid support technology [149] and the Sloning building block technique [150], which was exploited in order to create high-quality saturation mutagenesis libraries [94a]. Most recently, the Twist Platform was demonstrated to be the best approach currently to utilize solid-state gene synthesis as a means to eliminate amino acid bias in saturation mutagenesis [48, 151].

3.2.7.2 **The Sloning Approach to Solid-Phase Gene Synthesis of a Mutant Library: Comparison with the Respective Molecular Biological Saturation Mutagenesis Library**

The Sloning procedure is based on solid phases with bound biotin-modified oligonucleotides. In order to generate a defined gene sequence, a certain number of chemically synthesized building blocks ("splinkers") that contain self-complementary regions are ligated to the anchor, immobilized, washed, and cleaved with the formation of sub-fragments of 18 base pairs in each cycle. Full-length or large gene fragments are thus accessible in a fully automated manner. Sloning libraries have been prepared in order to incorporate the non-natural amino acid *O*-methyl-L-tyrosine in proteins [152] and engineer binding affinity in anticalins [153] and in antibodies [154].

The previously studied selective oxidative hydroxylation of testosterone [155] served as the model reaction in a Sloning study aimed at controlling regio- and stereo-selectivity of oxidative hydroxylation of steroids using cytochrome P450-BM3 [48, 94a] (Figure 3.14a). *The purpose was to compare the quality of saturation mutagenesis libraries generated by the conventional molecular biological technique with that of the synthetic Sloning libraries.* Three CAST libraries were designed at Site A (R47/T49/Y51), B (V78/A82), and C (M185/L188) lining or near the binding pocket (Figure 3.14b). The goal was to compare the traditional PCR-based saturation mutagenesis libraries with the respective Sloning libraries. Sites A and C were randomized conventionally using NDC codon degeneracy encoding 12 amino acids (R, D, N, C, G, H, I, L, S, V, F, and Y) (Figure 3.14c). In the Sloning experiments, the codon choice for the same 12 amino acids differed only in the last nucleotide of 5 codons due to a codon optimization algorithm for *E. coli* (Figure 3.14d). The randomization scheme for library B includes all 20 canonical amino acids, which in the case of PCR-based saturation mutagenesis is covered by NNK codon degeneracy. It encodes 32 defined codons, including one stop codon, but shows redundancy for amino acids A, G, P, T, V and R, L, S with two and three codons, respectively. In contrast, Sloning gene synthesis does not involve degenerate codons, which means that full randomization is ensured with a non-redundant set of 20 codons (Figure 3.14d) [48, 94a].

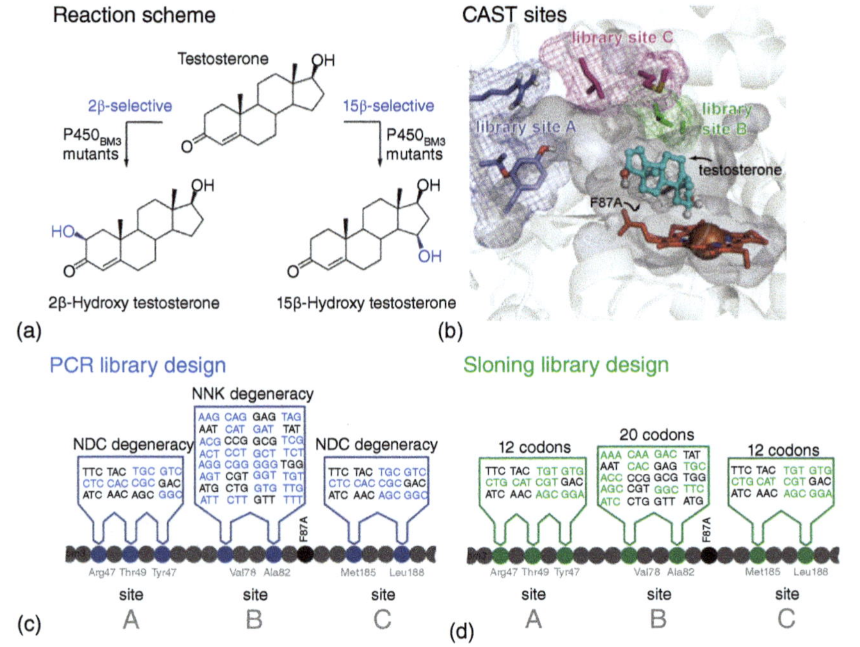

Figure 3.14 Model reaction and library design for comparing traditional PCR-based saturation mutagenesis libraries with Slone libraries [48, 94a]. (a) Testosterone hydroxylation by P450BM3 mutants. (b) Active site of P450BM3 mutant F87A. The three CAST sites and the F87A residue are highlighted. The structure was modeled by docking computations using the Schrödinger software and the picture was created with PyMol. (c) Diversity design of the combinatorial P450BM3-F87A libraries used in this study. Library A consists of three simultaneously randomized positions, whereas library B and C consist of two. PCR-based libraries use either the nonredundant NDC codon (library A + C) or the redundant NNK codon (library B). (d) Sloning-based libraries encode the same set of amino acids using the displayed codon usages. Gray codons are present in both designs. Source: Ref. [94a]/American Chemical Society.

All three PCR-based libraries were generated by the improved MegaPrimer PCR protocol (Scheme 3.7) [75]. Upon adjusting the annealing temperature and increasing the number of amplification cycles, libraries with sufficient numbers of colonies per transformation (more than 10 000) were obtained. Following pooling [99a] and plasmid extraction, sequencing was performed. As judged by the QQC [99a], acceptable quality was observed [94a]. All of the designed libraries were obtained commercially (MorphoSys, Munich, Germany). Gene fragments of 683 base pairs (954 bp with flanking sequences) were cloned into the target plasmid, followed by transformation, leading to: 5000 clones for library A (A-SLO), 11 000 clones for library B (B-SLO), and 5000 clones for library C (C-SLO). All libraries were transformed into *E. coli* BL21-Gold (DE3). Library screening was carried out using automated HPLC. Following random sequencing, statistical analyses were performed [94a]. The screening results proved to be encouraging. This study showed for the first time that the quality of designed DNA libraries produced by solid-phase synthesis, specifically of the synthetic Sloning libraries at the DNA and subsequently at the protein level, is higher than that of the respective conventional PCR-based saturation mutagenesis libraries. This correlates with less screening. *However, the weakness of this study was the failure to perform a sufficiently high degree of gene sequencing and oversampling, needed for a more quantitative statistical analysis* [48, 94a].

3.2.7.3 The Twist Approach to Solid-Phase Gene Synthesis of a Mutant Library: Comparison with Molecular Biological Saturation Mutagenesis Library

When attempting to exploit solid-phase gene synthesis as a means to produce the highest-quality saturation mutagenesis libraries with minimal amino acid bias, the Twist procedure proved to be distinctly more successful [48, 151]. It utilizes the so-called Twist-Platform, offered commercially by Twist Bioscience [156]. The proprietary technology is based on DNA synthesis that occurs on silicon chips in minute amounts, enabling the synthesis of at least 100 000 unique variant oligonucleotide sequences. These are extracted from the respective plates, and then equimolar primers are mixed and assembled into a full-length gene by overlap extension PCR, ready to be cloned. Finally, this leads to the desired variants as designed by the customer (Figure 3.15) [48, 151].

In this study, limonene epoxide hydrolase (LEH) *from Rhodococcus erythropolis* was used as the enzyme, and the stereoselective desymmetrization of cyclohexene oxide (**1**) served as the model reaction (Figure 3.16) [151]. The DNAs were synthesized with flanking sequences containing restriction enzyme sites (Ndel and HindIII), which were ligated into the digested plasmid pET22b with corresponding restriction enzymes. The resulting plasmids having LEH gene mutants were then transformed into *E. coli* BL21(DE3). *In order to identify the degree of fidelity and to check whether uniform incorporation of planned diversity had occurred, the colonies were harvested, followed by DNA sequencing.* Screening at the protein level was carried out by GC, which allowed conversion and determination of the relative amounts of the enantiomeric (*R,R*)- and (*S,S*)-products.

In order to construct a "model library," an LEH site composed of four CAST residues (Met78, Ile80, Leu114, and Ile116) was subjected to randomization using a highly reduced amino acid alphabet comprising valine, phenylalanine, and tyrosine (V–F–Y). The principle of such Triple Code Saturation Mutagenesis (TCSM) [157] is discussed in Chapter 4, but for the purpose of the present study, this particular choice plays no role. Theoretically, the library generated by conventional PCR-based saturation mutagenesis and the Twist counterpart should both harbor 256 different mutants, at DNA as well as protein level. Massive DNA sequencing with oversampling factors (OFs) of 1, 2, and 3 was performed for comparison. Statistically, the results are most meaningful in the case of OF = 3 [48, 151]. The data in Table 3.3 and Figure 3.17 allowed for several unambiguous conclusions, including the assessment of genetic diversity, which was defined as the observed ratio of the number of mutants obtained after sequencing to the number of theoretically expected mutants (in the present case 256). It can be

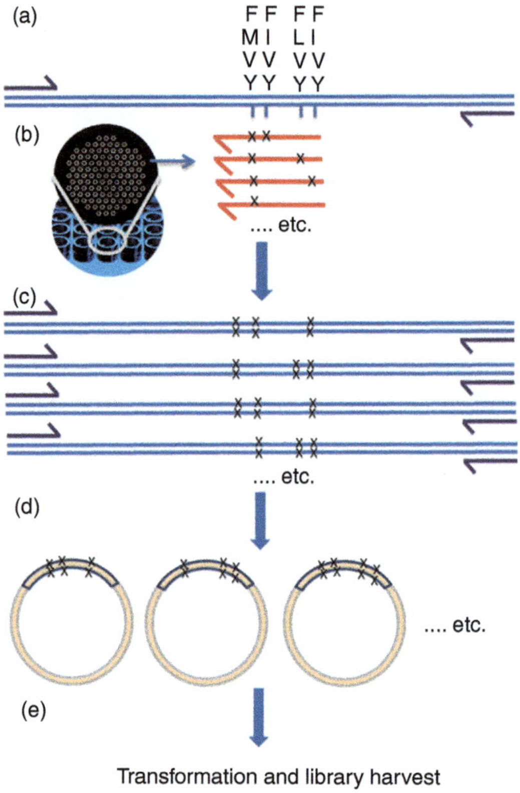

Figure 3.15 Combinatorial mutant library fabrication via the Twist system. Source: Ref. [151]/John Wiley & Sons.

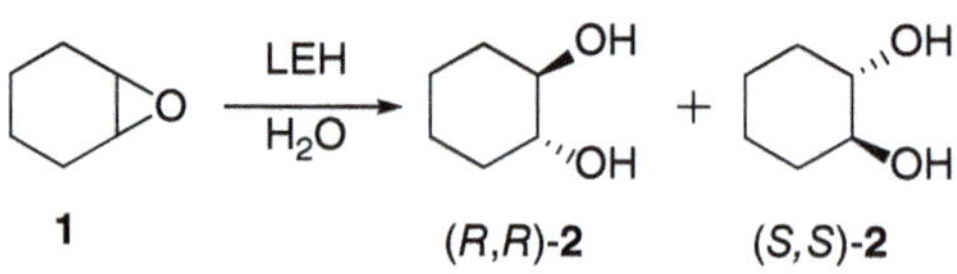

Figure 3.16 LEH-catalyzed hydrolytic desymmetrization of prochiral cyclohexene oxide (**1**). Source: Ref. [151]/John Wiley & Sons.

seen that at OF = 3, the library generated by conventional saturation mutagenesis contains only 144 of the expected 256 mutants, corresponding to a mere 56.3% genetic diversity. In sharp contrast, *the Twist library was found to harbor 249 of the 256 theoretically possible mutants, which means that a genetic diversity of 97.3% was achieved!* Furthermore, the number of variants characterized by an incomplete sequence was more than twice as high as in the PCR-based saturation mutagenesis library. The Twist library also contained distinctly more undesired WT members, e.g. 11 versus 4 at an OF of 3. Therefore, the consequence of efficient synthetic gene library production is: "*You get what you designed*"*!*

Table 3.3 Comparison of PCR-based SM library with the on-chip gene synthesis library on DNA level.

Oversampling factor	No. of colonies sequenced	Libraries	No. of different mutants found	No. of mutants with incomplete sequence[a)]	No. of WT	Genetic diversity (%)[b)]
1	276	PCR-based SM-library	109	36	4	42.5
	276	Silicon-based gene synthesis library	164	18	2	64.1
2	552	PCR-based SM-library	135	88	8	52.7
	552	Silicon-based gene synthesis library	218	47	2	85.2
3	828	PCR-based SM-library	144	130	11	56.3
	828	Silicon-based gene synthesis library	249	65	4	97.3

a) Number of colonies containing deletions/insertions or those in which sequencing was not possible for various reasons.
b) Ration of the number of mutants obtained after sequencing to the number of theoretically possible mutants expected (256 mutants) expressed in percentage.
Source: Ref. [151]/John Wiley & Sons.

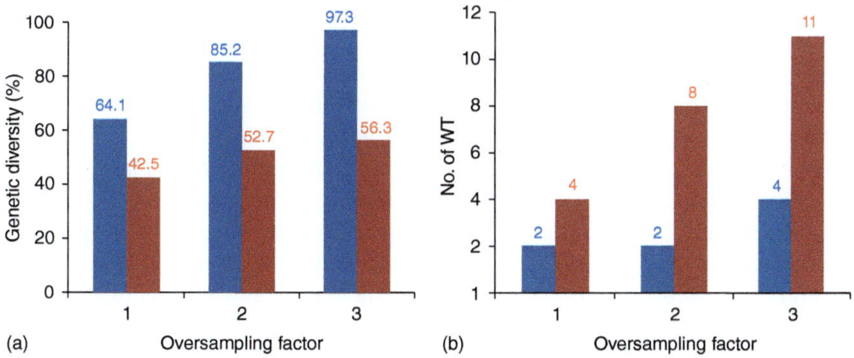

Figure 3.17 Comparison of the Twist-based library (blue) and the conversional PCR-based library (red) in genetic diversity (a) and the wild-type numbers (b). Source: Ref. [151]/John Wiley & Sons.

A different, although somewhat crude way to assess library quality is to analyze the percentage distribution of different residues at all targeted positions, in the present case at the four positions 78, 80, 114, and 116 (Figure 3.18) [48, 151]. In the PCR-based saturation mutagenesis library, the appearance frequency of different residues at each position of the mutants was found to vary considerably, and the parent amino acid at each position correlates with the highest appearance frequency (Figure 3.17a). This indicates suboptimal library

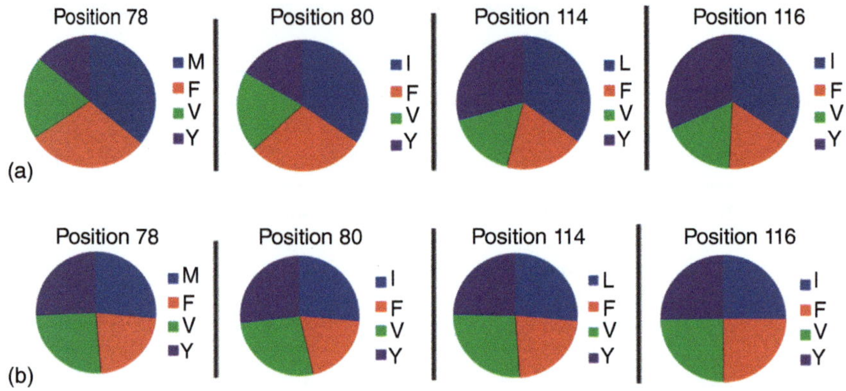

Figure 3.18 Percentage distribution of different residues at positions 78, 80, 114, and 116 of LEH for (a) combinatorial PCR-based SM library and (b) on-chip gene synthesis library. An oversampling factor of 3 was employed in both cases. Source: Ref. [151]/John Wiley & Sons.

quality. In contrast, a nearly perfect pattern was observed upon analyzing the Twist library (Figure 3.17b).

Although not directly relevant to the purpose of the Twist study, the two libraries were also compared at the protein level by screening for enantioselective variants as catalysts in the desymmetrization of cyclohexene oxide (**1**) with the formation of (*R*,*R*)- and (*S*,*S*)-**2** (Figure 3.16) [151]. Both libraries contained enantioselective hits, as expected on the basis of the CAST design, but the synthetic Twist library was found to harbor a number of highly enantioselective variants that were not discovered in the conventional saturation mutagenesis library. *If the prices of commercial mutant libraries continue to tumble, researchers may be advised to spend more of their time optimizing the actual library design, which will consequently correlate with less screening in the laboratory.*

As already indicated, this study [151] has important ramifications for the future of directed evolution, as emphasized in a review [48]. When researchers design, generate, and screen a saturation mutagenesis library conventionally, they can no longer be sure that all of the planned variants have in fact been formed. The screening results at protein level may be acceptable, but it is likely that considerably better results were lost. This conclusion is supported in Section 4.5.

3.2.8 Computational Tools and the Role of Machine Learning (ML) in Directed Evolution and Rational Enzyme Design

3.2.8.1 Introductory Remarks

Achieving maximum efficiency in directed evolution not only requires the use of advanced mutagenesis methods but also guidance by appropriate computational tools which accompany the experimental work [158]. Depending upon the specific goal of a directed evolution project, different types of computational methods and computer aids are available. Two types of tools have been developed, those that concentrate on the optimal use of degenerate codons, library size, and degree of oversampling; and those that analyze protein structures in the quest to identify functionally important residues (hot spots) that can then be varied by site-specific mutagenesis or focused randomization via saturation mutagenesis. It is important to note that essentially the same tools are currently used in rational enzyme design (Section 3.3). See also Chapter 4 for additional details. A number into investigations of rational design for enhancing protein stability have continued to appear, sometimes in combination with semi-rational methods. These will not be reviewed here because protein stabilization is treated in Chapter 6.

Today, a number of *de novo* protein design algorithms are available for a variety of applications, the Rosetta computer package [159] and the HotSpot Wizard and related algorithms [160] being two particularly prominent examples, which were recently extended to include machine learning (ML)/deep learning (DL) techniques. Rosetta involves a multi-step computational procedure based on:

- Protein structure prediction.
- Quantum mechanical (QM) energy refinement.
- Sequence design.
- Additional guidance in decision making by ML/DL and other techniques.

3.2.8.2 **Designing Mutant Libraries and Estimating Library Completeness**

When rationally designing small libraries (collections of mutants) and estimating completeness as a function of oversampling, several metrics are available (which can also be used in semi-rational directed evolution), including:

- GLUE and GLUE-IT (**GLUE**-**I**ncluding **T**ranslation) [30a, b] (http://guinevere.otago.ac.nz/cgi-bin/aef/glue-IT.pl)
- CASTER [70b] (http://www.kofo.mpg.de/en/research/biocatalysis)
- TopLib [91c] (http://stat.haifa.ac.il/~yuval/toplib)
- SwiftLib (Swift Library) [161] (http://rosettadesign.med.unc.edu/SwiftLib)
- DYNAMCC_D [162] (http://www.dynamcc.com/dynamcc_d)

The CASTER is a user-friendly computer aid, which is based on the Patrick/Firth metric GLUE, but it also provides the user with a great deal of additional helpful information such as electronic and steric properties of amino acids, the genetic code, and other useful tips for fast guidance [70b]. In contrast to SwiftLib, it does not tell the user which degenerate codons to use. In the case of the second type of computational tools, several programs are available, which help in identifying and evaluating residues at the protein level that could influence catalytic properties, these hot spots then being positions at which rational site-specific mutagenesis (or saturation mutagenesis) can be attempted. Typical software packages are listed below:

- HotSpot Wizard [163] (http://loschmidt.chemi.muni.cz/hotspotwizard)
- CASTp (Computed Atlas of Surface Topography of proteins) [164] (http://cast.engr.uic.edu)

- 3DM (3D Molecular class-specific information systems) [165] (http://3dmcsis.systemsbiology.nl)
- ConSurf (Conservation Surface mapping) [166] (http://consurf.tau.ac.il)
- MBLOSUM (Mutagenesis Blosum) [167] (http://apps.cbu.uib.no./mblosum)
- Scorecons (Score residue conservation) [168] (www.ebi.ac.uk/thornton-srv/databases/cgi-bin/valdar/scorecons_server.pl)
- FamClash (Family sequence residue-residue clashes) [169]
- SIRCH (Second-order mean-field Identification of Residue–residue Clashes in protein Hybrids) [170]
- IPRO (Iterative Protein Redesign and Optimization) [171]
- CorNet (Correlated mutation analyses Network) [172a, b]
- ANT (Ambiguous Nucleotide Tool) [172c]
- Funclib [173] (https://funclib.weizmann.ac.il/bin/steps)
- Flex_ddG [174]
- Cartesian_ddG [175]
- dTERMen (design with tertiary motifs energies) [176]

When aiming for enhanced protein thermostability (Chapter 6) and or activity, the versatile HotSpot Wizard can be used in addition to the following computational guides:

- B-FITTER [70b] (http://www.kofo.mpg.de/en/research/biocatalysis)
- CUPSAT (Cologne University Protein Stability Analysis Tool) [177] (http://cupsat.tu-bs.de)
- CAVER [178] (http://www.caver.cz)
- FOLDEF (Fold-X Energy Function) [179] (http://foldx.crg.es)
- PoPMuSiC [180] (http://babylone.ulb.ac.be/popmusic)
- FRESCO (Framework for Rapid Enzyme Stabilization by Computational libraries) [181]
- FireProt [182]
- PROSS (Protein Repair One Stop Shop) (http://pross.weizmann.ac.il) [183]

The optimal choice of a computational tool depends upon the particular problem at hand. Other computational tools for different purposes, such as identifying binding sites in proteins, have been reviewed [184].

Several other approaches are presented here in more detail, one of them being SCHEMA [185]. It is a structure-guided method utilizing recombinant processes (Scheme 3.13). Blocks of sequences are first identified that minimize structural disruption when recombination with the formation of chimeric proteins occurs. Pairs of interacting residues within 4.5 A of each other are identified, which are then utilized as a basis for constructing contact matrices. The program includes an optimization algorithm that selects optimal crossovers that minimize the average disruption of the library. All interactions that break upon recombination contribute to a disruption score, which is used in designing shuffling experiments. In order to compute the average disruption, high-resolution structural data of at least one of the proteins is required. SCHEMA has been applied to

P450 monooxygenases [185a], ß-lactamases [185b], and cellulases [185c], all by the Arnold group. A type of hybrid between FamClash and SCHEMA has been claimed to be particularly efficient [186].

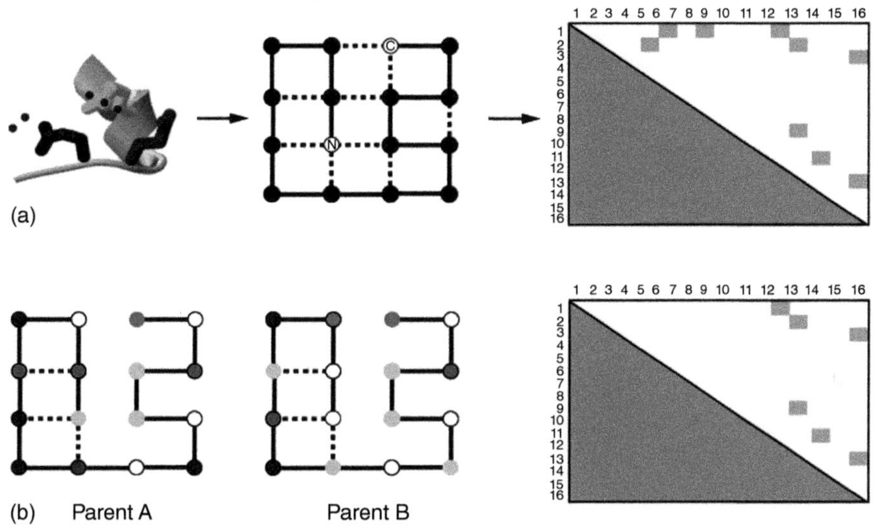

Scheme 3.13 SCHEMA disruption based upon a contact matrix representing interactions between amino acids in the three-dimensional structure of a protein (illustrated here with a simplified model). (a) Disruptions in a simplified model; (b) Contact matrix to be adjusted for the sequence identity of the parent enzymes. Source: Ref. [185]/John Wiley & Sons.

In analogy to quantitative structure–activity relationships (QSARs) used in therapeutic drug discovery, an algorithm based on protein sequence–activity relationships (ProSAR) has been developed by an industrial group for application in directed evolution [187]. As each evolutionary cycle is traversed, information from sequence-activity data is utilized. The best mutant is then used as a template for programming diversity in the subsequent round by inferring the contributions of mutational effects on the catalytic profile of the enzyme (Scheme 3.14). At any point, about 50 mutations as variables are evaluated in the combinatorial libraries, and the hits as well as some less improved mutants are sequenced. As part of the ProSAR analysis, mutations are assigned to four classes: (i) beneficial, which are fixed in the population in the next round; (ii) potentially beneficial, which are sent back for retesting; (iii) deleterious, which are discarded; and (iv) neutral, which are also discarded. Diversity, in addition to shuffling, needs to be increased by applying other mutagenesis methods such as rational design or saturation mutagenesis. ProSAR has been applied in the directed evolution of several enzymes, with activity and stereoselectivity being the catalytic parameters of interest. Examples include a halohydrin dehalogenase as the catalyst in the production of the cholesterol-lowering drug Lipitor [187] and a transaminase in the production of the anti-diabetic drug Sitagliptin [188]. The latter study utilizes, inter alia, ISM (see Scheme 3.9 and Chapter 4).

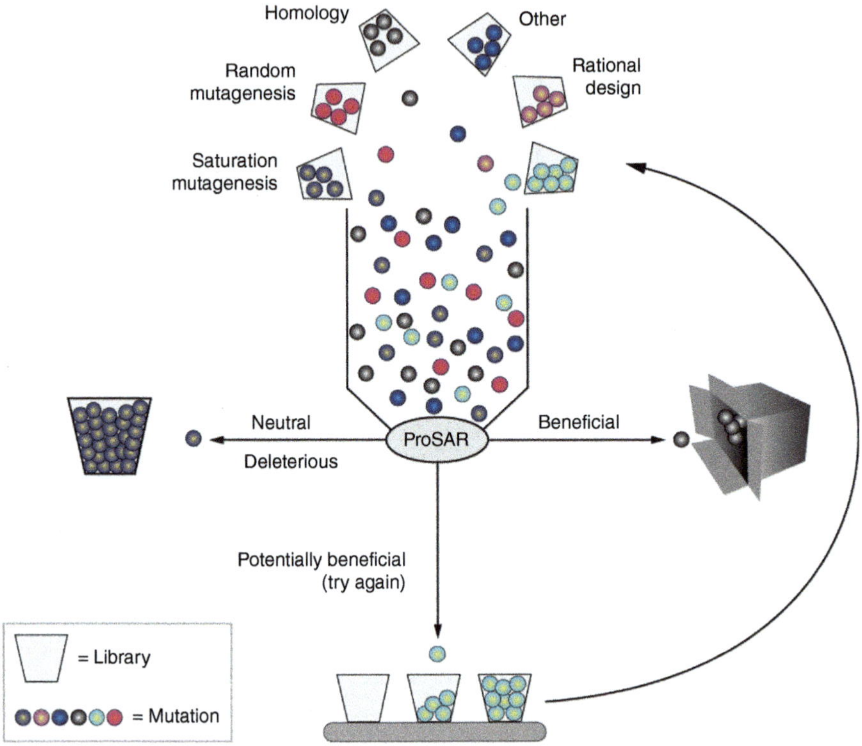

Scheme 3.14 Formal representation of ProSAR. Source: Adapted from Refs. [187, 188].

Since stereoselectivity stands at the heart of modern organic and pharmaceutical chemistry, it was of acute interest to test other approaches, especially those that actively utilize ML and DL as forms of Artificial Intelligence (AI). A seminal study regarding these techniques is based on the Adaptive Substituent Reordering Algorithm (ASRA) [189], specifically as a means to predict stereoselectivity improvements when applying ISM at sites lining the enzyme binding pocket [190]. It was tested computationally and experimentally in the enhancement of enantioselectivity of the epoxide hydrolase from *Aspergillus niger* (ANEH) as the catalyst in the hydrolytic kinetic resolution of a chiral epoxide. Two substitution positions, inter alia, were chosen as mutation targets, and each amino acid at every position was assigned a random distinct integer between 1 and 20, with the total number of possible mutants being 400 in this case. In step 1, a small subset of the 400 mutants containing substitutions in both positions were synthesized and their respective enantioselectivity (E-value) measured (Scheme 3.15). Due to the random integer assignment, the initial selectivity landscape is irregular and provides no predictive power. In step 2, the optimal integer assignment for each amino acid at each position was identified so that the property landscape is as regular as possible. It should be noted that when an amino acid at position 1 (or position 2) "moves," meaning that its integer assignment is changed from a certain value to another, all 20 amino acids at the other position will move along with it in order to maintain consistent indexing. In step 3, the location of the best mutants was predicted based on the geometric features of the reordered selectivity landscape. For example, the circle should be a desired area because of the monotonic landscape geometry. The identified mutants are relevant in the next ISM round (return to step 1) [190].

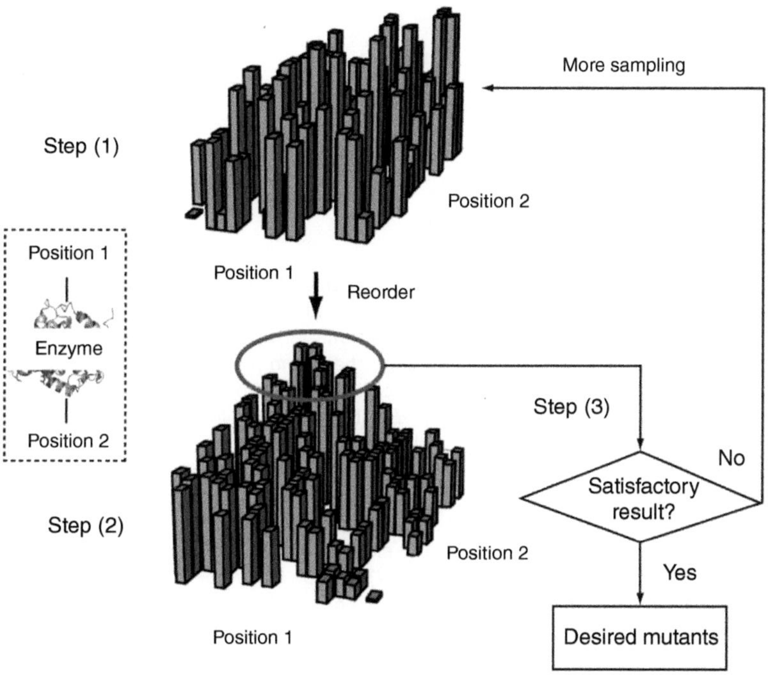

Scheme 3.15 Steps when applying ASRA to directed evolution. Source: Ref. [190]/John Wiley & Sons.

A typical result of reordering is shown in Figure 3.19. When applying ASRA, no assumptions regarding linearity, additivity, or any structure–property relationship are made. Rather, the algorithm allows the identification of the underlying regularity of the protein-property landscape, in the present case, stereoselectivity. Mutants with new sequences showing stereoselectivity factors of $E > 50$ were predicted, which was substantiated experimentally [190].

In a different ML/DL-based approach, the so-called Innov'SAR metric deserves mention [1n, 191]. Accordingly, by using an amino acid index, the

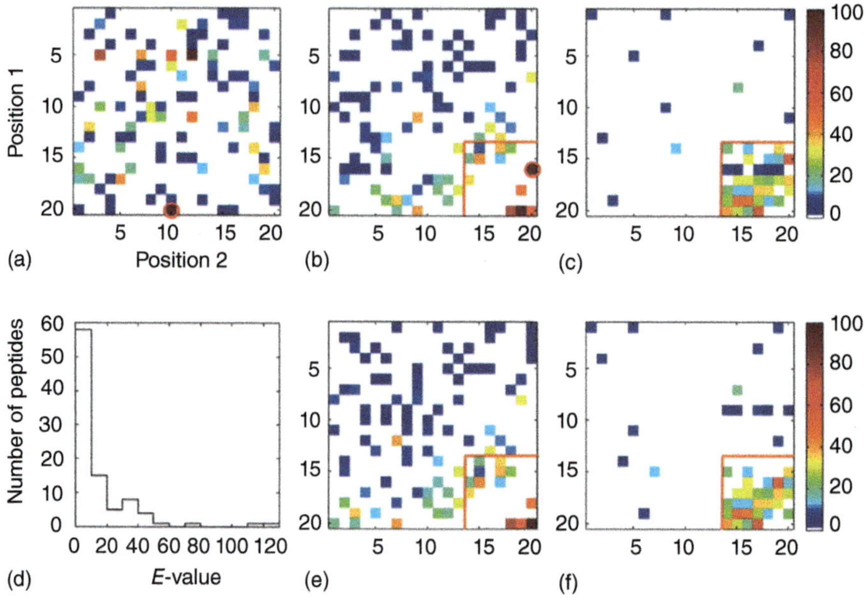

Figure 3.19 Optimal reordering of the E-value enantioselectivity landscapes with 60 minutes reaction time [190]. (a) Color heat map for the enantioselectivity landscape (E-values) of 95 randomly sampled mutants plotted with a random amino acid ordering. Each color square represents one mutant with red indicating a high E-value and blue corresponding to a low E-value (see color x^x bar on the far right). White squares are unsampled proteins. (b) E-value landscape of the 95 mutants using the ASRA-identified optimal amino acid ordering. The result predicts that proteins with high E-values are most likely located in the lower right corner. The mutant at position 16/20 (circled in red in both A and B) of the reordered landscape turned out to be the same as the mutant at position 20/19; the wrong protein was accidentally placed in this position in the experiment. (c) E-value landscape for 45 newly sampled mutants, guided by the ordering in (b). (d) E-value distribution for the 95 initial random mutants. (e) Reordered E-value landscape for the 94 mutants (excluding the erroneous mutant at position 16/20 in (b)). (f) E-value landscape for the 45 newly sampled mutants, based on the ordering in the enantioselectivity factor E. Source: Ref. [190]/John Wiley & Sons.

alphabetical protein sequence is encoded into a numerical sequence. Different from ASRA, each index then corresponds to a numerical value associated with one or more physicochemical properties of each amino acid. Finally, a digital signal processing technique called Fast Fourier Transform (FFT) converts numerical signals with the construction of an energy versus frequency power spectrum (dubbed the protein spectrum). The Innov'SAR technique was applied on the same directed evolution ANEH experimental platform as in the above ASRA-study [192]. Innov'SAR requires the protein sequence information to complete the encoding phase (Figure 3.20a), and the resultant numerical sequence coupled with the labeled data was recognized by the modeling process. Eventually, the regression model was used in the predictive phase (Figure 3.20b).

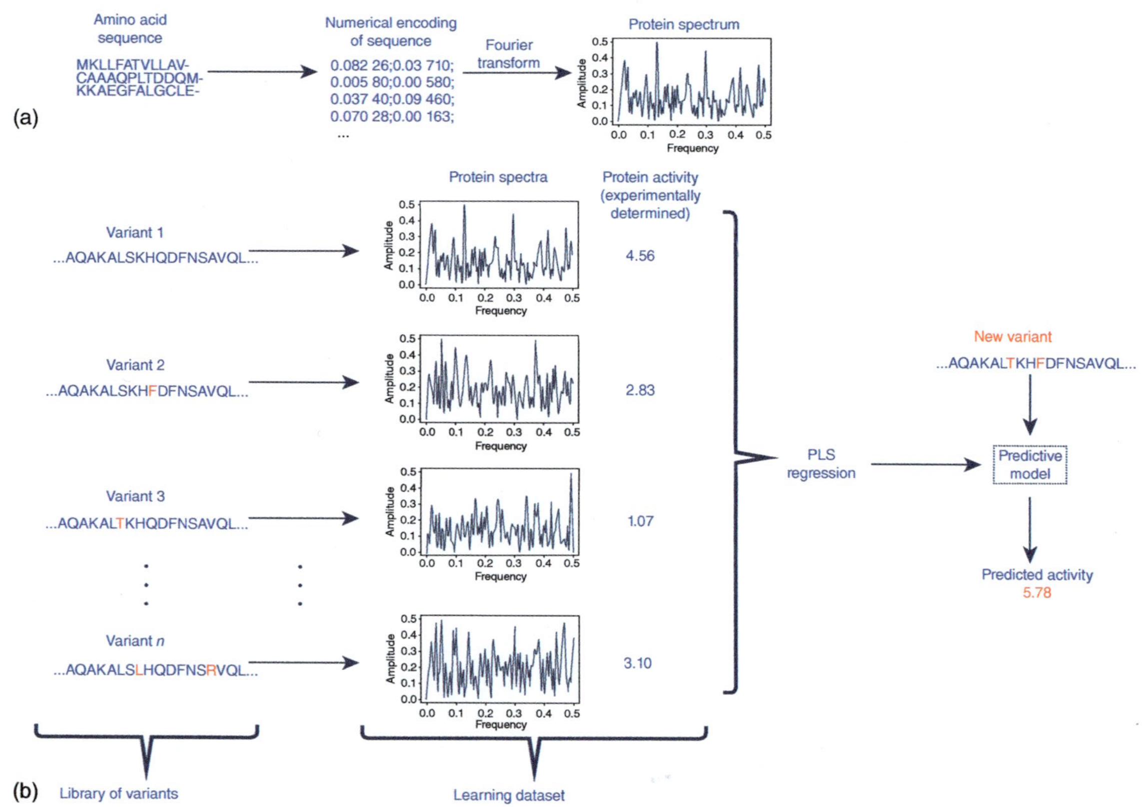

Figure 3.20 Workflow of the Innov'SAR method. (a) A protein sequence is encoded into a protein spectrum; (b) The primary sequences are transformed into protein spectra, and constructed as a regression model that is used to predict properties of variants such as activity and enantioselectivity. Source: Ref. [192]/Springer Nature/CC BY 4.0.

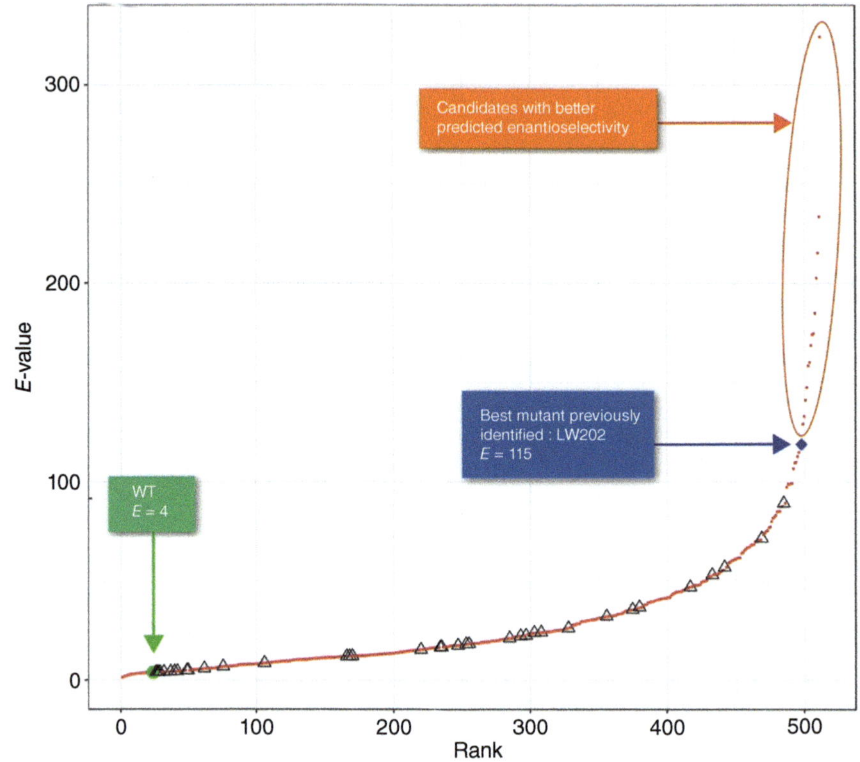

Figure 3.21 Ranking of the enantioselectivity *E*-values for the 512 possible variants of ANEH using the innov'SAR metric. Source: Ref. [192]/Springer Nature/CC BY 4.0.

Practically, 512 variants were obtained from combinations of the nine single point mutations (2^9). Mutants with enhanced activity and enantioselectivity were indeed harvested [192]. Several variants were predicted to have highly improved *E*-values relative to the earlier best variant obtained by CAST/ISM (Figure 3.21). One of the best ones, characterized by a never before reported DNA sequence, was predicted to display an enantioselectivity factor of $E = 250$, in line with the subsequently determined experimental value of $E = 253$ [192]. Innov'SAR has also been applied successfully in an attempt to predict enhanced enzyme stability [193] (see Chapter 6). An analysis of the relative merits and limitations of ASRA versus Innov'SAR, the first two ML-based studies for improving stereoselectivity, has appeared [1n], but more work is necessary before a final assessment can be made.

The third example of applying ML for evolving enzyme stereoselectivity concerns a Fe/ß-ketoglutarate dependent halogenase for late-stage functionalization of pharmaceutically important soraphens [194]. Nevertheless, it must be remembered that ML by itself is a technique that can be exploited as a helpful guide when engaging in directed enzyme evolution or rational enzyme design, not more nor less. As a result, machine learning has pervaded all topics of protein science as well, starting from the first report in 1992 [195] and continuing to the present day, as critically analyzed by a comprehensive review with emphasis on the actual techniques and algorithms. Baker et al have applied ML/DL to the challenge of rational protein design [196]. This ML-version computationally identifies amino acid sequences that encode very low energy target structures and importantly, provides the energy difference between the folded state and the lowest-lying alternative states. A convolutional neural network called trRosetta is essential. The captured features are general properties of the folding energy landscape, which predicts residue–residue distances and orientations from input sets of aligned sequences. Upon combining these predictions with Rosetta energy minimization, structure predictions are possible. The network was used to directly design sequences that fold into a desired structure by maximizing the probability of the observed residue-residue distances and orientations versus all others. Epistatic phenomena was not observed.

ML-assisted mutagenesis has also been successfully applied for the directed evolution of fluorescent proteins [197]. In this case study, a green fluorescent protein was chosen as a starting template, and a number of engineered proteins were obtained, showing yellow fluorescence. The overview of this mutagenesis approach is represented in Figure 3.22. In general, the procedure consists of four steps. Firstly, an initial small library of GFP variants was constructed via point

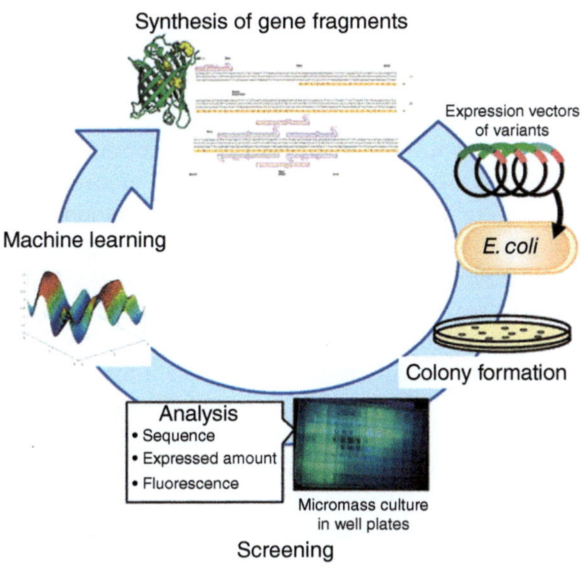

Figure 3.22 Schematic illustration of the machine-learning-guided mutagenesis method as applied for evolving GFP to YFP. Source: Saito et al. [197]/with permission of American Chemical Society.

saturation mutagenesis and random mutagenesis. Secondly, the variants in the library were experimentally analyzed to access the sequence and functional data. Next, labeled-data from 155 proteins was used to train a Gaussian process, thereby proposing the second-round mutagenesis library. Eventually, after screening 80 variants in the second-round library, 12 of them showed enhanced yellow fluorescence [197].

Apart from playing a possible role in directed evolution, ML can also be used in rational design or *de novo* design. In a prominent example, the algorithm ProteinGAN was developed based on the generative adversarial network (GAN), which enables the generation of novel functional protein sequences by "learning" natural protein sequence diversity [198]. In the case study of using malate dehydrogenase as a model enzyme, 13 out of 55 sequences generated by ProteinGAN proved to be soluble with catalytic activity. Remarkably, a highly mutated variant of 106 amino-acid substitutions was harvested in this study [198]. Other ML frameworks, e.g. Unirep [199], ECNet [200], PyPEF [201], CLADE [202], etc., also deserve to be mentioned for guiding protein engineering. Finally, the Rosetta algorithms [203] provide a computational means to access, inter alia, rationally designed enzymes showing promiscuous catalytic behavior, which can be improved by directed evolution in subsequent steps (see Chapter 7). Further reviews of ML and DL in protein design are available [204].

3.3 Diverse Approaches to Rational Enzyme Design

3.3.1 Introductory Remarks

As discussed in Chapter 1, several contributions in directed evolution during the period 1986–1997 marked the beginning of a new era in biocatalysis, evidenced by seminal papers on increasing enzyme thermostability (kanamycin nucleotidyltransferase) [205a], robustness toward hostile organic solvents (protease) [205b], resistance to antibiotics (ß-lactamase) [103a], and especially enhancing enantioselectivity (lipase) [205c]. Prior to these historical developments, Michael Smith had developed the molecular biological method of genetically substituting the amino acid at any residue of a protein for any one of the other 19 canonical amino acids [206], which in principle set the stage for so-called rational design of enzymes as an alternative to directed evolution. According to this concept, a limited number of predictions of point mutations are made and then the small collection of variants is tested experimentally. This exciting new research field originally focused on thermostability, while it was known that enzyme immobilization as part of process engineering is another technique for stabilization. Rational enzyme design for increasing enzyme thermostability and resistance to hostile organic solvents was reviewed in 2004 [207]. It was guided by the idea of introducing salt bridges on the protein surface for sufficient rigidification, constructing disulfide bonds likewise for rigidification, resorting to molecular dynamics (MD) simulations, and applying phylogenetic analyses (consensus) [208]. As more and more studies appeared, the accumulating knowledge of mutational effects provided further support for new projects in rational design. In parallel to these developments, increasing the robustness of enzymes was studied by directed evolution as an alternative [209]. It needs to be pointed out that organic chemists completely ignored these advances because they were interested in *stereoselectivity*, for which many synthetic transitions metal catalysts were being developed with great success [210]. Indeed, they were skeptical concerning enzymes in general, not believing that they could be part of their toolbox. With the advent of directed evolution of stereoselective enzymes [31], this attitude slowly began to change. Enhancing stereoselectivity by rational enzyme design was considered for a long time to be a daunting task [211]. Predicting the necessary exchange of an amino acid at a given residue appeared to be a mission impossible, not to speak of predicting several mutations. What residues should be rationally chosen, and which amino acid exchange should then be implemented at each residue? As will be shown below, the advantages of semi-rational directed evolution for influencing stereoselectivity and the alternative rational enzyme design did not develop on separate tracks. Indeed, the two approaches have merged (Section 3.4).

Due to the apparent difficulties at the time, not rational design was chosen for manipulating stereoselectivity, but instead directed evolution, originally utilizing *random* mutagenesis methods. The hydrolytic kinetic resolution of 2-methyl decanoic acid *p*-nitrophenyl ester, catalyzed by the PAL, served as the model experimental platform and the selectivity factor E as the degree of enantioselectivity [205c] (see Chapters 1 and 4 for details). Following four rounds of epPCR, the enantioselectivity factor increased 10-fold from $E = 1.4$ (S enantiomer) to $E = 11$, at the time a breakthrough, but far from industrially relevant. Since subsequent epPCR experiments proved to be disappointing ($E = 13$), the doctoral student Steffi Wilensek, working in the Reetz group, suggested, inter alia, focused saturation mutagenesis at sites lining the binding pocket using so-called NNK codon degeneracy (20 canonical amino acids as building blocks). *It was the first time that this technique was applied to enzyme stereoselectivity*. Today, some researchers wonder why it took so long to realize that this approach is the way to go. It worked indeed, but the best results ($E = 594$) [212] came later, following the systematization in the form of the CAST [87]. It was a crucial move from "blind" to semi-rational directed evolution of stereoselectivity. Importantly, ISM was also established, which simply meant that the best mutant of the first generation was used as a template for saturation mutagenesis at a different site in the second cycle, and so on [70, 88, 89]. Thus, CAST/ISM as a form of *focused* mutagenesis is different from iterative epPCR, iterative mutator strain-based mutagenesis, or iterative DNA shuffling, which are all *random*. This difference is of fundamental importance in protein engineering and pervades throughout this monograph.

3.4 Merging Semi-rational Directed Evolution and Rational Enzyme Design by Focused Rational Iterative Site-Specific Mutagenesis (FRISM)

A strict separation line between advanced semi-rational directed evolution, such as new-generation versions of CAST/ISM [1a, n, 89, 95] *and rational enzyme design* [83, 84, 158] *does not exist*. The development of Focused Rational Iterative Site-specific Mutagenesis (FRISM) goes even one step further, because it constitutes a successful and *systematized* fusion of the two approaches [213]. This development has been summarized in a review [3]. Prior to this development, several other rational enzyme design studies deserve mention, which are related to FRISM. An early example is a stereoselective Diels–Alder cycloaddition in which the Rosetta algorithm package [203, 214] was employed (Figure 3.23) [215]. Based on QM computations, over 200 Rosetta predictions were first made for possible mutants with sufficient activity. Then, 84 were expressed and tested in the model reaction. Mutations that enable H-bond interactions at the acceptor and donor residues were designed for improving binding and others were added for increasing activity. Two variants showed Diels–Alder activity, and one of them corresponded to the predicted high enantioselectivity (>95% ee; ee = enantiomeric excess) and complete endo-selectivity. The researchers concluded that the relatively low success rate calls for further improvement in Rosetta [215]. Designing exo-selectivity was not attempted, which would have been particularly interesting because it would violate standard "rules" in organic chemistry.

A scrutiny of the literature shows that Rosetta undergoes continuous improvements. As an example, protein backbone manipulation by RosettaRemodel is based on a unified interface to the Rosetta modeling suite, with which control over numerous aspects of flexible backbone protein design computations is possible [203a]. Loop insertion/deletion, disulfide engineering, domain assembly, loop remodeling, motif drafting, and *de novo* structure modeling belong to the challenges covered by this approach (Figure 3.24).

Rational enzyme design has also been applied to the standard P450-BM3 monooxygenase using (4*R*)-limonene as a chiral substrate [216]. Since WT was found to deliver a mixture of products at low conversion, the purpose was to

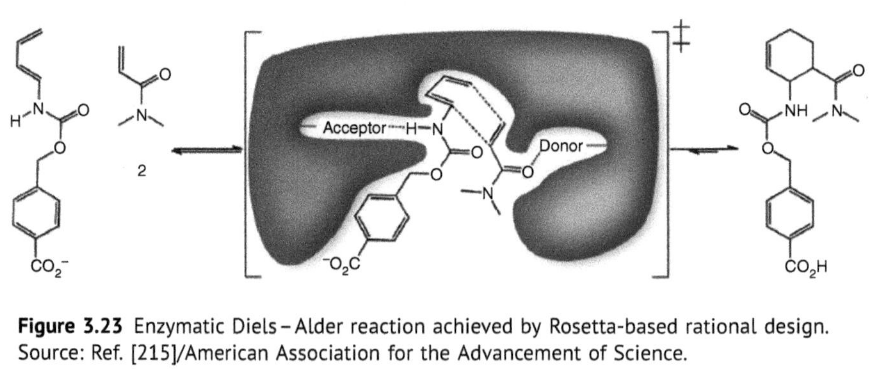

Figure 3.23 Enzymatic Diels–Alder reaction achieved by Rosetta-based rational design. Source: Ref. [215]/American Association for the Advancement of Science.

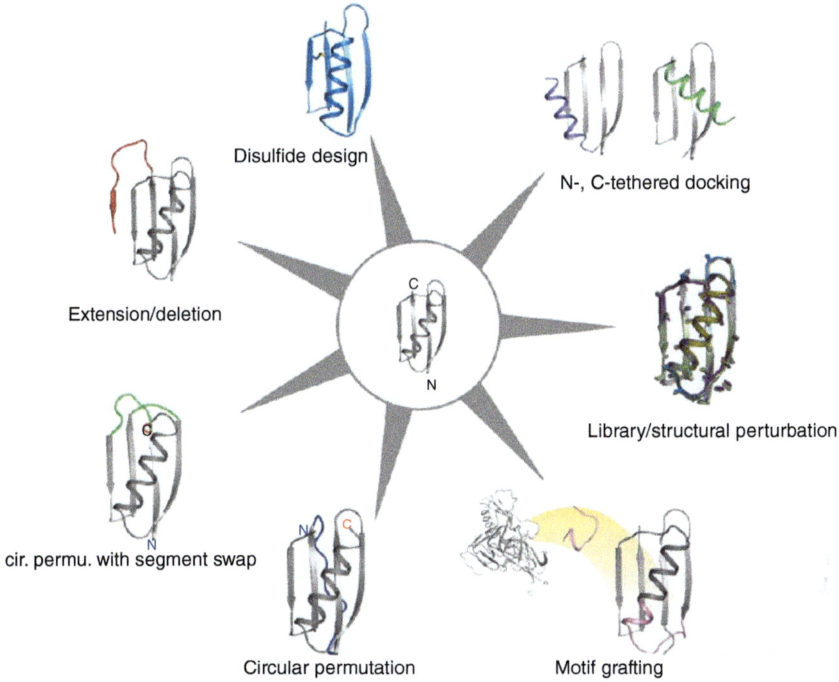

Figure 3.24 RosettaRemodel effective for a variety of protein backbone manipulations based on the crystal structure of the model protein G (PDB ID: 1PGA). The colored regions indicate the spots where predictions were made for various purposes. Source: Ref. [203a]/PLOS/CC BY.

increase activity for C7-hydroxylation with the formation of perillyl alcohol. The rational design campaign involved several steps that were involved in the rational design procedure, beginning with the screening of a previous mutant library and performing MD computations. This led the researchers to test the introduction of hydrophobic amino acids alanine, valine, phenylalanine, leucine, and isoleucine at various residues. A collection of 48 single and double mutants at residues 87 and 328 was constructed and tested, whereby variant A328V formed 27% of the desired perillyl alcohol. Using this mutation, rational mutagenesis at positions 87 and 328 was checked again with the introduction of hydrophobic amino acids. Since no improvements set in, more MD simulations were carried out, and mutations involving hydrophobic amino acids were added to mutant A328V, leading to 60% of the desired product. In the final round of rational design, a similar strategy resulted in the further addition of A328V/L43F, which enabled 97% conversion to perillyl alcohol [216]. Like in CAST/ISM, the overall strategy is *focused*, not *random* iterative mutagenesis, except that only small collections of mutants have to be generated and screened.

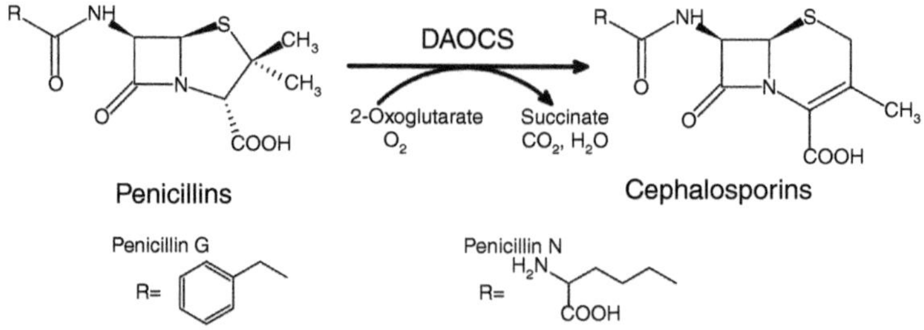

Figure 3.25 Conversion of penicillin G and N into cephalosporins catalyzed by mutants of scDAOCS. Source: Ref. [214]/American Society for Microbiology.

Similarly, rational enzyme design was employed for the generation of more active mutants of deacetoxycephalosporin C synthase from *Streptomyces clavuligerus* (scDAOCS) [214], which also preceded FRISM. As the authors noted themselves, they were inspired by ISM and therefore dubbed the technique Iterative Combinatorial Mutagenesis (ICM), which led to distinctly enhanced activity toward penicillin G and N (Figure 3.25). The template used as a start was a known quadruple mutant C155Y/Y184H/V275I/C281Y, evolved earlier by the same authors as a result of random epPCR and DNA shuffling followed by screening of ~10 500 clones, which had shown a 41-fold increase in k_{cat}/K_m relative to WT. As part of the ICM study, three rounds of targeted mutagenesis were performed in which a total of only 24 new mutants had to be screened. The best mutant resulted in an 87-fold increase in k_{cat}/K_m relative to WT [214]. The criteria (guidelines) for predicting the 24 new mutants were not clearly explained. MD computations were also not performed, but reference to earlier mutations was made, which was probably helpful. The practical results are nonetheless of notable significance in the pharmaceutical industry. A related rational design strategy was recently used in biocatalytic Pictet–Spengler reactions with the formation of pharmaceutically important 1-aryltetrahydro-ß-carbolines [217].

In addition to Rosetta and Hotspot Wizard, which can be combined with ML, other important techniques of utility in rational enzyme design include *alanine scanning* (or other amino acid scans), in which all or a limited number of residues are subjected to such amino acid exchanges [218]. This allows the identification of potential hotspots at which further mutagenesis can be performed, while occasionally some mutations already lead to an improved catalytic profile which can then be combined. Related is the concept of *mutability scanning*, according to which not just one amino acid but several are introduced at a select number of hotspots [219]. A fingerprint of positive, neutral, and detrimental effects on each residue becomes visible. However, this can be quite labor intensive. Therefore, only a limited number of such amino acid exchange events are implemented, based on consensus data, previous mutational experience and ML. In an impressive application, amorpha-4,11-diene synthase served as the enzyme, which is responsible for the first (rate-determining) step in the biosynthesis of antimalarial artemisinin [220]. Up to 16 residues surrounding the binding pocket were first considered, but this was reduced drastically by mutability landscaping, guided by consideration of sequences of other terpene synthases (consensus). Among the 258 produced variants which were screened for activity, the double mutant T399S/H448A enhanced k_{cat} fivefold [220]. In principle, this could be improved even more if the sequence and mutational information were to be used in a follow-up ISM-study. Amino acid scanning and mutability landscaping are related to the idea of performing NNK-based saturation mutagenesis at every residue of an enzyme.

The low number of predicted and tested mutants in the above study is also a central issue in advanced CAST/ISM since the generation of smarter and smaller mutant libraries continues to be a crucial goal. Along this line, a revealing study showed that only 384 transformants had to be screened for enantioselectivity in order to ensure 95% coverage of the respective CAST library [221]. This low number is comparable to the screening effort in many studies that rely on rational enzyme design, which again suggests that the two approaches to protein engineering are merging.

The first FRISM study was aimed at enhancing and inverting enantio- *and* diastereo-selectivity in an unusual lipase-catalyzed kinetic resolution [213a], which was then systematically classified [1a]. The new technique was inspired by CAST/ISM and nicely merges the two approaches of protein engineering. The first step is the same as in conventional rational enzyme design, i.e. predictions of a small set of variants are made and then experimentally tested. The best mutant is subsequently used as a template for a second cycle of prediction/screening, and so on. *One wonders why this systematization had not been suggested earlier.* The case of two mutational sites, as illustrated in Figure 3.26, pictures two FRISM sites together with the respective two pathways. Each site comprises either one or more residues. Since the two pathways do not necessarily terminate with the identical final variant, the scheme in Figure 3.26a is not symmetrical.

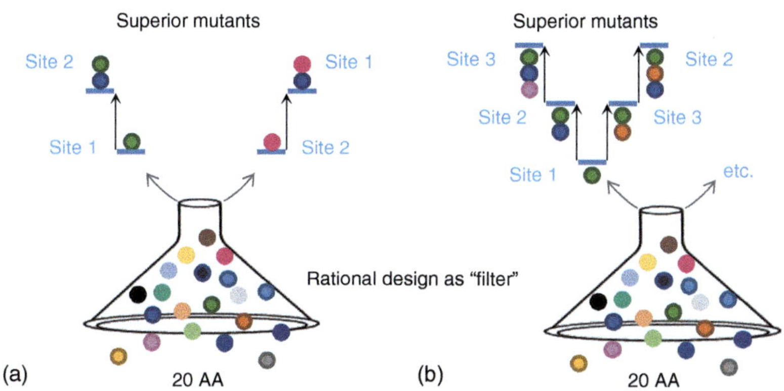

Figure 3.26 Adapted illustration of Focused Rational Iterative Site-specific Mutagenesis (FRISM). (a) FRISM with two mutational sites; (b) FRISM with three mutational sites, two of the six possible pathways being shown. Source: Adapted from Ref. [1a, 213].

When three sites are chosen, 3! = 6 evolutionary pathways are possible, two of which are illustrated in Figure 3.26b. The kinetic resolution employing a racemic alcohol and a racemic acid ester was attempted, catalyzed by the CALB. Consequently, two diastereomeric products are possible, each in two enantiomeric forms (enantio-preference), forming four different stereoisomers. It is worthwhile to look at a few details of this ground-breaking study.

As noted above, the lipase CALB was used as the catalyst in the challenging goal of generating four different variants for the four theoretically possible stereoisomers of a product compound 1-phenylethyl 2-phenylpropanoate bearing two centers of chirality (Figure 3.27) [213]. An unusual type of kinetic resolution in a transacylating process was necessary. Following the docking of substrate 2-phenylpropionic acid *p*-nitrophenyl ester into the binding pocket of CALB, seven possible residues at the acid and alcohol pockets were identified for site-specific mutagenesis. Then a limited number of amino acids with different steric properties were chosen to replace the WT amino acids at each of the CAST hotspots. Alanine (A) or glycine (G) were considered as small, and leucine (L) and phenylalanine (F) as large. Thereafter, the predicted variants were tested in kinetic resolution in order to identify the best one for each of the four pathways. These limited exploratory experiments revealed a small number of hits, which formed the basis for extending the list by considering a small number of additional mutations with similarly sized amino acids (A^+, L^+, and F^+), including G, V/C/I/M, and Y/W, respectively. As further guides for the best predictions concerning possibly improved mutants, the known (*R*)-selectivity

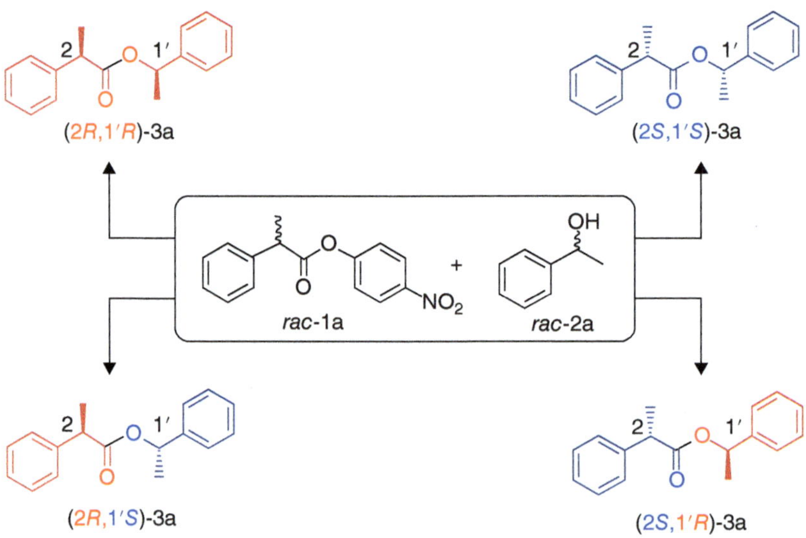

Figure 3.27 Stereoconvergent transacylation between 2-phenylpropionic acid *p*-nitrophenyl ester (*rac*-**1**) and 1-phenylethanol (*rac*-**2a**) catalyzed by CALB mutants. Source: Ref. [213]/American Chemical Society.

of WT CALB toward chiral alcohols and previous mutations for inverting to (*S*)-selectivity were considered. Along similar lines, the optimal first-generation variant(s) for each of the four pathways were predicted and tested, which indeed worked well, although still far away from 90% stereoselectivity, which did not surprise since only the first half of FRISM had been implemented.

The subsequent second half of the overall FRISM process was guided by similar considerations. The reader interested in applying the technique should read the details of the original publications (that requires a great deal of concentration!). Suffice it to say that indeed, four stereoconvergent variants were designed, which enabled more than 95% stereoselectivity in favor of the four stereoisomers, respectively (Figure 3.28). In this project, not just one stereoselective variant of an enzyme was obtained, as in traditional directed evolution studies, but four mutants, yet a total of only 100 variants had to be assessed for selectivity! The screening of traditional mutant libraries was circumvented. For mechanistic purposes, this study included kinetics and QM/MM computations, which unveiled the source of selectivities.

In a rapid succession of reports, FRISM was applied and extended using different types of enzymes [222]. Prominent examples involve a photodecarboxylase

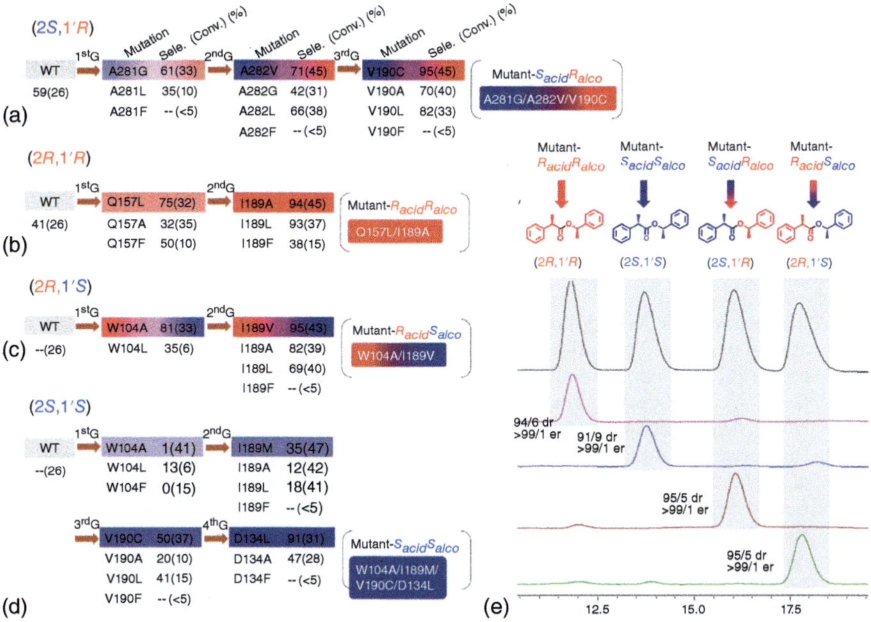

Figure 3.28 Focused Rational Iterative Site-specific Mutagenesis (FRISM) applied to the model CALB-catalyzed transacylating kinetic resolution (Figure 3.27) resulting in completely diastereo- and enantio-selective variants A, B, C, and D. Abbreviations: G, generation; Sele, selectivity. Conversion data are shown in parentheses after selectivity data according HPLC analysis (OJ-H, hexane:isopropanol = 80 : 20). Source: Ref. [213]/American Chemical Society.

[222a], and a glycosyltransferase for synthetically difficult regioselective glucosylation of poly-hydroxy compounds [222b]. A noteworthy study appeared around the same time in which a glycosyltransferase was also engineered for substrate acceptance, but in this case using CAST/ISM [223]. The authors reported that the use of FRISM would not have provided their best mutants. *Clearly, well-focused CAST/ISM and FRISM are complementary rational approaches* [3].

3.5 Conclusions and Perspectives

This chapter clearly demonstrates (i) the crucial role of methodology development in directed evolution [1a] and rational enzyme design [57, 83, 84, 158], and (ii) that the two approaches have merged. Information on the advantages and disadvantages of the various molecular biological gene mutagenesis techniques has been offered, including analyses of possible pitfalls. Nevertheless, the choice of a given method when starting a new protein engineering project depends on other factors as well, such as economic issues [101]. All these factors together determine the final strategy to be used. With the aid of case studies, Chapter 4 illustrates how such decisions are optimally made by providing further guidelines. In most cases, it will be seen that evolving pronounced stereo- and/or regio-selectivity as well as activity, which touches on the heart of organic and pharmaceutical chemistry, is best achieved by semi-rational saturation mutagenesis (CAST/ISM). *The latest improved versions of CAST/ISM can therefore be called "rational" rather than "semi-rational."* Chapter 4 also emphasizes the emergence of FRISM [3].

References

1. Recent reviews covering different aspects of directed evolution (see reference [2] for older citations): (a) Qu, G., Li, A., Acevedo-Rocha, C.G. et al. (2020). The crucial role of methodology development in directed evolution of selective enzymes. *Angew. Chem. Int. Ed.* **59**: 13204–13231. (b) Arnold, F.H. (2019). Innovation by evolution: bringing new chemistry to life (nobel lecture). *Angew. Chem. Int. Ed.* **58**: 14420–14426. (c) Chen, K. and Arnold, F.H. (2020). Engineering new catalytic activities in enzymes. *Nat. Catal.* **3**: 203–213. (d) Wang, Y., Xue, P., Cao, M. et al. (2021). Directed evolution: methodologies and applications. *Chem. Rev.* https://doi.org/10.1021/acs.chemrev.1c00260. (e) Klein, A.S. and Zeymer, C. (2021). Design and engineering of artificial metalloproteins: from de novo metal coordination to catalysis. *Protein Eng. Des. Sel.* **34**: gzab003. (f) Widersten, M. (2020). Engineering aldolases for asymmetric synthesis. *Methods Enzymol.* **644**: 149–167. (g) Wang, Z., Sekar, B.S., and Li, Z. (2021). Recent advances in artificial enzyme cascades for the production of value-added chemicals. *Bioresour. Technol.* **323**: 124551. (h) Bunzel, H.A., Garrabou, X., Pott, M., and Hilvert, D. (2018). Speeding up enzyme discovery and engineering with ultrahigh-throughput methods. *Curr. Opin. Struct. Biol.* **48**: 149–156. (i) Dinmukhamed, T., Huang, Z., Liu, Y. et al. (2021). Current advances in

design and engineering strategies of industrial enzymes. *Syst. Microbiol. Biomanuf.* **1**: 15–23. (j) Sandoval, B.A. and Hyster, T.K. (2020). Emerging strategies for expanding the toolbox of enzymes in biocatalysis. *Curr. Opin. Chem. Biol.* **55**: 45–51. (k) Chang, C.-Y., Vila, J.C.C., Bender, M. et al. (2021). Engineering complex communities by directed evolution. *Nat. Ecol. Evol.* **5**: 1011–1023. (l) Xiong, W., Liu, B., Shen, Y. et al. (2021). Protein engineering design from directed evolution to de novo synthesis. *Biochem. Eng. J.* **174**: 108096. (m) Vornholt, T., Christoffel, F., Pellizzoni, M.M. et al. (2021). Systematic engineering of artificial metalloenzymes for new-to-nature reactions. *Sci. Adv.* **7**: eabe4208. (n) Li, G., Dong, Y., and Reetz, M.T. (2019). Can machine learning revolutionize directed evolution of selective enzymes? *Adv. Synth. Catal.* **361**: 2377–2386. (o) Alejaldre, L., Pelletier, J.N., and Quaglia, D. (2021). Methods for enzyme library creation: which one will you choose? *BioEssays* **43**: e2100052. (p) Hanefeld, U., Hollmann, F., and Paul, C.E. (2022). Biocatalysis making waves in organic chemistry. *Chem. Soc. Rev.* **51**: 594–627. (q) Edwardson, T.G.W., Levasseur, M.D., Tetter, S. et al. (2022). Protein cages: from fundamentals to advanced applications. *Chem. Rev.* **122** (9): 9145–9197. (r) Ospina, F., Schülke, K.H., and Hammer, S.C. (2022). Biocatalytic alkylation chemistry: building molecular complexity with high selectivity. *ChemPlusChem* **87**: e202100454.

2. Selected reviews of directed evolution of enzymes in the period 2009–2015: (a) Gillam, E.M.J., Copp, J.N., and Ackerley, D.F. (ed.) (2014). Directed evolution library creation. In: *Methods in Molecular Biology*, 1179. Totowa, NJ: Humana Press. (b) Bommarius, A.S., Blum, J.K., and Abrahamson, M.J. (2011). Status of protein engineering for biocatalysts: how to design an industrially useful biocatalyst. *Curr. Opin. Chem. Biol.* **15** (2): 194–200. (c) Jäckel, C. and Hilvert, D. (2010). Biocatalysts by evolution. *Curr. Opin. Biotechnol.* **21** (6): 753–759. (d) Brustad, E.M. and Arnold, F.H. (2011). Optimizing non-natural protein function with directed evolution. *Curr. Opin. Chem. Biol.* **15** (2): 201–210. (e) Goldsmith, M. and Tawfik, D.S. (2012). Directed enzyme evolution: beyond the low-hanging fruit. *Curr. Opin. Struct. Biol.* **22** (4): 406–412. (f) Reetz, M.T. (2012). Directed evolution of enzymes. In: *Enzyme Catalysis in Organic Synthesis*, 3e (ed. K. Drauz, H. Gröger and O. May), 119–190. Weinheim: Wiley-VCH. (g) Widersten, M. (2014). Protein engineering for development of new hydrolytic biocatalysts. *Curr. Opin. Chem. Biol.* **21**: 42–47. (h) Denard, C.A., Ren, H., and Zhao, H. (2015). Improving and repurposing biocatalysts via directed evolution. *Curr. Opin. Chem. Biol.* **25**: 55–64. (i) Currin, A., Swainston, N., Day, P.J., and Kell, D.B. (2015). Synthetic biology for the directed evolution of protein biocatalysts: navigating sequence space intelligently. *Chem. Soc. Rev.* **44**: 1172–1239. (j) Lutz, S. and Bornscheuer, U.T. (ed.) (2009). *Protein Engineering Handbook*. Weinheim: Wiley-VCH. (k) Reetz, M.T. (2012). Laboratory evolution of stereoselective enzymes as a means to expand the toolbox of organic chemists. *Tetrahedron* **68**: 7530–7548.
3. Reetz, M.T. (2022). Making enzymes suitable for organic chemistry by rational protein design. *ChemBioChem* https://doi.org/10.1002/cbic.202200049.
4. Studies based on mutator strains: (a) Selifonova, O. and Schellenberger, V. (2003). Evolution of microorganisms using mutator plasmids. In: *Directed Evolution Library Creation: Methods and Protocols*, Methods in Molecular Biology, vol. **231** (ed. F.H. Arnold and G. Georgiou), 45–52. Totowa, NJ: Humana Press Inc. (b) Muteeb, G. and Sen, R. (2010). Random mutagenesis using a mutator strain. *Methods Mol. Biol.* **634**: 411–419. (c) Carr, R., Alexeeva, M., Enright, A. et al. (2003). Directed evolution of an amine oxidase possessing both broad substrate

specificity and high enantioselectivity. *Angew. Chem. Int. Ed.* **42** (39): 4807–4810. (d) Serero, A., Jubin, C., Loeillet, S. et al. (2014). Mutational landscape of yeast mutator strains. *Proc. Natl. Acad. Sci. U.S.A.* **111** (5): 1897–1902. (e) Agilent Technologies, Inc. (2015). XL1-Red Competent Cells, Agilent Instruction Manual, Catalog 200129. (f) Badran, A. and Liu, D. (2015). Development of potent in vivo mutagenesis plasmids with broad mutational spectra. *Nat. Commun.* **6**: 8425.

5. (a) Leung, D.W., Chen, E., and Goeddel, D.V. (1989). A method for random mutagenesis of a defined DNA segment using a modified polymerase chain reaction. *Technique* **1**: 11–15. (b) Cadwell, R.C. and Joyce, G.F. (1994). Mutagenic PCR. *PCR Methods Appl.* **3**: S136–S140.
6. (a) Jiang, W., Zhuang, Y., Wang, S., and Fang, B. (2015). Directed evolution and resolution mechanism of 1,3-propanediol oxidoreductase from *Klebsiella pneumoniae* toward higher activity by error-prone PCR and bioinformatics. *PLoS One* **10**: e0141837. (b) Qian, W., Ou, L., Li, C. et al. (2020). Evolution of glucose dehydrogenase for cofactor regeneration in bioredox processes with denaturing agents. *ChemBioChem* **21**: 2680–2688. (c) Liu, Y., Xu, G., Zhou, J. et al. (2020). Structure-guided engineering of D-carbamoylase reveals a key loop at substrate entrance tunnel. *ACS Catal.* **10**: 12393–12402. (d) Burke, A.J., Lovelock, S.L., Frese, A. et al. (2019). Design and evolution of an enzyme with a non-canonical organocatalytic mechanism. *Nature* **570**: 219–223.
7. McCullum, E., Williams, B.R., Zhang, J., and Chaput, J. (2010). Random mutagenesis by error-prone PCR. *Methods Mol. Biol.* **634**: 103–109.
8. Kaur, J. and Sharma, R. (2006). Directed evolution: an approach to engineer enzymes. *Crit. Rev. Biotechnol.* **26** (3): 165–199.
9. Zaccolo, M., Williams, D.M., Brown, D.M., and Gherardi, E. (1996). An approach to random mutagenesis of DNA using mixtures of triphosphate derivatives of nucleoside analogues. *J. Mol. Biol.* **255** (4): 589–603.
10. Abou-Nader, M. and Benedik, M.J. (2010). Rapid generation of random mutant libraries. *Bioeng. Bugs* **1** (5): 337–340.
11. Yang, J., Ruff, A.J., Arlt, M., and Schwaneberg, U. (2017). Casting epPCR (cepPCR): a simple random mutagenesis method to generate high quality mutant libraries. *Biotechnol. Bioeng.* **114** (9): 1921–1927.
12. Tee, K.L. and Wong, T.S. (2013). Polishing the craft of genetic diversity creation in directed evolution. *Biotechnol. Adv.* **31** (8): 1707–1721.
13. Ye, J., Wen, F., Xu, Y. et al. (2015). Error-prone PCR-based mutagenesis strategy for rapidly generating high-yield influenza vaccine candidates. *Virology* **482**: 234–243.
14. Cirino, P.C., Mayer, K.M., and Umeno, D. (2003). Generating mutant libraries using error-prone PCR. In: *Directed Evolution Library Creation*, vol. **231** (ed. F. Arnold and H. Georgiou), 3–9. Humana Press.
15. Reetz, M.T. (2016). *Directed Evolution of Selective Enzymes: Catalysts for Organic Chemistry and Biotechnology*. Weinheim: Wiley-VCH.
16. Eggert, T., Reetz, M.T., and Jaeger, K.-E. (2004). Directed evolution by random mutagenesis: a critical evaluation. In: *Enzyme Functionality – Design, Engineering, and Screening* (ed. A. Svendsen), 375–390. New York: Marcel Dekker.
17. (a) Neylon, C. (2004). Chemical and biochemical strategies for the randomization of protein encoding DNA sequences: library construction methods for directed evolution. *Nucleic Acids Res.* **32** (4): 1448–1459. (b) Lutz, S. and Patrick, W.M. (2004). Novel methods for directed evolution of enzymes: quality, not quantity. *Curr. Opin. Biotechnol.* **15** (4): 291–297.
18. (a) Weiss, G. and von Haeseler, A. (1995). Modeling the polymerase chain reaction. *J. Comput. Biol.* **2**: 49–61. (b) Moore, G.L. and Maranas, C.D. (2000). Modeling DNA mutation and

recombination for directed evolution experiments. *J. Theor. Biol.* **205** (3): 483–503. (c) Pritchard, L., Corne, D., Kell, D. et al. (2005). A general model of error-prone PCR. *J. Theor. Biol.* **234** (4): 497–509. (d) Sylvestre, J., Chautard, H., Cedrone, F., and Delcourt, M. (2006). Directed evolution of biocatalysts. *Org. Process Res. Dev.* **10**: 562–571. (e) Wong, T.S., Roccatano, D., and Schwaneberg, U. (2007). Challenges of the genetic code for exploring sequence space in directed protein evolution. *Biocatal. Biotransform.* **25**: 229–241. (f) Zhao, J., Kardashliev, T., Joëlle Ruff, A. et al. (2014). Lessons from diversity of directed evolution experiments by an analysis of 3,000 mutations. *Biotechnol. Bioeng.* **111** (12): 2380–2389.

19. Claveau, S., Sasseville, M., and Beauregard, M. (2004). Alcohol-mediated error prone PCR. *DNA Cell Biol.* **23**: 789–795.

20. Minamoto, T., Wada, E., and Shimizu, I. (2012). A new method for random mutagenesis by error-prone polymerase chain reaction using heavy water. *J. Biotechnol.* **157** (1): 71–74.

21. Vanhercke, T., Ampe, C., Tirry, L., and Denolf, P. (2005). Reducing mutational bias in random protein libraries. *Anal. Biochem.* **339**: 9–14.

22. (a) Biles, B.D. and Connolly, B.A. (2004). Low-fidelity *Pyrococcus furiosus* DNA polymerase mutants useful in error-prone PCR. *Nucleic Acids Res.* **32** (22): e176. (b) Kardashliev, T., Ruff, A., Zhao, J., and Schwaneberg, U. (2014). A high-throughput screening method to reengineer DNA polymerases for random mutagenesis. *Mol. Biotechnol.* **56** (3): 274–283.

23. Vartanian, J.P., Henry, M., and Wain-Hobson, S. (1996). Hypermutagenic PCR involving all four transitions and a sizeable proportion of transversions. *Nucleic Acids Res.* **24** (14): 2627–2631.

24. (a) Miyazaki, K. and Takenouchi, M. (2002). Creating random mutagenesis libraries using megaprimer PCR of whole plasmid. *Biotechniques* **33** (5): 1033, 1034, 1036–1038. (b) Miyazaki, K. (2003). Creating random mutagenesis libraries by megaprimer PCR of whole plasmid (MEGAWHOP). *Methods Mol. Biol.* **231**: 23–28. (c) Miyazaki, K. (2011). MEGAWHOP cloning: a method of creating random mutagenesis libraries via megaprimer PCR of whole plasmids. *Methods Enzymol.* **498**: 399–406.

25. (a) Yu, H., Li, J., Zhang, D. et al. (2009). Improving the thermostability of *N*-carbamyl-D-amino acid amidohydrolase by error-prone PCR. *Appl. Microbiol. Biotechnol.* **82** (2): 279–285. (b) Gleichmann, T., Diensthuber, R.P., and Möglich, A. (2013). Charting the signal trajectory in a light-oxygen-voltage photoreceptor by random mutagenesis and covariance analysis. *J. Biol. Chem.* **288** (41): 29345–29355. (c) Liang, C., Gui, X., Zhou, C. et al. (2015). Improving the thermoactivity and thermostability of pectate lyase from *Bacillus pumilus* for ramie degumming. *Appl. Microbiol. Biotechnol.* **99** (6): 2673–2682. (d) Jiang, P., Mu, S., Li, H. et al. (2015). Design and application of a novel high-throughput screening technique for 1-deoxynojirimycin. *Sci. Rep.* **5**: 8563.

26. (a) Zhang, D., Zhu, F., Fan, W. et al. (2011). Gradually accumulating beneficial mutations to improve the thermostability of *N*-carbamoyl-D-amino acid amidohydrolase by step-wise evolution. *Appl. Microbiol. Biotechnol.* **90** (4): 1361–1371. (b) Sandström, A.G., Wikmark, Y., Engström, K. et al. (2012). Combinatorial reshaping of the *Candida antarctica* lipase a substrate pocket for enantioselectivity using an extremely condensed library. *Proc. Natl. Acad. Sci. U.S.A.* **109** (1): 78–83. (c) Saß, S., Kadow, M., Geitner, K. et al. (2012). A high-throughput assay method to quantify Baeyer–Villiger monooxygenase activity. *Tetrahedron* **68** (37): 7575–7580.

27. Agudo, R., Roiban, G.D., and Reetz, M.T. (2012). Achieving regio- and enantioselectivity of P450-catalyzed oxidative CH activation of small functionalized molecules by structure-guided directed evolution. *ChemBioChem* **13** (10): 1465–1473.

28. (a) Copp, J.N., Hanson-Manful, P., Ackerley, D.F., and Patrick, W.M. (2014). Error-prone PCR and effective generation of gene variant libraries for directed evolution. *Methods Mol. Biol.* **1179**: 3–22. (b) Bill, A., Rosethorne, E.M., Kent, T.C. et al. (2014). High throughput mutagenesis for identification of residues regulating human prostacyclin (hiP) receptor expression and function. *PLoS One* **9** (6): e97973.
29. (a) Packer, M. and Liu, D. (2015). Methods for the directed evolution of proteins. *Nat. Rev. Genet.* **16**: 379–394. (b) Shao, W., Ma, K., Le, Y. et al. (2017). Development and use of a novel random mutagenesis method: in situ error-prone PCR (is-epPCR). *Methods Mol. Biol.* **1498**: 497–506.
30. (a) Firth, A.E. and Patrick, W.M. (2005). Statistics of protein library construction. *Bioinformatics* **21**: 3314–3315. (b) Firth, A.E. and Patrick, W.M. (2008). GLUE-IT and PEDEL-AA: new programmes for analyzing protein diversity in randomized libraries. *Nucleic Acids Res.* **36** (Web Server issue): W281–W285. (c) Denault, M. and Pelletier, J.N. (2007). Protein library design and screening: working out the probabilities. In: *Protein Engineering Protocols* (ed. K.M. Arndt and K.M. Müller), 127–154. Totowa, NJ: Humana Press. (d) Bosley, A.D. and Ostermeier, M. (2005). Mathematical expressions useful in the construction, description and evaluation of protein libraries. *Biomol. Eng.* **22**: 57–61. (e) Nov, Y. (2012). When second best is good enough: another probabilistic look at saturation mutagenesis. *Appl. Environ. Microbiol.* **78** (1): 258–262.
31. Reetz, M.T., Zonta, A., Schimossek, K. et al. (1997). Creation of enantioselective biocatalysts for organic chemistry by *in vitro* evolution. *Angew. Chem. Int. Ed. Engl.* **36** (24): 2830–2832.
32. (a) Arnold, F.H. (1998). Enzyme engineering reaches the boiling point. *Proc. Natl. Acad. Sci. U.S.A.* **95** (5): 2035–2036. (b) Bloom, J.D., Silberg, J.J., Wilke, C.O. et al. (2005). Thermodynamic prediction of protein neutrality. *Proc. Natl. Acad. Sci. U.S.A.* **102** (3): 606–611. (c) Tracewell, C.A. and Arnold, F.H. (2009). Directed enzyme evolution: climbing fitness peaks one amino acid at a time. *Curr. Opin. Chem. Biol.* **13** (1): 3–9.
33. Reetz, M.T., Wilensek, S., Zha, D., and Jaeger, K.-E. (2001). Directed evolution of an enantioselective enzyme through combinatorial multiple-cassette mutagenesis. *Angew. Chem. Int. Ed.* **40** (19): 3589–3591.
34. Zha, S., Wilensek, S., Hermes, M., and Jaeger, K.-E. (2001). Complete reversal of an enzyme-catalyzed reaction by directed evolution. *Chem. Commun.* 2664–2665.
35. (a) Zaccolo, M. and Gherardi, E. (1999). The effect of high-frequency random mutagenesis on *in vitro* protein evolution: a study on TEM-1 β-lactamase. *J. Mol. Biol.* **285** (2): 775–783. (b) Daugherty, P.S., Chen, G., Iverson, B.L., and Georgiou, G. (2000). Quantitative analysis of the effect of the mutation frequency on the affinity maturation of single chain Fv antibodies. *Proc. Natl. Acad. Sci. U.S.A.* **97** (5): 2029–2034.
36. Drummond, D.A., Iverson, B.L., Georgiou, G., and Arnold, F.H. (2005). Why high-error-rate random mutagenesis libraries are enriched in functional and improved proteins. *J. Mol. Biol.* **350** (4): 806–816.
37. (a) Holland, E.G., Buhr, D.L., Acca, F.E. et al. (2013). AXM mutagenesis: an efficient means for the production of libraries for directed evolution of proteins. *J. Immunol. Methods* **394**: 55–61. (b) Holland, E.G., Acca, F.E., Belanger, K.M. et al. (2015). *In vivo* elimination of parental clones in general and site-directed mutagenesis. *J. Immunol. Methods* **417**: 67–75.
38. (a) Fujii, R., Kitaoka, M., and Hayashi, K. (2006). Error-prone rolling circle amplification: the simplest random mutagenesis protocol. *Nat. Protoc.* **1**: 2493–2497. (b) Fujii, R., Kitaoka, M., and Hayashi, K. (2004). One-step random mutagenesis by error-prone

rolling circle amplification. *Nucleic Acids Res.* **32**: e145. (c) Fujii, W., Kano, K., Sugiura, K., and Naito, K. (2013). Repeatable construction method for engineered zinc finger nuclease based on overlap extension PCR and TA-cloning. *PLoS One* **8** (3): e59801. (d) Fujii, R., Kitaoka, M., and Hayashi, K. (2014). Error-prone rolling circle amplification greatly simplifies random mutagenesis. *Methods Mol. Biol.* **1179**: 23–29.

39. (a) Wong, T.S., Tee, K.L., Hauer, B., and Schwaneberg, U. (2004). Sequence saturation mutagenesis (SeSaM): a novel method for directed evolution. *Nucleic Acids Res.* **32** (3): e26. (b) Wong, T.S., Roccatano, D., Loakes, D. et al. (2008). Transversion-enriched sequence saturation mutagenesis (SeSaM-Tv+): a random mutagenesis method with consecutive nucleotide exchanges that complements the bias of error-prone PCR. *Biotechnol. J.* **3**: 74–82. (c) Shivange, A.V., Marienhagen, J., Mundhada, H. et al. (2009). Advances in generating functional diversity for directed protein evolution. *Curr. Opin. Chem. Biol.* **13**: 19–25. (d) Ruff, A.J., Kardashliev, T., Dennig, A., and Schwaneberg, U. (2014). The sequence saturation mutagenesis (SeSaM) method. *Methods Mol. Biol.* **1179**: 45–68.

40. (a) Ali, M.M., Li, F., Zhang, Z. et al. (2014). Rolling circle amplification: a versatile tool for chemical biology, materials science and medicine. *Chem. Soc. Rev.* **43** (10): 3324–3341. (b) Kobori, T. and Takahashi, H. (2014). Expanding possibilities of rolling circle amplification as a biosensing platform. *Anal. Sci.* **30** (1): 5–64.

41. Shahsavarian, M.A., Le Minoux, D., Matti, K.M. et al. (2014). Exploitation of rolling circle amplification for the construction of large phage-display antibody libraries. *J. Immunol. Methods* **407**: 26–34.

42. Huovinen, T., Brockmann, E.-C., Akter, S. et al. (2012). Primer extension mutagenesis powered by selective rolling circle amplification. *PLoS One* **7** (2): e31817.

43. Koyanagi, T., Yoshida, E., Minami, H. et al. (2008). A rapid, simple, and effective method of constructing a randomly mutagenized plasmid library free from ligation. *Biosci. Biotechnol., Biochem.* **72** (4): 1134–1137.

44. Brands, S., Brass, H.U.C., Klein, A.S. et al. (2020). A colourimetric high-throughput screening system for directed evolution of prodigiosin ligase PigC. *Chem. Commun.* **56** (61): 8631–8634.

45. Yang, G., Miton, C.M., and Tokuriki, N. (2020). A mechanistic view of enzyme evolution. *Protein Sci.* **29**: 1724–1747.

46. (a) Janahan, P., Hibbert, E.G., Russell, A.J., and Dalby, P.A. (2009). Distributions of enzyme residues yielding mutants with improved substrate specificities from two different directed evolution strategies. *Protein Eng. Des. Sel.* **22** (7): 401–411. (b) Morley, K.L. and Kazlauskas, R.J. (2005). Improving enzyme properties: when are closer mutations better? *Trends Biothechnol.* **23** (5): 231–237. (c) Hibbert, E.G. and Dalby, P.A. (2005). Directed evolution strategies for improved enzymatic performance. *Microb. Cell Fact.* **4** (1): 29.

47. (a) Chothia, C., Gough, J., Vogel, C., and Teichmann, S. (2003). Evolution of the protein repertoire. *Science* **300** (5626): 1701–1703. (b) Savino, S., Desmet, T., and Franceus, J. (2022). Insertions and deletions in protein evolution and engineering. *Biotechnol. Adv.* **60**: 108010.

48. Li, A., Sun, Z., and Reetz, M.T. (2018). Solid-phase gene synthesis for mutant library construction: the future of directed evolution? *ChemBioChem* **19** (19): 2023–2032.

49. Sondek, J. and Shortle, D. (1992). A general strategy for random insertion and substitution mutagenesis: substoichiometric coupling of trinucleotide phosphoramidites. *Proc. Natl. Acad. Sci. U.S.A.* **89** (8): 3581–3585.

50. Murakami, H., Hohsaka, T., and Sisido, M. (2002). Random insertion and deletion of arbitrary number of bases for codonbased random mutation of DNAs. *Nat. Biotechnol.* **20**: 76–81.

51. (a) Forloni, M., Liu, A.Y., and Wajapeyee, N. (2018). Creating insertions or deletions using overlap extension polymerase chain reaction (PCR) mutagenesis. *Cold Spring Harbor Protoc.* **2018** (8): https://doi.org/10.1101/pdb-prot097758. (b) Zeng, F., Zhang, Y., Zhang, Z. et al. (2016). Multiple-site fragment deletion, insertion and substitution mutagenesis by modified overlap extension PCR. *Biotechnol. Biotechnol. Equip.* **31** (2): 339–348.
52. (a) Heim, R. and Tsien, R.Y. (1996). Engineering green fluorescent protein for improved brightness, longer wavelengths and fluorescence resonance energy transfer. *Curr. Biol.* **6** (2): 178–182. (b) Tsien, R.Y. (2009). Constructing and exploiting the fluorescent protein paintbox (nobel lecture). *Angew. Chem. Int. Ed.* **48** (31): 5612–5626.
53. Cormack, B.P., Valdivia, R.H., and Falkow, S. (1996). FACS-optimized mutants of the green fluorescent protein (GFP). *Gene* **173** (1): 33–38.
54. (a) Budisa, N. (2013). Expanded genetic code for the engineering of ribosomally synthetized and post-translationally modified peptide natural products (RIPPS). *Curr. Opin. Biotechnol.* **24** (4): 591–598. (b) Lang, K. and Chin, J.W. (2014). Cellular incorporation of unnatural amino acids and bioorthogonal labeling of proteins. *Chem. Rev.* **114** (9): 4764–4806. (c) Sun, S.B., Schultz, P.G., and Kim, C.H. (2014). Therapeutic applications of an expanded genetic code. *ChemBioChem* **15** (12): 1721–1729.
55. Zheng, S. and Kwon, I. (2012). Manipulation of enzyme properties by noncanonical amino acid incorporation. *Biotechnol. J.* **7** (1): 47–60.
56. (a) Pikkemaat, M.G. and Janssen, D.B. (2002). Generating segmental mutations in haloalkane dehalogenase: a novel part in the directed evolution toolbox. *Nucleic Acids Res.* **30**: e35. (b) Jones, D.D. (2005). Triplet nucleotide removal at random positions in a target gene: the tolerance of TEM-1β-lactamase to an amino acid deletion. *Nucleic Acids Res.* **33**: e80.
57. Bocola, M., Schulz, F., Leca, F. et al. (2005). Converting phenylacetone monooxygenase into phenylcyclohexanone monooxygenase by rational design: towards practical Baeyer–Villiger monooxygenases. *Adv. Synth. Catal.* **347** (7–8): 979–986.
58. Reetz, M.T. and Wu, S. (2008). Greatly reduced amino acid alphabets in directed evolution: making the right choice for saturation mutagenesis at homologous enzyme positions. *Chem. Commun.* **43**: 5499–5501.
59. Erdogan, E., Jones, R.J., Matzlin, P. et al. (2005). A novel mutagenesis method generating high yields of closed circular mutant DNA with one primer per mutant. *Mol. Biotechnol.* **30**: 21–30.
60. Emond, S., Petek, M., Kay, E.J. et al. (2020). Accessing unexplored regions of sequence space in directed enzyme evolution via insertion/deletion mutagenesis. *Nat. Commun.* **11** (1): 3469.
61. Chaloupkova, R., Liskova, V., Toul, M. et al. (2019). Light-emitting dehalogenases: reconstruction of multifunctional biocatalysts. *ACS Catal.* **9** (6): 4810–4823.
62. Schenkmayerova, A., Pinto, G.P., Toul, M. et al. (2021). Engineering the protein dynamics of an ancestral luciferase. *Nat. Commun.* **12** (1): 3616.
63. Long, Y. and Cech, T.R. (2021). Targeted mutagenesis in human iPSCs using CRISPR genome-editing tools. *Methods* **191**: 44–58.
64. Miyazaki, K. and Arnold, F.H. (1999). Exploring nonnatural evolutionary pathways by saturation mutagenesis: rapid improvement of protein function. *J. Mol. Evol.* **49** (6): 716–720.
65. Liebeton, K., Zonta, A., Schimossek, K. et al. (2000). Directed evolution of an enantioselective lipase. *Chem. Biol.* **7** (9): 709–718.
66. Weiß, M.S., Pavlidis, I.V., Spurr, P. et al. (2017). Amine transaminase engineering for spatially bulky substrate

acceptance. *ChemBioChem* **18** (11): 1022–1026.

67. Estell, D.A., Graycar, T.P., and Wells, J.A. (1985). Engineering an enzyme by site-directed mutagenesis to be resistant to chemical oxidation. *J. Biol. Chem.* **260** (11): 6518–6521.

68. (a) Steipe, B. (2004). Consensus-based engineering of protein stability: from intrabodies to thermostable enzymes. *Methods Enzymol.* **388**: 176–186. (b) Lehmann, M., Loch, C., Middendorf, A. et al. (2002). The consensus concept for thermostability engineering of proteins: further proof of concept. *Protein Eng. Des. Sel.* **15** (5): 403–411. (c) Polizzi, K.M., Chaparro-Riggers, J.F., Vazquez-Figueroa, E., and Bommarius, A.S. (2006). Structure-guided consensus approach to create a more thermostable penicillin G acylase. *Biotechnol. J.* **1** (5): 531–536.

69. Wu, S., Acevedo, J.P., and Reetz, M.T. (2010). Induced allostery in the directed evolution of an enantioselective Baeyer–Villiger monooxygenase. *Proc. Natl. Acad. Sci. U.S.A.* **107** (7): 2775–2780.

70. (a) Reetz, M.T., Carballeira, J.D., and Vogel, A. (2006). Iterative saturation mutagenesis on the basis of B-factors as a strategy for increasing protein thermostability. *Angew. Chem. Int. Ed.* **45** (46): 7745–7751. (b) Reetz, M.T. and Carballeira, J.D. (2007). Iterative saturation mutagenesis (ISM) for rapid directed evolution of functional enzymes. *Nat. Protoc.* **2** (4): 891–903. (c) Reetz, M.T., Soni, P., Fernandez, L. et al. (2010). Increasing the stability of an enzyme toward hostile organic solvents by directed evolution based on iterative saturation mutagenesis using the B-FIT method. *Chem. Commun.* **46**: 8657–8658.

71. Hogrefe, H.H., Cline, J., Youngblood, G.L., and Allen, R.M. (2002). Creating randomized amino acid libraries with the Quikchange multisite-directed mutagenesis kit. *Biotechniques* **33** (5): 1158–1160, 1162, 1164, 1165.

72. Sarkar, G. and Sommer, S.S. (1990). The "megaprimer" method of site-directed mutagenesis. *Biotechniques* **8** (4): 404–407.

73. (a) Ho, S.N., Hunt, H.D., Horton, R.M. et al. (1989). Site-directed mutagenesis by overlap extension using the polymerase chain-reaction. *Gene* **77** (1): 51–59. (b) Zhang, H.Y., Zhang, J., Lin, L. et al. (1993). Enhancement of the stability and activity of aspartase by random and site-directed mutagenesis. *Biochem. Biophys. Res. Commun.* **192** (1): 15–21. (c) Heckman, K.L. and Pease, L.R. (2007). Gene splicing and mutagenesis by PCR-driven overlap extension. *Nat. Protoc.* **2** (4): 924–932.

74. Reikofski, J. and Tao, B.Y. (1992). Polymerase chain reaction (PCR) techniques for site-directed mutagenesis. *Biotechnol. Adv.* **10** (4): 535–547.

75. Sanchis, J., Fernández, L., Carballeira, J.D. et al. (2008). Improved PCR method for the creation of saturation mutagenesis libraries in directed evolution: application to difficult-to-amplify templates. *Appl. Microbiol. Biotechnol.* **81** (2): 387–397.

76. Sadler, J.C., Green, L., Swainston, N. et al. (2018). Fast and flexible synthesis of combinatorial libraries for directed evolution. *Methods Enzymol.* **608**: 59–79.

77. (a) Kirsch, R.D. and Joly, E. (1998). An improved PCR-mutagenesis strategy for two-site mutagenesis or sequence swapping between related genes. *Nucleic Acids Res.* **26** (7): 1848–1850. (b) Zheng, L., Baumann, U., and Reymond, J.-L. (2004). An efficient one-step site-directed and site-saturation mutagenesis protocol. *Nucleic Acids Res.* **32** (14): e115.

78. (a) Ono, A., Matsuda, A., Zhao, J., and Santi, D.V. (1995). The synthesis of blocked triplet-phosphoramidites and their use in mutagenesis. *Nucleic Acids Res.* **23** (22): 4677–4682. (b) Neuner, P., Cortese, R., and Monaci, P. (1998). Codon-based mutagenesis using dimer-phosphoramidites. *Nucleic Acids Res.* **26** (5): 1223–1227. (c) Gaytán, P. and Roldán-Salgado, A. (2013). Elimination of redundant and stop codons during the chemical synthesis of degenerate oligonucleotides.

Combinatorial testing on the chromophore region of the red fluorescent protein mKate. *ACS Synth. Biol.* **2** (8): 453–462.

79. Hughes, M.D., Nagel, D.A., Santos, A.F. et al. (2003). Removing the redundancy from randomised gene libraries. *J. Mol. Biol.* **331** (5): 973–979.

80. Ashraf, M., Frigotto, L., Smith, M.E. et al. (2013). Proximax randomization: a new technology for non-degenerate saturation mutagenesis of contiguous codons. *Biochem. Soc. Trans.* **41** (Pt 5): 1189–1194.

81. Li, A., Acevedo-Rocha, C.G., and Reetz, M.T. (2018). Boosting the efficiency of site-saturation mutagenesis for a difficult-to-randomize gene by a two-step PCR strategy. *Appl. Microbiol. Biotechnol.* **102**: 6095–6103.

82. (a) Dennig, A., Shivange, A.V., Marienhagen, J., and Schwaneberg, U. (2011). OmniChange: the sequence independent method for simultaneous site-saturation of five codons. *PLoS One* **6**: e26222. (b) Hidalgo, A., Schließmann, A., Molina, R. et al. (2008). A one-pot, simple methodology for cassette randomisation and recombination for focused directed evolution. *Protein Eng. Des. Sel.* **21** (9): 567–576. (c) Hidalgo, A., Schließmann, A., and Bornscheuer, U.T. (2014). One-pot simple methodology for cassette randomization and recombination for focused directed evolution. *Methods Mol. Biol.* **1179**: 207–212. (d) Püllmann, P., Ulpinnis, C., Marillonnet, S. et al. (2019). Golden mutagenesis: an efficient multi-site-saturation mutagenesis approach by Golden Gate cloning with automated primer design. *Sci. Rep.* **9**: 10932.

83. Ema, T., Kamata, S., Takeda, M. et al. (2010). Rational creation of mutant enzyme showing remarkable enhancement of catalytic activity and enantioselectivity toward poor substrates. *Chem. Commun.* **46** (30): 5440–5442.

84. Pleiss, J. (2012). Rational design of enzymes. In: *Enzyme Catalysis in Organic Synthesis*, 3e (ed. K. Drauz, H. Gröger and O. May), 89–117. Weinheim: Wiley-VCH.

85. (a) Fischer, E. (1894). Einfluss der Configuration auf die Wirkung der Enzyme. *Deutsch. Chem. Ges.* **27** (3): 2985–2993. (b) Frieder, W. (1995). 100 Years "Schlüssel-Schloss-Prinzip": what made Emil Fischer use this analogy? *Angew. Chem. Int. Ed. Engl.* **33** (2324): 2364–2374.

86. (a) Pauling, L. (1948). Nature of forces between large molecules of biological interest. *Nature* **161** (4097): 707–709. (b) Amyes, T.L. and Richard, J.P. (2013). Specificity in transition state binding: the Pauling model revisited. *Biochemistry* **52** (12): 2021–2035.

87. Reetz, M.T., Bocola, M., Carballeira, J.D. et al. (2005). Expanding the range of substrate acceptance of enzymes: combinatorial active-site saturation test. *Angew. Chem. Int. Ed.* **44** (27): 4192–4196.

88. (a) Reetz, M.T., Wang, L.-W., and Bocola, M. (2006). Directed evolution of enantioselective enzymes: iterative cycles of CASTing for probing protein-sequence space. *Angew. Chem. Int. Ed.* **45** (8): 1236–1241. (b) Reetz, M.T. (2005). Evolution im Reagenzglas: Neue Perspektiven für die Weiße Biotechnologie. Tätigkeitsberichte der Max-Planck-Gesellschaft. pp. 327–331. (c) Acevedo-Rocha, C.G., Sun, Z., and Reetz, M.T. (2018). Clarifying the difference between iterative saturation mutagenesis as a rational guide in directed evolution and omnichange as a gene mutagenesis technique. *ChemBioChem* **19** (24): 2542–2544.

89. Acevedo-Rocha, C., Höbenreich, S., and Reetz, M.T. (2014). Iterative saturation mutagenesis: a powerful approach to engineer proteins by systematically simulating Darwinian evolution. *Methods Mol. Biol.* **1179**: 103–128.

90. (a) Reetz, M.T., Kahakeaw, D., and Lohmer, R. (2008). Addressing the numbers problem in directed evolution. *ChemBioChem* **9** (11): 1797–1804. (b) Clouthier, C.M., Kayser, M.M., and Reetz, M.T. (2006). Designing new Baeyer–Villiger monooxygenases using

restricted casting. *J. Org. Chem.* **71** (22): 8431–8437.

91. (a) Nov, Y. (2013). Fitness loss and library size determination in saturation mutagenesis. *PLoS One* **8** (7): e68069. (b) Nov, Y. (2014). Probabilistic methods I directed evolution: library size, mutation rate, and diversity. *Methods Mol. Biol.* **1179**: 261–278. (c) Nov, Y., Fulton, A., and Jaeger, K.-E. (2013). Optimal scanning of all single-point mutants of a protein. *J. Comput. Biol.* **20**: 990–997.

92. (a) Regan, L. and DeGrado, W. (1988). Characterization of a helical protein designed from first principles. *Science* **241** (4868): 976–978. (b) Osawa, S., Jukes, T.H., Watanabe, K., and Muto, A. (1992). Recent evidence for evolution of the genetic code. *Microbiol. Rev.* **56** (1): 229–264. (c) Davidson, A.R., Lumb, K.J., and Sauer, R.T. (1995). Cooperatively folded proteins in random sequence libraries. *Nat. Struct. Mol. Biol.* **2** (10): 856–864. (d) Walter, K.U., Vamvaca, K., and Hilvert, D. (2005). An active enzyme constructed from a 9-amino acid alphabet. *J. Biol. Chem.* **280** (45): 37742–37746.

93. Li, A., Qu, G., Sun, Z., and Reetz, M.T. (2019). Statistical analysis of the benefits of focused saturation mutagenesis in directed evolution based on reduced amino acid alphabets. *ACS Catal.* **9** (9): 7769–7778.

94. (a) Hoebenreich, S., Zilly, F.E., Acevedo-Rocha, C.G. et al. (2015). Speeding up directed evolution: combining the advantages of solid-phase combinatorial gene synthesis with statistically guided reduction of screening effort. *ACS Synth. Biol.* **4** (3): 317–331. (b) Nett, N., Hoebenreich, S., Lingnau, J.B. et al. (2018). P450-catalyzed regio- and diastereoselective steroid hydroxylation: efficient directed evolution enabled by mutability landscaping. *ACS Catal.* **8** (4): 3395–3410.

95. Review of directed evolution of stereoselective enzymes with emphasis on iterative saturation mutagenesis: Reetz, M.T. (2011). Laboratory evolution of stereoselective enzymes: a prolific source of catalysts for asymmetric reactions. *Angew. Chem. Int. Ed.* **50** (1): 138–174.

96. Miller, D.C., Lal, R.G., Marchetti, L.A., and Arnold, F.H. (2022). Biocatalytic one-carbon ring expansion of aziridines to azetidines via a highly enantioselective [1,2]-Stevens rearrangement. *J. Am. Chem. Soc.* **144** (11): 4739–4745.

97. Wu, Q., Soni, P., and Reetz, M.T. (2013). Laboratory evolution of enantiocomplementary *Candida antarctica* lipase B mutants with broad substrate scope. *J. Am. Chem. Soc.* **135** (5): 1872–1881.

98. García-Marquina, G., Núñez-Franco, R., Peccati, D.F. et al. (2022). Deconvoluting the directed evolution pathway of engineered acyltransferase LovD. *ChemCatChem* **14**: e202101349.

99. (a) Bougioukou, D.J., Kille, S., Taglieber, A., and Reetz, M.T. (2009). Directed evolution of an enantioselective enoate-reductase: testing the utility of iterative saturation mutagenesis. *Adv. Synth. Catal.* **351** (18): 3287–3305. (b) Sullivan, B., Walton, A.Z., and Stewart, J.D. (2013). Library construction and evaluation for site saturation mutagenesis. *Enzyme Microb. Technol.* **53** (1): 70–77.

100. (a) Tang, L., Gao, H., Zhu, X. et al. (2012). Construction of "small-intelligent" focused mutagenesis libraries using well-designed combinatorial degenerate primers. *Biotechniques* **52** (3): 149–158. (b) Kille, S., Acevedo-Rocha, C.G., Parra, L.P. et al. (2013). Reducing codon redundancy and screening effort of combinatorial protein libraries created by saturation mutagenesis. *ACS Synth. Biol.* **2** (2): 83–92.

101. Acevedo-Rocha, C.G., Reetz, M.T., and Nov, Y. (2015). Economical analysis of saturation mutagenesis experiments. *Sci. Rep.* **5**: 10654.

102. (a) Xiao, H., Nasertorabi, F., Choi, S.-H. et al. (2015). Exploring the potential impact of an expanded genetic code on protein function. *Proc. Natl. Acad. Sci. U.S.A.* **112**: 6961–6966. (b) Hoesl, M.G. and Budisa, N. (2011).

In vivo incorporation of multiple noncanonical amino acids into proteins. *Angew. Chem. Int. Ed.* **50**: 2996–2902. (c) Chin, J.W. (2014). Expanding and reprogramming the genetic code of cells and animals. *Annu. Rev. Biochem.* **83**: 379–408. (d) Bae, J.H., Rubini, M., Jung, G. et al. (2003). Expansion of the genetic code enables design of a novel "gold" class of green fluorescent proteins. *J. Mol. Biol.* **328** (5): 1071–1081. (e) Kwon, I. and Lim, S.I. (2013). Non-natural amino acids for protein engineering and new protein chemistries. *Macromol. Chem. Phys.* **214**: 1295–1301. (f) Drienovská, I. and Roelfes, G. (2020). Expanding the enzyme universe with genetically encoded unnatural amino acids. *Nat. Catal.* **3**: 193–202.

103. (a) Stemmer, W.P. (1994). Rapid evolution of a protein *in vitro* by DNA shuffling. *Nature* **370** (6488): 389–391. (b) Crameri, A., Raillard, S.-A., Bermudez, E., and Stemmer, W.P.C. (1998). DNA shuffling of a family of genes from diverse species accelerates directed evolution. *Nature* **391** (6664): 288–291. (c) Crameri, A. and Stemmer, W.P.C. (1995). Combinatorial multiple cassette mutagenesis creates all the permutations of mutant and wild-type sequences. *BioTechniques* **18**: 194–196.

104. Joern, J. (2003). DNA shuffling. In: *Directed Evolution Library Creation*, vol. **231** (ed. F. Arnold and G. Georgiou), 85–89. Humana Press.

105. (a) Kikuchi, M., Ohnishi, K., and Harayama, S. (1999). Novel family shuffling methods for the *in vitro* evolution of enzymes. *Gene* **236** (1): 159–167. (b) Kikuchi, M., Ohnishi, K., and Harayama, S. (2000). An effective family shuffling method using single-stranded DNA. *Gene* **243** (1–2): 133–137. (c) Zha, W., Zhu, T., and Zhao, H. (2003). Family shuffling with single-stranded DNA. In: *Directed Evolution Library Creation*, vol. **231** (ed. F. Arnold and G. Georgiou), 91–97. Humana Press.

106. Behrendorff, J.Y.H., Johnston, W., and Gillam, E.J. (2014). Restriction enzyme-mediated DNA family shuffling. In: *Directed Evolution Library Creation*, vol. **1179** (ed. E.M.J. Gillam, J.N. Copp and D. Ackerley), 175–187. Springer New York.

107. Ouyang, F. and Zhao, M. (2019). Enhanced catalytic efficiency of CotA-laccase by DNA shuffling. *Bioengineered* **10** (1): 182–189.

108. Chronopoulou, E.G., Papageorgiou, A.C., Ataya, F. et al. (2018). Expanding the plant GSTome through directed evolution: DNA shuffling for the generation of new synthetic enzymes with engineered catalytic and binding properties. *Front. Plant Sci.* **9**: 1737.

109. (a) Zhao, H., Giver, L., Shao, Z. et al. (1998). Molecular evolution by staggered extension process (StEP) *in vitro* recombination. *Nat. Biotechnol.* **16**: 258–261. (b) Garcia-Ruiz, E., Mate, D., Ballesteros, A. et al. (2010). Evolving thermostability in mutant libraries of ligninolytic oxidoreductases expressed in yeast. *Microb. Cell Fact.* **9** (1): 17.

110. Coco, W.M., Levinson, W.E., Crist, M.J. et al. (2001). DNA shuffling method for generating highly recombined genes and evolved enzymes. *Nat. Biotechnol.* **19** (4): 354–359.

111. Gibbs, M.D., Nevalainen, K.M.H., and Bergquist, P.L. (2001). Degenerate oligonucleotide gene shuffling (DOGS): a method for enhancing the frequency of recombination with family shuffling. *Gene* **271**: 13–20.

112. Bergquist, P.L., Reeves, R.A., and Gibbs, M.D. (2005). Degenerate oligonucleotide gene shuffling (DOGS) and random drift mutagenesis (RNDM): two complementary techniques for enzyme evolution. *Biomol. Eng.* **22**: 63–72.

113. Ostermeier, M., Shim, J.H., and Benkovic, S.J. (1999). A combinatorial approach to hybrid enzymes independent of DNA homology. *Nat. Biotechnol.* **17**: 1205–1209.

114. Lutz, S., Ostermeier, M., and Benkovic, S.J. (2001). Rapid generation of incremental truncation libraries for protein engineering using α-phosphothioate nucleotides. *Nucleic Acids Res.* **29**: e16.

115. Kawarasaki, Y., Griswold, K.E., Stevenson, J.D. et al. (2003). Enhanced crossover SCRATCHY: construction and high-throughput screening of a combinatorial library containing multiple non-homologous crossovers. *Nucleic Acids Res.* **31**: e126.

116. Sieber, V., Martinez, C.A., and Arnold, F.H. (2001). Libraries of hybrid proteins from distantly related sequences. *Nat. Biotechnol.* **19**: 456–460.

117. Higara, K. and Arnold, F.H. (2003). General method for sequence-independent site-directed chimeragenesis. *J. Mol. Biol.* **330**: 287–296.

118. Lee, S.H., Ryu, E.J., Kang, M.J. et al. (2003). A new approach to directed gene evolution by recombined extension on truncated templates (RETT). *J. Mol. Catal. B: Enzym.* **26**: 119–129.

119. Ikeuchi, A., Kawarasaki, Y., Shinbata, T., and Yamane, T. (2003). Chimeric gene library construction by a simple and highly versatile method using recombination-dependent exponential amplification. *Biotechnol. Progr.* **19**: 1460–1467.

120. O'Maille, P.E., Bakhtina, M., and Tsai, M.D. (2002). Structure-based combinatorial protein engineering (SCOPE). *J. Mol. Biol.* **321**: 677–691.

121. Hamamatsu, N., Aita, T., Nomiya, Y. et al. (2005). Biased mutation-assembling: an efficient method for rapid directed evolution through simultaneous mutation accumulation. *Protein Eng. Des. Sel.* **18**: 265–271.

122. Stutzman-Engwall, K., Conlon, S., Fedechko, R. et al. (2005). Semi-synthetic DNA shuffling of *aveC* leads to improved industrial scale production of doramectin by *Streptomyces avermitilis*. *Metab. Eng.* **7**: 27–37.

123. (a) Herman, A. and Tawfik, D.S. (2007). Incorporating synthetic oligonucleotides via gene reassembly (ISOR): a versatile tool for generating targeted libraries. *Protein Eng. Des. Sel.* **20**: 219–226. (b) Rockah-Shmuel, L., Tawfik, D.S., and Goldsmith, M. (2014). Generating targeted libraries by the combinatorial incorporation of synthetic oligonucleotides during gene shuffling (ISOR). *Methods Mol. Biol.* **1179**: 129–137.

124. Coco, W.M., Encell, L.P., Levinson, W.E. et al. (2002). Growth factor engineering by degenerate homoduplex gene family recombination. *Nat. Biotechnol.* **20**: 1246–1250.

125. Ness, J.E., Kim, S., Gottman, A. et al. (2002). Synthetic shuffling expands functional protein diversity by allowing amino acids to recombine independently. *Nat. Biotechnol.* **20**: 1251–1255.

126. Zha, D., Eipper, A., and Reetz, M.T. (2003). Assembly of designed oligonucleotides as an efficient method for gene recombination: a new tool in directed evolution. *ChemBioChem* **4** (1): 34–39.

127. Acevedo-Rocha, C.G. and Reetz, M.T. (2014). Assembly of designed oligonucleotides: a useful tool in synthetic biology for creating high quality combinatorial DNA libraries. *Methods Mol. Biol.* **1179**: 189–206.

128. Gibson, D.G., Young, L., Chuang, R.Y. et al. (2009). Enzymatic assembly of DNA molecules up to several hundred kilobases. *Nat. Methods* **6**: 343–345.

129. Chao, R., Yuan, Y., and Zhao, H. (2015). Recent advances in DNA assembly technologies. *FEMS Yeast Res.* **15** (1): 1–9.

130. Ramirez, L.C., Tusell, C.C., Stoffel, G. et al. (2020). In vivo selection for formate dehydrogenases with high efficiency and specificity towards NADP. *ACS Catal.* **10** (14): 7512–7525.

131. Herrmann, A.K., Bender, C., Kienle, E. et al. (2019). A robust and all-inclusive pipeline for shuffling of adeno-associated viruses. *ACS Synth. Biol.* **8** (1): 194–206.

132. Hill, D., O'Connor, P.M., Altermann, E. et al. (2020). Extensive bacteriocin gene shuffling in the *Streptococcus bovis/Streptococcus equinus* complex reveals gallocin D with activity against vancomycin resistant enterococci. *Sci. Rep.* **10** (1): 13431.

133. Mittal, P., Brindle, J., Stephen, J. et al. (2018). Codon usage influences fitness through RNA toxicity. *Proc. Natl. Acad. Sci. U.S.A.* **115** (34): 8639–8644.

134. Agmon, N., Temple, J., Tang, Z. et al. (2020). Phylogenetic debugging of a complete human biosynthetic pathway transplanted into yeast. *Nucleic Acids Res.* **48** (1): 486–499.

135. Yu, Y. and Lutz, S. (2011). Circular permutation: a different way to engineer enzyme structure and function. *Trends Biotechnol.* **29**: 18–25.

136. (a) Cunningham, B.A., Hemperly, J.J., Hopp, T.P., and Edelman, G.M. (1979). Favin versus concanavalin A: circularly permuted amino acid sequences. *Proc. Natl. Acad. Sci. U.S.A.* **76** (7): 3218–3222. (b) Lo, W.-C., Lee, C.-C., Lee, C.-Y., and Lyu, P.-C. (2009). Cpdb: a database of circular permutation in proteins. *Nucleic Acids Res.* **37** (Database issue): D328–D332.

137. (a) Hennecke, J., Sebbel, P., and Glockshuber, R. (1999). Random circular permutation of dsba reveals segments that are essential for protein folding and stability. *J. Mol. Biol.* **286** (4): 1197–1215. (b) Butler, J.S., Mitrea, D.M., Mitrousis, G. et al. (2009). Structural and thermodynamic analysis of a conformationally strained circular permutant of barnase. *Biochemistry* **48** (15): 3497–3507. (c) Cheltsov, A.V., Barber, M.J., and Ferreira, G.C. (2001). Circular permutation of 5-aminolevulinate synthase: mapping the polypeptide chain to its function. *J. Biol. Chem.* **276** (22): 19141–19149.

138. (a) Qian, Z. and Lutz, S. (2005). Improving the catalytic activity of *Candida antarctica* lipase B by circular permutation. *J. Am. Chem. Soc.* **127** (39): 13466–13467. (b) Yu, Y. and Lutz, S. (2010). Improved triglyceride transesterification by circular permuted *Candida antarctica* lipase B. *Biotechnol. Bioeng.* **105** (1): 44–50. (c) Reitinger, S., Yu, Y., Wicki, J. et al. (2010). Circular permutation of *Bacillus circulans* xylanase: a kinetic and structural study. *Biochemistry* **49** (11): 2464–2474. (d) Qian, Z., Fields, C.J., and Lutz, S. (2007). Investigating the structural and functional consequences of circular permutation on lipase B from *Candida antarctica. ChemBioChem* **8** (16): 1989–1996. (e) Qian, Z., Horton, J.R., Cheng, X., and Lutz, S. (2009). Structural redesign of lipase B from *Candida antarctica* by circular permutation and incremental truncation. *J. Mol. Biol.* **393** (1): 191–201.

139. Pierre, B., Shah, V., Xiao, J., and Kim, J.R. (2015). Construction of a random circular permutation library using an engineered transposon. *Anal. Biochem.* **474**: 16–24.

140. Mehta, M.M., Liu, S., and Silberg, J.J. (2012). A transposase strategy for creating libraries of circularly permuted proteins. *Nucleic Acids Res.* **40** (9): e71.

141. (a) Bennett, M.J., Choe, S., and Eisenberg, D. (1994). Domain swapping: entangling alliances between proteins. *Proc. Natl. Acad. Sci. U.S.A.* **91** (8): 3127–3131. (b) Wulff, B.B.H., Thomas, C.M., Smoker, M. et al. (2001). Domain swapping and gene shuffling identify sequences required for induction of an Avr-dependent hypersensitive response by the tomato Cf-4 and Cf-9 proteins. *Plant Cell* **13** (2): 255–272. (c) Park, S.-H., Park, H.-Y., Sohng, J.K. et al. (2009). Expanding substrate specificity of GT-B fold glycosyltransferase via domain swapping and high-throughput screening. *Biotechnol. Bioeng.* **102** (4): 988–994. (d) Yan, C. and Sack, J.S. (2022). X-ray structure of a human cardiac muscle troponin C/troponin I chimera in two crystal forms. *Acta Crystallogr.* **F78**: 17–24.

142. Ostermeier, M. and Benkovic, S.J. (2000). Evolution of protein function by domain swapping. *Adv. Protein Chem.* **55**: 29–77.

143. Further key papers on domain swapping: (a) Golczak, M., Sears, A.E., Kiser, P.D., and Palczewski, K. (2015). Lrat-specific domain facilitates vitamin a metabolism by domain swapping in HRASLS3. *Nat. Chem. Biol.* **11** (1): 26–32. (b) Zhou, X., Wang, H., Zhang, Y. et al. (2012). Alteration of substrate specificities of thermophilic α/β hydrolases through domain swapping and domain interface optimization. *Acta Biochim. Biophys. Sin.* **44** (12): 965–973. (c) Yamanaka, M., Nagao,

S., Komori, H. et al. (2015). Change in structure and ligand binding properties of hyperstable cytochrome c555 from aquifex aeolicus by domain swapping. *Protein Sci.* **24** (3): 366–375. (d) Roach, C.R., Hall, D.E., Zerbe, P., and Bohlmann, J. (2014). Plasticity and evolution of (+)-3-carene synthase and (−)-sabinene synthase functions of a Sitka spruce monoterpene synthase gene family associated with weevil resistance. *J. Biol. Chem.* **289** (34): 23859–23869. (e) Pardo, I. and Camarero, S. (2015). Laccase engineering by rational and evolutionary design. *Cell. Mol. Life Sci.* **72** (5): 897–910. (f) Kang, J.-Y., Ryu, S.H., Park, S.-H. et al. (2014). Chimeric cytochromes p450 engineered by domain swapping and random mutagenesis for producing human metabolites of drugs. *Biotechnol. Bioeng.* **111** (7): 1313–1322. (g) Evans, B.S., Chen, Y., Metcalf, W.W. et al. (2011). Directed evolution of the nonribosomal peptide synthetase admk generates new andrimid derivatives *in vivo*. *Chem. Biol.* **18** (5): 601–607. (h) Shingate, P. and Sowdhamini, R. (2012). Analysis of domain-swapped oligomers reveals local sequence preferences and structural imprints at the linker regions and swapped interfaces. *PLoS One* **7** (7): e39305. (i) Chang, C., Huang, R., Yan, Y. et al. (2015). Uncovering the formation and selection of benzylmalonyl-CoA from the biosynthesis of splenocin and enterocin reveals a versatile way to introduce amino acids into polyketide carbon scaffolds. *J. Am. Chem. Soc.* **137** (12): 4183–4190. (j) van Beek, H.L., de Gonzalo, G., and Fraaije, M.W. (2012). Blending Baeyer–Villiger monooxygenases: using a robust BVMO as a scaffold for creating chimeric enzymes with novel catalytic properties. *Chem. Commun.* **48**: 3288–3290. (k) Ghanbarpour, A., Santos, E.M., Pinger, C. et al. (2020). Human cellular retinol binding protein II forms a domain-swapped trimer representing a novel fold and a new template for protein engineering. *ChemBioChem* **21**: 3192–3196.

144. Xiong, A.S., Peng, R.H., Zhuang, J. et al. (2008). Non-polymerase-cyclingassembly-based chemical gene synthesis: strategies, methods, and progress. *Biotechnol. Adv.* **26**: 121–134.

145. (a) Carothers, J.M., Goler, J.A., and Keasling, J.D. (2009). Chemical synthesis using synthetic biology. *Curr. Opin. Biotechnol.* **20**: 498–503. (b) Leprince, A., van Passel, M.W., and dos Santos, V.A. (2012). Streamlining genomes: toward the generation of simplified and stabilized microbial systems. *Curr. Opin. Biotechnol.* **23**: 651–658.

146. Xiong, A.S., Peng, R.H., Zhuang, J. et al. (2008). Chemical gene synthesis: strategies, softwares, error corrections, and applications. *FEMS Microbiol. Rev.* **32**: 522–540.

147. (a) Melnikov, A., Murugan, A., Zhang, X. et al. (2012). Systematic dissection and optimization of inducible enhancers in human cells using a massively parallel reporter assay. *Nat. Biotechnol.* **30**: 271–277. (b) Kwasnieski, J.C., Mogno, I., Myers, C.A. et al. (2012). Complex effects of nucleotide variants in a mammalian cis-regulatory element. *Proc. Natl. Acad. Sci. U.S.A.* **109**: 19498–19503. (c) Patwardhan, R.P., Lee, C., Litvin, O. et al. (2009). High-resolution analysis of DNA regulatory elements by synthetic saturation mutagenesis. *Nat. Biotechnol.* **27**: 1173–1175. (d) Sharon, E., Kalma, Y., Sharp, A. et al. (2012). Inferring gene regulatory logic from high-throughput measurements of thousands of systematically designed promoters. *Nat. Biotechnol.* **30**: 521–530.

148. LeProust, E.M., Peck, B.J., Spirin, K. et al. (2010). Synthesis of highquality libraries of long (150mer) oligonucleotides by a novel depurination controlled process. *Nucleic Acids Res.* **38**: 2522–2540.

149. Mulligan, J.T. and Parker, H.-Y. (2008). Solid phase methods for polynucleotide production. US Patent 7,482,119, filed

23 October 2004 and issued 27 January 2009.

150. Van den Brulle, J., Fischer, M., Langmann, T. et al. (2008). A novel solid phase technology for high-throughput gene synthesis. *BioTechniques* **45**: 340–343.

151. Li, A., Acevedo-Rocha, C.G., Sun, Z. et al. (2017). Beating bias in directed evolution of proteins: combining high-fidelity on-chip solid-phase gene synthesis with efficient gene assembly for combinatorial library construction. *ChemBioChem* **19** (3): 221–228.

152. Kuhn, S.M., Rubini, M., Fuhrmann, M. et al. (2010). Engineering of an orthogonal aminoacyl-tRNA synthetase for efficient incorporation of the non-natural amino acid *O*-methyl-L-tyrosine using fluorescence-based bacterial cell sorting. *J. Mol. Biol.* **404**: 70–87.

153. Gebauer, M., Schiefner, A., Matschiner, G., and Skerra, A. (2013). Combinatorial design of an Anticalin directed against the extradomain b for the specific targeting of oncofetal fibronectin. *J. Mol. Biol.* **425**: 780–802.

154. Bowers, P.M., Neben, T.Y., Tomlinson, G.L. et al. (2013). Humanization of antibodies using heavy chain complementarity-determining region 3 grafting coupled with *in vitro* somatic hypermutation. *J. Biol. Chem.* **288**: 7688–7696.

155. Kille, S., Zilly, F.E., Acevedo, J.P., and Reetz, M.T. (2011). Regio- and stereoselectivity of P450-catalysed hydroxylation of steroids controlled by laboratory evolution. *Nat. Chem.* **3** (9): 738–743.

156. Banyai, W., Peck, B.J., Fernandez, A. et al. De novo synthesized gene libraries. WO 2015021080 A2, filed 05 August 2014 and issued 12 February 2015.

157. Sun, Z., Lonsdale, R., Wu, L. et al. (2016). Structure-guided triple code saturation mutagenesis: efficient tuning of the stereoselectivity of an epoxide hydrolase. *ACS Catal.* **6**: 1590–1597.

158. Reviews of computational techniques in directed evolution: (a) Sebestova, E., Bendl, J., Brezovsky, J., and Damborsky, J. (2014). Computational tools for designing smart libraries. *Methods Mol. Biol.* **1179**: 291–314. (b) Zaugg, J., Gumulya, Y., Gillam, E.J., and Bodén, M. (2014). Computational tools for directed evolution: a comparison of prospective and retrospective strategies. *Methods Mol. Biol.* **1179**: 315–333. (c) Swiderek, K., Tunon, I., Moliner, V., and Bertran, J. (2015). Computational strategies for the design of new enzymatic functions. *Arch. Biochem. Biophys.* **582**: 68–79. (d) Ebert, M.C. and Pelletier, J.N. (2017). Computational tools for enzyme improvement: why everyone can – and should – use them Maximilian. *Curr. Opin. Chem. Biol.* **37**: 89–96. (e) Marques, S.M., Planas-Iglesias, J., and Damborsky, J. (2021). Web-based tools for computational enzyme design. *Curr. Opin. Struct. Biol.* **69**: 19–34. (f) Sharrock, A.V., Mulligan, T.S., Hall, K.R. et al. (2022). NTR 2.0: a rationally engineered prodrug-converting enzyme with substantially enhanced efficacy for targeted cell ablation. *Nat. Methods* **19**: 205–215. (g) Osuna, S. (2021). The challenge of predicting distal active site mutations in computational enzyme design. *WIREs Comput. Mol. Sci.* **11** (3): e1502.

159. (a) Alford, R.F., Leaver-Fay, A., Jeliazkov, J.R. et al. (2017). The Rosetta all-atom energy function for macromolecular modeling and design. *J. Chem. Theory Comput.* **13** (6): 3031–3048. (b) Anishchenko, I., Pellock, S.J., Chidyausiku, T.M. et al. (2021). De novo protein design by deep network hallucination. *Nature* **600** (7889): 547–552. (c) Du, Z., Su, H., Wang, W. et al. (2021). The trRosetta server for fast and accurate protein structure prediction. *Nat. Protoc.* **16** (12): 5634–5651.

160. Reviews and key articles on HotSpot Wizard, FireProt and related rational design techniques: (a) Sequeiros-Borja, C.E., Surpeta, B., and Brezovsky, J. (2021). Recent advances in user-friendly computational tools to engineer protein function. *Brief. Bioinform.* **22** (3): bbaa150. (b) Musil, M., Khan, R.T., Beier, A. et al. (2021). FireProtASR: a web server for fully automated ancestral sequence reconstruction. *Briefings Bioinf.* **22** (4): 1–11. (c) Musil, M., Stourac, J., Bendl, J. et al. (2017). FireProt: web server for automated design of thermostable proteins. *Nucleic Acids Res.* **45** (W1): W393–W399. (d) Xia, Y., Li, X., Yang, L. et al. (2021). Development of thermostable sucrose phosphorylase by semi-rational design for efficient biosynthesis of alpha-D-glucosylglycerol. *Appl. Microbiol. Biotechnol.* **105** (19): 7309–7319. (e) Sumbalova, L., Stourac, J., Martinek, T. et al. (2018). HotSpot Wizard 3.0: web server for automated design of mutations and smart libraries based on sequence input information. *Nucleic Acids Res.* **46** (W1): W356–W362. (f) Wei, R., von Haugwitz, G., Pfaff, L. et al. (2022). Mechanistic-based design of efficient PET hydrolases. *ACS Catal.* **12**: 3382–3396.

161. Jacobs, T.M., Yumerefendi, H., Kuhlman, B., and Leaver-Fay, A. (2015). SwiftLib: rapid degenerate-codon-library optimization through dynamic programming. *Nucleic Acids Res.* **43** (5): e34.

162. Pines, G., Pines, A., and Eckert, C.A. (2022). Highly efficient libraries design for saturation mutagenesis. *Synth. Biol.* https://doi.org/10.1093/synbio/ysac006.

163. Pavelka, A., Chovancova, E., and Damborsky, J. (2009). HotSpot wizard: a web server for identification of hot spots in protein engineering. *Nucleic Acids Res.* **37**: W376–W383.

164. Dundas, J., Ouyang, Z., and Miteva, M.A. (2006). CASTp: computed atlas of surface topography of proteins with structural and topographical mapping of functionally annotated residues. *Nucleic Acids Res.* **34**: W116–W118.

165. Kuipers, R.K., Joosten, H.-J., van Berkel, W.J.H. et al. (2010). 3DM: systematic analysis of heterogeneous superfamily data to discover protein functionalities. *Proteins* **78** (9): 2101–2113.

166. Ashkenazy, H., Abadi, S., Martz, E. et al. (2016). ConSurf 2016: an improved methodology to estimate and visualize evolutionary conservation in macromolecules. *Nucleic Acids Res.* **44**: W344–W350.

167. Ma, B.-G. and Berezovsky, I.N. (2010). The mblosum: a server for deriving mutation targets and position-specific substitution rates. *J. Biomol. Struct. Dyn.* **28** (3): 415–419.

168. Valdar, W.S.J. (2002). Scoring residue conservation. *Proteins* **48**: 227–241.

169. Saraf, M.C., Horswill, A.R., Benkovic, S.J., and Maranas, C.D. (2004). FamClash: a method for ranking the activity of engineered enzymes. *Proc. Natl. Acad. Sci. U.S.A.* **101**: 4142–4147.

170. Moore, G.L. and Maranas, C.D. (2003). Identifying residue–residue clashes in protein hybrids by using a second-order mean-field approach. *Proc. Natl. Acad. Sci. U.S.A.* **100** (9): 5091–5096.

171. Saraf, M.C., Moore, G.L., Goodey, N.M. et al. (2006). IPRO: an iterative computational protein library redesign and optimization procedure. *Biophys. J.* **90**: 4167–4180.

172. (a) Kuipers, R.K.P., Joosten, H.-J., Verwiel, E. et al. (2009). Correlated mutation analyses on super-family alignments reveal functionally important residues. *Proteins Struct. Funct. Bioinf.* **76** (3): 608–616. (b) Nobili, A., Tao, Y., Pavlidis, I.V. et al. (2015). Simultaneous use of *in silico* design and a correlated mutation network as a tool to efficiently guide enzyme engineering. *ChemBioChem* **16** (5): 805–810. (c) Engqvist, M.K.M. and Nielsen, J. (2015). ANT: software for generating and evaluating degenerate codons for natural and expanded genetic codes. *ACS Synth. Biol.* **4**: 935–938.

173. Khersonsky, O., Lipsh, R., Avizemer, Z. et al. (2018). Automated design of efficient and functionally diverse enzyme repertoires. *Mol. Cell* **72** (1): 178–186.e5.

174. Barlow, K.A., Conchúir, Ó., Thompson, S. et al. (2018). Flex DdG: Rosetta ensemble-based estimation of changes in protein–protein binding affinity upon mutation. *J. Phys. Chem. B* **122** (21): 5389–5399.

175. Park, H., Bradley, P., Greisen, P. et al. (2016). Simultaneous optimization of biomolecular energy functions on features from small molecules and macromolecules. *J. Chem. Theory Comput.* **12**: 6201–6212.

176. Zhou, J., Panaitiu, A.E., and Grigoryan, G. (2020). A general-purpose protein design framework based on mining sequence–structure relationships in known protein structures. *Proc. Natl. Acad. Sci. U.S.A.* **117** (2): 1059–1068.

177. Parthiban, V., Gromiha, M.M., and Schomburg, D. (2006). Cupsat: prediction of protein stability upon point mutations. *Nucleic Acids Res.* **34** (Web Server issue): W239–W242.

178. Chovancova, E., Pavelka, A., Benes, P. et al. (2012). Caver 3.0: a tool for the analysis of transport pathways in dynamic protein structures. *PLoS Comput. Biol.* **8** (10): e1002708.

179. Guerois, R., Nielsen, J.E., and Serrano, L. (2002). Predicting changes in the stability of proteins and protein complexes: a study of more than 1000 mutations. *J. Mol. Biol.* **320** (2): 369–387.

180. Dehouck, Y., Kwasigroch, J., Gilis, D., and Rooman, M. (2011). Popmusic 2.1: a web server for the estimation of protein stability changes upon mutation and sequence optimality. *BMC Bioinf.* **12** (1): 151.

181. Wijma, H.J., Floor, R.J., Jekel, P.A. et al. (2014). Computationally designed libraries for rapid enzyme stabilization. *Protein Eng. Des. Sel.* **27** (2): 49–58.

182. Bednar, D., Beerens, K., Sebestova, E. et al. (2015). FireProt: energy- and evolution-based computational design of thermostable multiple-point mutants. *PLoS Comput. Biol.* **11** (11): e1004556.

183. (a) Campeotto, I., Goldenzweig, A., Davey, J. et al. (2017). One-step design of a stable variant of the malaria invasion protein RH5 for use as a vaccine immunogen. *PNAS* **114** (5): 998–1002. (b) Yang, L., Peplowski, L., Shen, Y. et al. (2022). Enhancing thermostability and activity of sucrose phosphorylase for high-level production of 2-*O*-α-D-glucosylglycerol. *Syst. Microbiol. Biomanuf.* https://doi.org/10.1007/s43393-022-00090-y.

184. (a) Henrich, S., Salo-Ahen, O.M.H., Huang, B. et al. (2010). Computational approaches to identifying and characterizing protein binding sites for ligand design. *J. Mol. Recognit.* **23** (2): 209–219. (b) Lin, Y., Yoo, S., and Sanchez, R. (2012). SiteComp: a server for ligand binding site analysis in protein structures. *Bioinformatics* **28** (8): 1172–1173. (c) Zhao, J., Cao, Y., and Zhang, L. (2020). Exploring the computational methods for protein-ligand binding site prediction. *Comput. Struct. Biotechnol.* **18**: 417–426.

185. (a) Saab-Rincon, G., Li, Y., Meyer, M. et al. (2009). Protein engineering by structure-guided SCHEMA recombination. In: *Protein Engineering Handbook* (ed. S. Lutz and U.T. Bornscheuer), 481–492. Weinheim: Wiley-VCH Verlag GmbH. (b) Heinzelman, P., Snow, C.D., Smith, M.A. et al. (2009). SCHEMA recombination of a fungal cellulase uncovers a single mutation that contributes markedly to stability. *J. Biol. Chem.* **284**: 26229–26233. (c) Heinzelman, P., Romero, P.A., and Arnold, F.H. (2013). Efficient sampling of SCHEMA chimera families to identify useful elements. *Methods Enzymol.* **523**: 351–368.

186. Pantazes, R.J., Saraf, M.C., and Maranas, C.D. (2007). Optimal protein library design using recombination or point mutations based on sequence based scoring functions. *Protein Eng. Des. Sel.* **20**: 361–373.

187. Fox, R.J., Davis, S.C., Mundorff, E.C. et al. (2007). Improving catalytic function by ProSAR-driven enzyme evolution. *Nat. Biotechnol.* **25**: 338–344.

188. Savile, C.K., Janey, J.M., Mundorff, E.C. et al. (2010). Biocatalytic asymmetric synthesis of chiral amines from ketones applied to Sitagliptin manufacture. *Science* **329** (5989): 305–309.

189. (a) Liang, F., Feng, X.-j., Lowry, M., and Rabitz, H. (2005). Maximal use of minimal libraries through the adaptive substituent reordering algorithm. *J. Phys. Chem. B* **109** (12): 5842–5854. (b) Carvalho, S.A., da Silva, E.F., Santa-Rita, R.M. et al. (2004). Synthesis and antitrypanosomal profile of new functionalized 1,3,4-thiadiazole-2-arylhydrazone derivatives, designed as non-mutagenic megazol analogues. *Bioorg. Med. Chem. Lett.* **14** (24): 5967–5970.

190. Feng, X., Sanchis, J., Reetz, M.T., and Rabitz, H. (2012). Enhancing the efficiency of directed evolution in focused enzyme libraries by the adaptive substituent reordering algorithm. *Chem. Eur. J.* **18** (18): 5646–5654.

191. Cadet, F., Fontaine, N., Vetrivel, I. et al. (2018). Application of Fourier transform and proteochemometrics principles to protein engineering. *BMC Bioinf.* **19** (1): 382.

192. Cadet, F., Fontaine, N., Li, G. et al. (2018). A machine learning approach for reliable prediction of amino acid interactions and its application in the directed evolution of enantioselective enzymes. *Sci. Rep.* **8**: 16757.

193. Li, G., Qin, Y., Fontaine, N.T. et al. (2021). Machine learning enables selection of epistatic enzyme mutants for stability against unfolding and detrimental aggregation. *ChemBioChem* **22** (5): 904–914.

194. Büchler, J., Malca, S.H., Patsch, D. et al. (2022). Algorithm-aided engineering of aliphatic halogenase WelO5* for the asymmetric late-stage functionalization of soraphens. *Nat. Commun.* **13** (1): 371.

195. Muggleton, S., King, R.D., and Sternberg, M.J. (1992). Protein secondary structure prediction using logic-based machine learning. *Protein Eng.* **5** (7): 647–657.

196. Norn, C., Wicky, B., Juergens, D. et al. (2021). Protein sequence design by conformational landscape optimization. *Proc. Natl. Acad. Sci. U.S.A.* **118** (11): e2017228118.

197. Saito, Y., Oikawa, M., Nakazawa, H. et al. (2018). Machine-learning-guided mutagenesis for directed evolution of fluorescent proteins. *ACS Synth. Biol.* **7** (9): 2014–2022.

198. Repecka, D., Jauniskis, V., Karpus, L. et al. (2021). Expanding functional protein sequence spaces using generative adversarial networks. *Nat. Mach. Intell.* **3**: 324–333.

199. Alley, E.C., Khimulya, G., Biswas, S. et al. (2019). Unified rational protein engineering with sequence-based deep representation learning. *Nat. Methods* **16** (12): 1315–1322.

200. Luo, Y., Jiang, G., Yu, T. et al. (2021). ECNet is an evolutionary context-integrated deep learning framework for protein engineering. *Nat. Commun.* **12** (1): 1–14.

201. Siedhoff, N.E., Illig, A.M., Schwaneberg, U., and Davari, M.D. (2021). PyPEF—an integrated framework for data-driven protein engineering. *J. Chem. Inf. Model.* **61** (7): 3463–3476.

202. Qiu, Y., Hu, J., and Wei, G.W. (2021). Cluster learning-assisted directed evolution. *Nat. Comput. Sci.* **1** (12): 809–818.

203. (a) Huang, P.-S., Ban, Y.-E.A., Richter, F. et al. (2011). RosettaRemodel: a generalized framework for flexible backbone protein design. *PLoS One* **6** (8): e24109. (b) Kiss, G., Çelebi-Ölçüm, N., Moretti, R. et al. (2013). Computational enzyme design. *Angew. Chem. Int. Ed.* **52** (22): 5700–5725. (c) Lovelock, S.L., Crawshaw, R., Basler, S. et al. (2022). The road to fully programmable protein catalysis. *Nature* **606** (7912): 49–58.

204. (a) Narayanan, P.S. and Runthala, A. (2022). Accurate computational evolution of proteins and its dependence on deep learning and machine learning strategies. *Biocatal. Biotransform.* https://doi.org/10.1080/10242422.2022.2030317. (b) Ding, W., Nakai, K., Gong, H. (2022) Protein design via deep learning. *Briefings Bioinf.*, https://doi.org/10.1093/bib/bbac102. (c) Sapoval, N., Aghazadeh, A., Nute, M.G. et al. (2022). Current progress and open challenges for applying deep learning across the biosciences. *Nat. Commun.* **13** (1): 1728.

205. (a) Liao, H., Mckenzie, T., and Hageman, R. (1986). Isolation of a thermostable enzyme variant by cloning and selection in a thermophile. *Proc. Natl. Acad. Sci. U.S.A.* **83** (3): 576–580. (b) Chen, K.Q. and Arnold, F.H. (1993). Tuning the activity of an enzyme for unusual environments – sequential random mutagenesis of subtilisin-E for catalysis in dimethylformamide. *Proc. Natl. Acad. Sci. U.S.A.* **90** (12): 5618–5622. (c) Reetz, M.T., Zonta, A., Schimossek, K. et al. (1997). Creation of enantioselective biocatalysts for organic chemistry by *in vitro* evolution. *Angew. Chem. Int. Ed.* **36** (24): 2830–2832.

206. (a) Smith, M. (1985). In vitro mutagenesis. *Annu. Rev. Genet.* **19**: 423–462. (b) Smith, M. (1994). Synthetic DNA and biology (nobel lecture). *Angew. Chem. Int. Ed. Engl.* **33** (12): 1214–1221.

207. Eijsink, V.G.H., Bjørk, A., Gåseidnes, S. et al. (2004). Rational engineering of enzyme stability. *J. Biotechnol.* **113** (1–3): 105–120.

208. (a) Steipe, B., Schiller, B., Plückthun, A., and Steinbacher, S. (1994). Sequence statistics reliably predict stabilizing mutations in a protein domain. *J. Mol. Biol.* **240** (3): 188–192. (b) Janda, J.O., Busch, M., Kück, F. et al. (2012). CLIPS-1D: analysis of multiple sequence alignments to deduce for residue-positions a role in catalysis, ligand-binding, or protein structure. *BMC Bioinf.* **13**: 55. (c) Sternke, M., Tripp, K.W., and Barrick, D. (2019). Consensus sequence design as a general strategy to create hyperstable, biologically active proteins. *Proc. Natl. Acad. Sci. U.S.A.* **116** (23): 11275–11284.

209. (a) Arnold, F.H. (1998). Design by directed evolution. *Acc. Chem. Res.* **31** (3): 125–131. (b) Romero, P.A. and Arnold, F.H. (2009). Exploring protein fitness landscapes by directed evolution. *Nat. Rev. Mol. Cell Biol.* **10** (12): 866–876.

210. (a) Knowles, W.S. (2002). Asymmetric hydrogenations (nobel lecture). *Angew. Chem. Int. Ed. Engl.* **41** (12): 1999–2007. (b) Noyori, R. (2002). Asymmetric catalysis: science and opportunities (nobel lecture). *Angew. Chem. Int. Ed. Engl.* **41** (12): 2008–2022. (c) Sharpless, K.B. (2002). Searching for new reactivity (nobel lecture). *Angew. Chem. Int. Ed. Engl.* **41** (12): 2024–2032.

211. Wilensek, S. (2001). Dissertation, *Gerichtete Evolutionals ein Mittel zur Erzeugung enantioselektiver Enzyme für die organische Synthese*, Ruhr-Universität-Bochum/Germany.

212. Reetz, M.T., Prasad, S., Carballeira, J.D. et al. (2010). Iterative saturation mutagenesis accelerates laboratory evolution of enzyme stereoselectivity: rigorous comparison with traditional methods. *J. Am. Chem. Soc.* **132** (26): 9144–9152.

213. (a) Xu, J., Cen, Y., Singh, W. et al. (2019). Stereodivergent protein engineering of a lipase to access all possible stereoisomers of chiral esters with two stereocenters. *J. Am. Chem. Soc.* **141** (19): 7934–7945. (b) Li, D., Wu, Q., and Reetz, M.T. (2020). Focused rational iterative site-specific mutagenesis (FRISM). *Methods Enzymol.* **643**: 225–242.

214. Ji, J., Fan, K., Tian, X. et al. (2012). Iterative combinatorial mutagenesis as an effective strategy for generation of deacetoxycephalosporin C synthase with improved activity toward penicillin G. *Appl. Environ. Microbiol.* **78** (21): 7809–7812.

215. Siegel, J.B., Zanghellini, A., Lovick, H.M. et al. (2010). Computational design of an enzyme catalyst for a stereoselective bimolecular Diels–Alder reaction. *Science* **329** (5989): 309–313.

216. Seifert, A., Antonovici, M., Hauer, B., and Pleiss, J. (2011). An efficient route to selective bio-oxidation catalysts: an iterative approach comprising modeling, diversification, and screening, based on CYP102A1. *ChemBioChem* **12** (9): 1346–1351.

217. Eger, E., Schrittwieser, J.H., Wetzl, D. et al. (2020). Asymmetric biocatalytic synthesis of

1-aryltetrahydro-β-carbolines enabled by "substrate walking". *Chem. Eur. J.* **26** (69): 16281–16285.

218. Morrison, K.L. and Weiss, G.A. (2001). Combinatorial alanine-scanning. *Curr. Opin. Chem. Biol.* **5** (3): 302–307.

219. (a) van der Meer, J.Y., Biewenga, L., and Poelarends, G.J. (2016). The generation and exploitation of protein mutability landscapes for enzyme engineering. *ChemBioChem* **17** (19): 1792–1799. (b) van der Meer, J.Y., Poddar, H., Baas, B.J. et al. (2016). Using mutability landscapes of a promiscuous tautomerase to guide the engineering of enantioselective Michaelases. *Nat. Commun.* **7**: 10911.

220. Abdallah, I.I., van Merkerk, R., Klumpenaar, E., and Quax, W.J. (2018). Catalysis of amorpha-4,11-diene synthase unraveled and improved by mutability landscape guided engineering. *Sci. Rep.* **8** (1): 9961.

221. Acevedo-Rocha, C.G., Li, A., D'Amore, L. et al. (2021). Pervasive cooperative mutational effects on multiple catalytic enzyme traits emerge via long-range conformational dynamics. *Nat. Commun.* **12** (1): 1621.

222. (a) Xu, J., Fan, J., Lou, Y. et al. (2021). Light-driven decarboxylative deuteration enabled by a divergently engineered photodecarboxylase. *Nat. Commun.* **12** (1): 3983. (b) Li, J., Qu, G., Shang, N. et al. (2021). Near-perfect control of the regioselective glucosylation enabled by rational design of glycosyltransferases. *Green Synth. Catal.* **2** (1): 45–53. (c) Li, D., Han, T., Xue, J. et al. (2021). Engineering fatty acid photodecarboxylase to enable highly selective decarboxylation of trans fatty acids. *Angew. Chem. Int. Ed.* **60** (38): 20695–20699. (d) Ma, N., Fang, W., Liu, C. et al. (2021). Switching an artificial P450 peroxygenase into peroxidase via mechanism-guided protein engineering. *ACS Catal.* **11** (14): 8449–8455. (e) Tian, C., Yang, J., Liu, C. et al. (2022). Engineering substrate specificity of HAD phosphatases and multienzyme systems development for the thermodynamic-driven manufacturing sugars. *Nat. Commun.* **13**: 3582.

223. Wen, Z., Zhang, Z.M., Zhong, L. et al. (2021). Directed evolution of a plant glycosyltransferase for chemo- and regioselective glycosylation of pharmaceutically significant flavonoids. *ACS Catal.* **11**: 14781–14790.

4 Guidelines for Applying Gene Mutagenesis Methods in Organic Chemistry, Pharmaceutical Applications, and Biotechnology

4.1 Some General Tips

Useful and general tips when applying rational design (Section 4.1.1) and directed evolution (Section 4.1.2) are given here. Thereafter, a description of comparative studies as well as recent developments follows. Protein engineers specializing in rational design or directed evolution have learned from each other in recent years, and the two fields have started to merge with the development of Focused Rational Iterative Site-specific Mutagenesis (FRISM) and related approaches, as explained in Section 3.10. The latest examples are highlighted here in Section 4.6.

4.1.1 Rational Design

For a long time, the rational design of stereoselective enzyme mutants as biocatalysts in organic chemistry and biotechnology remained beyond the imagination of protein engineers (see Section 1.3). This does not surprise, because even the directed evolution of enzyme stereoselectivity was a daunting task. Fortunately, this is slowly changing. Some impressive successes have already been reported using the Rosetta algorithms (see Section 3.3) in the area of promiscuous artificial metalloenzymes (see Chapter 7). In a more general assessment concerning the majority of known enzyme types, examples of purely rational design without the inclusion of saturation mutagenesis, error-prone polymerase chain reaction (epPCR), or DNA shuffling are rare. Semi-rational directed evolution of the kind combinatorial active-site saturation test (CAST)/iterative saturation mutagenesis (ISM) and advanced rational versions thereof (Section 3.5) can be augmented by adding a well-designed single mutation at an appropriate residue in the enzyme. In principle, to avoid pitfalls, rational design needs to be guided by appropriate data, techniques, and methods, including:

- Structural and mechanistic data
- Phylogenetic analysis (consensus technique)
- Docking and molecular dynamics (MD) computations

Enzyme Engineering: Selective Catalysts for Applications in Biotechnology, Organic Chemistry, and Life Science, First Edition. Manfred T. Reetz, Zhoutong Sun, and Ge Qu.

- Quantum mechanical (QM) calculations
- Previous mutational data
- The Rosetta computational package (which includes some of the above)
- Machine learning

In 2013, it was stated in a comprehensive review of computational enzyme design that rational design of such enzyme parameters as stereo- and/or regio-selectivity should be attempted only if an efficient medium- or high-throughput assay is *not* available [1]. Since then, this view has not changed much, as is evident in newer reviews [2]. A recent example of rationally manipulating substrate scope (activity) while maintaining high stereoselectivity without the aid of saturation mutagenesis, epPCR or DNA shuffling concerns a ω-transaminase from *Pseudomonas jessenii* (ω-TA), belonging to the class III transaminases [3]. This enzyme catalyzes the reaction of prochiral ketones with the formation of chiral amines, a biocatalytic transformation of great importance in the pharmaceutical industry [4] (Scheme 4.1). While strict (*S*)-enantioselectivity is generally observed with enzymes of this type, substrate scope is narrow, which greatly limits applications. This study is highlighted here as an important step forward, although directed evolution of transaminases (or amine dehydrogenases) based on semi-rational CAST/ISM has been shown *not* to require excessive or expensive ee-screening [5].

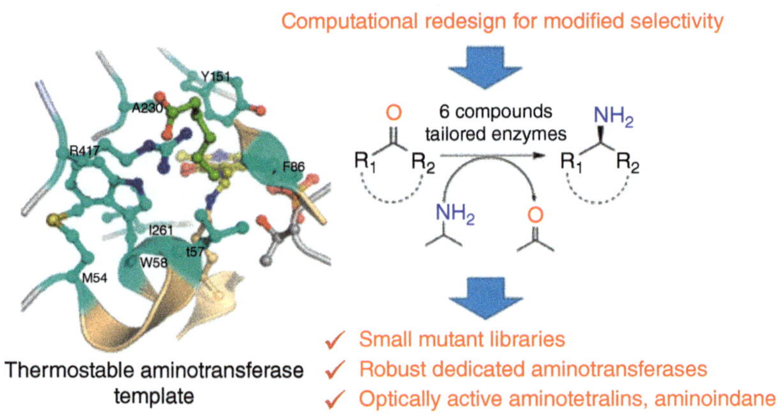

Scheme 4.1 Utility of computationally designed mutants of ω-transaminase from *Pseudomonas jessenii* (ω-TA) in the synthesis of pharmaceutically important chiral amines. Source: [3]/American Chemical Society.

In order to proceed rationally, the authors of the ω-TA study first considered the enzyme mechanism (Scheme 4.2). The catalytic cycle is described by two half-reactions with the occurrence of several intermediates. Then the Rosetta enzyme design software [1, 6] was applied. It is based on a search algorithm that computationally mutates selected residues randomly and identifies low-energy solutions by considering residue identities and their rotamers as well as ligand

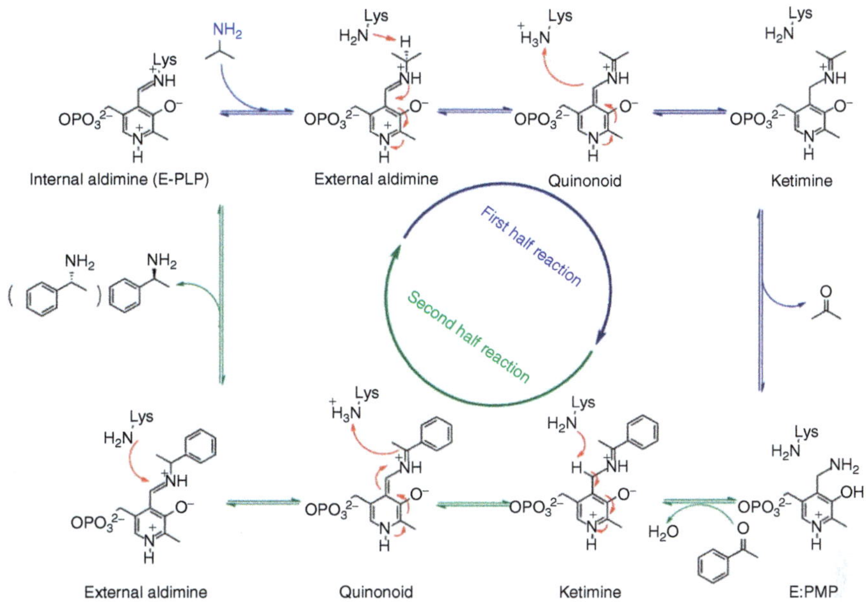

Scheme 4.2 Reaction mechanism for the transamination of acetophenone to enantiopure 1-phenylethylamine with isopropylamine as an amino donor. Source: [3]/American Chemical Society.

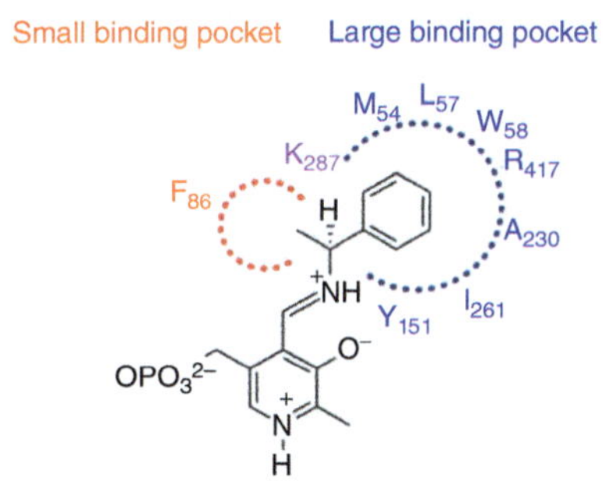

Figure 4.1 Key residues lining the large and small binding pockets of ω-TA PjTA-R6. Source: Meng et al. [3]/American Chemical Society.

conformations [7]. In this study, the Rosetta interface energy was employed as the primary metric, which makes extensive MD computations unnecessary in the early ranking phase. When performing docking experiments, the substrates were covalently bound in the form of the aldimine intermediate (Figure 4.1). This limits the search space and speeds up the process because degrees of freedom are greatly reduced.

Six structurally different bulky amines with an (*S*)-configuration were considered. In a dock-and-design step, mutations that sterically allow the accommodation of the respective aldimine intermediates were identified. Small virtual mutant libraries were constructed containing amino acid exchanges at the eight positions (Figure 4.1), and these were produced by site-specific mutagenesis and

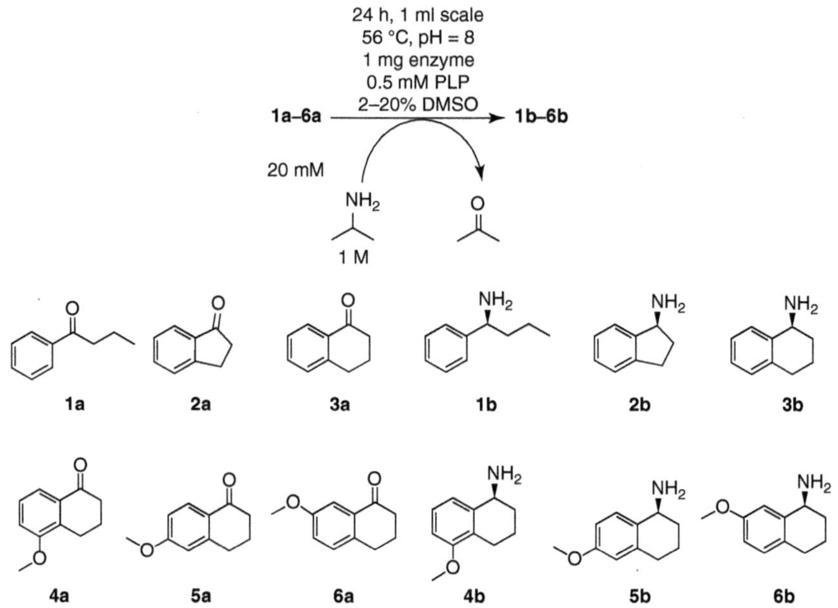

Figure 4.2 Ketone substrates (left half) and (*S*)-configurated chiral amines as products (right half). Source: Meng et al. [3]/American Chemical Society.

tested as catalysts in the laboratory. Eventually, 40 designed top-ranking variants based on the computed interface energy were constructed and experimentally assayed. Two of the designs showed the desired catalytic activity towards the six ketone substrates with high enantioselectivity (ee > 99%) (Figure 4.2). Moreover, several crystal structures of improved variants were analyzed in this study, which are in line with the Rosetta predictions. The results proved to be impressive, because exceedingly bulky substrates were readily accepted (Figure 4.2). In terms of thermostability, 12 out of the 40 variants displayed comparable or slightly increased stability. In the case of reduced thermostability, the water displacement upon substrate binding may be the cause, which leads to an increased energy barrier that is not considered in the Rosetta calculations [3].

Finally, *it should be noted that in this study, enantioselectivity itself was not designed, since WT and mutants are all strictly (S)-selective*. The reader may ask: How would one proceed if (*R*)-selectivity were the goal? The answer to this pressing question may lie in the optimal fusion of rational design and directed evolution based on advanced CAST/ISM. As outlined in Section 3.10, a major step forward is the development of FRISM, with Section 4.6 highlighting new examples.

In yet another study featuring rational enzyme design, chemoselectivity in oxidation was addressed by employing a difficult transformation of a substrate for which the toolbox of synthetic organic chemists offers no ecologically viable solution. It concerns the P450-BM3 catalyzed reaction of benzene with the

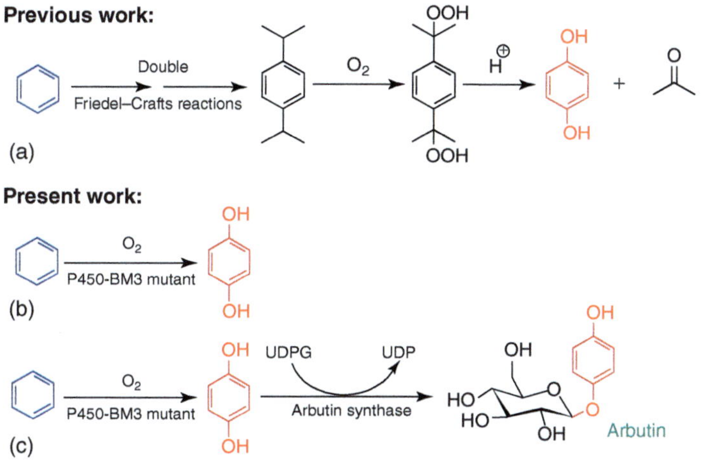

Scheme 4.3 The conventional procedure of synthesizing hydroquinone (HQ) from benzene (a), the newly developed biocatalytic route (b), and the biocatalytic cascade synthesis of arbutin from benzene using designed *E. coli* cells harboring a P450-BM3 mutant in a practical application fashion (c). Source: [8]/John Wiley & Sons.

formation of phenol and dihydroquinone (HQ), respectively (Scheme 4.3) [8]. If chemists try to oxidize benzene using synthetic reagents in the absence of enzymes, the initially formed phenol is more reactive than the starting material and therefore forms various dihydroxy benzenes and other products in an uncontrollable manner. The industrial synthetic procedure of dihydroquinone therefore requires two Friedel–Crafts reactions, oxidation with the formation of a di-peroxy compound followed by acid catalyzed fragmentation, all under harsh conditions (Scheme 4.3a).

The P450-BM3 based process could be stopped at the HQ product, formed regioselectively (Scheme 4.3b), which was also exploited in the one-pot transformation of benzene to arbutin, a known natural product used in the treatment of skin ailments. This was achieved by including it in the *Eshcherichia coli* cells with the enzyme arbutin synthase (Scheme 4.3c). The crucial question was how to rationally design mutants that enable these amazing transformations. In an earlier study in which P450-BM3 mutants had been evolved for the regio- and stereo-selective conversion of cyclohexanone to the acyloins (*R*)- and (*S*)-2-hydroxy cyclohexanone, exploratory NNK based saturation mutagenesis had been performed at eight CAST residues [9]. The mutational fingerprint was then analyzed, which showed that, especially at residues A78, A82, and V328, the introduction of phenylalanine signalled notably higher activity. Consequently, in the HQ-study, this amino acid was introduced not just at these three residues by site-specific mutagenesis, but also as double mutants A82F/A328F and V78F/A328F and triple mutant A82F/A328F/V78F. These six mutants were expected to reduce the size of the large P450-BM3 binding pocket. Indeed, these few variants were first tested in the reaction of phenol, which led

Table 4.1 Conversion of cyclohexanone to stereoisomeric cyclohexane-1,2-diols using *E. coli* cells co-expressing P450-BM3 mutants and 2,3-butanediol dehydrogenases.

Entry	Catalysts	Conv. (%)[a)]	*trans/ meso*	er (%)[c)]	Favored stereo-isomer	Product distribution (%)[b)]		
						Cyclo-hexanol	2-Hydroxycy-clohexan-1-one	Cyclo-hexane-1,2-diol
1	*E. coli* (P450ATC06 + BDHA)	90	98 : 2	>99 : 1	(*R,R*)	0	1	99
2	*E. coli* (P450ATD04 + BDHA)	98	5 : 95	—[d)]	*meso*	0	1	99
3	*E. coli* (P450ATC06 + BUDC)	84	8 : 92	—[d)]	*meso*	17	1	82
4	*E. coli* (P450ATD04 + BUDC)	98	95 : 5	>99:1	(*S,S*)	3	0	97
5	*E. coli* (P450ATC06 + LBDHA)	91	51 : 49	—[d)]	N.A.	29	2	69
6	*E. coli* (P450ATD04 + LBDHA)	97	96 : 4	98 : 2	(*S,S*)	8	1	91

a) Conditions: 10 mM substrate, 30 °C and 200 rpm, five hours. The conversion was determined by GC analysis and is based on the amount of converted substrate.
b) Relative amounts based on the peak areas in the GC chromatograms.
c) Determined for *trans*-cyclohexane-1,2-diols.
d) Values not determined.
N.A., not available.
Source: Ref. [9]/John Wiley & Sons.

chemo- and regio-selectively to HQ with no overoxidation and only traces of the regioisomer catechol (Table 4.1). As expected, the mutants also catalyzed the conversion of benzene itself to HQ. Finally, using the triple mutant and arbutin synthase in *E. coli* allowed benzene to be converted to arbutin in a single whole cell cascade system in excellent yield [9]. *The study is an illustration of how previous mutagenic information can be exploited in tackling challenging problems by rational design without the need to screen mutant libraries.* At the same time, it points to the current limitations of rational design, since similar scenarios using other substrates and enzyme types are likely to be rare. However, as time goes on, more data will become available, which makes rational enzyme design easier.

Apart from engineering CASTing sites, manipulating enzyme function by altering allosteric interactions while conserving the active site is also an attractive goal. In 2013, the Sanchez-Ruiz group constructed several ancient β-lactamases via ancestral sequence reconstruction [10a]. One of the resurrected β-lactamases, called GNCA, shares a common 3D structure with the extant TEM-1 β-lactamase (RMSD ~0.6 Å), and ~50% conservation in their amino acid sequence. Nevertheless, GNCA and TEM-1 β-lactamase showed a notable

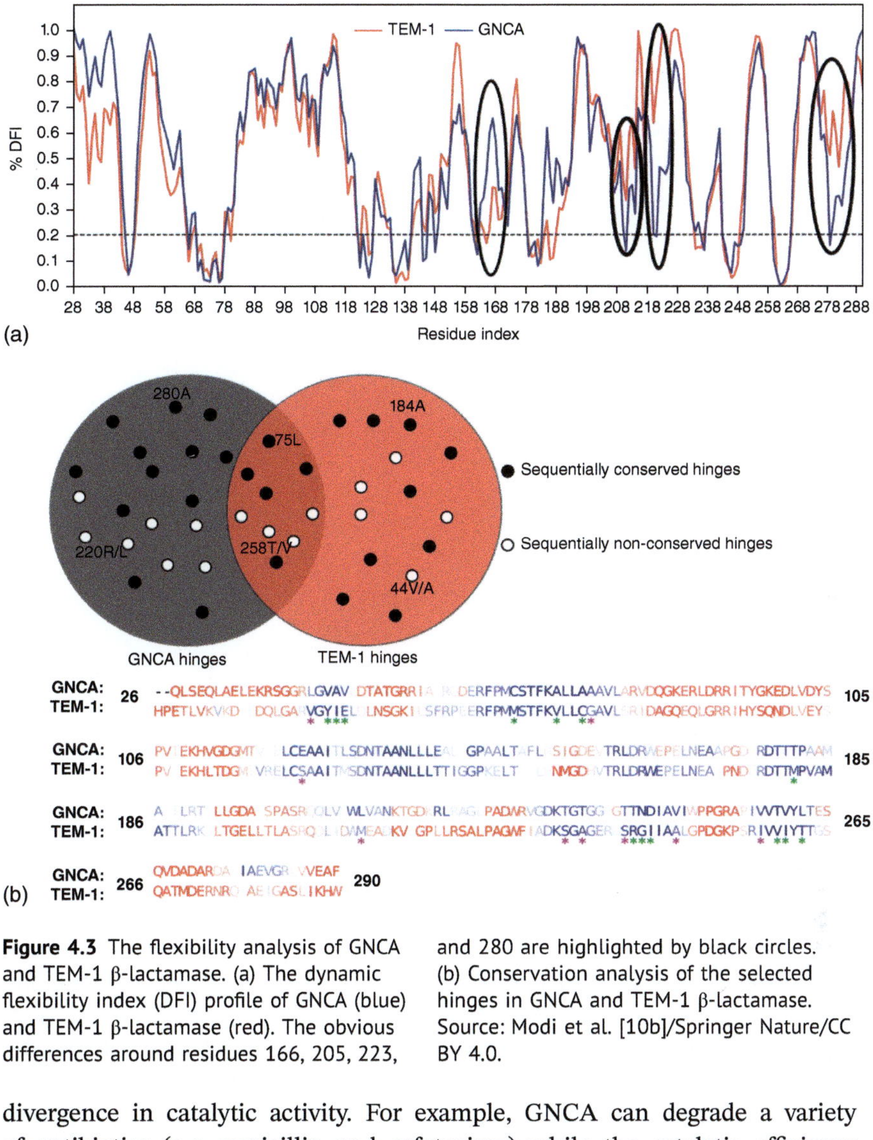

Figure 4.3 The flexibility analysis of GNCA and TEM-1 β-lactamase. (a) The dynamic flexibility index (DFI) profile of GNCA (blue) and TEM-1 β-lactamase (red). The obvious differences around residues 166, 205, 223, and 280 are highlighted by black circles. (b) Conservation analysis of the selected hinges in GNCA and TEM-1 β-lactamase. Source: Modi et al. [10b]/Springer Nature/CC BY 4.0.

divergence in catalytic activity. For example, GNCA can degrade a variety of antibiotics (e.g. penicillin and cefotaxime) while the catalytic efficiency is moderate. In contrast, TEM-1 β-lactamase only displays resistance toward penicillins, but the activity is about two magnitudes higher than that of GNCA [10]. Guided by a conformational dynamics computational approach, the same group and their collaborators recently continued engineering GNCA, aiming to tailor its substrate promiscuity and activity [10b]. The flexibility of hinge regions in the two enzymes was thoroughly analyzed by performing molecular dynamics simulations (MDs) in conjunction with the dynamic flexibility index (DFI) calculations (Figure 4.3).

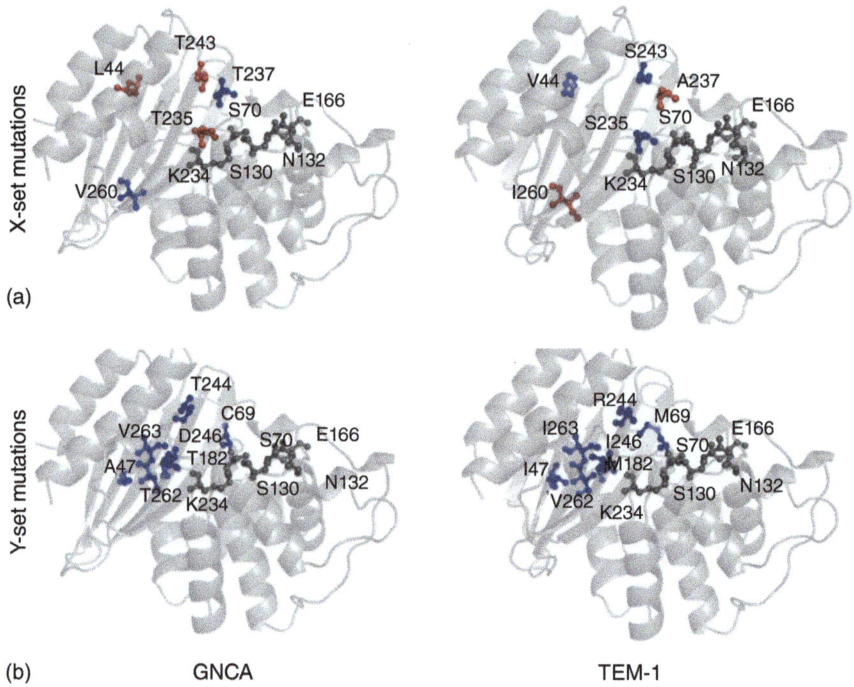

Figure 4.4 The substituting residues selected by flexibility profiles. (a) Non-common and sequentially non-conserved residues substituted in set X. (b) Common and sequentially non-conserved residues substituted in set Y. The substituting residues with low dynamic flexibility index (DFI) values are colored in blue, while the ones with high DFI are shown in red. Active site is shown in gray. Source: Modi et al. [10b]/Springer Nature/CC BY 4.0.

As a result of a consensus analysis (Figure 4.3b), the residues in the hinge regions were divided into two groups: non-common and sequentially non-conserved residues as X-set mutations (Figure 4.4a), and common and sequentially non-conserved residues as Y-set mutations (Figure 4.4b). By shifting the substituting residues from TEM-1 β-lactamase to GNCA, along with fine-tuning the dynamic allosteric interactions, the resultant GNCA variant bearing 21 substitutions showed a threefold increase in activity toward benzylpenicillin and a 10 000-fold decrease for degrading cefotaxime, a remarkable selectivity [10b].

The above representative example highlights the significant roles of allostery and conformational dynamics when tailoring enzyme properties in rational design procedures, which are also demonstrated by many other studies [11]. Other cases of rational design can be found in Table 5.2.

4.1.2 Directed Evolution

In general, application of any one of the gene mutagenesis techniques is likely to provide improved enzyme variants (see Chapter 3), but choosing the best method for a given study coupled with the optimal strategy on how to apply it ensures efficiency needed in practical (industrial) applications [12, 13]. The optimal choice(s) depends on the catalytic parameter(s) to be engineered. Thermostability and robustness in hostile organic solvents will be treated separately in Chapter 6. The focus here is on the directed evolution of stereo- and regio-selective enzymes as catalysts in organic chemistry and biotechnology.

Quite often, more than one enzyme parameter needs to be improved, and then two strategic options are possible [13]: (i) engineer the parameters separately in two sequential projects, or (ii) optimize both parameters simultaneously in a single project. For practical reasons, it is useful to choose robust enzymes for directed evolution projects, since protein stability promotes evolvability [14]. Consequently, when considering a not so robust enzyme due to ready availability or if patent reasons are involved, enhancing thermostability should precede the engineering of other parameters such as activity or selectivity (Scheme 4.4a). However, it may also be possible to improve both parameters simultaneously by designing appropriate experimental platforms. This can be achieved by heating microtiter plates at a given temperature for a defined length of time prior to high-throughput screening for activity (or selectivity). All variants of inferior thermostability would be denatured by such a heat treatment, and only those in which thermostabilizing mutations have been introduced will "survive" (Scheme 4.4b) [13]. Today, the simultaneous optimization of two parameters seems particularly appealing, but also challenging [15] (see Chapters 8 and 9).

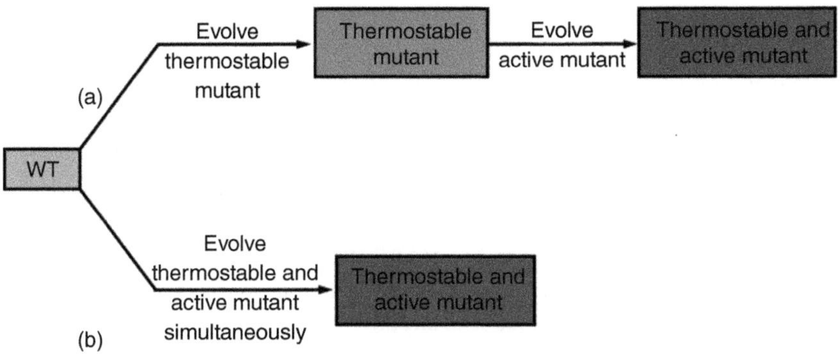

Scheme 4.4 Two choices when attempting to optimize thermostability and activity of an enzyme [13]. (a) Engineer thermostability first and then activity; (b) Engineer both thermostability and activity simultaneously. Source: [13]/John Wiley & Sons.

In other directed evolution endeavors, activity and/or enantioselectivity constitute the catalytic parameters of interest [15, 16]. In such cases, a given library may contain variants with the highest enantioselectivity (property A) but low activity (property B), or vice versa, variants with the highest activity but low enantioselectivity. It has been demonstrated that the very best hits with respect to one parameter (enantioselectivity) should not be used as templates in the subsequent cycle of mutagenesis/expression/screening when seeking higher activity as well as stereoselectivity, and vice versa [16]. Rather, a compromise is recommended in which several medium- or higher-quality hits, but not the very best ones in terms of stereoselectivity, are not discarded as often done in the past. Rather, they can be used as templates in the next directed evolution round, which focuses on the other catalytic parameter (property B, e.g. activity). This strategy is illustrated in Scheme 4.5 [16]. These kinds of non-discarded variants are reminiscent of neutral drift mutants [17], but the term "lateral hits" may be more appropriate [16]. A remote relationship to the Eigen/Schuster notion of quasi-species [18] may also be noted, as invoked in other directed evolution studies [19]. An opposing view is to evolve maximally high activity first, and then to tune stereoselectivity. Comparative studies are needed in order to make final conclusions [13].

Irrespective of the type of genetic method used in a directed evolution project, the crucial question concerning the choice of the optimal number of point mutations to occur in each mutagenesis/expression/screening cycle should be routinely addressed. Different opinions have been voiced on this important issue, and certainly successful examples of single-point mutations, but also of multiple point mutations, have been reported numerous times [12, 13]. In 2009, it was strongly recommended that climbing the hill in a fitness landscape is best achieved when only single mutations are allowed to accumulate one by one in evolutionary cycles [20]. However, this conclusion did not consider previous studies in which the simultaneous introduction of more than one-point mutation was shown to be highly efficient [21–23], with success being due to cooperative effects (more than traditional additivity) within a set of point mutations and between sets of mutations [24–27]. The occurrence of pronounced mutational cooperativity in directed evolution has been interpreted as a sign of efficacy [27].

Whenever an enzyme's substrate scope needs to be shifted or enlarged because a given compound of interest is not accepted (lack of notable activity), two different directed evolution strategies can be applied [13]:

- Proceed conventionally by applying CAST/ISM using the substrate of interest directly in mutagenesis/screening rounds, aided by structural, consensus sequence data and/or computational support.
- Apply the *in vitro* coevolutionary strategy initially developed for manipulating binding affinity [28] to directed evolution [29].

The *in vitro* coevolutionary strategy is a type of "substrate walking," in which a compound structurally more closely related to the natural substrate, but reacting slower, is first subjected to directed evolution for enhanced activity. This is followed by several similar iterative steps using structurally altered compounds

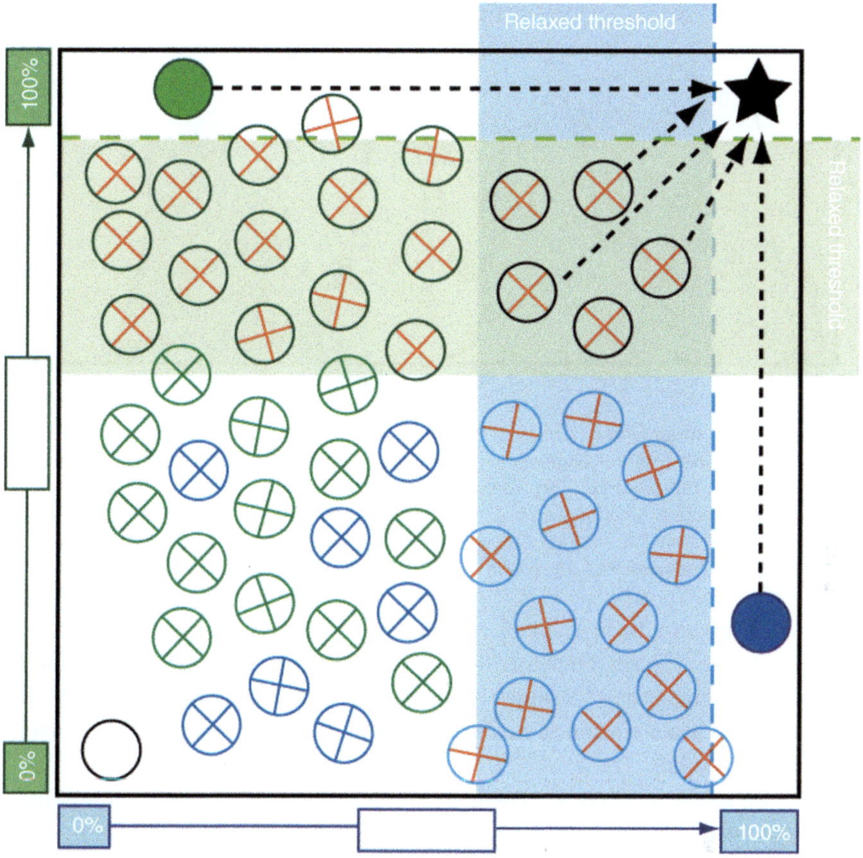

Scheme 4.5 Preferred approach for the simultaneous optimization of two catalyst properties A and B [13, 16]. Black star indicates the desired variant; blue and green dashed lines: stringent thresholds; blue and green rectangles: relaxed thresholds; blue and green filled circles: best mutant for property A and B, respectively, which are not used in further mutagenesis; red-crossed blue and green circles: variants with improved property A or B; red-crossed black circles: mutants with improved A and B property. Black dashed arrows: second round of mutagenesis. Source: Adapted from Refs. [13, 16].

until the final substrate of actual interest is targeted (Scheme 4.6). The latter option, although conceptionally intriguing, requires several steps that may not be necessary due to recent developments in dramatically improved directed evolution methods and strategies. However, should such advanced techniques fail, for whatever reason, the approach based on *in vitro* coevolution may be the best way to solve the problem.

A crucial issue concerns the choice of the mutagenesis method. In attempts to manipulate stereoselectivity, activity, and/or substrate scope, a variety of different molecular biological techniques such as epPCR, saturation mutagenesis, and

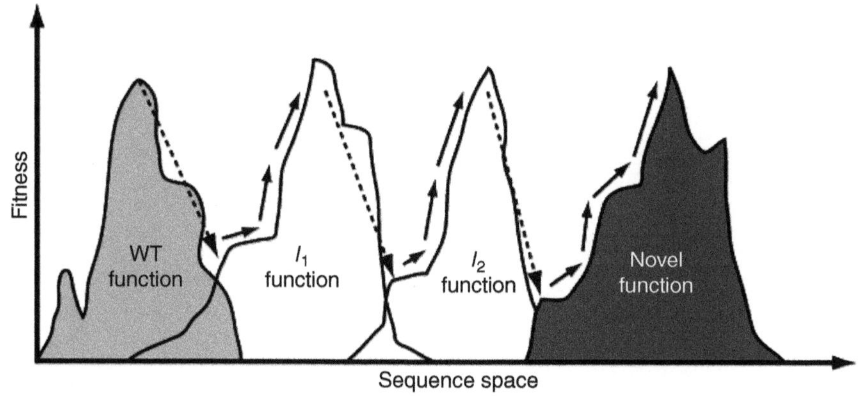

Scheme 4.6 The strategy of *in vitro* coevolution (substrate walking) for engineering novel protein functions [28, 29]. The wild-type (WT) protein function and the novel protein function are separated by an inactive region of sequence space, which may be filled by two intermediate functions (I_1 and I_2) that are amenable to conventional directed evolution. The arrows illustrate a potential evolutionary path leading to the novel protein function. Source: [28]/With permission of Elsevier.

DNA shuffling have been invoked. As already explained in Section 3.5, it took a number of years of research before the preferred strategy emerged: semi-rational saturation mutagenesis at residues lining or near the binding pocket [12, 13]. In Section 4.2, this conclusion is further analyzed by highlighting and analyzing comparative studies.

As emphasized in Chapter 3, whenever saturation mutagenesis is applied, at least a crude check on the quality of the mutant library is mandatory. In other words, the experimenter poses the question: Has the expected diversity been approximately achieved? This can be done fairly fast by the Quick Quality Control (QQC), in which a limited number of transformants are pooled and sequenced. *The cost of sequencing has been reduced drastically in recent years, which means that such an investment is worthwhile because it helps to prevent the screening of mutants that do not exist.* Alternatively, the Stewart method of generating quantitative, so-called *Q*-values is more exact but a little more labor-intensive (see Chapter 2).

4.2
Rare Cases of Comparative Directed Evolution Studies

4.2.1
Converting a Galactosidase into a Fucosidase

In an early comparative study, DNA shuffling [30] versus saturation mutagenesis [31] in the directed evolution of substrate acceptance was investigated, which

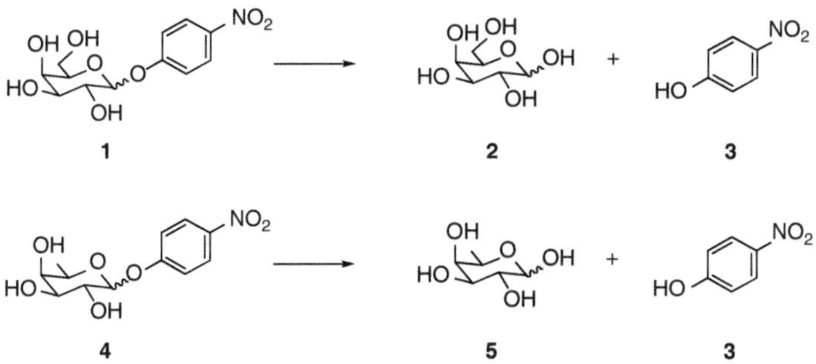

Scheme 4.7 Model reactions in the directed evolution of a fucosidase from a galactosidase. Source: Adapted from Refs. [30, 31].

involved the evolution of a fucosidase starting from a galactosidase. The model reactions utilized the hydrolysis of the “natural” synthetic substrate **1** → **2** + **3** catalyzed by *E. coli lacZ* ß-galactosidase (BGAL) and the respective reaction of the “nonnatural” substrate **4** → **5** + **3** to be catalyzed by an evolved variant (Scheme 4.7). The two substrates differ by a single hydroxyl group at position C6 ($-CH_2OH$ versus $-CH_3$), but otherwise the stereochemical features of the compounds are identical. In the reaction of substrate **4**, wild-type (WT) BGAL is a poor catalyst in a sluggish reaction, which the authors wanted to improve.

The steps involved in the earlier DNA shuffling study are summarized in Scheme 4.8 [30]. A total of seven DNA shuffling cycles were performed, each time about 10 000 colonies were screened using a crude but convenient on-plate color test for activity. About 2–5% of the transformants displayed enhanced activity as judged by the pretest, and typically 20–40 variants were then isolated and studied separately by kinetics.

The best mutant from the seventh and final round of shuffling showed a 66-fold increase in fucosidase activity and a reduction in galactosidase activity, but it still displayed a preference for the “natural” substrate **1** by a factor of 2.7 [30]. Thus, the desired reversal of substrate preference was not achieved. The best variant is characterized by eight-point mutations, only two of which are near the active site. Apparently, subtle effects are involved, which were not unambiguously uncovered. It has been pointed out that since deconvolution experiments of multi-mutational variants [27] were not performed, the conclusion that all eight-point mutations are really necessary for the change in substrate acceptance is not fully conclusive [13].

In the second study, saturation mutagenesis was employed rather than DNA shuffling, the purpose being to compare the two gene mutagenesis methods by employing the identical substrates, the same enzyme BGAL, and the previous on-plate screening assay [31]. Based on X-ray data of BGAL [32], saturation mutagenesis was applied at a site comprising three residues 201, 540, and 604, which bind sodium ions (Figure 4.5). It was speculated that simultaneous randomization at these positions as part of the binding site would provide variants that no longer coordinate Na^+, nor interact with the hydroxyl group at C6. In contrast,

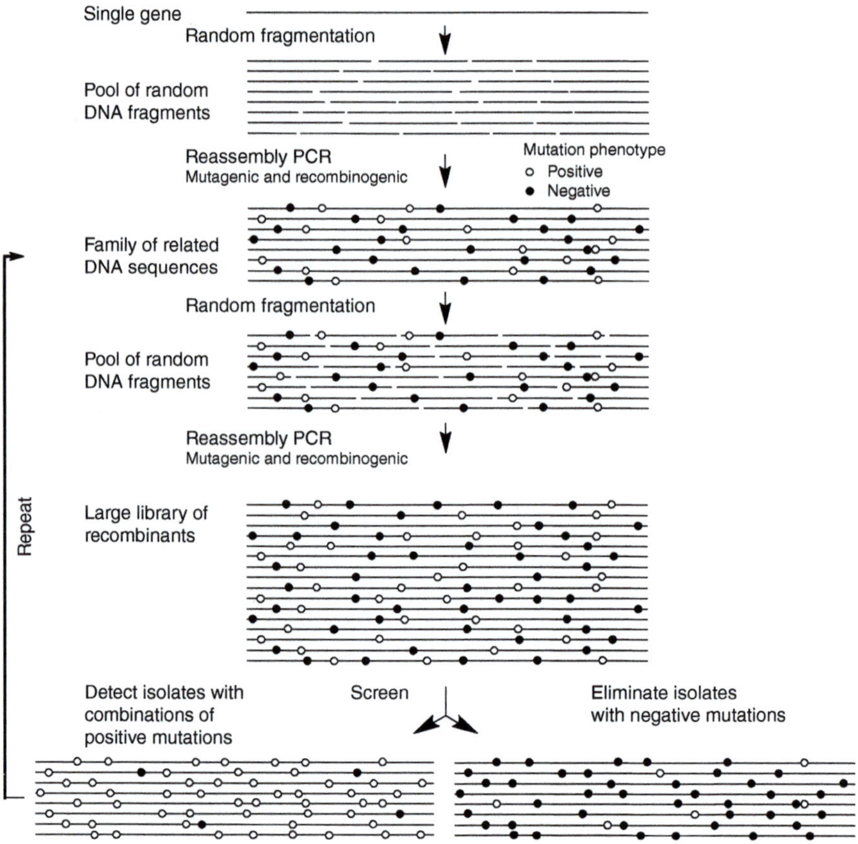

Scheme 4.8 DNA shuffling process used in the directed evolution of a fucosidase starting from a galactosidase. Source: [30]. Copyright (1997) National Academy of Sciences, U.S.A.

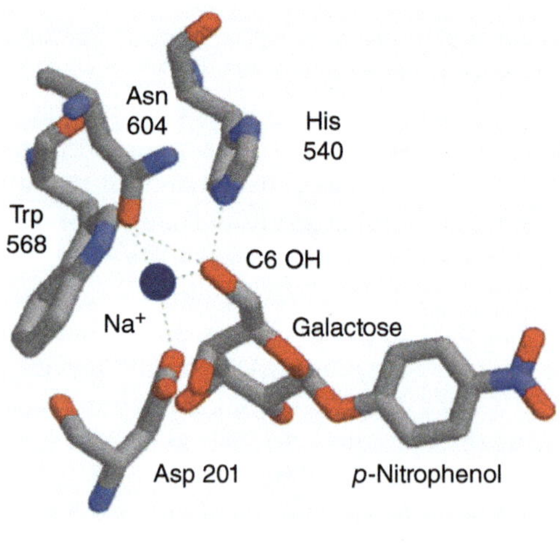

Figure 4.5 Structure of BGAL active site [32] used as a guide in designing saturation mutagenesis at amino acid positions 201, 540, and 604. Source: [31]/With permission of Elsevier.

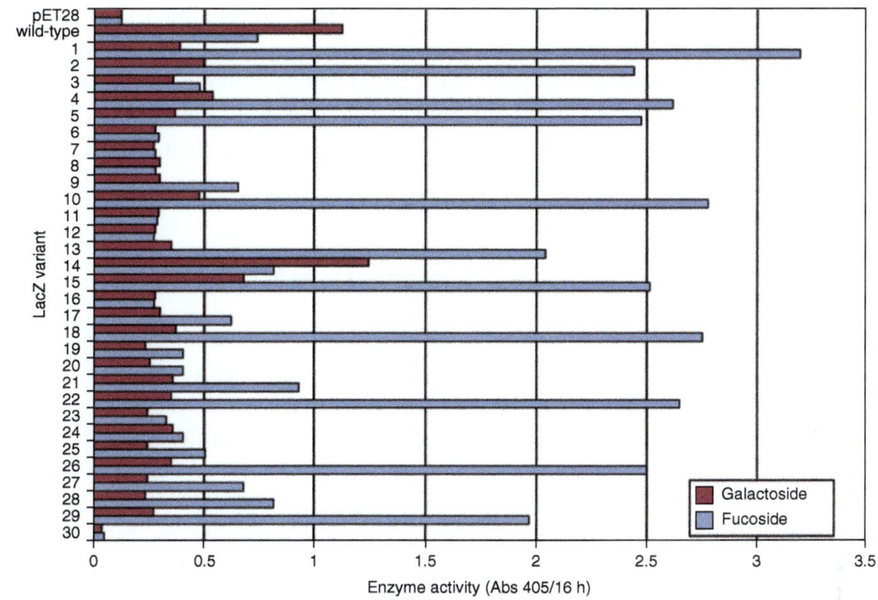

Figure 4.6 Selected BGAL variants resulting from saturation mutagenesis at a site composed of amino acid positions 201, 540, and 604. Source: [31]/With permission of Elsevier.

interaction directly with the methyl group of substrate **4** was expected, which should shift substrate selectivity in favor of the "fuco-substrate" **4** and less so for the "galacto-substrate" **1**.

Saturation mutagenesis was performed using NNK codon degeneracy encoding all 20 canonical amino acids, followed by screening with the on-plate pretest. In this way, about 10 000 transformants were assayed, leading to the discovery of a handful of active variants showing pronounced degrees of preference for substrate **4** (Figure 4.6) [31].

It was found that in the reaction of substrate **4**, the best variant, H540V/N604T, is considerably more active (180-fold increase in k_{cat}/K_M), while favoring its transformation relative to that of the "natural" synthetic substrate **1**. The observed 7000-fold switch is dramatic. Although the interpretation of the results on a molecular level was difficult because Asp201 was retained, the overall conclusion regarding the pros and cons of DNA shuffling versus saturation mutagenesis was convincing: *Focused library generation by structure-guided saturation mutagenesis is superior to DNA shuffling*. The generation and screening of a single saturation mutagenesis library (10 000 transformants) leading to these excellent results contrasts with the formation and screening of seven DNA shuffling libraries (total 70 000 transformants) with the evolution of notably inferior variants. Nevertheless, the authors were careful not to generalize their conclusion [31].

From today's perspective, an important aspect of this study concerns the issue of library coverage. A 3-residue site would require for 95% library coverage the screening of about 98 000 transformants (Table 3.3). Since only 10 000 were actually assayed, it is clear that full or nearly full library coverage is not necessary for obtaining acceptable results [31]. Even lower library coverage in saturation mutagenesis for enhanced stereoselectivity and activity may be sufficient in some cases [33]. However, the best variants will be missed by such a procedure. Nonetheless, for practical applications, the *n*th best hit may be acceptable [34].

4.2.2 Enhancing and Inverting the Enantioselectivity of the Lipase from *Pseudomonas aeruginosa* (PAL)

Comparing different mutagenesis techniques in directed evolution is best done over an extended period of time, since each of them undergoes improvements [12a]. The most systematically performed comparison of the different mutagenesis methods and strategies is based on a series of studies over a period of more than a decade [13]. The model reaction involves the hydrolytic kinetic resolution of *rac*-**6** with a slight preference for (*S*)-**7** ($E = 1.1$), catalyzed by mutants of the lipase from *Pseudomonas aeruginosa* (PAL) (Scheme 4.9) [26, 33–38]. As shown in the introductory Chapter 1, this system was used in the original proof-of-principle study, showing for the first time that directed evolution can be used to control the enantioselectivity of enzymes [35].

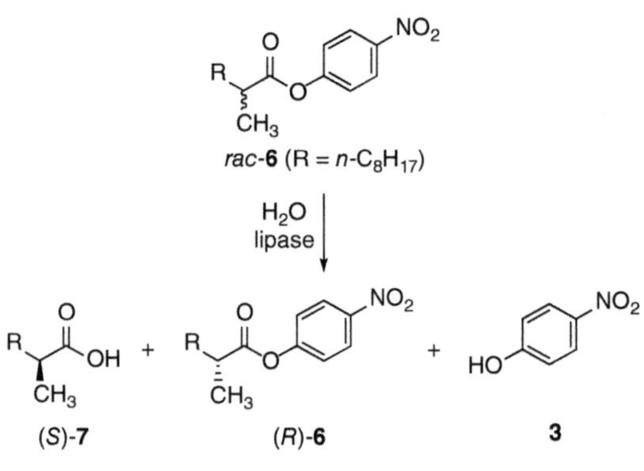

Scheme 4.9 Model reaction used in the directed evolution of the enzyme PAL. Source: Adapted from Refs. [26, 33, 35–38].

Initially, four cycles of epPCR at low mutation averaging one amino acid exchange per enzyme were transversed. In each round, a single mutation was introduced randomly, leading to the accumulation of the (*S*)-selective variant S149G/S155L/V47G/F259L characterized by a selectivity factor of $E = 11.3$ (see Scheme 1.11) [35]. Two further epPCR rounds failed to provide notably improved variants, with *E*-values of only 13–15 being achieved [39]. It is a clear indication that recursive epPCR is not ideally suited for enhancing stereoselectivity, certainly not in this particular enzyme system. Moreover, most of the hot spots appeared to be remote from the active site as judged by a homology model (the crystal structure of PAL was not published until later). Therefore, methodology development was initiated by exploring other strategies.

At the time, it was assumed that the four observed mutations S149G/S155L/V47G/F259L occurred at hot spots, but that the newly introduced amino acids might not be optimal. Consequently, saturation mutagenesis was applied at these positions [36]. In principle, wild-type (WT) or first-, second-, third-, or fourth-generation mutants can be used as templates. In some cases, this provided variants displaying higher enantioselectivity, but in others, no improvements were detected. For example, using the third generation mutant S149G/S155L/V47G as the template, saturation mutagenesis using NNK codon degeneracy encoding all 20 canonical amino acids was performed at position 155. Variant S149G/S155F/V47G was identified in which leucine was exchanged for phenylalanine, leading to a notable increase in enantioselectivity ($E \approx 20$). Randomization at this position using other templates likewise led to the same exchange S155F. However, focusing on other positions failed to provide better variants. Therefore, alternating saturation mutagenesis at different positions and epPCR were tested, with the best variant V47G/V55G/S149G/S155F/S164G displaying a selectivity factor of $E = 26$ (Scheme 4.10) [36]. This can be viewed as a forerunner of ISM (for a detailed discussion of ISM, Section 3.5).

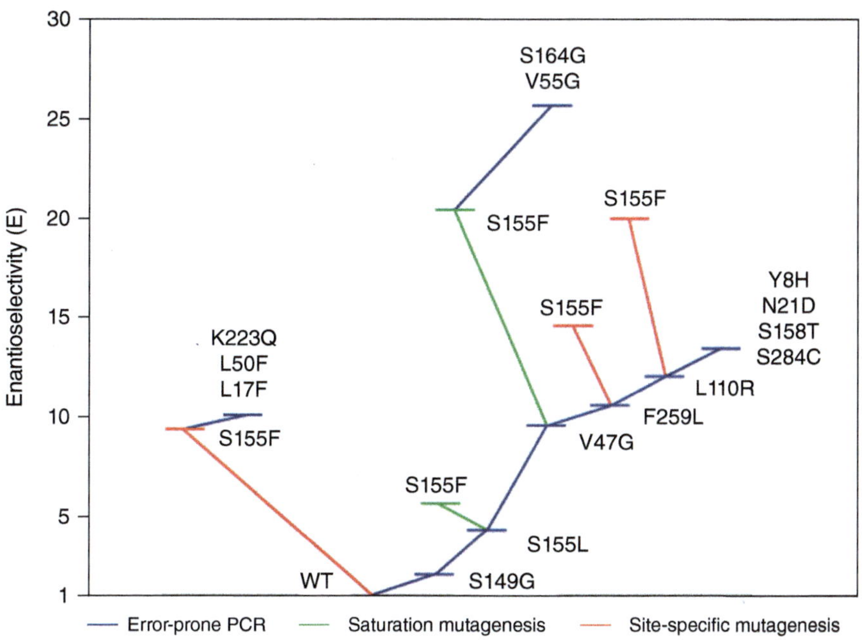

Scheme 4.10 Alternating saturation mutagenesis at different positions with epPCR in the quest to enhance the enantioselectivity of PAL in the model reaction *rac*-**6** → (*S*)-**7** + **3**. Source: Ref. [36]/With permission of Elsevier.

While these strategies ensured partial success, the observed enantioselectivities were not high enough for practical applications. Therefore, other approaches were tested, the first involving saturation mutagenesis (NNK codon degeneracy) at a site comprising four amino acid positions 160–163 at the binding pocket of PAL [33] (see Scheme 1.11). This experiment was guided by the PAL crystal structure [40] *and constitutes the first example of focused saturation mutagenesis at a site lining the binding pocket of an enzyme with the aim of enhancing stereoselectivity*. It was essentially an attempt to reshape the binding pocket of an enzyme, keeping Emil Fischer's lock-and-key hypothesis in mind. After screening only 5000 transformants, the quadruple mutant G160A/S161D/L162G/N163F showing a selectivity factor of $E = 30$ was identified [33]. In those early days, oversampling statistics were not considered, but subsequently it became clear that for 95% library coverage, about 3×10^6 transformants should have been screened, far beyond practical possibilities. The result was later interpreted as an indication that aiming for full library coverage in saturation mutagenesis is not absolutely necessary [13].

In the same study [33], another saturation mutagenesis experiment likewise led to notable success. A 2-residue site Ser155/Leu162 was randomized, again using NNK codon degeneracy, which provided two variants (labeled VIII and IX) of similar enantioselectivity, $E = 34$ and $E = 30$, respectively, in favor of (*S*)-**7**. These positive results did not lead to the realization that saturation mutagenesis at sites lining the binding pocket is superior to epPCR and DNA shuffling. Fortunately, this conclusion was finally made several years later with the systematization of saturation mutagenesis at sites lining the binding pocket in the form of the CAST [38] and the emergence of ISM [21] (see Section 3.5).

In the PAL study describing saturation mutagenesis at the 2- and 4-residue sites, a second strategy was then tested in which the genes of two previous mutants (labeled IV and V) were subjected to DNA shuffling [33]. In order to increase diversity in the DNA shuffling experiment, a modified form called Combinatorial Multiple-Cassette Mutagenesis (CMCM) [41] was applied (for a detailed description, see Section 3.5). Accordingly, a mutagenic oligocassette was included in the shuffling, which allows parallel randomization at positions 155 and 162 (Scheme 4.11) [33]. This procedure provided an enzyme (PAL variant

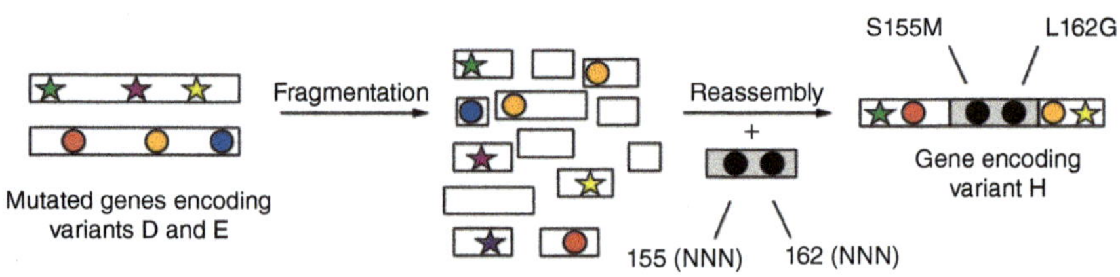

Scheme 4.11 Extended CMCM in the evolution of an (*S*)-selective variant X in the hydrolytic kinetic resolution of *rac*-**6** [33]. Green star: position 20; purple star: position 161; yellow star: position 234; red circle: position 53; orange circle: position 180; blue circle: position 272. Source: Ref. [33]/John Wiley & Sons.

X) characterized by six-point mutations D20N/S53P/S155M/L162G/180I/T234S and the highest enantioselectivity ($E = 51$) in the model reaction observed up to that point (Scheme 4.11) [33]. Interestingly, only L162G was found to be next to the binding pocket, the other five-point mutations being more or less remote.

On the basis of these exploratory experiments, it was provisionally concluded that the combination of epPCR and DNA shuffling is the optimal strategy for evolving stereoselective enzymes. Indeed, at the time other groups joined efforts in generalizing directed evolution of stereoselectivity using other types of enzymes along similar lines as summarized in an early review [37]. However, a subsequent mechanistic and QM/MM study suggested that only two of the six-point mutations in the best variant X are necessary for high enantioselectivity [42]. Indeed, the predicted double mutant S53P/L162G was generated and shown to have a selectivity factor of $E = 63$ in the model reaction, demonstrating that four of the six-point mutations are superfluous and actually reduce stereoselectivity [42]. This was a clear signal that the strategy, which includes epPCR and DNA shuffling is successful but not overly efficient [13]. A total of 50 000 transformants had to be screened in all of these experiments, as summarized in Scheme 4.12. Similar screening efforts were made in order to invert stereoselectivity favoring (*R*)-**7** [43].

Following two crucial advances in developing more efficient saturation mutagenesis, namely the structure-guided use of reduced amino acid alphabets [23] and ISM [21, 22] (see Section 3.5), the PAL model system was re-visited in order to allow for a more rigorous comparison of mutagenesis methods and strategies [26]. The same model of reaction was employed. This crystal structure of PAL [40] served as a guide, together with ISM as the mutagenetic strategy [26].

It is known that lipases have "two" binding pockets, one harboring the acid and the other the alcohol part of the ester. Therefore, it was decided to focus saturation mutagenesis on the respective CAST sites. Since the chiral center in substrate *rac*-**6** is in the acid part of the molecule with catalytically active S82 being between the two segments, six residues were judged as being important for mutagenesis, namely M16, L17, L159, L162, L231, and V232, all surrounding the acid part of the binding pocket (Figure 4.7) [26]. The distance values of the α-C-atom of all these six residues to the α-C-atom of the acid moiety (C4) in *rac*-**6** were estimated to be: M16 (5.3 Å); L17 (8.4 Å); L159 (8.5 Å); L162 (6.5 Å); L231 (7.0 Å); and V232 (7.3 Å).

How to group the six chosen CAST residues into appropriately sized multi-residue randomization sites, if at all, was the next task, in addition to choosing the right reduced amino acid alphabet for minimizing the screening effort (see Chapters 2 and 3). These strategic questions are addressed in a general way in Section 4.3. In the PAL study, three 2-residue randomization sites were defined, namely A (M16/L17), B (L159/L162), and C (L231/V232) (Figure 4.7). Each library would require the screening of about 3000 transformants for 95% coverage, which adds up to 9000. This is much more than the screening effort when generating and screening six separate single-residue libraries, but this conclusion is reversed in the subsequent iterative steps if the complete ISM

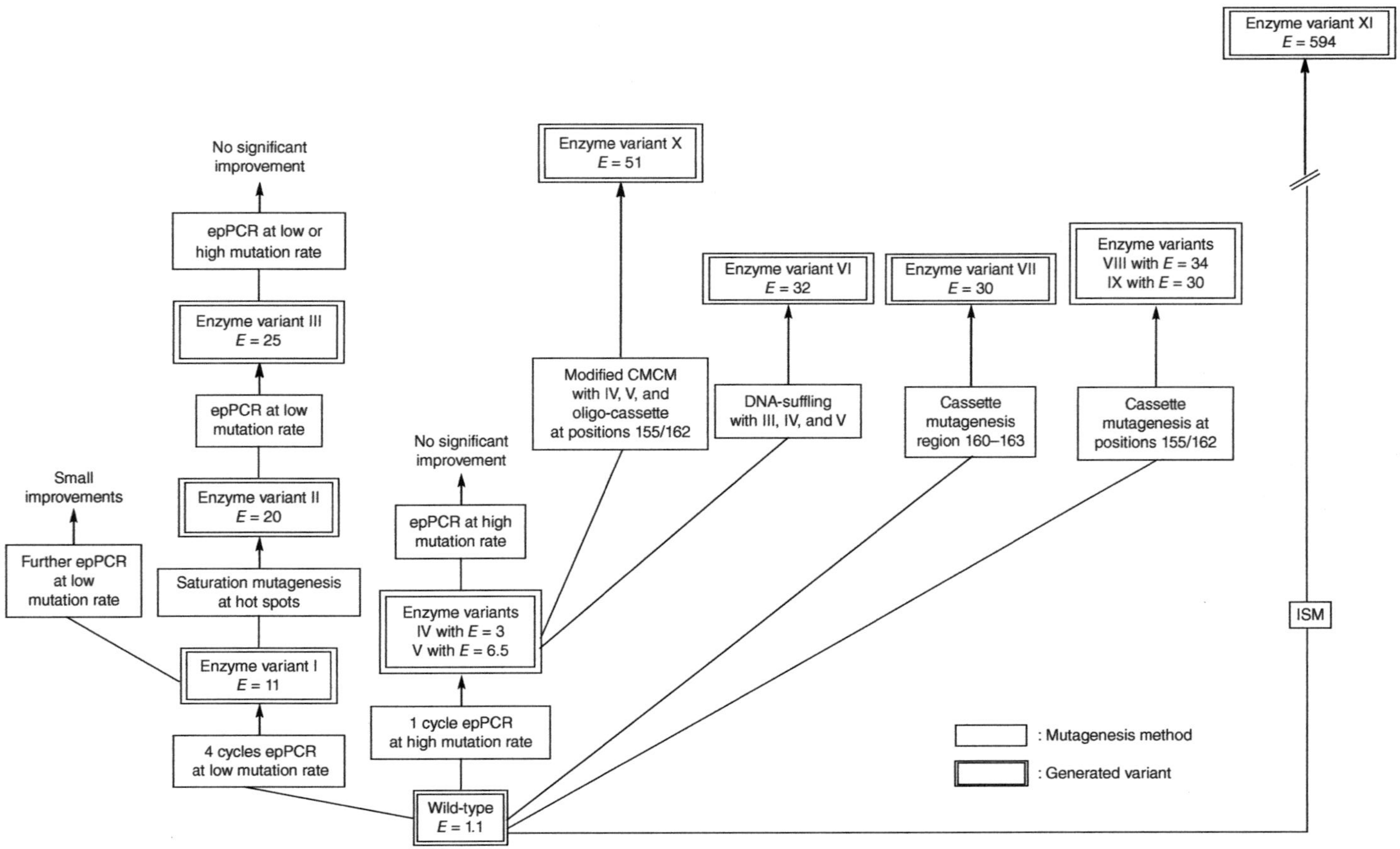

Scheme 4.12 Summary of all comparative studies of PAL as a catalyst in the hydrolytic kinetic resolution of *rac*-**6**, including the result of the final study based on ISM (far right) [26] in a 12-year effort. Source: Adapted from Ref. [26].

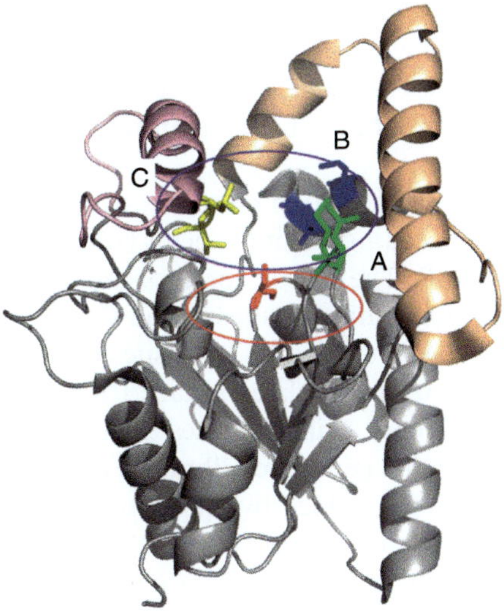

Figure 4.7 Schematic representation of amino acid residues considered for saturation mutagenesis [26], based on the X-ray structure of WT-PAL [40]: sites A (Met16/Leu17, green), B (Leu159/Leu162, blue), and C (Leu231/Val232, yellow) around the active site Ser82 (stick representation in red) in the acid-binding pocket (purple circle). The red circle marks the alcohol-binding pocket, in the case at hand, harboring the *p*-nitrophenyl moiety of *rac*-**6**. At the top of the picture, helix and loop in wheat (right, Asp113-Leu156) and light pink (left, Pro203-Asn228) represent lid 1 and lid 2, respectively. Source: Adapted from [26, 40].

Table 4.2 Statistical consequences as a function of grouping single CAST residues into randomization sites.

Type of grouping	No. of pathways in complete ISM scheme	No. of libraries in complete ISM scheme	Total number of transformants screened in complete ISM scheme for 95% coverage	
			Using NNK codon	Using NDT or DNT codon
Six single-residue sites	720	1956	183 864 (94 per lib.)	66 504 (34 per lib.)
Three double-residue sites	6	15	45 990 (3066 per lib.)	6450 (430 per lib.)
Two triple-residue sites	2	4	392 652 (98 163 per lib.)	20 700 (5 175 per lib.)
One six-residue site	1	1	3.21×10^9	8.95×10^6

Source: Ref. [26]/American Chemical Society.

scheme is fully investigated (Table 4.2) [26]. Importantly, randomization at sites comprising two or three residues can lead to double or triple mutants, respectively, in which the respective point mutations may interact with one another cooperatively (more than additively) [27].

In the case of NDT codon degeneracy encoding 12 amino acids (Phe, Leu, Ile, Val, Tyr, His, Asn, Asp, Cys, Arg, Ser, and Gly), only 430 transformants would have to be screened (a total of ≈1300 for all three libraries). Experimentally, libraries A, B, and C were first produced using NNK codon degeneracy, and in each case ≈3000 transformants were screened. In libraries A and C, slightly improved variants were discovered, but the best hit originated in library B: variant 1F8 was characterized by a single-point mutation L162N with $E = 8$ in favor of (*S*)-**7**. At this point, a decision had to be made on how to proceed in the upward climb. A complete 3-site ISM scheme involves 6 evolutionary pathways and a total of 15 saturation mutagenesis libraries (Scheme 3.9b). Experience has shown that it is not necessary to explore all upward pathways in an ISM study, but some are more productive than others [44].

Mutant 1F8 was then used as the template for saturation mutagenesis at sites A and C employing DNT codon degeneracy encoding 11 amino acids (Ala, Asn, Asp, Cys, Gly, Ile, Phe, Ser, Thr, Tyr, and Val). The reason for choosing DNT rather than NDT for second generation randomization at sites A and C in this particular case is simple: DNT does not encode leucine, and both sites already harbor leucine, thereby reducing the amount of template mutant ("starting mutant," in this case 1F8) appearing in the library. This increases the library's quality somewhat. Upon screening the two second generation libraries, a highly active and stereoselective variant was identified in which two-point mutations accumulated, M16A/L17F, leading to the final triple mutant M16A/L17F/L162N (enzyme variant XI) characterized by a selectivity factor of $E = 594$ (Scheme 4.12) [26]. Due to the dramatically improved catalytic profile, further ISM at site C was not necessary. The triple mutant is also a good catalyst for the hydrolytic kinetic resolution of other chiral esters. The study includes kinetics, molecular dynamics (MD) computations, and deconvolution experiments, which uncover unusually large cooperative effects existing between the second mutational set (M16A/L17F) and the initial point mutation (L162N) (see Chapter 8 for further details).

Less than 10 000 transformants were screened using an on-plate pretest for activity followed by automated gas chromatography (GC) analysis of the active hits. This is much less than the invested effort in the older approach based on epPCR and DNA shuffling (50 000 transformants). At the time of the ISM study [26], the full significance of using reduced amino acid alphabets in the quest to reduce the screening effort was not as clear as it currently is. Today, a researcher would not select NNK codon degeneracy for the initial libraries, but NDT or DNT codon degeneracy, which encode balanced sets of polar/nonpolar, charged/non-charged, and hydrophobic/hydrophilic amino acids [13]. In such a scenario, less than 1900 transformants would have to be screened for optimizing PAL. As highlighted in Section 4.3, even smaller amino acid alphabets can be considered, provided structural data is available when designing the mutant libraries.

Up to this point in the analysis, all other comparative studies [12, 31], although not as extensive as the PAL investigations, show that epPCR and DNA shuffling can in principle be successful to some extent when wanting to enhance or invert stereoselectivity, rate, and substrate scope of enzymes. These are the most important catalytic parameters when applying enzymes as catalysts in organic chemistry and/or biotechnology [45, 46]. *But today, when aiming to manipulate selectivity and/or activity, the superiority of semi-rational saturation mutagenesis at sites lining the binding pocket over epPCR and DNA shuffling is evident as indicated by many studies, and it was even suspected years ago* [12a, e, g, h, 13, 45, 47].

Today, saturation mutagenesis at sites lining the enzyme binding pocket is being used by many groups [48]. As highlighted in Section 4.3, *choosing the best strategy for applying this technique is likewise of prime importance.*

4.3 Choosing the Best Strategy When Applying Saturation Mutagenesis

4.3.1 General Guidelines

Once the choice has been made in favor of saturation mutagenesis in a new protein engineering project focusing on stereo- and/or regio-selectivity, activity, or substrate scope, quite different strategies for applying this form of directed evolution are possible. The optimal approaches have emerged and continue to undergo further improvements. Some tips on how to apply saturation mutagenesis optimally have already been outlined in Section 3.5 and in this chapter, including the use of statistical metrics for estimating oversampling in relation to %-library coverage. These tips revolve around the challenge of maximizing library quality while minimizing the screening effort (a bottleneck of directed evolution in general). If an enzyme under study has not been characterized by an X-ray structure needed to define CAST sites, then a homology model must be resorted to, which usually works quite well in spite of introducing some uncertainty. If no structural data of any kind is available, which is rare, epPCR, DNA shuffling, or a mutator strain should be chosen, possibly followed by saturation mutagenesis at the identified "hot spots."

In older studies based on saturation mutagenesis, NNK, or NNS codon degeneracy encoding all 20 canonical amino acids was used, e.g. when targeting stereoselectivity [13, 21]. A crucial development for increasing the efficacy of this protein engineering technique was the utilization of reduced amino acid alphabets [23, 45], which lower structural diversity but provide a simple technique for reducing the degree of oversampling in the screening step [25]. Saturation mutagenesis at a multi-residue site can be performed in three fundamentally different ways [13], each resulting in vastly different screening efforts. These include:

- Conventional NNK (or NNS) codon degeneracy encoding all 20 canonical amino acids at all individual positions of a multi-residue randomization site in separate saturation mutagenesis experiments.
- A rationally chosen reduced amino acid alphabet at all positions of a multi-residue randomization site in a single procedure.
- A different codon degeneracy at each position of a multi-residue site in a single saturation mutagenesis experiment.

When using reduced amino acid alphabets, two fundamentally different strategies are possible [48a, b, d], which is part of the systematization of saturation mutagenesis (see Section 4.3.3 for details). Strategy 1 constitutes the traditional

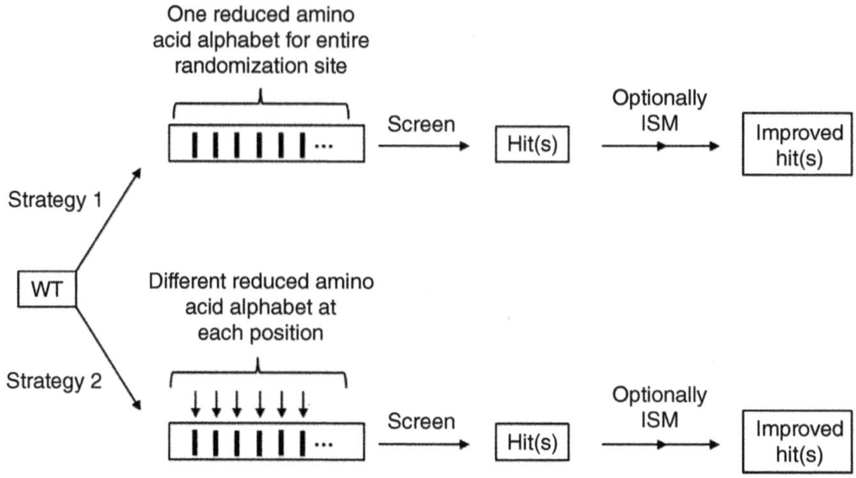

Scheme 4.13 Two different approaches to the use of reduced amino acid alphabets in saturation mutagenesis, if necessary, followed ISM. Source: Adapted from Refs. [12a, 23, 25, 48a, d].

way to perform saturation mutagenesis at a multi-residue randomization site, namely a defined codon degeneracy corresponding to the chosen amino acid alphabet. Strategy 2 involves a different codon degeneracy encoding a chosen reduced amino acid alphabet at each residue of a multi-residue randomization site also in a single saturation mutagenesis experiment (Scheme 4.13).

Here, two representative examples of the traditional approach are analyzed, in these cases based on NNK codon degeneracy without the use of a reduced amino acid alphabet. Shortcomings will be pointed out. In a study concerning the directed evolution of a stereoselective transaminase, ISM [21, 22] played a crucial role, but DNA shuffling and site-specific mutagenesis were also employed [29]. The goal was to increase the activity of the transaminase ATA-117 as a catalyst in the key step of the industrial production of sitagliptin phosphate (**10**), a therapeutic drug in the treatment of diabetes (Scheme 4.14).

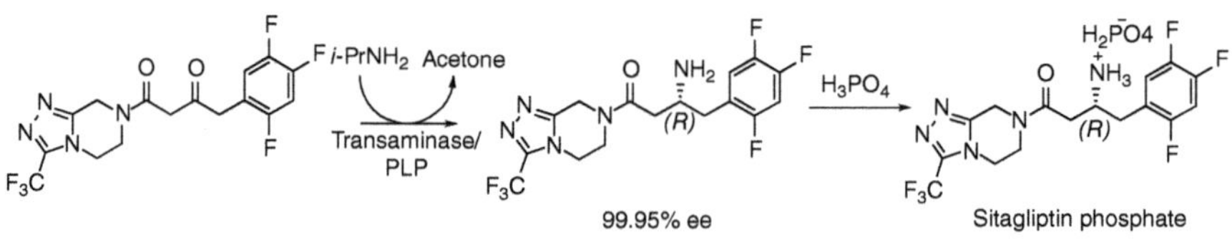

8 9 10

Scheme 4.14 Biocatalytic route to sitagliptin phosphate using a transaminase evolved by applying ISM, epPCR, and DNA shuffling. Source: Adapted from Ref. [29].

The enzyme ATA-117 is a homolog of the structurally well characterized transaminase from *Arthrobacter* sp., which is known to be (*R*)-selective in the reductive amination of methyl ketones and small cyclic ketones [49]. At the time, industrial researchers refrained from using the "real" substrate **8** in mutagenesis/screening experiments [29]. Instead, they first resorted to Zhao's *in vitro* coevolution concept according to Scheme 4.15 [28]. Substrate walking was performed by first focusing protein engineering on the truncated methyl ketone **11** as a substrate, with isopropylamine (**12**) serving as the amine-donor (Scheme 4.15). WT ATA-117 is a poor catalyst in this transformation, but nevertheless provided 4% conversion at 2 g l^{-1} substrate concentration.

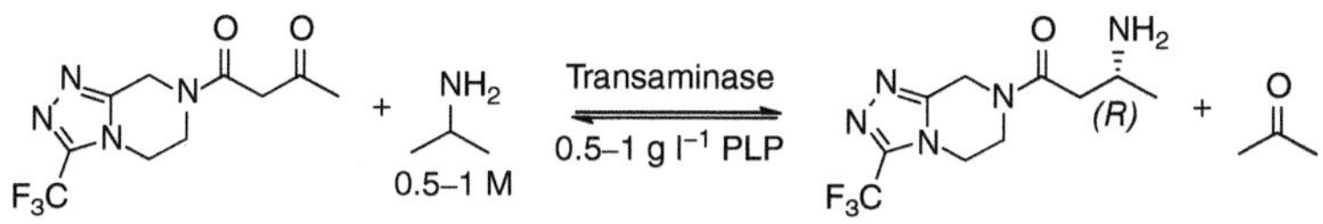

11 12 13 14

Scheme 4.15 Model compound (**11**) used in substrate walking based on *in vitro* coevolution. Source: Adapted from Ref. [29].

Using a homology model of ATA-117, docking computations were performed, which allowed reasonable choices of randomization sites lining the binding pocket (CAST sites, although not designated as such). Although extensive details were not reported, this procedure provided variant S223P with an 11-fold increase in activity in the model reaction of ketone **11** [29]. This mutant was then used as a template for ISM experiments, this time using the "real" substrate **8**. Docking experiments indicated that the trifluoromethyl group in the substrate could interact with residues V69, F122, T283, and A284. Four saturation mutagenesis libraries were created individually at these four positions, in addition to the generation of a combinatorial library using several residues simultaneously. Whereas randomization at the single residues failed to provide improved variants, the combinatorial library led to an active variant characterized by four-point mutations lining "small" and "large" parts of the binding pocket. Double mutants F122I/V69G, F122I/A284G, F122V/V69G, F122V/A284G, F122L/V69G, and F122L/A284G proved to be hits, all containing the parent mutation S223P. Activity was still quite low, but in the absence of point mutation S223P, no activity whatsoever was observed, as demonstrated by a deconvolution experiment. Full or partial deconvolution of multi-mutational variants constitutes an excellent way to identify cooperative (more than additive) mutations and/or deleterious amino acid exchange events [27]. The value of substrate walking as part of the *in vitro* coevolution approach is clear. However, the inclusion of position 223 in an extended combinatorial library was not explored, which may have made substrate walking superfluous.

Following these findings, the most active variant was then used as the parent for the next round of ISM (although this acronym was not used), and the beneficial mutations from the small-pocket and large-pocket saturation mutagenesis libraries were combined into a new library, which led to a variant having

12-point mutations and showing a 75-fold increase in activity. Unfortunately, nothing is mentioned regarding enantioselectivity, but it can be assumed that the researchers observed (*R*)-selectivity in all cases in favor of the desired amine **9**. Although the results at this stage are impressive, a number of practical problems still have to be solved, e.g. finding reaction conditions under which the substrate is soluble, increasing substrate and amine donor concentration, solvent tuning, pH optimization, as well as enhancing enzyme performance under operating conditions. Therefore, 11 further rounds of mutagenesis/screening were performed using DNA shuffling, epPCR-based random mutagenesis, rational design, and saturation mutagenesis at second-sphere sites from the binding pocket, all of this being done in parallel to process development. In some cases, mutagenesis cycles were guided by ProSAR (see Section 3.9.2). A total of 36 480 transformants were assayed using an LC/MS/MS screen in the early phase (low activity variants) and an automated achiral high-performance liquid chromatography (HPLC) system for all other transformants (≈2 minutes per sample), followed by chiral HPLC of the best hits. The best variant was reported to have 27 mutations [29]. In 50% dimethylsulfoxide (DMSO), it converts 200 g l^{-1} of the prositagliptin ketone **8** to sitagliptin (**9**) with >99.95% ee (*R*). One of the best mutants was also reported to be active and enantioselective in the reductive amination of several other structurally different ketones [29].

The overall results of this study regarding the catalytic performance of the best ATA-117 variant under operating conditions are impressive. However, since many experimental details are lacking, it is difficult to assess the actual efficiency of the reported mutagenesis strategy [13]. Was the order of the mutagenesis cycles in the multi-step process planned the way it was presented? If so, why was the particular order of mutagenesis events chosen? Or were local minima in the upward climb encountered (not reported), calling for different mutagenesis techniques? Thus, it is difficult to learn from this study. With the most recent emergence of improved ISM techniques based on large randomization sites in combination with designed reduced amino acid alphabets [12a], a saturation mutagenesis approach without the need to shuffle or to invoke site-specific mutagenesis may be an attractive alternative, certainly in future studies. Nevertheless, one of the interesting observations made in this study is the fact that saturation mutagenesis at single residue sites lining the binding pocket does not always result in the discovery of improved mutants (local minima?), whereas grouping them into larger randomization sites, which are then subjected to saturation mutagenesis seems to be the better strategy [29].

Prior to the above study, the actual first time that ISM was applied to an enzyme, in this case for boosting its enantioselectivity, concerned the hydrolytic kinetic resolution of epoxide *rac*-**15** with the formation of (*S*)-**16** [21]. The epoxide hydrolase from *Aspergillus niger* (ANEH) served as the biocatalyst. The successes and pitfalls of this early report and subsequent studies of the model reaction are analyzed here. WT ANEH was known to be slightly (*S*)-selective ($E = 4.6$) (Scheme 4.16).

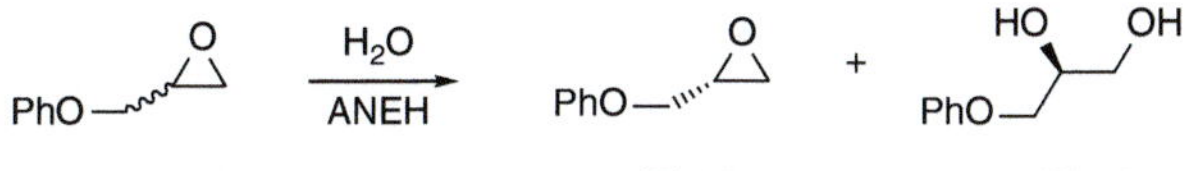

Scheme 4.16 Model hydrolytic kinetic resolution of the glycidyl ether *rac*-**15** catalyzed by ANEH. Source: Ref. [21]/John Wiley & Sons.

Guided by the crystal structure of WT ANEH [50], 15 amino acid positions lining the tunnel-like binding pocket were identified. Rather than generating 15 single residue randomization libraries at these CAST positions, they were grouped into six randomization sites comprising two or three residues: A (positions 193/195/196), B (215/217/219), C (329/330), D (349/350), E (317/318), and F (244/245/249) (Figure 4.8). All six initial libraries A–F were generated using NNK codon degeneracy and screened for enhanced enantioselectivity in the model reaction, with the best variant (L215F/A217D/R219S) originating from library B. This mutant was then used as the template in the otherwise arbitrarily chosen ISM pathway WT → B → C → D → F → E leading to the best variant LW202, characterized by 9-point mutations and showing notably improved enantioselectivity ($E = 115$) [21]. Due to the already high stereoselectivity, the last site A in the series was not visited. A total of about 20 000 transformants were screened using the Mülheim multiplexing MS-based high-throughput

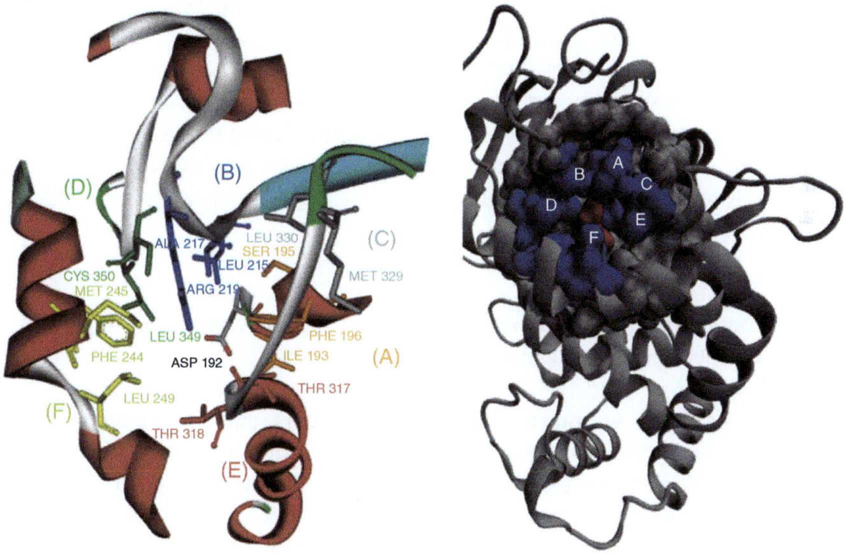

Figure 4.8 CAST sites A–E [21] of the epoxide hydrolase from *Aspergillus niger* (ANEH) chosen on the basis of the X-ray crystal structure of the WT [50]. Left: Defined randomization sites A (orange), B (blue), C (gray), D (green), and E (yellow); Right: Top view of the tunnel-like ANEH binding pocket showing sites A–E (blue) and the catalytically active Asp192 (red). Source: Adapted from [50].

ee-screening system [51] (see Chapter 2). Limited alternative ISM exploration was performed, with pathway WT → B → C → D → E leading to a variant with a selectivity factor of $E = 49$ [8b], but continuing to site A or F failed to produce improved variants. Apparently, a local minimum in the fitness pathway landscape was encountered, as we know today. It took several years of research to find a way out of such dead ends (see below), and to improve the ISM-based strategy so that the screening work is typically reduced to 2000–4000 transformants. With recent methodology developments [12a], in the most directed evolution studies considerably less mutants need to be screened.

While this study was the first to introduce ISM for manipulating the catalytic profile of an enzyme, several methodological aspects were neglected [21]. Firstly, a statistical analysis regarding screening/oversampling was not performed. Today, this fast and routine procedure is highly recommended [12a, 23]. Later calculations showed that 95% library coverage was not ensured in this study, again demonstrating that less screening can still be successful in finding improved variants (although missing possibly better variants). Secondly, no attempt was made to escape from a local minimum on the fitness landscape. Thirdly, further ISM pathways were not considered. Fourthly, library quality was not checked by the QQC [16] nor by the more quantitative *Q*-values [52], which were introduced a few years later. Therefore, one may consider the evolution of the best variant, LW202, to be a result of fortune. Subsequent and likewise successful ISM studies [12c, 26], which also rely on arbitrarily chosen pathways, speak against this hypothesis. Dozens of other successful ISM-based studies have appeared, as summarized in reviews [12a, 45, 52]; see also Chapter 5. In the case of B-FIT based thermostabilization of a lipase, highly improved variants were evolved by choosing arbitrary ISM pathways [22]. Nevertheless, some uncertainty remains. Even with today's most sophisticated theoretical QM/MM techniques, it is impossible to predict the optimal upward pathway in an ISM scheme. Nevertheless, some simple and practical guidelines have emerged, as noted in Section 4.3.2.

4.3.2 Choosing Optimal Pathways in Iterative Saturation Mutagenesis (ISM) and Escaping from Local Minima in Fitness Landscapes

Some ISM pathways are more productive than others, and some may be characterized by local minima, meaning the absence of any improved variants in a given library. Such "dead ends" are not restricted to ISM, they are a universal phenomenon when applying other gene mutagenesis techniques such as epPCR, DNA shuffling, or mutator strains [12]. In order to address this fundamental problem, a complete ISM landscape was experimentally constructed. All 24 trajectories of a 4-site system were explored, which meant that all 64 relevant mutant libraries were generated and screened for enantioselectivity [44]. It involved the epoxide hydrolase ANEH as the catalyst in the same model reaction (Scheme 4.16). However, rather than constructing the respective complete 5-site

ISM scheme [21], which would require excessive exploration of 120 pathways, a truncated version based on only four CAST sites in ANEH was designed [44]. It requires considerably less laboratory work, because only 24 pathways have to be constructed, involving a total of 64 mutant libraries. In order to reduce the experimental work even further, the original 3-residue sites B (215/217/219) and F (244/245/249) were truncated to 2-residue sites B* (215/219) and F* (244/249). Moreover, NDT codon degeneracy was chosen encoding only 12 amino acids (Phe, Leu, Ile, Val, Tyr, His, Asn, Asp, Cys, Arg, Ser, and Gly), which likewise requires less screening for 95% library coverage (only about 430 transformants) [44]. These practical measures reduce structural diversity quite a bit. Nevertheless, variants that are even better than the previous best mutant LW202 were identified. Keeping the goal of the study in mind, the complete fitness pathway landscape was analyzed in detail, which revealed several phenomena that are important when applying ISM.

The results of the 12 most productive ISM pathways are shown in Figure 4.9a, with the respective terminal variants showing E-values in the range of 78–159 favoring (*S*)-**16**. The results of the other set of 12 trajectories are pictured in Figure 4.9b, which features variants displaying selectivity factors in the range of $E = 28–78$, which is also considerably better than the performance of WT ANEH ($E = 4.6$). Thus, all 24 pathways provide respectable results, the best one WT → F* → B* → E → D leading to variant GUY-228 ($E = 159$) [44], which is superior to the originally best mutant LW202 ($E = 115$) [21]. These results help to explain why arbitrarily chosen pathways in numerous ISM have been successful, although better options may have been missed [44].

The results allow for the construction of a fitness-pathway landscape by considering the experimental data at all stages of a given pathway in a stacking mode that links WT ANEH with the respective final mutant in each case of all 24 trajectories (Figures 4.10 and 4.11).

This unique fitness landscape reveals several local minima, meaning that in the respective libraries no improved variants were found. *The evolutionary pathways point to an important result in this study, namely a simple way to escape from such dead ends.* Rather than abandoning a pathway characterized by a local minimum, or switching to epPCR or DNA shuffling (or giving up), the use of an inferior mutant showing *lower* enantioselectivity (or one similar to the one in the previous upward climb), as the template in the subsequent ISM step leads to a notably improved variant. One of several such pathways is WT ($E = 4.6$) → E ($E = 21$) → B* ($E = 36$) → D ($E = 32$) → F* ($E = 97$). The use of inferior or non-improved mutants is reminiscent but not identical to the concept of neutral drift [17a–c], and of the Eigen/Schuster concept of quasi-species in natural evolution [18], which has been invoked in other directed evolution studies [19, 53]. As the most general conclusion of the study, the use of non-improved or inferior mutants as templates in subsequent mutagenesis cycles is recommended whenever local minima are encountered in any directed evolution study, including those driven by epPCR or DNA shuffling. To date, this strategy has not been used very often in directed evolution because researchers

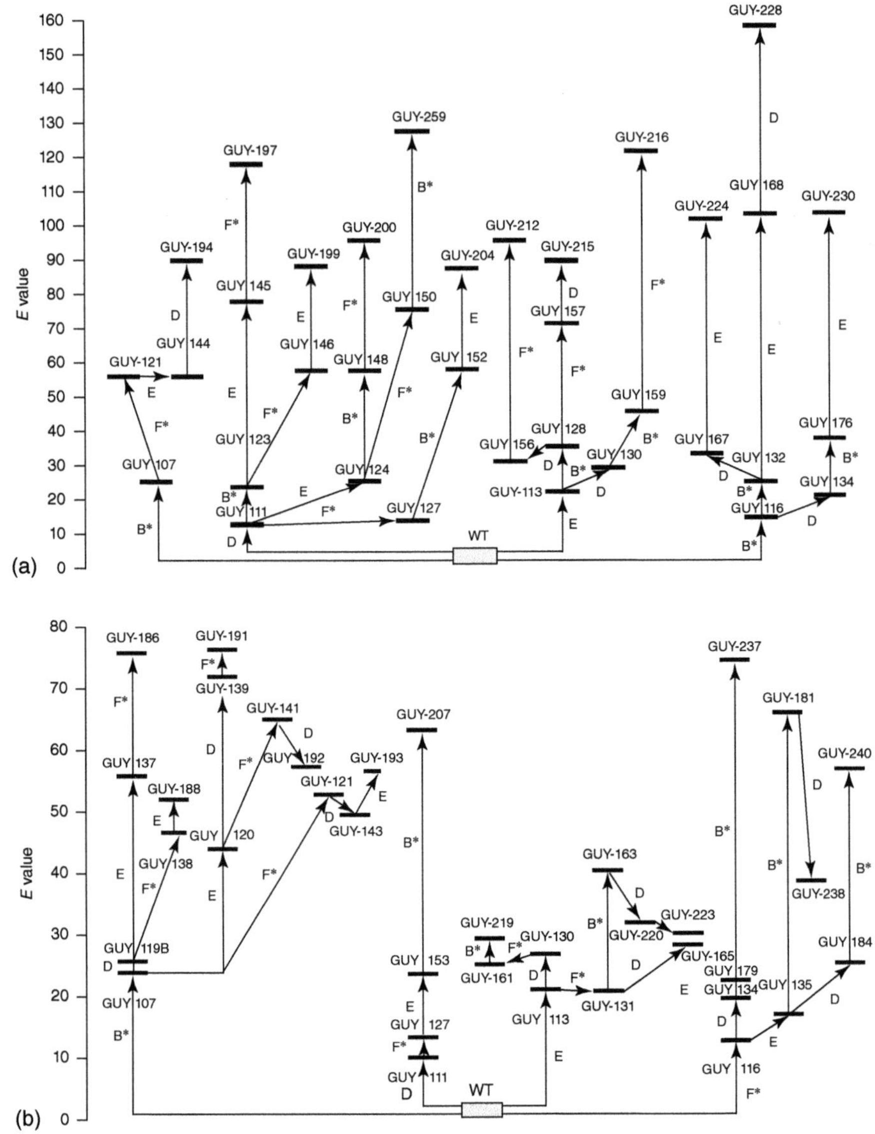

Figure 4.9 Complete experimental exploration of a 24-pathway ISM system involving the ANEH-catalyzed hydrolytic kinetic resolution of *rac*-**15** (Scheme 4.16). (a) Portion of the 24-pathway ISM scheme showing the 12 best pathways leading to ANEH variants displaying $E > 78$ (*S*); (b) Portion of the 24-pathway ISM scheme showing the 12 least productive pathways leading to ANEH variants with $E = 28-78$ (*S*). Source: [44]/John Wiley & Sons.

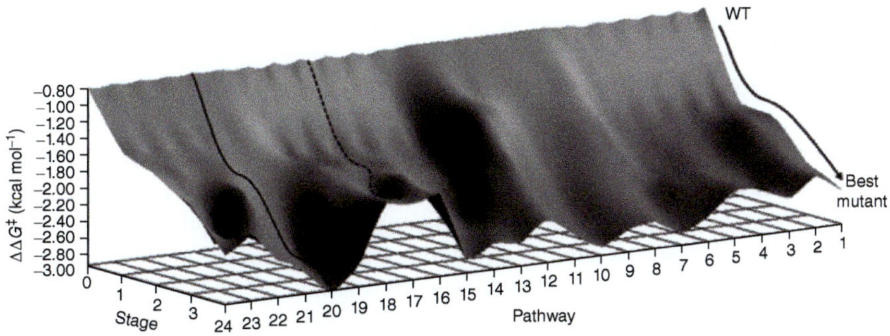

Figure 4.10 Fitness pathway landscape featuring the 24 trajectories leading from WT ANEH to the respective final variants with enhanced enantioselectivity at the end of each pathway as specified by the respective experimental $\Delta\Delta G^{\dagger}$ values [44]. Solid line: typical pathway in which each mutant library contains at least one variant displaying enhanced enantioselectivity; dotted line: typical pathway in which at least one library is devoid of an improved variant, in which case an inferior mutant was employed in the subsequent ISM step thereby escaping from the local minimum. Notice that WT is placed on top, and the final mutants on the bottom, a procedure, which was reversed in later studies [12a]. Source: [44]/John Wiley & Sons.

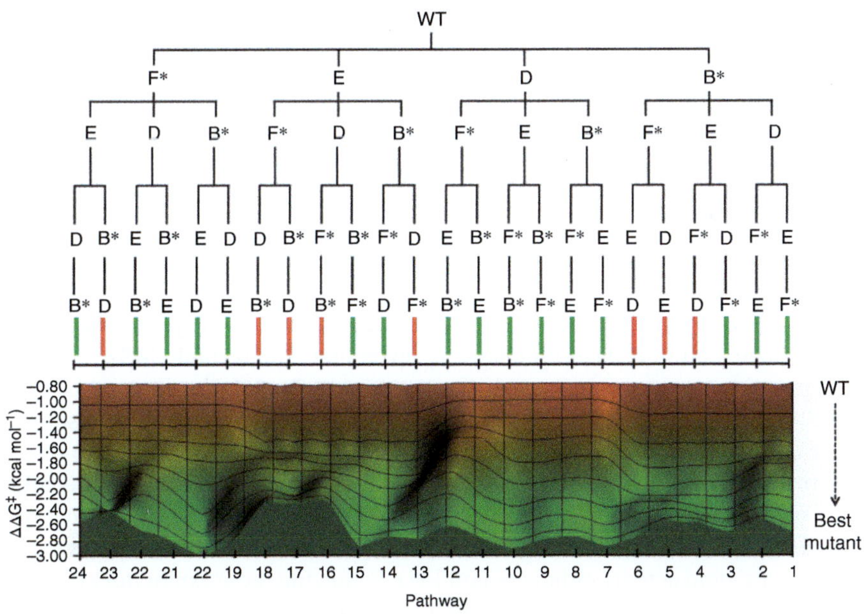

Figure 4.11 Free energy profiles of the 24 ISM pathways in the directed evolution of ANEH, as pictured in a front view of the fitness-pathway landscape [44]. In the green pathways all relevant saturation mutagenesis libraries contain improved variants (enhanced enantioselectivity) in the model reaction (Scheme 4.16); the eight red pathways denote those in which at least one library in the four-step evolutionary process is devoid of any improved variants (local minimum). Source: Adapted from [44].

usually prefer to abandon the respective experimental platform and switch to another mutagenesis method or give up [12a]. In a B-FIT study aimed at thermostabilization, it was shown that the least favorable starting points and inferior mutants can map superior evolutionary pathways [54]. When analyzing the experimental data of the epoxide hydrolase study, the first derivative of $\Delta\Delta G^{\dagger}$ at every stage of each of the 24 pathways is also an instructive way to picture a fitness landscape [44].

As Table 5.1 reveals, ISM is being increasingly used to manipulate stereo- and regio-selectivity as well as activity and substrate acceptance. The areas of application generally concern the biocatalytic synthesis of enantiopure intermediates needed in the preparation of chiral pharmaceuticals, plant protecting agents or fragrances [45]. The potential of ISM in the production of bio-fuels [55] is enormous. The technique has also been applied in completely different areas, e.g. in the directed evolution of *N*-oligosaccharyltransferases of relevance in vaccine production [56].

An intriguing application of ISM was reported in a study focusing on the potential production of universal blood by enzyme-catalyzed selective cleavage of the antigenic components [57]. Decades ago, a vision was put forward calling for selective and active glycoside hydrolases [58]. Unfortunately, all the presently known glycoside hydrolases that catalyze such challenging reactions are so slow that the original vision could not be implemented experimentally. The situation changed when ISM-based directed evolution was applied to the glycoside hydrolase from *Streptococcus pneumonia* SP3-BS, which cleaves selectively the entire terminal trisaccharide antigenic determinants of both A- and B-antigens from some of the linkages on the surface glycans of red blood cells [57]. In order to enhance enzyme activity for the cleavage of the Galß-1,3-GlcNAc linkage of type 1A antigens (Figure 4.12a–d), a high-throughput microtiter-based assay was first established in which a fluorogenic substrate was devised comprising the type 1A blood group pentasaccharide linked by a ß-glycosidic bond to a methylumbelliferone.

The crystal structure of Sp3GH98 harboring type 2A Lewis^{Y} pentassacharide [59] formed a rational basis for choosing seven first- and second-CAST residues for saturation mutagenesis, Tyr530, Asn559, Tyr560, Trp561, Ile562, Asn592, and Lys624 (Figure 4.12e). At a later stage of the ISM process, two additional residues were considered. All seven initial NNK libraries were screened, leading to three improved variants, Asn559Ser, Asn592Val, and Asn592Ser with a threefold enhanced hydrolytic activity. The mutations were combined with the formation of the double mutants Asn559Ser/Asn592Val and Asn559Ser/Asn592Ser, which showed a sixfold activity enhancement [57]. Following these exploratory experiments, ISM was initiated using the double mutants as templates for individual randomization at residues Tyr530, Tyr560, Trp561, Ile562, and Lys624. Further ISM exploration involving various 2-residue sites from the above individual amino acid positions and spatially close residues Glu630, Glu663, and Lys677 followed. This ISM exercise provided a variant characterized by five-point mutations, Tyr530His/Asn559Ser/Asn592Val/Glu630Leu/Lys677Leu with a

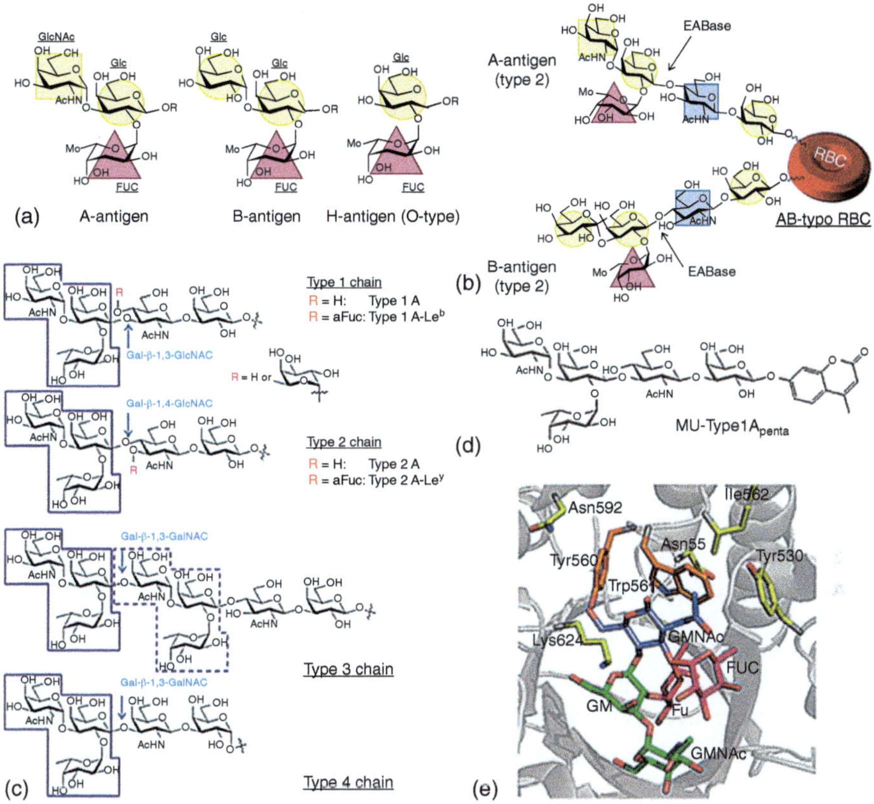

Figure 4.12 Toward universal blood using ISM [57]. (a) Carbohydrate antigenic determinants of A-, B-, and H-antigens. The H-antigen is present on glycans of the O blood-group and is typically nonantigenic except in rare cases. (b) Site of cleavage of A- and B-antigens by GH98 EABase enzymes from type 2 chains of erythrocytes. (c) Various chain types to which A-antigens are present on erythrocytes and other cell types. (d) Structure of the fluorogenic substrate MUType1Apenta. (e) First- and second-sphere randomization sites chosen for Iterative Saturation Mutagenesis (ISM), guided by the X-ray structure of Sp3GH98. First sphere: Tyr 560 and Trp561; second sphere: Tyr 530, Asn559, Ile 562, Asn592, and Lys624. Source: [57]/American Chemical Society.

120-fold increase in activity while maintaining the desired regioselectivity [57]. A final round of epPCR-based mutagenesis improved activity by only a small extent (Figure 4.13). It was shown experimentally that the complete removal of the antigens had been achieved. Moreover, antibody-based immunofluorescence control experiments were carried out with real substrates, thereby demonstrating efficient removal of type 1A antigens from the surface of red blood cells.

As analyzed in a highlight featuring this research, the overall strategy has opened a new door in this exciting research field [60]. It was suggested that further ISM experimentation is likely to provide even better results. This could

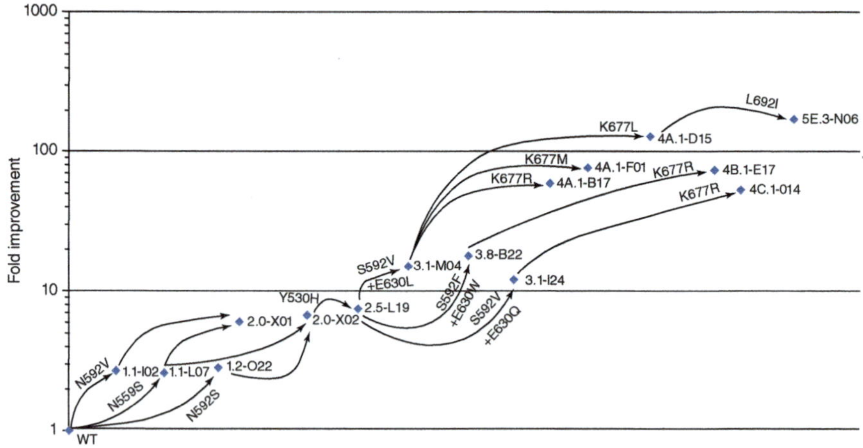

Figure 4.13 Evolutionary pathways of Sp3GH98 based on iterative saturation mutagenesis (ISM) and one final round of epPCR (upper right). Source: [57]/American Chemical Society.

involve the grouping of the individual amino acid positions into multi-residue sites followed by ISM on the basis of appropriately chosen reduced amino acid alphabets.

Interestingly, the Withers group continued their study by discovering new enzymes that convert A to universal O type blood [61]. A two-step catalytic pathway was established by using an α-1,3-linked-*N*-acetylgalactosamine deacetylase (FpGalNAcDeAc) and a galactosaminidase (FpGalNase) from *Flavonifractor plautii*. The former enzyme enables the deacetylation of the A antigen, while the latter one directs the cleavage of galactosamine to generate H antigen (Figure 4.14). The crystallographic structure of FpGalNAcDeAc was solved in this study, and the multiple active sites were determined by mutagenesis. However, application of advanced CAST/ISM has not yet been undertaken on this enzyme, which may further improve its activity and specificity for real clinical tests.

4.3.3
Systematization of Saturation Mutagenesis with Further Practical Tips

The systematization of semi-rational saturation mutagenesis requires guidelines for choosing residues to be mutagenized, how to group them into larger sites, and how to choose reduced amino acid alphabets for focused randomization. The guidelines are more or less the same as the ones suggested in Section 4.1.1 for rational design:

- X-ray data or homology model
- Mechanistic information
- Previous mutational data

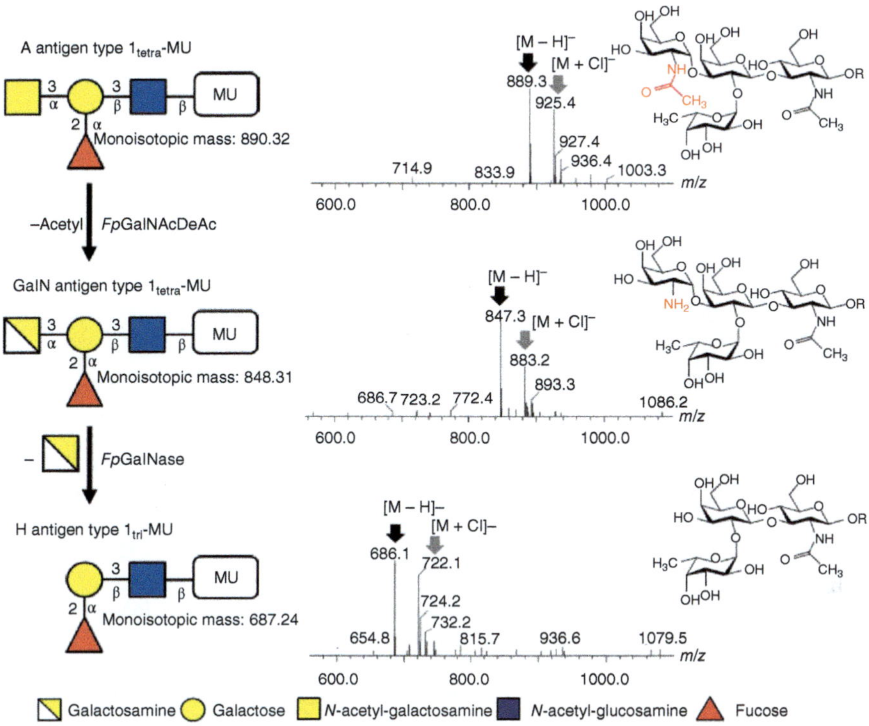

Figure 4.14 Schematic illustration of a unique mechanism to generate H antigen catalyzed by a deacetylase (FpGalNAcDeAc) and a galactosaminidase (FpGalNase), revealed by mass spectra data. Source: [61]/Springer Nature.

- Phylogenetic analysis (consensus technique)
- Docking and molecular dynamics (MD) computations
- Machine learning

At this point, the reader is reminded of the possibility of two different strategies when applying saturation mutagenesis (see Section 4.3.1): A single experiment with a given codon degeneracy being applied simultaneously to all of the chosen residues (Strategy 1), or a different codon degeneracy at each residue of a multi-residue randomization site (Strategy 2). The first example of Strategy 2 for enhancing stereoselectivity was the directed evolution of phenyl acetone monooxygenase (PAMO) [48a]. This robust Baeyer–Villiger monooxygenase readily accepts phenyl acetone but not such synthetically interesting compounds such as cyclohexanone or its derivatives [62]. For example, ketones **17a–b** are essentially inert to PAMO-catalyzed oxidation (Scheme 4.17) [48a]. The goal was to evolve substrate acceptance (rate) as well as enantioselectivity in the oxidative kinetic resolution of these compounds using saturation mutagenesis at each residue of a randomization site at the binding pocket, guided by X-ray

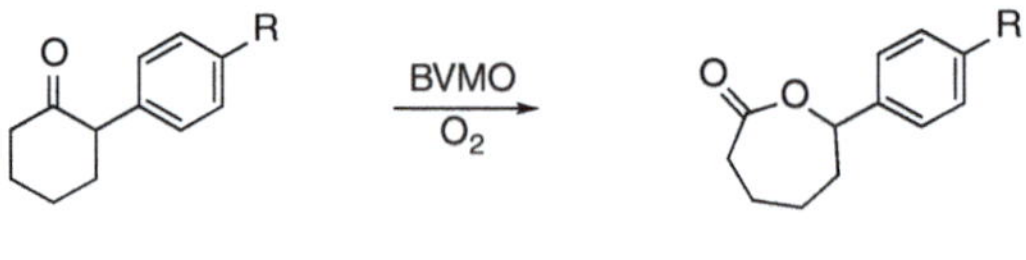

rac-**17a** R = H
b R = Cl

(*R*)- or (*S*)-**18a** R = H
b R = Cl

Scheme 4.17 Oxidative kinetic resolution catalyzed by PAMO mutants. Source: Adapted from Ref. [48a]/Royal Society of Chemistry.

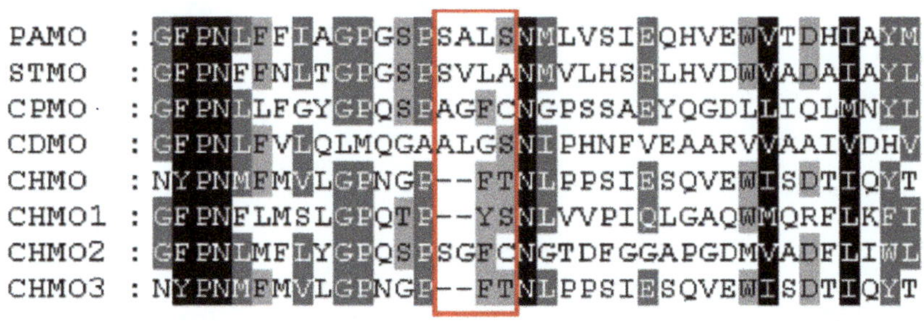

Scheme 4.18 Multiple Sequence Alignment (MSA) of eight BVMOs (441–444 loop in gray box). Source: Adapted from Ref. [48a]/Royal Society of Chemistry.

data and a phylogenetic analysis in the form of Multiple Sequence Alignment (MSA).

First, on the basis of the PAMO crystal structure [63], four residues in loop 441–444 next to the binding pocket were identified as possible CAST sites. Previous application of rational design using site-specific mutagenesis at positions in the loop was only partially successful [64]. NNK-based randomization of a 4-residue CAST site would require the screening of 3.1 million transformants for 95% library coverage, and even NDT codon degeneracy would still call for ≈62 000 screened clones (Table 3.2). Although considerably less library coverage may still provide improved variants [33], this was not tested in the study. Rather, eight Baeyer–Villiger monooxygenases were aligned with focus on the loop region (Scheme 4.18). As can be seen, only a limited number of amino acids are conserved at the four positions: Ser and Ala (position 441); Ala, Val, Gly, and Leu (position 442); Leu, Phe, Gly, and Tyr (position 443); and Ser, Ala, Cys, and Thr (position 444). The plan was to use these amino acids as building blocks at the respective positions of the 4-residue randomization site, the degree of oversampling being highly reduced by this strategy [48a].

Appropriate codon degeneracies were designed in order to match as extensively as possible the amino acids occurring at these four positions according to MSA, while also introducing a limited number of additional amino acids, as building blocks for slightly enhanced diversity (Table 4.3). At all positions, the WT amino acid is maintained as defined by the chosen codon degeneracy. At position 441, KCA codon degeneracy means the introduction of only one new

Table 4.3 Choice of codon degeneracies at each position in the 441–444 loop of PAMO.

Amino acid positions	Codon degeneracy	Encoded amino acids	Codons	Oversampling for 95% coverage
441	KCA	A, (S)	864	2587
442	KBG	S, (A), L, V, W, G		
443	BGC	F, H, (L), V, Y, G, D, R, C		
444	NSC	(S), A, P, T, R, G, C		

Degenerate codons: A (adenine); B (cytosine/guanine/thymine); C (cytosine); G (guanine); S (cytosine/guanine); K (guanine/thymine); N (adenine/cytosine/guanine/thymine) [48a]. In the column listing the encoded amino acids, the WT amino acids are shown in parentheses.
Source: Ref. [48a]/Royal Society of Chemistry.

amino acid (in this case, Ala), while structural diversity at the other positions is higher.

Screening of 2587 would be necessary for 95% library coverage, but only 1700 were assayed. Several active hits were identified, with PAMO variant Ser441Ala/Ala442Trp/Leu443Tyr/Ser444Thr showing the highest activity and enantioselectivity ($E = 70$ in favor of *R*-**18a**). This variant is an even better catalyst for the reaction of substrate **17b** ($E > 200$), which is also not accepted by WT PAMO. Two conclusions result from this proof-of-principle study: (i) A different reduced amino acid alphabet can be used effectively at each position within a multi-residue randomization site in a single saturation mutagenesis experiment; (ii) MSA-based bioinformatics constitutes a practical guide in this endeavor [48]. MSA-guided saturation mutagenesis using a defined amino acid alphabet for the entire randomization site has since been applied and extended to other directed evolution studies focusing on stereoselectivity [65].

Especially challenging is the directed evolution of PAMO as catalysts in the oxidative kinetic resolution of 2-alkyl substituted cyclohexanone derivatives that are completely inert. In these cases, the variant Ser441Ala/Ala442Trp/Leu443Tyr/Ser444Thr is inactive. Rather than resorting to *in vitro* coevolution by applying substrate walking (which could well be successful) [28], a completely different strategy was tested [66]. Further analysis of the sequence alignment (Scheme 4.19) showed that proline at positions 437 and 440 is highly conserved, with cyclododecanone monooxygenase (CDMO) being the only exception. This suggests that proline, known to impart some degree of rigidity to proteins [67], is necessary at these positions for maintaining stability and function. By standard logic, it should not be exchanged by another amino acid. Nevertheless, in the hope of evolving a new catalytic profile, the opposite strategy was implemented by performing NNK-based saturation mutagenesis at positions 437 and 440 (a second sphere CAST residue) [66]. This unorthodox MSA-guided approach proved to be successful, with the randomization library at position 440 containing several highly active and stereoselective single mutants with essentially no tradeoff in stability. Variants Pro440Leu, Pro440Ile, Pro440Asn, Pro440His, Pro440Tyr, and Pro440Trp are active and highly stereoselective in oxidative

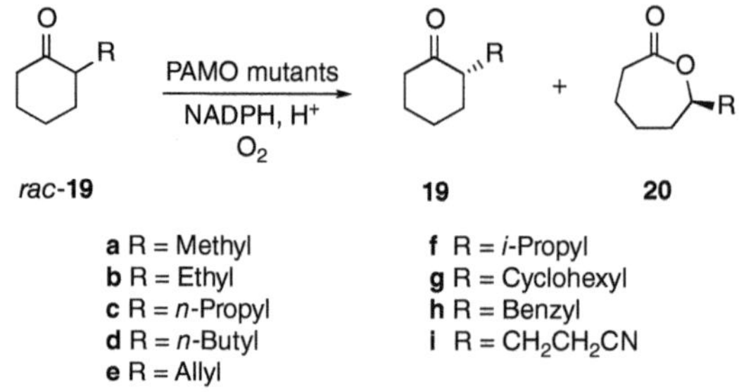

Scheme 4.19 PAMO-catalyzed oxidative kinetic resolution of 2-alkyl substituted cyclohexanone derivatives. Source: Ref. [66]/American Chemical Society.

kinetic resolution of the previous ketones **17a–b**, but also in reactions of a wide variety of structurally different 2-alkyl derivatives **19a–i** (Scheme 4.19), with selectivity factors of $E = 150-200$ being typical. Even the bulky substrate 2-cyclohexylcyclohexanone (**19g**) is readily accepted with high enantioselectivity and activity ($E > 200$).

Increased flexibility and conformational dynamics of the extended PAMO loop may be the reason for the dramatic effects, as recently shown in the study of a P450 monooxygenase [11c]. Long-range enzyme dynamics have also been revealed in other protein studies, as in the catalysis of isopenicillin *N* synthase, where the intermediate generation of an Fe(III) superoxide induces changes in active site volume and remote conformational effects, as shown by femtosecond crystallography [68]. In the PAMO case, a thorough theoretical analysis of the experimental results has yet to be performed. It is also uncertain how general the unusual MSA-based strategy is, i.e. whether amino acid exchange events at highly conserved residues remote or near the binding pocket a successful when evolving other enzymes. The opposite strategy using combinatorial proline scanning with the introduction of one or more prolines at CAST sites (or remote loops) also needs to be explored.

In a subsequent study, the approach based on the use of a different amino acid alphabet at each position of a multi-residue site (Strategy 2 in Scheme 4.13) was reported in the directed evolution of *Candida antarctica* A (CALA) as a biocatalyst in the hydrolytic kinetic resolution of α-substituted carboxylic acid esters [48b]. An earlier study by the same group had shown that CAST/ISM is successful in evolving a number of variants for asymmetric transformations of this kind [69], but extension to more bulky ibuprofen-type esters was not achieved. The new study concentrated on substrates **21–24** (Scheme 4.20) by first applying substrate walking [28]. A triple mutant (F149Y/I150N/F233G), which had proven to be highly active and enantioselective for substrate **18** using CAST/ISM but which was a poor catalyst for the larger substrate **21** (slow and stereorandom reactions), was used as the starting template for further saturation mutagenesis experiments [48b].

First, the bulky substrate **21** was docked inside the CALA binding pocket in the oxyanion form (tetrahedral intermediate at Ser184), leading to the

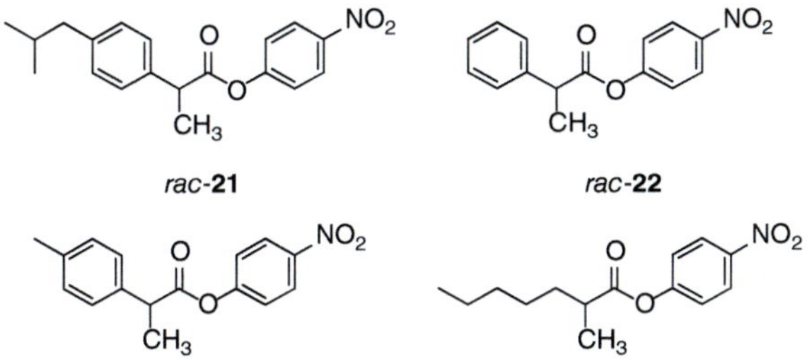

rac-**21** *rac*-**22**

rac-**23** *rac*-**24**

Scheme 4.20 Substrates investigated in the CAST-based directed evolution of the lipase CALA using mutant F149Y/I150N/F233G as template. Source: Ref. [48b]/Proceedings of the National Academy of Sciences.

conclusion that nine residues at the acyl binding region should be mutable (Figure 4.15). Then, other residues at this large CAST site were eliminated from further consideration because they proved to be highly conserved, as shown by an MSA (PSI-BLAST) analysis. In view of the PAMO study (Scheme 4.18) [66], this may not have been necessary, but nine otherwise well-chosen positions sufficed for reshaping the binding pocket. Since substrate **21** is too bulky to be readily accepted, small amino acids as building blocks in reduced amino acid alphabets were chosen for saturation mutagenesis. Phe149Tyr and Ile150Asn were included because in the earlier study this pair had been shown to be

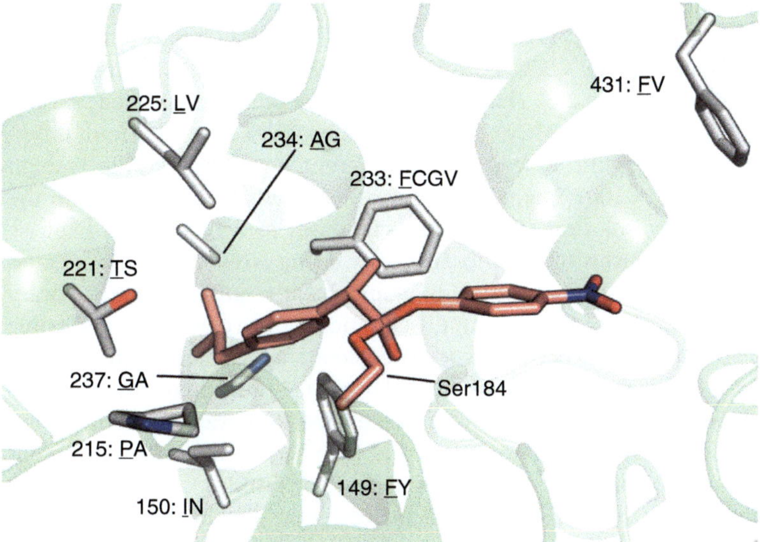

Figure 4.15 Binding pocket of CALA showing tetrahedral intermediate with substrate 21 and nine residues for potential saturation mutagenesis [48b]. The original WT residues are underlined. Source: [48b]. Copyright (2012) National Academy of Sciences, U.S.A.

Table 4.4 Combinatorial use of amino acids as building blocks employed in saturation mutagenesis at the 9-residue randomization site of CALA (Figure 4.15).

Position	WT residue	Alternative residue(s)
149	Phe	Tyr
150	Ile	Asn
215	Pro	Ala
221	Thr	Ser
225	Leu	Val
233	Phe	Cys/Gly/Val
234	Ala	Gly
237	Gly	Ala
431	Phe	Val

Source: Ref. [48b]. Copyright (2012) National Academy of Sciences, U.S.A.

essential in the evolution of high enantioselectivity toward a fairly broad range of similar substrates [69].

It was shown that sequence proximity in the nine chosen residues allowed ready clustering into four primer pairs. Together with primers for the ends of the gene, they were used in generating five partly overlapping fragments. These were then simultaneously assembled in a modified overlap extension polymerase chain reaction (OE-PCR) (see Chapter 3). The final set of amino acids as building blocks for simultaneous randomization at the 9-residue CAST site was chosen on the basis of structural information, mechanistic considerations, and previous mutational data (Table 4.4) [48b].

After screening about 2400 transformants in the highly condensed library corresponding to ≈90% coverage, only a few variants proved to be active toward substrate **21**. The best hit was a penta-substituted variant Thr221Ser/Leu225Val/Phe233Cys/Gly237Ala/Phe431Val in which four different amino acids were introduced at five different positions, leading to high stereoselectivity ($E = 100$). The best hit, as well as other CALA variants in the hydrolytic kinetic resolution of the other substrates in Scheme 4.20, likewise ensured acceptable levels of enantioselectivity. Using NNK codon degeneracy, encoding all 20 canonical amino acids would have required for 95% library coverage, the screening of 10^{14} potentially enantioselective clones, an impossible mission. Deconvolution experiments revealed cooperative (more than additive) effects [27], and suggested that the particular penta-substituted variant would not be accessible by normal ISM [48b]. Extensive ISM was not applied, but in principle such, exploration could provide many more mutants of different sequences that also feature highly improved catalytic profiles.

As pointed out in 2016 [13], the most interesting aspect in the CALA study [48b] was the use of codon degeneracies, which led to the combinatorial introduction of a single defined amino acid at eight positions (rather than just one position as in the PAMO study [48a]) and three amino acids at the ninth position (in addition to the respective WT amino acid) [48b]. A small and smart library resulted in one step, as in the PAMO study [48a], thereby avoiding the necessity to probe several cycles of mutagenesis/expression/screening. The option defined by strategy 2 (Scheme 4.13) was also chosen in an investigation of CALA as a catalyst in the acylating kinetic resolution of secondary alcohols [48c], in a study of the evolution of P450-BM3 mutants, which accept small alkanes such as propane [48d], and in other applications [48e–i].

The molecular biology in the two structure-guided utilizations of reduced amino acid alphabets in saturation mutagenesis is straightforward, but it is not clear whether strategy 1 [22, 23, 26, 52] or strategy 2 [48] is superior. They need to be studied more thoroughly before final assessments are possible. When comparing NNK with NDT codon degeneracy as part of strategy 1, the benefits of resorting to the reduced amino acid alphabet have been demonstrated [23a], but in that study, alternative reduced amino acid alphabets were not tested. In contrast, strategy 1 has been tested fairly systematically by studying the consequences of utilizing different reduced amino acid alphabets, with limonene epoxide hydrolase (LEH) from *Rhodococcus erythropolis* DCL14 serving as the catalyst in the hydrolytic desymmetrization of meso-type epoxides (Scheme 4.21) [70]. LEH is mechanistically different from other epoxide hydrolases because epoxide substrates undergo S_N2 reactions with water acting as the nucleophile (not an aspartate-residue), which has been modeled in a QM/MM study [71a] on the basis of the LEH crystal structure [71b]. In the initial report, the conventional CAST/ISM approach was successfully applied using NDT codon degeneracy in the randomization of four 2-residue sites: Met32/Leu35, Leu74/Ile80, Leu114/Ile116, and Met78/Val83 lining the LEH binding pocket (CASTing) [70]. Employing epoxide **25a** as the model substrate and arbitrarily chosen ISM pathways, both (*R*,*R*)- and (*S*,*S*)-selective mutants were evolved, enantiomeric ratios reaching er = 90 : 10 and er = 96 : 4, respectively. A total of 5000 transformants were screened.

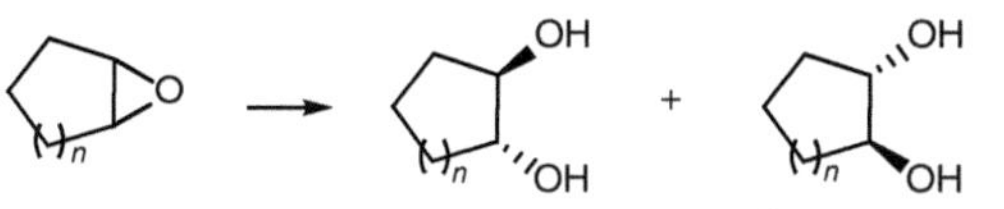

25 a *n* = 1
b *n* = 2
c *n* = 3

(*R*,*R*)-**26 a** *n* = 1
b *n* = 2
c *n* = 3

(*S*,*S*)-**26 a** *n* = 1
b *n* = 2
c *n* = 3

Scheme 4.21 Hydrolytic desymmetrization of meso-epoxides catalyzed by LEH and mutants thereof. Source: Adapted from Refs. [70–72a, b].

An interesting example of CAST/ISM concerns the synthetically important reductive desymmetrization of prochiral compounds to generate a single stereoisomer bearing two chirality centers [73]. It can be seen that in the case of the compound ethyl secodione, when either of the two carbonyl groups at 14- or 17-positions is reduced, two chiral centers are simultaneously formed (Figure 4.16a). The key challenge was how to obtain the desired single stereoisomer product (13*R*,17*S*)-ethyl secol, which is in high demand in the

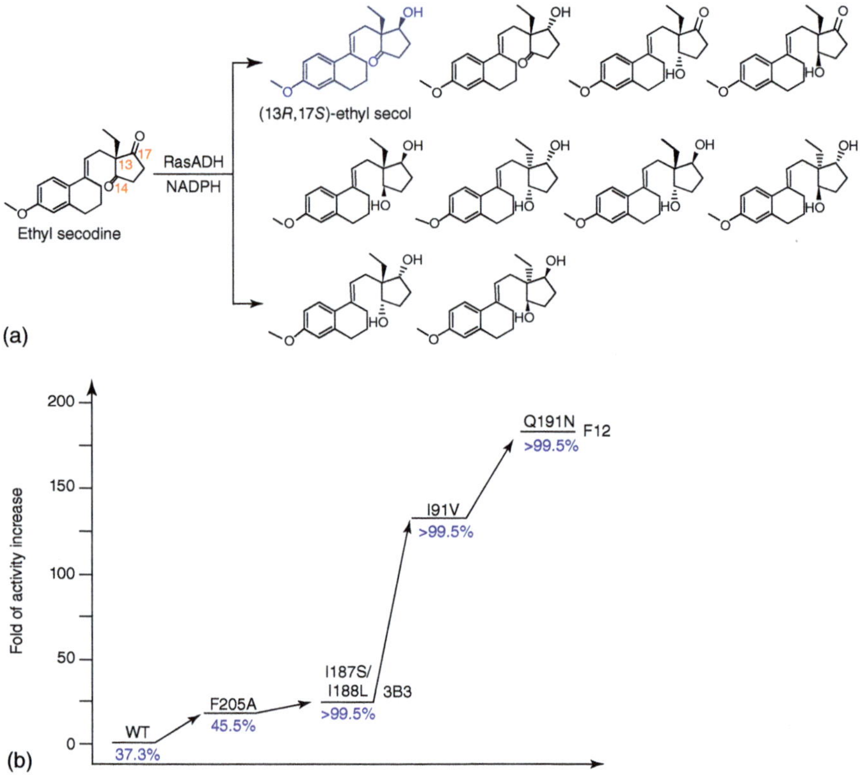

Figure 4.16 (a) All theoretically possible reductive products of ethyl secondine. The desired (13*R*,17*S*)-ethyl secol is marked in blue. (b) The evolution of RasADH in the transformation of ethyl secondine to (13*R*,17*S*)-ethyl secol using CAST/ISM. The percentage of (13*R*,17*S*)-ethyl secol in the product mixture is indicated in blue. Source: Adapted from [73].

synthesis of contraceptive hormones. To this end, an alcohol dehydrogenase from *Ralstonia* sp. (RasADH) was rationally engineered using ISM, based on structural information. After docking analysis, a handful of residues lining the substrate binding pocket were selected as hot spots and subjected to mutagenesis. In the first round of site-specific saturation mutagenesis (SSM), a variant with the point mutation F205A delivered an 18-fold increase in activity relative to the WT enzyme. The ratio of the desired (13*R*,17*S*)-product was improved from 37.3% for WT to 45.5% for the variant F205A (Figure 4.16b). Next, two additional sites, I187 and I188, were simultaneously mutated using F205 as a template. After library screening, a positive hit named 3B3 (I187S/I188L/F205A) was obtained, showing a 1.2-fold improvement in activity and a >99.5% percentage of the target (13*R*,17*S*)-product. On the basis of variant 3B3, five individual saturation mutagenesis libraries covering the sites I91/Q93, L144, H147, Y150, and Q191 were then constructed, respectively. Eventually, the resulting mutant

F12 (3B3/I91V/Q191N) was identified with notably enhanced activity (183-fold increase compared to the WT) with no tradeoff in stereoselectivity [73].

4.3.4 Single Code Saturation Mutagenesis (SCSM): Use of a Single Amino Acid as Building Block

In a curiosity-driven project, the question of whether the smallest amino acid alphabet, namely a single amino acid as a building block for randomization, can be used effectively in saturation mutagenesis at a large multi-residue site lining the binding pocket [72a]. It constitutes the extreme case of minimal building blocks in reshaping enzyme binding pockets and is therefore fundamentally different from the CALA-study (Strategy 2 in Scheme 4.13), in which up to four different amino acids were introduced in the final mutant (Table 4.4) [48b]. The hydrolytic desymmetrization of cyclohexene oxide (**25b**) with the formation of (*R*,*R*)-**26b** and (*S*,*S*)-**26b** served as the model reaction. Ten CAST residues were identified for saturation mutagenesis (Leu74, Phe75, Met78, Ile80, Leu103, Leu114, Ile116, Phe134, Phe139, and Leu147), which were grouped into a single randomization site (Figure 4.17) [72a, b]. WT LEH is only slightly (*S*,*S*)-selective (er = 52 : 48; ee = 4%). If such a 10-residue site were to be randomized using NNK codon degeneracy (20 amino acid alphabet) or NDT codon degeneracy (12 amino acid alphabet), then one would have to screen for 95% library coverage of 10^{15} or 10^{11} transformants, respectively. In contrast, when employing the smallest amino acid alphabet, a single amino acid, only about 3000 transformants would have to be screened for the same essentially complete library coverage. However, structural diversity would be dramatically reduced. It was speculated that such a strategy could be successful if right decisions were made regarding the choice of the amino acid in single codon saturation mutagenesis (SCSM). The crystal structure of LEH reveals that most of the amino acids surrounding the binding pocket are hydrophobic [71b].

Therefore, valine was chosen as the sole building block in saturation mutagenesis at the 10-residue randomization site. This procedure is reminiscent of combinatorial alanine scanning used in changing the binding properties of proteins [74a, b]. In another case, alanine scanning was applied in order to increase the activity of a P450 monooxygenase (although activity was measured using lysates and not isolated mutants, which causes some uncertainty in the assessment of activity) [74c]. In the LEH study, the reason for choosing valine as opposed to alanine has to do with the expectation that the sterically more demanding side-chain of this amino acid would induce a greater steric effect, possibly compensating for the low structural diversity imposed by the use of a single amino acid. Primer design is shown in Scheme 4.22 [72a].

Using the adrenaline pre-test for epoxide activity [75], 35 microtiter plates of 96-well format were screened (96 − 2 positive controls − 2 negative controls = 92), corresponding to 3220 transformants [72a]. A large number of active hits (533) were discovered, which is an indication that this strategy is productive, certainly

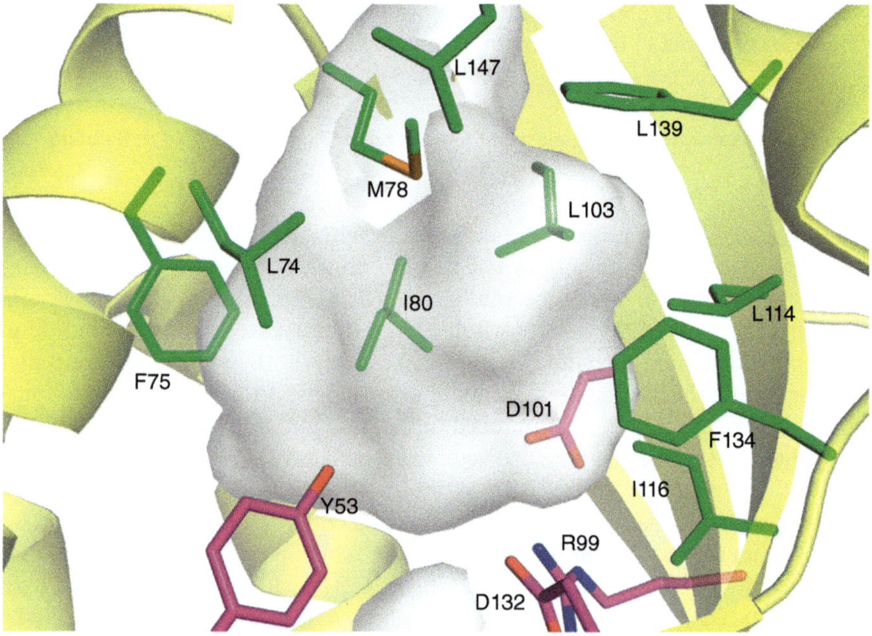

Figure 4.17 Large randomization site defined by 10 amino acid positions (green) chosen on the basis of the crystal structure of LEH [71b] with the catalytic residues being shown in pink. Source: [72a, b]/John Wiley & Sons.

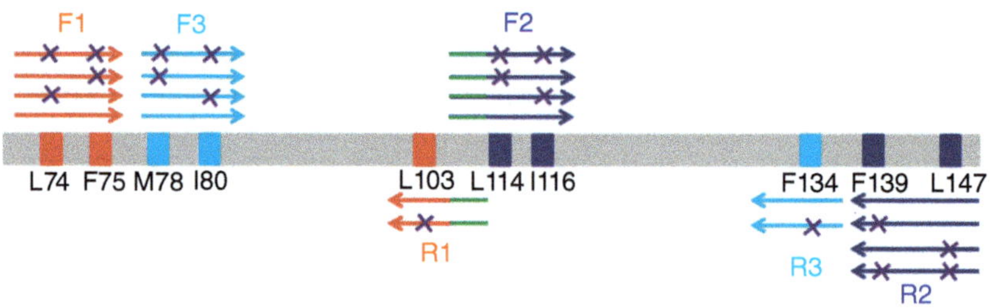

Scheme 4.22 Primer design and library construction using valine as the sole building block and the 10 randomization positions in LEH according to Figure 4.17. Source: Ref. [72a]/American Chemical Society.

in this particular case. These were then assessed for enantioselectivity by automated chiral GC. The result is shown in Figure 4.18, which features seven of the best hits discovered in this single mutant library, four of them favoring (*S*,*S*)-**26b** and three being (*R*,*R*)-selective, with enantiomeric ratios up to er = 97 : 3 and er = 12 : 88, respectively. It can be seen that three to five valines were introduced at different positions of the 10-residue site [72a].

The use of phenylalanine as the sole amino acid as building block in an analogous manner also worked nicely, although the degree of enantioselectivity was generally somewhat lower. In order to increase enantioselectivity even more,

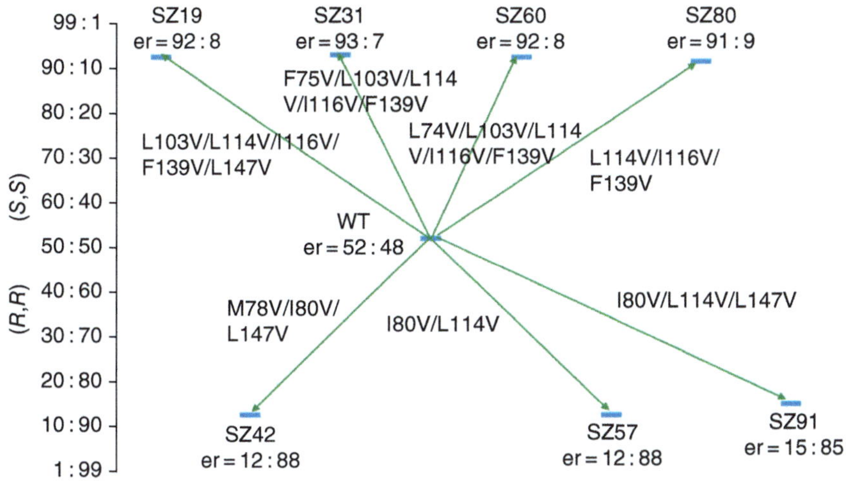

Figure 4.18 Best hits discovered in a mutant library created by a single saturation mutagenesis experiment using valine as the sole building block at a 10-residue CAST-randomization site in LEH serving as the catalyst in the hydrolytic desymmetrization of epoxide 25b (Scheme 4.21). Source: [72a]/John Wiley & Sons.

ISM [76] employing valine and phenylalanine in successive randomization steps led to the highest enantioselectivities (enantiomeric ratios up to 98 : 2) [72a]. In important control experiments, other amino acids such as serine or proline were used as the sole building blocks, anticipating that this would not provide any improved mutants. Indeed, this was observed, which is of mechanistic significance because it demonstrates that the rationale behind choosing valine was correct. (*R*,*R*)- and (*S*,*S*)-selective mutants were characterized by X-ray structural analyses, which revealed the distinct changes in the shape of the respective binding pockets [72a].

Although the apparent success when using the smallest reduced amino acid alphabet as part of strategy 1 (Scheme 4.13) is indisputable, it cannot be expected to be general. The choice of the best single amino acid in addition to WT was fairly straightforward in the epoxide hydrolase case because the binding pocket of WT is surrounded by hydrophobic amino acids. However, in the case of many (if not most) other enzymes, the structural situation is more complex, making the correct choice of a single amino acid as the sole building block at a large randomization site more difficult. This problem also applies to strategy 2, when using a single amino acid in addition to WT amino acid.

4.3.5 Triple Code Saturation Mutagenesis (TCSM): A Viable Compromise When Choosing Optimal Reduced Amino Acid Alphabets in CAST/ISM

Keeping the limitations of using a single amino acid as the smallest reduced amino acid alphabet in mind, while still being far away from NNK codon degeneracy, an option is to choose two amino acids as building blocks (in addition to WT). This can be called "double codon saturation mutagenesis" (DCSM) [77], but similar problems can arise as in the case of a single amino acid. The total structural diversity can be increased further by using three amino acids as building blocks in a process that has been dubbed "triple code saturation mutagenesis" (TCSM) when considering large multi-residue randomization sites [72b]. This option has emerged as a viable compromise between reduced structural diversity and excessive screening. Three options are then possible while still keeping screening to a minimum (typically 2000–3000 transformants or less):

- Screen less than 95% library coverage.
- Split the multi-residue randomization site into two smaller sites, A and B, and then apply ISM along pathway A → B or B → A (or both if one of them is not as unproductive as desired).
- Split the multi-residue randomization site into three smaller ones, A, B, and C, and proceed with ISM exploration.

When choosing strategy 1 according to Scheme 4.13, one and the same three-membered reduced amino acid alphabet can be employed at all (smaller) sites derived from splitting a large multi-residue site, or a different triple code can be chosen at each of the split randomization sites. TCSM proved to be surprisingly successful, especially when guided by X-ray structural data, consensus sequence alignment and computational aids [72b, c]. In the first study, which focused on LEH as the catalyst in the model hydrolytic desymmetrization of cyclohexene oxide (**25b**) (Scheme 4.21), a triple code comprising Val–Phe–Tyr was chosen on structural and mechanistic grounds, which led to high (*S*,*S*)-selectivity (99% ee) and (*R*,*R*)-selectivity (89% ee) without resorting to ISM. (*R*,*R*)-selectivity was not quite as high but was boosted to 97% by a single ISM step [72b]. Therefore, when comparing SCSM with TCSM in the same model reaction, the latter using three amino acids as building blocks is clearly superior to the former based on a single amino acid.

In a follow-up study, TCSM was applied to the thermally robust ADH from *Thermoethanolicus brockii* (TbSADH) as a catalyst in the asymmetric reduction of difficult-to-reduce ketones such as tetrahydrofuran-3-one [72c]. WT TbSADH shows low (*R*)-selectivity (23% ee). X-ray data and docking computations revealed five residues to be in contact with the substrate: A85, I86, W110, L294, and C295 [72c]. In order to make a rational choice regarding the triple code, NNK-based saturation mutagenesis at all five positions was first performed, with the observed amino acid substitutions that ensure higher (*R*)- or (*S*)-selectivity then being used as building blocks in subsequent TCSM. Based on the data from the NNK-experiments and X-ray structural information, two randomization sites were designed: Site A (A85/I86/L294/C295) and Site B (A85/I86/W110/L294). At site A, triple code Val–Asn–Leu was chosen, and based on all of the available data, mainly (*R*)-selective variants were expected. At site B, triple code Val–Gln–Leu was suggested to be optimal, leading to the

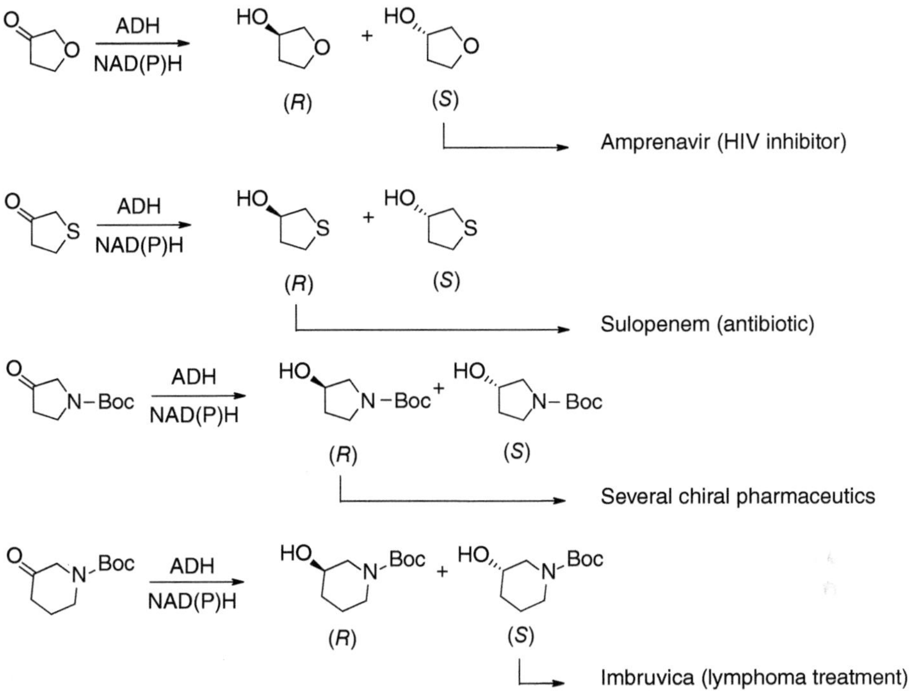

Scheme 4.23 Application of best variants of alcohol dehydrogenase TbSADH as catalysts in the asymmetric reduction of difficult-to-reduce ketones, evolved by application of triple code saturation mutagenesis (TCSM). Source: Ref. [72c]/American Chemical Society.

expectation that mainly (*S*)-selective variants should appear in the subsequent saturation mutagenesis library. Indeed, the respective TCSM libraries harbored several highly (*R*)-selective variants (97–99% ee) and (*S*)-selective counterparts (94–95% ee), respectively [72c]. The best variants were used as catalysts in the asymmetric reduction of other challenging substrates, their products being synthons for the preparation of a number of important therapeutic drugs (Scheme 4.23). Therefore, based on these initial studies, TCSM is emerging as the method of choice for evolving stereo- and regio-selectivity, substrate scope, and/or activity.

4.4 Techno-economical Analysis of Saturation Mutagenesis Strategies

The above discussions focus on the use of saturation mutagenesis in protein engineering of enzymes in the quest to minimize laboratory work, especially the amount of screening (bottleneck of directed evolution) [13]. An issue that has been neglected in directed evolution in general is the question of the number and respective cost of primers needed when implementing a given mutagenesis strategy. Therefore, it is advisable to perform an economic analysis before initiating a directed evolution project [78]. The following points need attention:

- Apply a library quality control, which is cheaper and faster than in the past.
- Consider the role of primer purity and costs according to supplier information with and without redundancy.
- Compare library quality, yield, randomization efficiency, and annealing bias using traditional and emergent randomization techniques based on mixtures of mutagenic primers.
- Choose the most cost-effective saturation mutagenesis scheme while also considering the screening costs and other experimental work, in a "compromise" procedure.

Only a few of the major highlights of this study are delineated here. The experimental results from which the techno-economical analysis was performed refer to saturation mutagenesis of P450-BM3 as the catalyst in the regio- and stereo-selective oxidative hydroxylation of steroids [76]. A total of 12 libraries were generated by different versions of saturation mutagenesis: (i) Traditional use of NNK codon degeneracy; (ii) Traditional use of NNS codon degeneracy; (iii) 22c-trick [79]; and (iv) Tang-approach [80]. In the comparative study, these different embodiments of saturation mutagenesis aim for the same goal regarding catalyst improvement, but require different primers and therefore lead to different costs. Moreover, primer quality and prices vary according to the supplier. In all cases, the QuikChange protocol (Section 3.3) was used, but employing a polymerase different from the traditional recipe [78]. The results of the QQC [16] and the respective quantitative Q-values [52] were obtained, revealing to what extent the desired diversity was actually introduced in the codon using a pooled DNA sequence electropherogram [76]. Table 4.5 summarizes the results of this type of analysis as applied to the given case. The sequence results are summarized in Table 4.6.

Upon comparing the six NNK libraries, it became clear that those using primers from supplier 3 led to the highest mean Q_{pool} value relative to the results obtained when using primers from suppliers 1 and 2. In the techno-economical study, only primers from supplier 3 were used thereafter, specifically when comparing the traditional codon degeneracy, namely the Tang [80] and 22c-trick [79] approaches [78]. Figure 4.19 reflects the total expenditure, as a function of screening cost when randomizing a single position using five different saturation mutagenesis schemes [78].

The data in Figure 4.19 is based on the assumption that the primer cost equals the fixed cost. However, the economically optimal randomization scheme is influenced by changes in both primer cost and screening cost, relative to the fixed cost of the transformation of interest.

Figure 4.20 outlines how cost space, shown in the 2-dimensional plane with axes as primer and screening costs, becomes partitioned into mutually exclusive regions. Thus, each of them corresponds to a different optimal choice of randomization procedure. It turns out, for example, that NNK is economically the

Table 4.5 Quick quality control and Q-values.

Library	Randomization Scheme	Primer		QQC charts and Q-values			
		Supplier	Purity	Theoretical Q-value	Solid Q_{pool}	Liquid Q_{pool}	Experimental Q_{exptl}
1	NNK	IDT Technologies	Desalted	1.000	0.470	0.515	0.798
2	NNK	IDT Technologies	HPLC	1.000	0.721	0.706	0.837
3	NNK	Life Technologies	Desalted	1.000	0.484	0.438	0.800
4	NNK	Life Technologies	HPLC	1.000	0.486	0.620	0.773
5	NNK	Metabion International	Desalted	1.000	0.770	0.657	0.863
6	NNK	Metabion International	HPLC	1.000	0.766	0.759	0.888
7	NNS	Metabion International	Desalted	1.000	0.643	0.549	0.872
8	NNS	Metabion International	HPLC	1.000	0.614	0.592	0.844
9	22c-trick	Metabion International	Desalted	1.000	0.557	0.575	0.979
10	22-trick	Metabion International	HPLC	1.000	0.590	0.579	0.798
11	Tang	Metabion International	Desalted	1.000	0.668	0.543	0.834
12	Tang	Metabion International	HPLC	1.000	0.581	0.516	0.742

Library specifications with resulting QQC charts and Q-values [78]. The three pie charts in each column/row correspond to the three positions in a codon. Black = guanidine; green = adenosine; red = threonine; blue = cytosine.

Source: Ref. [78]/Springer Nature/CC BY 4.0.

first choice when the screening cost is lower than the primer cost by a factor of at least 4.25, as given by the slope of the line separating the NNK and the 22c-trick areas. When the screening cost increases, then the 22c-trick becomes optimal, followed by the Tang approach, and finally by the individual generation of all 20 variants. Generally, the 22c-trick and Tang approaches are fairly similar. In further analytical refinements, schemes such as the one depicted in Figure 4.20 were modified [78].

These and other analyses, e.g. of annealing probabilities, culminated in several general conclusions and guidelines. When the screening costs are high, then

Table 4.6 Summary of P450-BM3 sequencing results obtained from 96 single colonies formed on agar plates per library.

Library	Successfully randomized	Yield (%)	>1 base per position	Non-target mutations	Primer misinsertions	Suboptimal sequencing	Missed amino acids
1	68	72.3	19	6	1	2	Met, Asp
2	65	68.4	24	6	—	1	Lys, Asn, His
3	55	59.8	29	5	3	4	Met, Lys, Asn, Phe
4	54	58.1	37	2	—	3	Ile
5	50	52.1	42	4	—	—	—
6	64	66.7	27	5	—	—	—
7	39	41.9	50	4	—	3	Met, Ile, Gln, Trp
8	64	67.4	24	6	1	1	Phe, Tyr
9	57	59.4	37	1	1	—	—
10	67	69.8	24	4	1	—	Asp, Tyr
11	51	53.7	41	1	2	1	Asp, Tyr
12	65	69.1	23	6	—	2	Lys, Asp, Tyr

Source: Ref. [78]/Springer Nature/CC BY 4.0.

approaches such as the 22c-trick [79] or the Tang-method [80] are economically superior. A faster and more economical method for performing the QQC is to use liquid cultures, also presented in this study [78]. It is highly recommended in order to save time and money when designing directed evolution experiments using saturation mutagenesis [76, 78]. Of significant practical importance is the conclusion that QQC should be combined with the Q-values in order to assess the overall quality of libraries.

It was also shown that primer purity has a notable effect on library yield [78]. Some suppliers offer primers of higher quality than others without additional purification steps. Researchers are advised to read the details of this study, which provides guidelines for choosing a randomization scheme optimally as a function of the screening costs and other experimental parameters [78]. These guidelines are also of help when applying any PCR-based method for library creation such as epPCR or DNA shuffling, including combinatorial gene preparation [81], gene assembly [82], and overlap extension PCR [83]. Finally, yet another cost aspect that needs to be considered concerns a different approach to mutant library construction [13], namely combinatorial solid-phase gene synthesis (Section 4.5).

4.5 Generating Mutant Libraries by Combinatorial Solid-Phase Gene Synthesis: The Future of Directed Evolution?

All of the methods of mutant library generation utilizing epPCR, saturation mutagenesis, or DNA shuffling are PCR-based [13, 83]. As outlined in Section 3.8, a principally different approach is based on combinatorial solid-phase gene

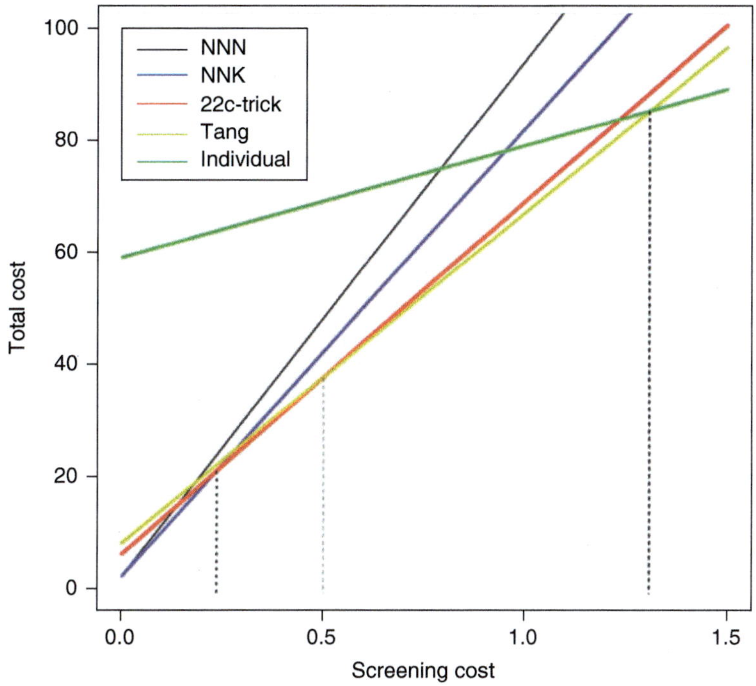

Figure 4.19 Total cost as a function of screening cost, when randomizing a single position using five randomization schemes. Primer cost is cprimer = 1. Source: [78]/Springer Nature/CC BY 4.0.

synthesis, which can be used to create mutant libraries at DNA level [84]. In the seminal study [84a], it was demonstrated that the quality of such synthetic gene libraries can be considerably higher than that of their respective PCR-based counterparts, which reduces the screening effort drastically. Nowadays, just two to three years following the appearance of the best study [84a], researchers in general can design their saturation mutagenesis libraries, send the genetic information to an appropriate company, which offers high quality combinatorial gene synthesis, and receive a few weeks later the respective libraries [84c, 85].

Gene synthesis on micro-chips appears to be the fastest and ultimately cheapest [84b]. When applied to mutant library creation for the purpose of directed evolution, new perspectives emerge. If prices continue to go down, this approach to library construction can be expected to revolutionize future directed evolution because the protein engineer can depend on the second law of directed evolution [13]: *You get what you designed* (Figure 4.21). This means that all of the time traditionally invested when applying the molecular biological tools for preparing saturation mutagenesis libraries can now be invested in the better design of such mutants. Moreover, due to the distinctly higher quality of synthetic libraries, less screening is required, and better catalytic profiles can be expected.

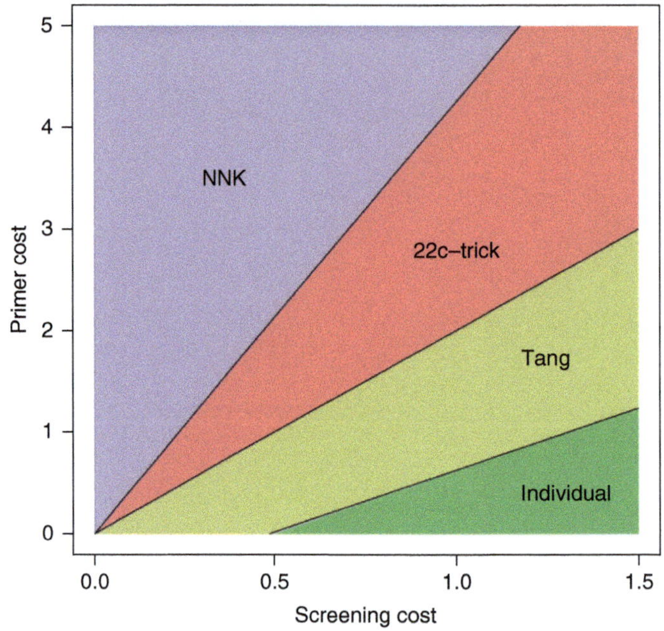

Figure 4.20 Cost space partitioned into regions according to the optimal randomization scheme (a single randomized position, assuming 100% yield, and no WT bias). Source: [78]/Springer Nature/CC BY 4.0.

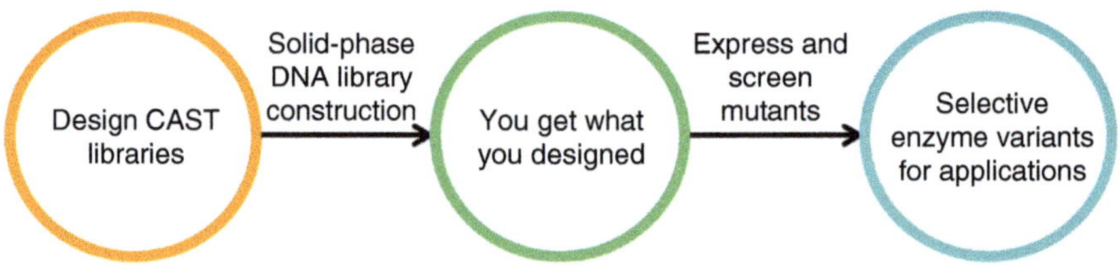

Figure 4.21 How to proceed when choosing solid-phase gene for mutant library construction and evolving selective and active variants.

4.6 Fusing Directed Evolution and Rational Design: New Examples of Focused Rational Iterative Site-Specific Mutagenesis (FRISM)

As pointed out in a review of methodology development in directed evolution [12a], and highlighted in Chapter 3, the emergence of FRISM constitutes a significant advance in protein engineering of selective and active enzymes as catalysts in organic chemistry and biotechnology [86].

While the results of the initial FRISM-study are indisputably impressive, it remains to be demonstrated that FRISM is a powerful tool of great generality. Therefore, it was tested using other enzyme types. For example, FRISM was

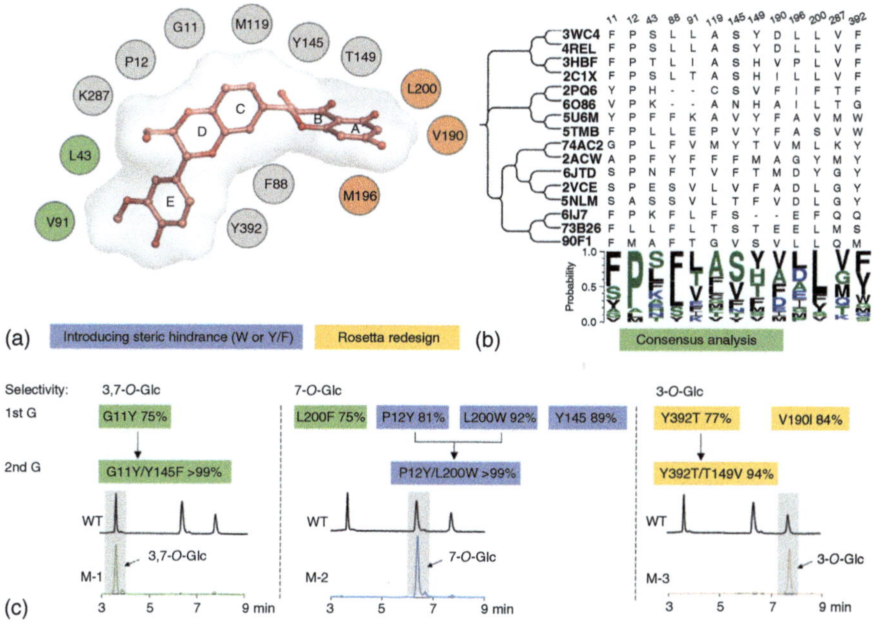

Figure 4.22 Engineering the regioselectivity of UGT74AC2 guided by Focused Rational Iterative Site-specific Mutagenesis (FRISM). (a) Key residues lining the acceptor binding pocket. (b) Consensus analysis. (c) HPLC profiles of the three best mutants G11Y/Y145F (M1), P12Y/L200W (M2), and Y392T/T149V (M3). Source: [87a]/With permission of Elsevier.

applied to a glycosyltransferase from *Siraitia grosvenorii* (UGT74AC2), with the aim of tailoring its glycosylation regioselectivity. Taking the polyhydroxy compound silybin as the model substrate, WT UGT74AC2 was found to deliver a product mixture with 22%, 39%, and 39% selectivity on the 3-OH, 7-OH, and 3,7-*O*-diglycoside, respectively [87a]. Based on the crystallographic structure of UGT74AC2, and docking analysis, several residues were assigned as hot spots (Figure 4.22a). Subsequently, each individual residue was substituted by a limited number of predictive suggestions, accessed from phylogenetic analysis (Figure 4.22b), structural information (e.g. introducing steric hindrance), as well as application of Baker's Rosetta computational package. After two rounds of computational mutagenesis and testing less than 100 variants, three double mutants Y392T/T149V, P12Y/L200W, and G11Y/Y145F were obtained, which enable 94%, >99%, and >99% selectivity on the 3-OH, 7-OH, and 3,7-*O*-diglycoside, respectively (Figure 4.22c). This work clearly supports the original claim that the application of FRISM in enzyme regio- and stereo-selective engineering is a viable approach [87b].

FRISM was also applied and extended using other types of enzymes [88]. As noted in Section 3.10, *optimally executed CAST/ISM and FRISM are complementary rational approaches, the difference being small* [89]. In both options, the Rosetta algorithms [6, 7] or the Damborsky metrics [2a] can be applied as an aid [12a, 89].

References

1. Kiss, G., Çelebi-Ölçüm, N., Moretti, R. et al. (2013). Computational enzyme design. *Angew. Chem. Int. Ed.* **52** (22): 5700–5725.
2. (a) Marques, S.M., Planas-Iglesias, J., and Damborsky, J. (2021). Web-based tools for computational enzyme design. *Curr. Opin. Struct. Biol.* **69**: 19–34. (b) Bunzel, H.A., Anderson, J.L.R., and Mulholland, A.J. (2021). Designing better enzymes: Insights from directed evolution. *Curr. Opin. Struct. Biol.* **67**: 212–218. (c) Lin, Y.W. (2020). Rational design of heme enzymes for biodegradation of pollutants toward a green future. *Biotechnol. Appl. Biochem.* **67** (4): 484–494. (d) Dinmukhamed, T., Huang, Z., Liu, Y. et al. (2020). Current advances in design and engineering strategies of industrial enzymes. *Systems Microbiol. Biomanuf.* **1**: 15–23. (e) Moulis, C., Guieysse, D., Morel, S. et al. (2021). Natural and engineered transglycosylases: green tools for the enzyme-based synthesis of glycoproducts. *Curr. Opin. Chem. Biol.* **61**: 96–106. (f) Mariz, B.P., Carvalho, S., Batalha, I.L., and Pina, A.S. (2021). Artificial enzymes bringing together computational design and directed evolution. *Org. Biomol. Chem.* **19** (9): 1915–1925. (g) Coates, T.L., Young, N., Jarrett, A.J. et al. (2021). Current computational methods for enzyme design. *Mod. Phys. Lett. A* **35**: 2150155.
3. Meng, Q., Ramírez-Palacios, C., Capra, N. et al. (2021). Computational redesign of an ω-transaminase from *Pseudomonas jessenii* for asymmetric synthesis of enantiopure bulky amines. *ACS Catal.* **11** (17): 10733–10747.
4. (a) Gomm, A. and O'Reilly, E. (2018). Transaminases for chiral amine synthesis. *Curr. Opin. Chem. Biol.* **43**: 106–112. (b) Li, G., Wang, J., and Reetz, M.T. (2018). Biocatalysis for the pharmaceutical industry created by structure-guided directed evolution of stereoselective enzymes. *Bioorg. Med. Chem.* **26**: 1241–1251.
5. (a) Abrahamson, M.J., Vázquez-Figueroa, E., Woodall, N.B. et al. (2012). Development of an amine dehydrogenase for synthesis of chiral amines. *Angew. Chem. Int. Ed.* **51** (16): 3969–3972. (b) Abrahamson, M.J., Wong, J.W., and Bommarius, A.S. (2013). The evolution of an amine dehydrogenase biocatalyst for the asymmetric production of chiral amines. *Adv. Synth. Catal.* **355** (9): 1780–1786. (c) Li, J.Y., Hui, H.T., Yi, Y. et al. (2015). Engineering of amine dehydrogenase for asymmetric reductive amination of ketone by evolving *Rhodococcus* phenylalanine dehydrogenase. *ACS Catal.* **5** (2): 1119–1122. (d) Weiß, M.S., Pavlidis, I.V., Spurr, P. et al. (2016). Protein-engineering of an amine transaminase for the stereoselective synthesis of a pharmaceutically relevant bicyclic amine. *Org. Biomol. Chem.* **14** (43): 10249–10254. (e) Weiß, M.S., Pavlidis, I.V., Spurr, P. et al. (2017). Amine transaminase engineering for spatially bulky substrate acceptance. *ChemBioChem* **18** (11): 1022–1026. (f) Yasukawa, K., Nakano, S., and Asano, Y. (2014). Tailoring D-amino acid oxidase from the pig kidney to *R*-stereoselective amine oxidase and its use in the deracemization of α-methylbenzylamine. *Angew. Chem. Int. Ed.* **53** (17): 4428–4431. (g) Chen, F., Cosgrove, S.C., Birmingham, W.R. et al. (2019). Enantioselective synthesis of chiral vicinal amino alcohols using amine dehydrogenases. *ACS Catal.* **9** (12): 11813–11818. (h) Land, H., Ruggieri, F., Szekrenyi, A. et al. (2020). Engineering the active site of an (*S*)-selective amine

transaminase for acceptance of doubly bulky primary amines. *Adv. Synth. Catal.* **362** (4): 812–821.

6. Richter, F., Leaver-Fay, A., Khare, S.D. et al. (2011). De novo enzyme design using Rosetta3. *PLoS One* **6** (5): e19230.
7. (a) Das, R. and Baker, D. (2008). Macromolecular modeling with Rosetta. *Annu. Rev. Biochem.* **77**: 363–382. (b) Baek, M. and Baker, D. (2022). Deep learning and protein structure modeling. *Nat. Methods* **19**: 11–26. (c) Lovelock, S.L., Crawshaw, R., Basler, S. et al. (2022). The road to fully programmable protein catalysis. *Nature* **606**: 449–458.
8. Zhou, H., Wang, B., Wang, F. et al. (2019). Chemo- and regioselective dihydroxylation of benzene to hydroquinone enabled by engineered cytochrome P450 monooxygenase. *Angew. Chem. Int. Ed.* **58** (3): 764–768.
9. Li, A., Ilie, A., Sun, Z. et al. (2016). Whole-cell-catalyzed multiple regio- and stereoselective functionalizations in cascade reactions enabled by directed evolution. *Angew. Chem. Int. Ed.* **55** (39): 12026–12029.
10. (a) Risso, V.A., Gavira, J.A., Mejia-Carmona, D.F. et al. (2013). Hyperstability and substrate promiscuity in laboratory resurrections of Precambrian β-lactamases. *J. Am. Chem. Soc.* **135** (8): 2899–2902. (b) Modi, T., Risso, V.A., Martinez-Rodriguez, S. et al. (2021). Hinge-shift mechanism as a protein design principle for the evolution of β-lactamases from substrate promiscuity to specificity. *Nat. Commun.* **12** (1): 1852.
11. (a) Crean, R.M., Gardner, J.M., and Kamerlin, S.C.L. (2020). Harnessing conformational plasticity to generate designer enzymes. *J. Am. Chem. Soc.* **142** (26): 11324–11342. (b) Kaczmarski, J.A., Mahawaththa, M.C., Feintuch, A. et al. (2020). Altered conformational sampling along an evolutionary trajectory changes the catalytic activity of an enzyme. *Nat. Commun.* **11**: 5945. (c) Acevedo-Rocha, C.G., Li, A., D'Amore, L. et al. (2021). Pervasive cooperative mutational effects on multiple catalytic enzyme traits emerge via long-range conformational dynamics. *Nat. Commun.* **12** (1): 1621. (d) Qu, G., Bi, Y., Liu, B. et al. (2022). Unlocking the stereoselectivity and substrate acceptance of enzymes: proline-induced loop engineering test. *Angew. Chem. Int. Ed.* **61** (1): e202110793. (e) Karamitros, C.S., Murray, K., Winemiller, B. et al. (2022). Leveraging intrinsic flexibility to engineer enhanced enzyme catalytic activity. *Proc. Natl. Acad. Sci. U.S.A.* **119** (23): e2118979119.
12. Selected reviews of directed evolution of enzymes as catalysts in organic and pharmaceutical chemistry and biotechnology covering mainly stereo- and/or regioselectivity and activity: (a) Qu, G., Li, A., Acevedo-Rocha, C.G. et al. (2020). The crucial role of methodology development in directed evolution of selective enzymes. *Angew. Chem. Int. Ed.* **59**: 13204–13231. (b) Arnold, F.H. (2019). Innovation by evolution: bringing new chemistry to life (nobel lecture). *Angew. Chem. Int. Ed.* **58**: 14420–14426. (c) Chen, K. and Arnold, F.H. (2020). Engineering new catalytic activities in enzymes. *Nat. Catal.* **3**: 203–213. (d) Wang, Y., Xue, P., Cao, M. et al. (2021). Directed evolution: methodologies and applications. *Chem. Rev.* https://doi.org/10.1021/acs.chemrev.1c00260. (e) Currin, A., Swainston, N., Day, P.J., and Kell, D.B. (2015). Synthetic biology for the directed evolution of protein biocatalysts: navigating sequence space intelligently. *Chem. Soc. Rev.* **44**: 1172–1239. (f) Denard, C.A., Ren, H., and Zhao, H. (2015). Improving and repurposing biocatalysts via directed evolution. *Curr. Opin. Chem. Biol.* **25**: 55–64. (g) Siloto, R.M.P. and Weselake, R.J. (2012). Site saturation mutagenesis: methods and application in protein engineering. *Biocatal. Agric. Biotechnol.* **1**: 181–189. (h) Reetz, M.T. (2011). Laboratory evolution of stereoselective enzymes: a prolific source of catalysts for asymmetric reactions. *Angew. Chem. Int. Ed.* **50** (1): 138–174. (i) Reetz, M.T. (2012). *Enzyme Catalysis in Organic Synthesis*, 3e (ed. K. Drauz, H. Gröger and O. May), 119–190. Weinheim: Wiley-VCH Verlag GmbH.

13. Reetz, M.T. (2016). *Directed Evolution of Selective Enzymes Catalysts for Organic Chemistry and Biotechnology*. Weinheim: Wiley-VCH.
14. (a) Bloom, J.D., Labthavikul, S.T., Otey, C.R., and Arnold, F.H. (2006). Protein stability promotes evolvability. *Proc. Natl. Acad. Sci. U.S.A.* **103**: 5869–5874.
15. Li, G., Maria-Solano, M.A., Romero-Rivera, A. et al. (2017). Inducing high activity of a thermophilic enzyme at ambient temperatures by directed evolution. *Chem. Commun.* **53** (68): 9454–9457.
16. Bougioukou, D.J., Kille, S., Taglieber, A., and Reetz, M.T. (2009). Directed evolution of an enantioselective enoate-reductase: testing the utility of iterative saturation mutagenesis. *Adv. Synth. Catal.* **351** (18): 3287–3305.
17. (a) Peisajovich, S.G. and Tawfik, D.S. (2007). Protein engineers turned evolutionists. *Nat. Methods* **4** (12): 991–994. (b) DePristo, M.A. (2007). The subtle benefits of being promiscuous: adaptive evolution potentiated by enzyme promiscuity. *HFSP J.* **1** (2): 94–98. (c) Rockah-Shmuel, L., Toth-Petroczy, A., and Tawfik, D.S. (2015). Systematic mapping of protein mutational space by prolonged drift reveals the deleterious effects of seemingly neutral mutations. *PLoS Comput. Biol.* **11** (8): e1004421. (d) Bloom, J.D. and Arnold, F.H. (2009). In the light of directed evolution: pathways of adaptive protein evolution. *Proc. Natl. Acad. Sci. U.S.A.* **106** (Suppl 1): 9995–10000.
18. Eigen, M., McCaskill, J., and Schuster, P. (1988). Molecular quasi-species. *J. Phys. Chem.* **92** (24): 6881–6891.
19. Kurtovic, S. and Mannervik, B. (2009). Identification of emerging quasi-species in directed enzyme evolution. *Biochemistry* **48** (40): 9330–9339.
20. Tracewell, C.A. and Arnold, F.H. (2009). Directed enzyme evolution: climbing fitness peaks one amino acid at a time. *Curr. Opin. Chem. Biol.* **13**: 3–9.
21. (a) Reetz, M.T., Wang, L.-W., and Bocola, M. (2006). Directed evolution of enantioselective enzymes: Iterative cycles of CASTing for probing protein-sequence space. *Angew. Chem. Int. Ed.* **45** (8): 1236–1241. (b) Initial study reporting four cycles of ISM for enhancing the stereoselectivity of the epoxide hydrolase ANEH: Reetz, M.T. (2005). Evolution im Reagenzglas: Neue Perspektiven für die Weiße Biotechnologie. *Tätigkeitsberichte der Max-Planck-Gesellschaft*. pp. 327–331.
22. Reetz, M.T. and Carballeira, J.D. (2007). Iterative saturation mutagenesis (ISM) for rapid directed evolution of functional enzymes. *Nat. Protoc.* **2** (4): 891–903.
23. (a) Reetz, M.T., Kahakeaw, D., and Lohmer, R. (2008). Addressing the numbers problem in directed evolution. *ChemBioChem* **9** (11): 1797–1804. (b) Clouthier, C.M., Kayser, M.M., and Reetz, M.T. (2006). Designing new Baeyer–Villiger monooxygenases using restricted casting. *J. Org. Chem.* **71** (22): 8431–8437.
24. Reetz, M.T. and Sanchis, J. (2008). Constructing and analyzing the fitness landscape of an experimental evolutionary process. *ChemBioChem* **9** (14): 2260–2267.
25. Reetz, M.T., Kahakeaw, D., and Sanchis, J. (2009). Shedding light on the efficacy of laboratory evolution based on iterative saturation mutagenesis. *Mol. BioSyst.* **5** (2): 115–122.
26. Reetz, M.T., Prasad, S., Carballeira, J.D. et al. (2010). Iterative saturation mutagenesis accelerates laboratory evolution of enzyme stereoselectivity: rigorous comparison with traditional methods. *J. Am. Chem. Soc.* **132** (26): 9144–9152.
27. Reetz, M.T. (2013). The importance of additive and non-additive mutational effects in protein engineering. *Angew. Chem. Int. Ed.* **52** (10): 2658–2666.
28. Chen, Z. and Zhao, H. (2005). Rapid creation of a novel protein function by *in vitro* coevolution. *J. Mol. Biol.* **348** (5): 1273–1282.
29. Savile, C.K., Janey, J.M., Mundorff, E.C. et al. (2010). Biocatalytic asymmetric synthesis of chiral amines from ketones applied to sitagliptin manufacture. *Science* **329** (5989): 305–309.
30. Zhang, J.-H., Dawes, G., and Stemmer, W.P.C. (1997). Directed evolution of a fucosidase from a galactosidase by

DNA shuffling and screening. *Proc. Natl. Acad. Sci. U.S.A.* **94** (9): 4504–4509.

31. Parikh, M.R. and Matsumura, I. (2005). Site-saturation mutagenesis is more efficient than DNA shuffling for the directed evolution of ß-fucosidase from ß-galactosidase. *J. Mol. Biol.* **352**: 621–628.

32. Juers, D.H., Heightman, T.D., Vasella, A. et al. (2001). A structural view of the action of *Escherichia coli (lacZ)* β-galactosidase. *Biochemistry* **40** (49): 14781–14794.

33. Reetz, M.T., Wilensek, S., Zha, D., and Jaeger, K.-E. (2001). Directed evolution of an enantioselective enzyme through combinatorial multiple-cassette mutagenesis. *Angew. Chem. Int. Ed.* **40** (19): 3589–3591.

34. (a) Nov, Y. (2012). When second best is good enough: another probabilistic look at saturation mutagenesis. *Appl. Environ. Microbiol.* **78** (1): 258–262. (b) Nov, Y. (2013). Fitness loss and library size determination in saturation mutagenesis. *PLoS One* **8** (7): e68069. (c) Nov, Y. (2014). Probabilistic methods in directed evolution: library size, mutation rate, and diversity. *Methods Mol. Biol.* **1179**: 261–278. (d) Nov, Y., Fulton, A., and Jaeger, K.-E. (2013). Optimal scanning of all single-point mutants of a protein. *J. Comput. Biol.* **20**: 990–997.

35. Reetz, M.T., Zonta, A., Schimossek, K. et al. (1997). Creation of enantioselective biocatalysts for organic chemistry by in vitro evolution. *Angew. Chem. Int. Ed.* **36** (24): 2830–2832.

36. Liebeton, K., Zonta, A., Schimossek, K. et al. (2000). Directed evolution of an enantioselective lipase. *Chem. Biol.* **7** (9): 709–718.

37. Reetz, M.T. (2004). Controlling the enantioselectivity of enzymes by directed evolution: practical and theoretical ramifications. *Proc. Natl. Acad. Sci. U.S.A.* **101** (16): 5716–5722.

38. Reetz, M.T., Bocola, M., Carballeira, J.D. et al. (2005). Expanding the range of substrate acceptance of enzymes: combinatorial active-site saturation test. *Angew. Chem. Int. Ed.* **44** (27): 4192–4196.

39. (a) Schimossek, K. (1998). Dissertation, *Neue Lipasen für die organische Chemie*, Ruhr-Universität Bochum. (b) Wilensek, S. (2001). Dissertation, *Gerichtete Evolution als ein Mittel zur Erzeugung enantioselektiver Enzyme für die organische Synthese*, Ruhr-Universität Bochum.

40. Nardini, M., Lang, D.A., Liebeton, K. et al. (2000). Crystal structure of *Pseudomonas aeruginosa* lipase in the open conformation: the prototype for family I.1 of bacterial lipases. *J. Biol. Chem.* **275** (40): 31219–31225.

41. Crameri, A. and Stemmer, W.P.C. (1995). Combinatorial multiple cassette mutagenesis creates all the permutations of mutant and wild-type sequences. *Biotechniques* **18**: 194–196.

42. (a) Bocola, M., Otte, N., Jaeger, K.-E. et al. (2004). Learning from directed evolution: theoretical investigations into cooperative mutations in lipase enantioselectivity. *ChemBioChem* **5** (2): 214–223. (b) Reetz, M.T., Puls, M., Carballeira, J.D. et al. (2007). Learning from directed evolution: further lessons from theoretical investigations into cooperative mutations in lipase enantioselectivity. *ChemBioChem* **8** (1): 106–112.

43. Zha, D., Wilensek, S., Hermes, M. et al. (2001). Complete reversal of enantioselectivity of an enzyme-catalyzed reaction by directed evolution. *Chem. Commun.* **24**: 2664–2665.

44. Gumulya, Y., Sanchis, J., and Reetz, M.T. (2012). Many pathways in laboratory evolution can lead to improved enzymes: how to escape from local minima. *ChemBioChem* **13** (7): 1060–1066.

45. Perspective on biocatalysts including the development of directed evolution: Reetz, M.T. (2013). Biocatalysis in organic chemistry and biotechnology: past, present, and future. *J. Am. Chem. Soc.* **135** (34): 12480–12496.

46. (a) Faber, K. (2011). *Biotransformations in Organic Chemistry*, 6e. Heidelberg: Springer. (b) Liese, A., Seelbach, K., and Wandrey, C. (ed.) (2006). *Industrial Biotransformations*. Weinheim: Wiley-VCH. (c) Gotor, V., Alfonso, I., and Garcia-Urdiales, E. (ed.) (2008). *Asymmetric Organic Synthesis with*

Enzymes. Weinheim: Wiley-VCH. (d) Tao, J., Lin, G.-Q., and Liese, A. (2009). *Biocatalysis for the Pharmaceutical Industry*. Singapore: Wiley. (e) Svendsen, A. (ed.) (2016). *Understanding Enzymes: Function, Design, Engineering, and Analysis*. Singapore: Pan Stanford Publishing.

47. (a) Paramesvaran, J., Hibbert, E.G., Russell, A.J., and Dalby, P.A. (2009). Distributions of enzyme residues yielding mutants with improved substrate specificities from two different directed evolution strategies. *Protein Eng. Des. Sel.* **22** (7): 401–411. (b) Morley, K.L. and Kazlauskas, R.J. (2005). Improving enzyme properties: when are closer mutations better? *Trends Biotechnol.* **23** (5): 231–237. (c) Hibbert, E.G. and Dalby, P.A. (2005). Directed evolution strategies for improved enzymatic performance. *Microb. Cell Fact.* **4** (1): 29.

48. (a) Reetz, M.T. and Wu, S. (2008). Greatly reduced amino acid alphabets in directed evolution: making the right choice for saturation mutagenesis at homologous enzyme positions. *Chem. Commun.* **43**: 5499–5501. (b) Sandström, A.G., Wikmark, Y., Engström, K. et al. (2012). Combinatorial reshaping of the *Candida antarctica* lipase a substrate pocket for enantioselectivity using an extremely condensed library. *Proc. Natl. Acad. Sci. U.S.A.* **109** (1): 78–83. (c) Wikmark, Y., Svedendahl Humble, M., and Bäckvall, J.E. (2015). Combinatorial library based engineering of *Candida antarctica* lipase A for enantioselective transacylation of *sec*-alcohols in organic solvent. *Angew. Chem. Int. Ed.* **54**: 4284–4288. (d) Sun, Z., Wikmark, Y., Bäckvall, J.-E., and Reetz, M.T. (2016). New concepts for increasing the efficiency in directed evolution of stereoselective enzymes. *Chem. Eur. J.* **22**: 5046–5054. (e) Chen, M.M.Y., Snow, C.D., Vizarra, C.L. et al. (2012). Comparison of random mutagenesis and semi-rational designed libraries for improved cytochrome P450 BM3-catalyzed hydroxylation of small alkanes. *Protein Eng. Des. Sel.* **25**: 171–178. (f) Evans, B.S., Chen, Y., Metcalf, W.W. et al. (2011). Directed evolution of the nonribosomal peptide synthetase AdmK generates new andrimid derivatives in vivo. *Chem. Biol.* **18** (5): 601–607. (g) Swe, P.M., Copp, J.N., Green, L.K. et al. (2012). Targeted mutagenesis of the *Vibrio fischeri* flavin reductase FRase I to improve activation of the anticancer prodrug CB1954. *Biochem. Pharmacol.* **84** (6): 775–783. (h) Dudek, H., Fink, M., Shivange, A. et al. (2014). Extending the substrate scope of a Baeyer–Villiger monooxygenase by multiple-site mutagenesis. *Appl. Microbiol.* **98** (9): 4009–4020. (i) Zhang, L., Lu, L., Fan, S. et al. (2015). One-step synthesis of α-Gal epitope and globotriose derivatives by an engineered α-galactosidase. *RSC Adv.* **5** (29): 22361–22364. (j) Chuang, H.-Y., Suen, C.-S., Hwang, M.-J., and Roffler, S.R. (2015). Toward reducing immunogenicity of enzyme replacement therapy: altering the specificity of human β-glucuronidase to compensate for α-iduronidase deficiency. *Protein Eng. Des. Sel.* **28** (11): 519–530.

49. (a) Truppo, M.D., Turner, N.J., and Rozzell, J.D. (2009). Efficient kinetic resolution of racemic amines using a transaminase in combination with an amino acid oxidase. *Chem. Commun.* **16**: 2127–2129. (b) Koszelewski, D., Clay, D., Rozzell, D., and Kroutil, W. (2009). Deracemisation of α-chiral primary amines by a one-pot, two-step cascade reaction catalysed by ω-transaminases. *Eur. J. Org. Chem.* **14**: 2289–2292.

50. Zou, J., Hallberg, B.M., Bergfors, T. et al. (2000). Structure of *Aspergillus niger* epoxide hydrolase at 1.8 Å resolution: implications for the structure and function of the mammalian microsomal class of epoxide hydrolases. *Structure* **8** (2): 111–122.

51. Reetz, M.T., Becker, M.H., Klein, H.-W., and Stöckigt, D. (1999). A method for high-throughput screening of enantioselective catalysts. *Angew. Chem. Int. Ed.* **38** (12): 1758–1761.

52. Sullivan, B., Walton, A.Z., and Stewart, J.D. (2013). Library construction and evaluation for site saturation mutagenesis. *Enzyme Microb. Technol.* **53** (1): 70–77.

53. Emrén, L.O., Kurtovic, S., Runarsdottir, A. et al. (2006). Functionally diverging molecular quasi-species evolve by crossing two enzymes. *Proc. Natl. Acad. Sci. U.S.A.* **103** (29): 10866–10870.
54. Gumulya, Y. and Reetz, M.T. (2011). Enhancing the thermal robustness of an enzyme by directed evolution: least favorable starting points and inferior mutants can map superior evolutionary pathways. *ChemBioChem* **12** (16): 2502–2510.
55. Bastian, S., Liu, X., Meyerowitz, J.T. et al. (2011). Engineered ketol-acid reductoisomerase and alcohol dehydrogenase enable anaerobic 2-methylpropan-1-ol production at theoretical yield in *Escherichia coli*. *Metab. Eng.* **13** (3): 345–352.
56. Ihssen, J., Haas, J., Kowarik, M. et al. (2015). Increased efficiency of *Campylobacter jejuni N*-oligosaccharyltransferase PglB by structure-guided engineering. *Open Biol.* **5** (4): 140227.
57. Kwan, D.H., Constantinescu, I., Chapanian, R. et al. (2015). Toward efficient enzymes for the generation of universal blood through structure-guided directed evolution. *J. Am. Chem. Soc.* **137** (17): 5695–5705.
58. Goldstein, J., Siviglia, G., Hurst, R. et al. (1982). Group B erythrocytes enzymatically converted to group O survive normally in A, B, and O individuals. *Science* **215** (4529): 168–170.
59. Higgins, M.A., Whitworth, G.E., El Warry, N. et al. (2009). Differential recognition and hydrolysis of host carbohydrate antigens by *Streptococcus pneumoniae* family 98 glycoside hydrolases. *J. Biol. Chem.* **284** (38): 26161–26173.
60. Sun, Z., Ilie, A., and Reetz, M.T. (2015). Towards the production of universal blood by structure-guided directed evolution of glycoside hydrolases. *Angew. Chem. Int. Ed.* **54** (32): 9158–9160.
61. Rahfeld, P., Sim, L., Moon, H. et al. (2019). An enzymatic pathway in the human gut microbiome that converts A to universal O type blood. *Nat. Microbiol.* **4** (9): 1475–1485.
62. (a) Fraaije, M., Wu, J., Heuts, D.H.M. et al. (2005). Discovery of a thermostable Baeyer–Villiger monooxygenase by genome mining. *Appl. Microbiol. Biotechnol.* **66** (4): 393–400. (b) de Gonzalo, G., Pazmiño, D.E.T., Ottolina, G. et al. (2005). Oxidations catalyzed by phenylacetone monooxygenase from *Thermobifida fusca*. *Tetrahedron: Asymmetry* **16** (18): 3077–3083.
63. Malito, E., Alfieri, A., Fraaije, M.W., and Mattevi, A. (2004). Crystal structure of a Baeyer–Villiger monooxygenase. *Proc. Natl. Acad. Sci. U.S.A.* **101** (36): 13157–13162.
64. Bocola, M., Schulz, F., Leca, F. et al. (2005). Converting phenylacetone monooxygenase into phenylcyclohexanone monooxygenase by rational design: towards practical Baeyer–Villiger monooxygenases. *Adv. Synth. Catal.* **347** (7–8): 979–986.
65. (a) Jochens, H. and Bornscheuer, U.T. (2010). Natural diversity to guide focused directed evolution. *ChemBioChem* **11** (13): 1861–1866. (b) Höhne, M., Schätzle, S., Jochens, H. et al. (2010). Rational assignment of key motifs for function guides in silico enzyme identification. *Nat. Chem. Biol.* **6** (11): 807–813. (c) Godinho, L.F., Reis, C.R., Rozeboom, H.J. et al. (2012). Enhancement of the enantioselectivity of carboxylesterase a by structure-based mutagenesis. *J. Biotechnol.* **158** (1–2): 36–43.
66. Reetz, M.T. and Wu, S. (2009). Laboratory evolution of robust and enantioselective Baeyer–Villiger monooxygenases for asymmetric catalysis. *J. Am. Chem. Soc.* **131** (42): 15424–15432.
67. (a) Fersht, A. (2000). *Structure and Mechanism in Protein Science*. New York: W.H. Freeman and Co. (b) Shen, B., Bai, J., and Vihinen, M. (2008). Physicochemical feature-based classification of amino acid mutations. *Protein Eng. Des. Sel.* **21**: 37–44.
68. Rabe, P., Kamps, J.J.A.G., Sutherlin, K.D. et al. (2021). X-ray free-electron laser studies reveal correlated motion during isopenicillin *N* synthase catalysis. *Sci. Adv.* **7** (34): eabh0250.

69. Engström, K., Nyhlén, J., Sandström, A.G., and Bäckvall, J.-E. (2010). Directed evolution of an enantioselective lipase with broad substrate scope for hydrolysis of α-substituted esters. *J. Am. Chem. Soc.* **132** (20): 7038–7042.
70. Zheng, H. and Reetz, M.T. (2010). Manipulating the stereoselectivity of limonene epoxide hydrolase by directed evolution based on iterative saturation mutagenesis. *J. Am. Chem. Soc.* **132** (44): 15744–15751.
71. (a) Hopmann, K.H., Hallberg, B.M., and Himo, F. (2005). Catalytic mechanism of limonene epoxide hydrolase, a theoretical study. *J. Am. Chem. Soc.* **127** (41): 14339–14347. (b) Arand, M., Hallberg, B., Zou, J. et al. (2003). Structure of *Rhodococcus erythropolis* limonene-1,2-epoxide hydrolase reveals a novel active site. *EMBO J.* **22** (11): 2583–2592.
72. (a) Sun, Z., Lonsdale, R., Kong, X.-D. et al. (2015). Reshaping an enzyme binding pocket for enhanced and inverted stereoselectivity: use of smallest amino acid alphabets in directed evolution. *Angew. Chem. Int. Ed.* **54**: 12410–12415. (b) Sun, Z., Lonsdale, R., Wu, L. et al. (2016). Structure-guided triple code saturation mutagenesis: efficient tuning of the stereoselectivity of an epoxide hydrolase. *ACS Catal.* **6**: 1590–1597. (c) Sun, Z., Lonsdale, R., Ilie, A. et al. (2016). Catalytic asymmetric reduction of difficult-to-reduce ketones: triple code saturation mutagenesis of an alcohol dehydrogenase. *ACS Catal.* **6**: 1598–1605.
73. Chen, X., Zhang, H., Maria-Solano, M.A. et al. (2019). Efficient reductive desymmetrization of bulky 1,3-cyclodiketones enabled by structure-guided directed evolution of a carbonyl reductase. *Nat. Catal.* **2**: 931–941.
74. (a) Morrison, K.L. and Weiss, G.A. (2001). Combinatorial alanine-scanning. *Curr. Opin. Chem. Biol.* **5** (3): 302–307. (b) Massova, I. and Kollman, P.A. (1999). Computational alanine scanning to probe protein–protein interactions: a novel approach to evaluate binding free energies. *J. Am. Chem. Soc.* **121** (36): 8133–8143. (c) Lewis, J.C., Mantovani, S.M., Fu, Y. et al. (2010). Combinatorial alanine substitution enables rapid optimization of cytochrome P450BM3 for selective hydroxylation of large substrates. *ChemBioChem* **11** (18): 2502–2505.
75. Wahler, D. and Reymond, J.-L. (2002). The adrenaline test for enzymes. *Angew. Chem. Int. Ed.* **41** (7): 1229–1232.
76. Acevedo-Rocha, C., Hoebenreich, S., and Reetz, M.T. (2014). Iterative saturation mutagenesis: a powerful approach to engineer proteins by systematically simulating Darwinian evolution. *Methods Mol. Biol.* **1179**: 103–128.
77. (a) Sun, Z., Lonsdale, R., Li, G., and Reetz, M.T. (2016). Comparing different strategies in directed evolution of enzyme stereoselectivity: single- versus double-code saturation mutagenesis. *ChemBioChem* **17** (19): 1865–1872. (b) Qu, G., Liu, B., Jiang, Y. et al. (2019). Laboratory evolution of an alcohol dehydrogenase towards enantioselective reduction of difficult-to-reduce ketones. *Bioresour. Bioprocess.* **6** (1): 18.
78. Acevedo-Rocha, C.G., Reetz, M.T., and Nov, Y. (2015). Economical analysis of saturation mutagenesis experiments. *Sci. Rep.* **5**: 10654.
79. Kille, S., Acevedo-Rocha, C.G., Parra, L.P. et al. (2013). Reducing codon redundancy and screening effort of combinatorial protein libraries created by saturation mutagenesis. *ACS Synth. Biol.* **2** (2): 83–92.
80. Tang, L., Gao, H., Zhu, X. et al. (2012). Construction of "small-intelligent" focused mutagenesis libraries using well-designed combinatorial degenerate primers. *Biotechniques* **52** (3): 149–158.
81. Currin, A., Swainston, N., Day, P.J., and Kell, D.B. (2014). Speedygenes: an improved gene synthesis method for the efficient production of error-corrected, synthetic protein libraries for directed evolution. *Protein Eng. Des. Sel.* **27** (9): 273–280.
82. Acevedo-Rocha, C.G. and Reetz, M.T. (2014). Assembly of designed oligonucleotides: a useful tool in synthetic

biology for creating high quality combinatorial DNA libraries. *Methods Mol. Biol.* **1179**: 189–206.

83. Williams, E.M., Copp, J.N., and Ackerley, D.F. (2014). Site-saturation mutagenesis by overlap extension PCR. *Methods Mol. Biol.* **1179**: 83–101.

84. (a) Li, A., Acevedo-Rocha, C.G., Sun, Z. et al. (2018). Beating bias in directed evolution of proteins: combining high-fidelity on-chip solid-phase gene synthesis with efficient gene assembly for combinatorial library construction. *ChemBioChem* **19** (3): 221–228. (b) Li, A., Sun, Z., and Reetz, M.T. (2018). Solid-phase gene synthesis for mutant library construction: the future of directed evolution? *ChemBioChem* **19** (19): 2023–2032. (c) Hoebenreich, S., Zilly, F.E., Acevedo-Rocha, C.G. et al. (2015). Speeding up directed evolution: combining the advantages of solid-phase combinatorial gene synthesis with statistically guided reduction of screening effort. *ACS Synth. Biol.* **4** (3): 317–331.

85. (a) Schmidt, T.L., Beliveau, B.J., Uca, Y.O. et al. (2015). Scalable amplification of strand subsets from chip-synthesized oligonucleotide libraries. *Nat. Commun.* **6**: 8634. (b) For a company specializing in microchip combinatorial gene synthesis, which was used in the evolution of P450-BM3 mutants as catalysts in C7-selective steroid hydroxylation: Li, A., Acevedo-Rocha, C.G., D'Amore, L. et al. (2020). Regio- and stereoselective steroid hydroxylation at C7 by cytochrome P450 monooxygenase mutants. *Angew. Chem. Int. Ed.* **59**: 12499–12505.

86. (a) Xu, J., Cen, Y., Singh, W. et al. (2019). Stereodivergent protein engineering of a lipase to access all possible stereoisomers of chiral esters with two stereocenters. *J. Am. Chem. Soc.* **141** (19): 7934–7945. (b) Li, D., Wu, Q., and Reetz, M.T. (2020). Focused Rational Iterative Site-specific Mutagenesis (FRISM). *Methods Enzymol.* **643**: 225–242.

87. (a) Li, J., Qu, G., Shang, N. et al. (2021). Near-perfect control of the regioselective glucosylation enabled by rational design of glycosyltransferases. *Green Synth. Catal.* **2**: 45–53. (b) Reetz, M.T. (2021). A breakthrough in protein engineering of a glycosyltransferase. *Green Synth. Catal.* **2**: 4–5.

88. (a) Xu, J., Fan, J., Lou, Y. et al. (2021). Light-driven decarboxylative deuteration enabled by a divergently engineered photodecarboxylase. *Nat. Commun.* **12** (1): 3983. (b) Wu, Q., Li, D., Han, T. et al. (2021). Engineering fatty acid photodecarboxylase to enable the highly selective decarboxylation of trans fatty acids. *Angew. Chem. Int. Ed.* **60** (38): 20695–20699. (c) Ma, N., Fang, W., Liu, C. et al. (2021). Switching an artificial P450 peroxygenase into peroxidase via mechanism-guided protein engineering. *ACS Catal.* **14**: 8449–8455. (d) Tian, C., Yang, J., Liu, C. et al. (2022). Engineering substrate specificity of HAD phosphatases and multienzyme systems development for the thermodynamic-driven manufacturing sugars. *Nat. Commun.* **13**: 3582.

89. Reetz, M.T. (2022). Making enzymes suitable for organic chemistry by rational design. *ChemBioChem* e202200049.

5
Tables of Selected Examples of Directed Evolution and Rational Design of Enzymes with Emphasis on Stereo- and Regio-selectivity, Substrate Scope and/or Activity

5.1
Introductory Explanations

The all-too-often observed insufficient activity, narrow substrate scope, and/or poor or wrong stereo- and regio-selectivity of enzymes constitute the major reasons why biocatalysts have not been employed more often in organic and pharmaceutical chemistry [1]. As noted in Chapters 1, 3, and 4 of the present monograph, directed evolution has heavily contributed to popularizing the use of enzymes while also opening new perspectives for biotechnology. The first study reporting the directed evolution of a stereoselective enzyme involved a lipase (Chapter 1) [2]. Since then, most of the major enzyme types have been subjected to directed evolution in successful attempts to enhance and/or invert enantioselectivity: hydrolases (e.g. lipases, esterases, nitrilases, epoxide hydrolases, and glycosidases), acylases (e.g. penicillin G), monooxygenases (e.g. P450 and Baeyer–Villiger enzymes, and monoamine oxidases), reductases (e.g. alcohol dehydrogenases and enoate reductases), ligases (e.g. aldolases, oxynitrilases, and thiamine diphosphate-dependent decarboxylases), and lyases (e.g. aryl malonate decarboxylases). These advances have been summarized in a previous 2016 monograph [1] and a 2020 review [3], in which tables of examples of directed evolution with emphasis on selectivity, substrate scope, and activity were presented, together with information on enzyme type and additional short comments. Here, we do something similar, which will help protein engineers when aiming to start new projects (Table 5.1). Selected examples of rational enzyme design are listed in Table 5.2 similarly, while promiscuous artificial metalloenzymes are featured in Chapter 7.

Table 5.1 includes typical studies from the recent literature, arranged according to enzyme type and kind of chemical transformation. The interested researcher can readily find selected studies focusing on the enzyme type of his/her interest, thereby enabling fast comparison of the different approaches. All major gene mutagenesis methods are considered, including epPCR, saturation mutagenesis (SM), iterative saturation mutagenesis (ISM), and DNA shuffling. Although all of these methods continue to be applied, especially with the aim of influencing activity, substrate scope, stereo- and regio-selectivity, saturation mutagenesis at

Enzyme Engineering: Selective Catalysts for Applications in Biotechnology, Organic Chemistry, and Life Science, First Edition. Manfred T. Reetz, Zhoutong Sun, and Ge Qu.

Table 5.1 Selected directed evolution studies of enzymes, many for enhanced stereo- and/or regio-selectivity, higher activity, and/or shifted substrate scope.

Enzyme	Evolved property	Mutagenesis method	Comment	References
α-2,6-Sialyltransferase (*Photobacterium damselae*)	Activity	SM, alanine scanning	Sialyllactose production	[4]
Alcohol dehydrogenase (*Thermoethanolicus brockii*)	Enantioselectivity	CAST/ISM	Induced axial chirality; RNG codon, eight-membered reduced amino acid alphabet for exploring effect of size	[5]
Alcohol dehydrogenase (*Thermoethanolicus brockii*)	Enantioselectivity; substrate scope; activity	Alanine scanning and site-specific mutagenesis; CAST/ISM	Conformational dynamics-guided; both enantiomers evolved for reduction of bulky ketones	[6]
Alcohol dehydrogenase (*Thermoanaerobacter ethanolicus*)	Enantioselectivity	SM	Three phenyl-substituted ketones as substrates	[7]
Alcohol dehydrogenase (*Lactobacillus brevis*)	Thermostability, activity	Combining earlier mutations obtained from epPCR; DNA shuffling	Preparation of precursor of atorvastatin; B-FITTER used in the interpretation	[8]
Alcohol dehydrogenase (*Candida parapsilosis*)	Enantioselectivity	ISM	Converted (*S*)-selectivity to be ambidextrous; amination of racemic alcohols to produce optically pure amines	[9]
Alcohol dehydrogenase (*Rhodococcus ruber DSM 44541*)	Activity, enantioselectivity	CAST/ISM	A degenerate codon set of NDT, VMA, ATG, and TGG was used; vicinal 1,2-diol and respective acyloin	[10]
Alcohol dehydrogenase (*R. ruber DSM 44541*)	Regio- and stereo-selectivity	CAST/ISM	Broad substrate acceptance toward compounds containing two secondary alcohols	[11]
Alcohol dehydrogenase (*Pseudomonas putida* KT2440)	Substrate scope	CAST	NNK codon; substrate channel evolution; oxidation of 5-(hydroxymethyl) furoic acid	[12]
Alcohol dehydrogenase (*Kluyveromyces polyspora*)	Activity, stereoselectivity	SM	18 CAST residues with NNK codon	[13]

Alcohol dehydrogenase (*S. stipites* CBS 6054)	Activity, stereoselectivity	SM	CAST/ISM, three rounds of mutagenesis	[14]
Amine dehydrogenase (*Lysinibacillus fusiformis*)	Activity	SM	NNK codon; chiral vicinal amino alcohols as products	[15]
Amine dehydrogenase (*Sporosarcina psychrophila*)	Activity	CAST/ISM	31 CAST residues with NNK codon; three rounds of mutagenesis	[16]
Amine dehydrogenase (*Petrotoga mobilis*)	Activity	epPCR; DNA shuffling	Two rounds of epPCR, followed by shuffling	[17]
Aminotransferase (*E. coli*)	Activity	Site-specific SM	SM performed on 28 residues, followed by combinatorial mutagenesis; valienamine synthesis	[18]
Amorpha-4,11-diene synthase (ADS)	Activity	SM	16 CAST residues with NDT, VMA, ATG, and TGG codon	[19]
Amylosucrase	Substrate scope in glycosylation reactions	ISM	Sucrose as donor	[20]
Amylosucrase (*Neisseria polysaccharea*)	Activity	SM	Thermostability also improved	[21]
Artificial retro-aldolase	Stereoselectivity	SM	Obtained four stereocomplementary catalysts	[22]
Arylmalonate decarboxylase (*Bordatella bronchoseptica*)	Racemase activity	Site-directed mutagenesis, SM	Cooperative mutational effects	[23]
Asparaginase (*Erwinia chrysanthemi*)	Activity	epPCR	Diminishes enzyme immunoreactivity important when treating acute lymphoblastic leukemia	[24]
Aspartase (*Bacillus* sp. YM55–1)	Promiscuous amino acid lyase activity	SM	NNK codon; cluster screening of 3×10^5 clones; single-site SM inefficient, three simultaneous mutations necessary; (*R*)-3-aminobutyrate from crotonic acid plus ammonia	[25]

(*continued overleaf*)

Table 5.1 (Continued)

Enzyme	Evolved property	Mutagenesis method	Comment	References
Baeyer–Villiger monooxygenase (PAMO)	Activity, cis/trans-diastereoselectivity	ISM	Five-membered reduced amino acid alphabet	[26]
Baeyer–Villiger monooxygenase (PAMO)	Activity, cis/trans-diastereoselectivity	ISM	Two 5-residue SM sites; four and five-membered reduced amino acid alphabets at different residues	[27]
Baeyer–Villiger monooxygenase (PAMO)	Reversal of enantioselectivity	CAST/ISM	Asymmetric sulfoxidation; extreme cooperative mutational effects revealed by fitness landscape via deconvolution	[28]
Baeyer–Villiger monooxygenase (CHMO)	Activity	SM; epPCR	NNK and NDT codons; SM followed by one round of epPCR	[29]
Baeyer–Villiger monooxygenase (CHMO)	Cofactor specificity	SM; epPCR	NNK codon; changed preference from NADPH to NADH	[30]
ß-Glycosidase Zm-p601 (maize)	Activity	SM, site-specific mutagenesis	NNM, NNK codon degeneracies	[31]
Carbonic anhydrase (*Desulfovibrio vulgaris*)	Activity, stability	ISM, DNA shuffling	CO_2 sequestration in *N*-methyl-diethanolamine	[32]
Carboxylic acid reductase (*Mycobacterium marinum*)	Substrate specificity	epPCR, site-directed SM	Coupled with a growth-assisted screening method	[33]
Carboxylic acid reductase (*Segniliparus rugosus*)	Activity	CAST	NDT, VMA, ATG, TGG codon; guided by molecular dynamics simulations-guided	[34]
Carotenoid synthase (*Staphylococcus aureus* C30)	Substrate scope	epPCR	C30 and C40 carotenoid synthesis; reading frame correction	[35]
CotA-laccases	Activity	DNA shuffling	Decolorization of azo dyes	[36]
D-Amino acid oxidase (from porcine kidney)	Enantioselectivity	ISM	Reversed enantioselectivity favoring (*R*)-product	[37]

Deacetoxycephalosporin C synthase (*Streptomyces clavuligerus*)	Activity	ISM	Combining point mutations combinatorially; ring-expansion of penicillin G to cephalosporin	[38]
Ene reductase (OYE; *Saccharomyces pastorianus*)	Enantioselectivity	SM	*Z*-ß-aryl-ß-cyanoacrylates as substrates; ß-lactams as final products	[39]
Ene reductase (*Pichia stipites*)	Enantioselectivity	ISM	Baylis–Hillman substrates	[40]
Endoglucanase (*E. coli*)	Activity	epPCR	Only one epPCR circle; cellulose hydrolysis	[41]
Endoglucanase (*Spirochaeta thermophila*)	Substrate scope	SM	Reduced amino acid alphabet; consensus; SM of loop next to active site; hydrolysis of cellulose derivative	[42]
Esterase (*Rhodobacter sphaeroides*)	Enantioselectivity	epPCR	3 cycles	[43]
Esterase (*Bacillus subtilis*)	Enantioselectivity	SM, site-specific mutagenesis	NDT codon; challenging substrate	[44]
Esterase (*Rhodococcus* sp. ECU1013)	Activity	epPCR, SM	Substrate channel evolution, cilastatin synthesis	[45]
Esterase (*Archaeoglobus fulgidus*)	Enantioselectivity, activity	epPCR, DNA shuffling, SM	Millions of mutants were screened using a "dual-channel" microfluidic droplet method	[46]
Epoxide hydrolase (metagenomic)	Enantioselectivity; activity	ISM	Several substrates	[47]
Epoxide hydrolase (*Solanum tuberosum*)	Stereoselectivity	ISM	Enantioconvergence	[48]
Epoxide hydrolase (*Vigna radiata*)	Stereoselectivity	SM	CAST; enantioconvergence	[49]
Epoxide hydrolase (*Aspergillus niger*)	Enantioselectivity	ISM	All 24 ISM pathways explored; how to escape from local minima	[1]
Flavin reductase (*Vibrio fisheri*)	Activity	SM	Different codon at different positions; prodrug activation	[50]
Formate dehydrogenases (*Candida methylica, Chaetomium thermophilum*)	Activity	ISM	Reduced amino acid alphabet; transformation of hydrogen carbonate into formate	[51]

(continued overleaf)

Table 5.1 (Continued)

Enzyme	Evolved property	Mutagenesis method	Comment	References
Formate dehydrogenase (*Chaetomium thermophilum*)	Activity	SM	NRT codon; CO_2 fixation	[52]
Galactase oxidase (*Fusarium* sp.)	Substrate scope	CAST/ISM	Useful for labeling glycoproteins	[53]
Geranylgeranyl diphosphate synthase (*Nicotiana tabacum*)	Activity	SM	NNK codon; CAST/ISM	[54]
Glucose dehydrogenase (*Bacillus megaterium*)	Organic solvent tolerance	epPCR; DNA shuffling; combinatorial mutagenesis	Two rounds of epPCR, followed by DNA shuffling and then one round of epPCR, eventually combinatorial mutagenesis	[55]
Glutamine dehydrogenase (*E. coli*)	Substrate scope	SM	NNK at 4-residue site; six million clones screened; L-homophenylalanine production	[56]
Glycosidase (ß-fructofuranosidase /*Schwanniomyces occidentalis*)	Transglycosylation activity, substrate selectivity	epPCR, DNA shuffling	6-kestose synthesis	[57]
Glycosidase (*N*-oligosaccharyl transferase; *Campylobacter jejuni*)	Activity	ISM	Application in vaccine	[58]
Glycosidase (Arabinofuranosyl hydrolase; *Clostridium thermocellum*)	Substrate scope in transglycosylation	epPCR	Aliphatic alcohols as acceptors	[59]
Glycosidase (cyclodextrin glycosyltransferase; *Paenibacillus macerans*)	Substrate acceptance	ISM	2-*O*-D-glucopyranosyl-L-ascorbic acid production	[60]

Glycosyltransferase (oled)	Substrate scope	SM	Sugar nucleotide synthesis	[61]
Glycosyltransferase UGT51	Regioselectivity	ISM	3-OH selectivity; anticancer ginsenoside Rh2 synthesis	[62]
Glycosyltransferase CaUGT3 (*Catharanthus roseus*)	Sugar-donor specificity	SM, site-specific mutagenesis	NDT, VMA, ATG, and TGG codons; converted from UDP-glucose to UDP-xylose	[63]
Glycosyltransferase (*Bacillus licheniformis* ZSP01)	Regioselectivity	ISM	Alanine scanning involved	[64]
Haloacid dehalogenase (HAD)-like phosphatase	Substrate specificity, activity	SM	B-factor and consensus guided	[65]
Halogenase (Tryptophan 7-halogenase RebH)	Activity, regioselectivity	epPCR	Chlorination; substrate walking	[66]
Halohydrin dehalogenase (*Agrobacterium radiobacter* HheC)	Enantioselectivity, activity	ISM	DC-analyzer applied	[67]
Haloalkane dehalogenase and ancestral luciferase	Promiscuity	Insertion–deletion	Backbone mutagenesis; B-factor analysis revealed enzyme dynamics	[68]
Halohydrin dehalogenase	Enantio- and regio-selectivity	ISM	NNK codon; kinetic resolution; azido derivatives of indoles	[69]
Halogenase WeI05	Regio- and diastereo-selectivity	CAST/ISM	Fe-dependent; NNK-based SM at 9 residues; also SKA and VNT codons at additional sites; consensus; late-stage chlorination of chiral terpenes	[70]
Halogenase WeL05*	Regio- and diastereo-selectivity	SM	Different codons VIG, AVP, TLA at different residues; machine learning; three randomization sites chosen based on docking and previous mutations; then ML; synthetic solid-phase library preparation (TWIST); chlorination of soraphens	[71]

(continued overleaf)

Table 5.1 (Continued)

Enzyme	Evolved property	Mutagenesis method	Comment	References
Halogenase Wi-WeL015	Regio- and diastereo-selectivity	CAST/ISM	SM using different codons at seven residues; consensus; late-stage halogenation of terpenoids	[72]
Human glutathione transferase (A2-2)	Substrate scope	SM	NDT and YHC codons; dietary isothiocyanates	[73]
Hydroxynitrilase (*Granulicella tundricola*)	Activity, enantioselectivity	epPCR and SM; combining mutations	Four substrates for pharmaceuticals production	[74]
Hydroxynitrile lyase (AtHNL)	Substrate scope, enantioselectivity	SM	99% ee (*S*); mechanism of promiscuous retro-nitroaldol reaction revealed	[75]
Hydroxynitrile lyase (*Prunus dulcis*)	Activity	SM	Also enhanced acidic tolerance	[76]
Human ß-glucuronidase	Substrate scope	SM	Different codon at different positions; α-iduronidase activity; reduced immunogenicity	[77]
Human-like L-asparaginase (*Homo sapiens*)	Activity	DNA shuffling	Potential clinical use	[78]
Imine reductase	Activity, thermostability	CAST, SM	Three CAST cycles, consensus	[79]
Laccase (*E. coli*)	Onset potential	epPCR, SM	Engineered more positive onset potential	[80]
Laccase (*Pycnoporus cinnabarinus*)	Activity	ISM	NNK codon; oxidation of sinapic acid of lignin	[81]
L-Aspartate β-decarboxylase (*Pseudomonas dacunhae*)	Activity	SM	Alanine scanning for identifying hot spots for SM	[82]
L-Aspartate-ß-semialdehyde dehydrogenase (*E. coli*)	Cofactor NADH acceptance	ISM	L-homoserine production	[83]
Limonene epoxide hydrolase	Enantioselectivity	ISM	Smallest amino acid alphabets	[84]
Limonene epoxide hydrolase	Enantioselectivity	ISM	Reduced amino acid alphabet	[85]
Lipase (*Rhizomucor miehei*)	Activity	epPCR	Four epPCR cycles	[86]

Lipase (*Pseudomonas aeruginosa*)	Activity, enantioselectivity	ISM	Reduced amino acid alphabet	[87]
Lipase (CALA)	Enantioselectivity	ISM	Reduced amino acid alphabet	[88]
Lipase (CALA)	Enantioselectivity	SM	Different codon degeneracy at each position; immobilization on Ni-coated 96-well microtiter plates	[89]
Lipase (CALA)	*Trans*/*cis* substrate selectivity	SM	Reduced amino acid alphabet	[90]
Lipase (CALA)	Substrate selectivity (the chain length)	SM	22c-trick [91]; SM performed on three residues and combined the beneficial hits	[92]
Lipase A (*B. subtilis*)	Enantioselectivity	SM and site-specific mutagenesis	Reduced amino acid alphabet; both enantiomers evolved	[93]
Lipase (CALB)	Enantioselectivity	ISM	Kinetic resolution, reduced amino acid alphabet, profen-esters	[94]
Lipase (CALB)	Enantioselectivity	SM and site-specific mutagenesis	Reduced amino acid alphabet	[95]
Lipase (pancreatic)	Activity, substrate-selectivity	epPCR	One cycle only	[96]
Lipase (*Pseudomonas* sp. Lip I.3)	Substrate scope, activity	SM	Entrance channel and lid mutations	[97]
Lipase (*Thermomyces lanuginosus*)	Stereoselectivity	epPCR, SM	Diastereoselective "kinetic resolution"	[98]
Lipase A (*B. subtilis*)	Enantioselectivity	ISM	Both enantiomers evolved	[93]
L-Threonine aldolase (*Bacillus nealsonii*)	Diastereoselectivity	CAST/ISM	NNK codon; in silico screening synthesis of ß-hydroxy-α-amino acids	[99]
Lysine decarboxylase (*Hafnia alvei* ASI.1009)	Substrate scope	epPCR, DNA shuffling	Cadaverine production	[100]
Metallo-ß-lactamase (MBL9)	Substrate scope	SM	Different codon at different sites; resistance to seven antibiotics	[101]
Monoamine oxidase (*A. niger*)	Substrate scope, enantioselectivity	ISM	(*R*)-mexiletine synthesis	[102]
Monoamine oxidase (*A. niger*)	Enantioselectivity	Mutator strain	Deracemization of alkaloids by mutant obtained earlier via mutator strain	[103]

(*continued overleaf*)

Table 5.1 (Continued)

Enzyme	Evolved property	Mutagenesis method	Comment	References
Myoglobin (*Physeter catodon*)	Activity, enantioselectivity	SM, ISM	NNK codon; bicyclic lactams as products	[104]
Nitrilase (*Pyrococcus abyssi*)	Activity, enantioselectivity	Complete SM at large region, ISM	Dynamic kinetic resolution	[105]
Nonheme diiron *N*-oxygenase AzoC	Substrate scope, activity	CAST, alanine scanning	Guidance by previous studies; azoxy six-membered nitrogen catenation synthesis	[106]
Oleic acid hydratase (*Elizabethkingia meningoseptica*)	Substrate scope, enantio- and regio-selectivity	SM	Asymmetric hydration of aliphatic alkenes	[107]
P450-bm3	Activity, regioselectivity	ISM	Bioorthogonal deprotection of caged compounds	[108]
P450-bm3	Regio-, diastereo-, and enantio-selectivity	SM	Six-membered reduced amino acid alphabet; simultaneous creation of two chirality centers	[109]
P450-bm3	Activity	ISM	Reduced amino acid alphabet; peroxide-driven hydroxylation of small alkanes	[110]
P450-bm3	Regioselectivity	ISM	NDT + GCG codon; CAST/ISM; C6 oxidation of sclareol	[111]
P450-pyr (*Sphingomonas* sp.-HXN-200)	Regio- and enantio-selectivity	ISM	Both enantiomers evolved	[112]
P450-pyr (*Sphingomonas* sp.-HXN-200)	Regioselectivity	ISM	*n*-Butanol to 1,4-butanediol reaction	[113]
P450 (*Streptomyces carbophilus*)	Activity	ISM	Mevastatin to pravastatin by regio- and stereo-selective hydroxylation	[114]
P450-pyr (*Sphingomonas* sp. HXN-200)	Regio- and enantio-selectivity	ISM	Subterminal hydroxylation of alkanes	[115]
P450 monooxygenase (*Labrenzia aggregata*)	Activity	epPCR	Four rounds of epPCR, followed by combinatorial mutagenesis	[116]

P450 (MycG)	Regioselectivity	epPCR	Sequential hydroxylation and epoxidation; accumulation of mycinamicin V	[117]
P450 monooxygenase (*Streptomyces antibioticus*)	Regio- and enantio-selectivity	Alanine scanning followed by site-specific mutagenesis	7ß-hydroxylation of lithocholic acid enabling ursodeoxycholic acid production	[118]
P450 CYP153A (*Marinobacter aquaeolei*)	Activity and regioselectivity	SM using 22c-trick, combining mutations	Terminal ω-hydroxylation of octanoic acid	[119]
P450 CYP154E1 (*Thermobifida fusca* YX)	Activity, selectivity	SM, 22c-trick	Seven individual site-SM	[120]
P450 CYP154E1 (*T. fusca* YX)	Activity, selectivity	In-house CYP mutants obtained earlier by SM were tested followed by site-specific mutagenesis	Production of anti-depressant (2*S*,6*S*)-hydroxynorketamine	[121]
Penicillin G acylase	Diastereoselectivity	SM	NDT codon; both diastereomers evolved	[122]
Peptiligase (*Bacillus amyloliquefaciens*)	Substrate scope	Site-directed mutagenesis, SM	In silico guidance (Rosetta); FRET-based substrate screening; peptide syntheses	[123]
Phenylalanine dehydrogenase (*Rhodococcus* sp. M4)	Substrate scope, enantioselectivity	ISM	Reductive amination, single enantiomer evolved	[124]
Phenylalanine dehydrogenase (*Bacillus badius*)	Substrate scope, enantioselectivity	ISM	Reduced amino acid alphabet, single enantiomer evolved	[125]
Phosphotriesterase (*Agrobacterium radiobacter*)	Substrate scope	SM	CAST; NNS codon; malathion acceptance	[126]
Phosphotriesterase (*Brevundimonas diminuta*)	Promiscuous activity	Insertion/deletion mutagenesis	Arylesterase/phosphotriesterase activity	[127]
Phosphotriesterase-like lactonase (*Geobacillus kaustophilus*)	Promiscuous activity	Insertion; loop SM	NNK codon; stepwise loop insertion strategy (StLois)	[128]

(*continued overleaf*)

Table 5.1 (Continued)

Enzyme	Evolved property	Mutagenesis method	Comment	References
Polyketide synthase (*Saccharopolyspora* erythraea)	Substrate scope	SM	Introducing non-native functional group into erythromycin	[129]
Polyethylene terephthalate hydrolase (*Ideonella sakaiensis*)	Thermostability, activity	SM	Guidance by CAST and disulfide bridge	[130]
Polysialyltransferase (*nmb*)	Size distribution of polymeric products	epPCR, SM	Potential medical applications	[131]
Prolidase (*Pseudoalteromonas* sp. SCSIO 04301)	Promiscuous activity	Site-directed mutagenesis, SM	CAST; NNK codon	[132]
Propanediol oxidoreductase	Activity, furfural tolerance	SM	Fourteen residues individually targeted; NNK	[133]
Pyrrolysyl-tRNA synthetase	Substrate acceptance	SM	Unnatural amino acid acceptance	[134]
Pyrrolysyl-tRNA synthetase	Substrate scope	ISM	NNK codon-based SM at five sites, three ISM cycles; modifying proteins by unnatural amino acids for Cu-free click chemistry	[135]
Reductive aminase (*Streptomyces purpureus*)	Substrate tolerance, activity	ISM	NNK codon; 96 residues individually targeted; abrocitinib JAK1 inhibitor	[136]
RNA polymerase (T7)	Substrate scope	SM	NNB and NNS codons; 2′-*O*-methyl-modified RNA synthesis	[137]
S-Allylcysteinyl-tRNA synthetase	Substrate scope	SM	NNK codon; acceptance of noncanonical amino acid S-allyl cysteine (Sac)	[138]
Self-labeling protein (HaloTag7)	Fluorescence lifetime multiplexing	SM	Ten residues at rhodamine binding site subjected to NNK-based SM; biosensor	[139]
Signal peptides of unspecific peroxygenases (UPOs)	Secretion	DNA shuffling	Improving UPO secretion in yeast	[140]

SRC Homology 3 (SH3)	Specificity	SM, domain shuffling	NNN codon; protein–protein interactions	[141]
Tagatose epimerase (*Pseudomonas cichorii*)	Substrate scope	ISM	C3-epimerization of D-fructose and L-sorbose	[142]
Transaminase (CV 2025)	Activity	SM	Phylogenetic analysis	[143]
Transaminase (*Vibrio fluvialis*)	Enantioselectivity, activity	ISM	Epoxidation–isomerization–amination enzyme cascades; formal anti-Markovnikov amination of olefins	[144]
Transaminase (*Vibrio fluvialis*)	Activity, robust, diastereoselectivity	ISM	11 rounds of directed evolution; preparation of chiral Sacubitril precursor	[145]
Transaminase (S6)	Substrate scope, stereoselectivity	epPCR; ISM	Random mutagenesis followed by several rounds of ISM; preparation of intermediate for rimegepant synthesis needed in migraine treatment	[146]
Transketolase (*E. coli*) and ω-transaminase (ATA117)	Stereoselectivity	CAST/ISM	NNK codon; one-pot two-step reaction; florfenicol synthesis	[147]
Transketolase (*E. coli*)	Substrate scope, stereoselectivity	CAST/ISM	Consensus-based reduced amino acid library; single diastereomer evolved	[148]
Transketolase (*Geobacillus stearothermophilus*)	Substrate scope, enantioselectivity	SM	NNS codon; different aldehydes; (*S*)-enantiomers formed	[149]
Unspecific peroxidase (*Myceliophthora thermophila*)	Activity, chemo- and regio-selective	SM; Golden mutagenesis/22c trick	Active site and entrance channel; benzylic hydroxylation	[150]
Xylanase (*Bacillus halodurans* S7)	Activity	Site-directed SM	NDT, VMA, ATG, TGG codon; entrance tunnel tuning; lignocellular hydrolysis	[151]
YfeX	Activity	DNA shuffling	Assisted by a versatile high-throughput GC–MS approach	[152]
YfeX	Activity	Alanine scanning	C–H functionalization of indoles with diazoacetonitrile; synthesis of cysmethynil	[153]

Abbreviations of mutational methods: epPCR (error-prone polymerase chain reaction); SM (saturation mutagenesis); ISM (iterative saturation mutagenesis); CAST (combinatorial active-site saturation test). In some cases, CAST or ISM was actually used, but the authors did not employ these acronyms. Then the column on mutagenesis method simply lists SM (saturation mutagenesis). ADH and carbonyl reductase represent essentially the same enzyme type.

Table 5.2 Selected studies of rational enzyme design (more publications are listed in Chapter 7 which features protein engineering of artificial metalloenzymes).

Enzyme	Evolved property	Mutagenesis method	Comment	References
4-Dimethylallyltryptophan synthase	Regioselectivity, substrate scope	Site-specific mutagenesis	Docking and steric hindrance-guided mutagenesis	[154]
4-Hydroxyphenylacetate 3-hydroxylase (4HPA3H)	Regioselectivity	Site-directed mutagenesis	Guidance by structural alignment	[155]
Acyltransferase MsAcT (*Mycobacterium smegmatis*)	Activity, substrate scope	Site-specific mutagenesis	Increased acyl transfer to hydrolysis ratio	[156]
Acyltransferase (EstA)	Selectivity 6-*O*-acetylation	Rationally combining known mutations by site-specific mutagenesis	Structural guided; promiscuous activity toward acetylating sugars	[157]
Acyltransferase (LovD)	Kinetics, stability	Rational deconvoluting known mutations	Improve the activity with 15 less mutations	[158]
Acyltransferase (PestE)	Activity	Rational site-specific mutagenesis	Enhance the hydrophobicity of active site	[159]
Alcohol dehydrogenase (*Candida parapsilosis*)	Stability/solvent resistance	SM and site-specific mutagenesis	Enantioselectivity not reported	[160]
Amidase (*Rhodococcus erythropolis* AJ270)	Enantioselectivity	Site-directed mutagenesis	Desymmetrization of meso O-heterocyclic dicarboxamides, both enantiomers evolved	[161]
Amino acid racemase (*P. putida*)	Organic solvent tolerance	Rational site-directed mutagenesis; DNA shuffling	Exchanging surface residues for arginine to enhance enzyme stability	[162]
Aminotransferase (AGT1)	Stability/frustration balance	Site-directed mutagenesis	Guidance by normalized B-factor	[163]
Ancestral β-lactamase GNCA	Substrate specificity	Site-specific mutagenesis	Conformational dynamics-guided; allosteric interactions; 21 amino acid exchanges; preference for penicillin degradation	[164]

Arylsulfatases and phosphonate monoester hydrolases	Promiscuous activity	Site-directed mutagenesis	catalytic promiscuity through single-point mutations	[165]
Cancer genes	Tumorigenesis	SM	568 mutational cancer genes; in silico SM using machine learning	[166]
Carboxylesterase (*Proteobacteria bacterium* SG_bin9)	Enantioselectivity	SM	Rational design at residues lining substrate binding pocket	[167]
Deacetoxycephalosporin C synthase (*Streptomyces clavuligerus*)	Substrate specificity, activity	FRISM	Rosetta-guided FRISM design	[168]
D-Threonine aldolase (*Filomicrobium marinum*)	Stereoselectivity	Site-specific mutagenesis	Catalytic mechanism-guided mutagenesis	[169]
Envelope glycoprotein E2 (Hepatitis C virus)	Stability, antigenicity	De novo design	Loop tuning; Monte Carlo sampling	[170]
Epoxide hydrolase (*Vigna radiata*)	Enantioselectivity	Site-directed mutagenesis	Enantioconvergent preparation of (*R*)-hexane-1,2-diol	[171]
Esterase (*Pseudomonas fluorescens*)	Substrate scope, enantioselectivity	Site-specific mutagenesis, SM, combining mutations	In silico guidance	[172]
Glycoside hydrolases	Activity	Rational site-specific mutagenesis	Consensus analysis; transglycosylation	[173]
Glycosyltransferase (*Oryza sativa*)	Regioselectivity	Rational site-specific mutagenesis	Enhanced β (1−2) glucosylation and eliminated β (1−6), consensus analysis	[174]
Glycosyltransferase UGT74AC2 (*Siraitia grosvenorii*)	Regioselectivity	FRISM	In silico design of the building blocks	[175]
Iron(III)−mimochrome VI (peroxidase)	Activity	Site-specific mutagenesis	Introduction of two 2-aminoisobutyric acids into artificial peptide-based peroxidase	[176]
Lipase (CALA)	Acyltransfer activity	Site-specific mutagenesis; 28 variants generated	Out of 28 rationally designed variants a single hit	[177]

(*continued overleaf*)

Table 5.2 (Continued)

Enzyme	Evolved property	Mutagenesis method	Comment	References
Lipase (CALB)	Diastereo- and enantio-selectivity	FRISM	Unconventional kinetic resolution using a racemic ester and a racemic alcohol; four FRISM pathways leading to four different mutants catalyzing four different stereoisomers, each with 95–99% selectivity	[178]
L-lactate dehydrogenase (*Plasmodium falciparum*)	Activity	Site-specific mutagenesis	Consensus-guided; cascade with production of (*S*)-2-hydroxy butyric acid	[179]
Methionine adenosyltransferase (*Homo sapiens*)	Substrate selectivity	Rational site-specific mutagenesis	Rational guidance by "bump-and-hole"; converting substrate recognition from L-Met to L-Metol	[180]
Peptiligase (*Bacillus amyloliquefaciens*)	Activity, substrate scope	Site-directed mutagenesis	Rational guidance by previous mutations/X-ray data; [14 + 14]-mer segment condensation	[181]
Photosensitizer protein (PSP)	Activity	Site-specific mutagenesis	Endowed photocatalytic CO_2-reducing ability with formation of CO	[182]
Serine ester hydrolase (EH102)	Substrate promiscuity	Site-specific mutagenesis	Guidance by structure, mechanism, and MD data; combining mutations; stereoselectivity not involved	[183]
Sortase A (*Streptococcus pyogenes*)	Substrate specificity; activity	Loop grafting	Guidance by both sequence and structural alignment	[184]
Squalene hopene cyclase (*Alicyclobacillus acidocaldarius*)	Diastereoselective cyclization	Rational site-directed mutagenesis	Combining point mutations combinatorially	[185]
Sucrose phosphorylase (*Leuconostoc mesenteroides*)	Activity, thermostability	Site-directed mutagenesis	FireProt-guided mutagenesis	[186]

amine dehydrogenase. *ACS Sustainable Chem. Eng.* **8** (46): 17054–17061.

18. Cui, L., Wei, X., Wang, X. et al. (2020). A validamycin shunt pathway for valienamine synthesis in engineered *streptomyces hygroscopicus* 5008. *ACS Synth. Biol.* **9** (2): 294–303.

19. Abdallah, I.I., van Merkerk, R., Klumpenaar, E., and Quax, W.J. (2018). Catalysis of amorpha-4,11-diene synthase unraveled and improved by mutability landscape guided engineering. *Sci. Rep.* **8** (1): 9961.

20. Champion, E., Guérin, F., Moulis, C. et al. (2012). Applying pairwise combinations of amino acid mutations for sorting out highly efficient glucosylation tools for chemo-enzymatic synthesis of bacterial oligosaccharides. *J. Am. Chem. Soc.* **134** (45): 18677–18688.

21. Daudé, D., Topham, C.M., Remaud-Siméon, M., and André, I. (2013). Probing impact of active site residue mutations on stability and activity of *Neisseria polysaccharea* amylosucrase. *Protein Sci.* **22** (12): 1754–1765.

22. Garrabou, X., Macdonald, D.S., Wicky, B.I.M., and Hilvert, D. (2018). Stereodivergent evolution of artificial enzymes for the Michael reaction. *Angew. Chem. Int. Ed.* **57** (19): 5288–5291.

23. Gaßmeyer, S.K., Yoshikawa, H., Enoki, J. et al. (2015). STD-NMR-based protein engineering of the unique arylpropionate-racemase AMDase G74C. *ChemBioChem* **16** (13): 1943–1949.

24. Costa, I.M., Moura, D.C., Lima, G.M. et al. (2022). Engineered asparaginase from Erwinia chrysanthemi enhances asparagine hydrolase activity and diminishes enzyme immunoreactivity – a new promise to treat acute lymphoblastic leukemia. *J. Chem. Technol. Biotechnol.* **97**: 228–239.

25. Vogel, A., Schmiedel, R., Hofmann, U. et al. (2014). Converting aspartase into a β-amino acid lyase by cluster screening. *ChemCatChem* **6** (4): 965–968.

26. Parra, L.P., Agudo, R., and Reetz, M.T. (2013). Directed evolution by using iterative saturation mutagenesis based on multiresidue sites. *ChemBioChem* **14** (17): 2301–2309.
27. Zhang, Z.G., Roiban, G.D., Acevedo, J.P. et al. (2013). A new type of stereoselectivity in Baeyer–Villiger reactions: access to *E*- and *Z*-olefins. *Adv. Synth. Catal.* **355** (1): 99–106.
28. Zhang, Z.G., Lonsdale, R., Sanchis, J., and Reetz, M.T. (2014). Extreme synergistic mutational effects in the directed evolution of a Baeyer–Villiger monooxygenase as catalyst for asymmetric sulfoxidation. *J. Am. Chem. Soc.* **136** (49): 17262–17272.
29. Ren, S.M., Liu, F., Wu, Y.Q. et al. (2021). Identification two key residues at the intersection of domains of a thioether monooxygenase for improving its sulfoxidation performance. *Biotechnol. Bioeng.* **118** (2): 737–744.
30. Maxel, S., Saleh, S., King, E. et al. (2021). Growth-based, high-throughput selection for NADH preference in an oxygen-dependent biocatalyst. *ACS Synth. Biol.* **10** (9): 2359–2370.
31. Turek, D., Klimeš, P., Mazura, P., and Brzobohatý, B. (2014). Combining rational and random strategies in β-glucosidase Zm-p60.1 protein library construction. *PLoS One* **9** (9): e108292.
32. Alvizo, O., Nguyen, L.J., Savile, C.K. et al. (2014). Directed evolution of an ultrastable carbonic anhydrase for highly efficient carbon capture from flue gas. *Proc. Natl. Acad. Sci. U.S.A.* **111** (46): 16436–16441.
33. Hu, Y., Zhu, Z., Gradischnig, D. et al. (2020). Engineering carboxylic acid reductase for selective synthesis of medium-chain fatty alcohols in yeast. *Proc. Natl. Acad. Sci. U.S.A.* **117** (37): 22974–22983.
34. (a) Qu, G., Liu, B., Zhang, K. et al. (2019). Computer-assisted engineering of the catalytic activity of a carboxylic acid reductase. *J. Biotechnol.* **306**: 97–104. (b) Qu, G., Fu, M., Zhao, L. et al. (2019). Computational insights into the catalytic mechanism of bacterial carboxylic acid reductase. *J. Chem. Inf. Model.* **59**: 832–841.
35. Furubayashi, M., Saito, K., and Umeno, D. (2014). Evolutionary analysis of the functional plasticity of *Staphylococcus aureus* C_{30} carotenoid synthase. *J. Biosci. Bioeng.* **117** (4): 431–436.
36. Ouyang, F. and Zhao, M. (2019). Enhanced catalytic efficiency of CotA-laccase by DNA shuffling. *Bioengineered* **10** (1): 182–189.
37. Yasukawa, K., Nakano, S., and Asano, Y. (2014). Tailoring D-amino acid oxidase from the pig kidney to *R*-stereoselective amine oxidase and its use in the deracemization of α-methylbenzylamine. *Angew. Chem. Int. Ed.* **53** (17): 4428–4431.
38. Ji, J., Fan, K., Tian, X. et al. (2012). Iterative combinatorial mutagenesis as an effective strategy for generation of deacetoxycephalosporin C synthase with improved activity toward penicillin G. *Appl. Environ. Microbiol.* **78** (21): 7809–7812.
39. (a) Brenna, E., Crotti, M., Gatti, F.G. et al. (2015). Opposite enantioselectivity in the bioreduction of (*Z*)-β-aryl-β-cyanoacrylates mediated by the tryptophan 116 mutants of old yellow enzyme 1: synthetic approach to (*R*)- and (*S*)-β-aryl-γ-lactams. *Adv. Synth. Catal.* **357** (8): 1849–1860. (b) Walton, A.Z., Conerly, W.C., Pompeu, Y. et al. (2011). Biocatalytic reductions of Baylis–Hillman adducts. *ACS Catal.* **1**: 989–993.
40. Walton, A.Z., Sullivan, B., Patterson-Orazem, A.C., and Stewart, J.D. (2014). Residues controlling facial selectivity in an alkene reductase and semirational alterations to create stereocomplementary variants. *ACS Catal.* **4** (7): 2307–2318.
41. Lin, L., Fu, C., and Huang, W. (2016). Improving the activity of the endoglucanase, Cel8M from *Escherichia coli* by error-prone PCR. *Enzyme Microb. Technol.* **86**: 52–58.
42. Hämäläinen, V., Barajas-López, J.D., Berlina, Y. et al. (2021). New thermostable endoglucanase from *Spirochaeta thermophila* and its

mutants with altered substrate preferences. *Appl. Microbiol. Biotechnol.* **105** (3): 1133–1145.

43. Ma, J., Wu, L., Guo, F. et al. (2013). Enhanced enantioselectivity of a carboxyl esterase from *Rhodobacter sphaeroides* by directed evolution. *Appl. Microbiol. Biotechnol.* **97** (11): 4897–4906.

44. Godinho, L.F., Reis, C.R., Rozeboom, H.J. et al. (2012). Enhancement of the enantioselectivity of carboxylesterase a by structure-based mutagenesis. *J. Biotechnol.* **158** (1, 2): 36–43.

45. Luan, Z.-J., Li, F.-L., Dou, S. et al. (2015). Substrate channel evolution of an esterase for the synthesis of cilastatin. *Catal. Sci. Technol.* **5** (5): 2622–2629.

46. Ma, F., Chung, M.T., Yao, Y. et al. (2018). Efficient molecular evolution to generate enantioselective enzymes using a dual-channel microfluidic droplet screening platform. *Nat. Commun.* **9** (1): 1030.

47. Kotik, M., Zhao, W., Iacazio, G., and Archelas, A. (2013). Directed evolution of metagenome-derived epoxide hydrolase for improved enantioselectivity and enantioconvergence. *J. Mol. Catal. B: Enzym.* **91**: 44–51.

48. Carlsson, Å.J., Bauer, P., Ma, H., and Widersten, M. (2012). Obtaining optical purity for product diols in enzyme-catalyzed epoxide hydrolysis: contributions from changes in both enantio- and regioselectivity. *Biochemistry* **51** (38): 7627–7637.

49. Li, F.L., Qiu, Y.Y., Zheng, Y.C. et al. (2020). Reprogramming epoxide hydrolase to improve enantioconvergence in hydrolysis of styrene oxide scaffolds. *Adv. Synth. Catal.* **362** (21): 4699–4706.

50. Swe, P.M., Copp, J.N., Green, L.K. et al. (2012). Targeted mutagenesis of the *Vibrio fischeri* flavin reductase FRase I to improve activation of the anticancer prodrug CB1954. *Biochem. Pharmacol.* **84** (6): 775–783.

51. Pala, U., Yelmazer, B., Çorbacioglu, M. et al. (2018). Functional effects of active site mutations in NAD^+-dependent formate dehydrogenases on transformation of hydrogen carbonate to formate. *Protein Eng. Des. Sel.* **31** (9): 327–335.

52. Çakar, M.M., Ruupunen, J., Mangas-Sanchez, J. et al. (2020). Engineered formate dehydrogenase from *Chaetomium thermophilum*, a promising enzymatic solution for biotechnical CO_2 fixation. *Biotechnol. Lett* **42** (11): 2251–2262.

53. Rannes, J.B., Ioannou, A., Willies, S.C. et al. (2011). Glycoprotein labeling using engineered variants of galactose oxidase obtained by directed evolution. *J. Am. Chem. Soc.* **133** (22): 8436–8439.

54. Dong, C., Qu, G., Guo, J. et al. (2022). Rational design of geranylgeranyl diphosphate synthase enhances carotenoid production and improves photosynthetic efficiency in *Nicotiana tabacum*. *Sci. Bull.* **67** (3): 315–327.

55. Qian, W.Z., Ou, L., Li, C.X. et al. (2020). Evolution of glucose dehydrogenase for cofactor regeneration in bioredox processes with denaturing agents. *ChemBioChem* **21** (18): 2680–2688.

56. Li, H. and Liao, J.C. (2014). Development of an NADPH-dependent homophenylalanine dehydrogenase by protein engineering. *ACS Synth. Biol.* **3** (1): 13–20.

57. de Abreu, M., Alvaro-Benito, M., Sanz-Aparicio, J. et al. (2013). Synthesis of 6-kestose using an efficient β-fructofuranosidase engineered by directed evolution. *Adv. Synth. Catal.* **355** (9): 1698–1702.

58. Ihssen, J., Haas, J., Kowarik, M. et al. (2015). Increased efficiency of *Campylobacter jejuni* *N*-oligosaccharyltransferase PglB by structure-guided engineering. *Open Biol.* **5** (4): 140227.

59. Pennec, A., Daniellou, R., Loyer, P. et al. (2015). Ara*f*51 with improved transglycosylation activities: one engineered biocatalyst for one specific acceptor. *Carbohydr. Res.* **402**: 50–55.

60. Han, R., Liu, L., Shin, H.-d. et al. (2013). Iterative saturation mutagenesis of −6 subsite residues in cyclodextrin glycosyltransferase from *Paenibacillus macerans* to improve maltodextrin specificity for

2-*O*-D-glucopyranosyl-L-ascorbic acid synthesis. *Appl. Environ. Microbiol.* **79** (24): 7562–7568.

61. Gantt, R.W., Peltier-Pain, P., Singh, S. et al. (2013). Broadening the scope of glycosyl transferase-catalyzed sugar nucleotide synthesis. *Proc. Natl. Acad. Sci. U.S.A.* **110** (19): 7648–7653.
62. Zhuang, Y., Yang, G.Y., Chen, X. et al. (2017). Biosynthesis of plant-derived ginsenoside Rh2 in yeast via repurposing a key promiscuous microbial enzyme. *Metab. Eng.* **42**: 25–32.
63. Bi, H., Qu, G., Wang, S. et al. (2022). Biosynthesis of a rosavin natural product in *Escherichia coli* by glycosyltransferase rational design and artificial pathway construction. *Metab. Eng.* **69**: 15–25.
64. Chen, T., Chen, Z., Wang, N. et al. (2021). Highly regioselective and efficient biosynthesis of polydatin by an engineered UGT_{BL}1-*At*SuSy cascade reaction. *J. Agric. Food. Chem.* **69** (31): 8695–8702.
65. Tian, C., Yang, J., Liu, C. et al. (2022). Engineering substrate specificity of HAD phosphatases and multienzyme systems development for the thermodynamic-driven manufacturing sugars. *Nat. Commun.* **13**: 3582.
66. Payne, J.T., Poor, C.B., and Lewis, J.C. (2015). Directed evolution of RebH for site-selective halogenation of large biologically active molecules. *Angew. Chem. Int. Ed.* **54** (14): 4226–4230.
67. Guo, C., Chen, Y., Zheng, Y. et al. (2015). Exploring the enantioselective mechanism of halohydrin dehalogenase from *Agrobacterium radiobacter* AD1 by iterative saturation mutagenesis. *Appl. Environ. Microbiol.* **81** (8): 2919–2926.
68. Schenkmayerova, A., Pinto, G.P., Toul, M. et al. (2021). Engineering the protein dynamics of an ancestral luciferase. *Nat. Commun.* **12** (1): 3616.
69. Zhang, F.R., Wan, N.W., Ma, J.M. et al. (2021). Enzymatic kinetic resolution of bulky spiro-epoxyoxindoles via halohydrin dehalogenase-catalyzed enantio- and regioselective azidolysis. *ACS Catal.* **11** (15): 9066–9072.
70. Hayashi, T., Ligibel, M., Sager, E. et al. (2019). Evolved aliphatic halogenases enable regiocomplementary C–H functionalization of a pharmaceutically relevant compound. *Angew. Chem. Int. Ed. Engl.* **58** (51): 18535–18539.
71. Büchler, J., Malca, S.H., Patsch, D. et al. (2022). Algorithm-aided engineering of aliphatic halogenase WelO5* for the asymmetric late-stage functionalization of soraphens. *Nat. Commun.* **13** (1): 371.
72. Duewel, S., Schmermund, L., Faber, T. et al. (2020). Directed evolution of an Feii-dependent halogenase for asymmetric C(sp^3)-H chlorination. *ACS Catal.* **10** (2): 1272–1277.
73. Zhang, W., Dourado, D.F.A.R., and Mannervik, B. (2015, 1850). Evolution of the active site of human glutathione transferase A2-2 for enhanced activity with dietary isothiocyanates. *Biochim. Biophys. Acta* **4**: 742–749.
74. Wiedner, R., Kothbauer, B., Pavkov-Keller, T. et al. (2015). Improving the properties of bacterial R-selective hydroxynitrile lyases for industrial applications. *ChemCatChem* **7**: 325–332.
75. Vishnu Priya, B., Sreenivasa Rao, D.H., Gilani, R. et al. (2022). Enzyme engineering improves catalytic efficiency and enantioselectivity of hydroxynitrile lyase for promiscuous retro-nitroaldolase activity. *Bioorg. Chem.* **120**: 105594.
76. Zheng, Y.C., Ding, L.Y., Jia, Q. et al. (2021). A high-throughput screening method for the directed evolution of hydroxynitrile lyase towards cyanohydrin synthesis. *ChemBioChem* **22** (6): 996–1000.
77. Chuang, H.-Y., Suen, C.-S., Hwang, M.-J., and Roffler, S.R. (2015). Toward reducing immunogenicity of enzyme replacement therapy: altering the specificity of human β-glucuronidase to compensate for α-iduronidase deficiency. *Protein Eng. Des. Sel.* **28** (11): 519–530.
78. Rigouin, C., Nguyen, H.A., Schalk, A.M., and Lavie, A. (2017). Discovery of human-like L-asparaginases with

potential clinical use by directed evolution. *Sci. Rep.* **7** (1): 10224.

79. Zhang, J., Liao, D., Chen, R. et al. (2022). Tuning an imine reductase for the asymmetric synthesis of azacycloalkylamines by concise structure-guided engineering. *Angew. Chem. Int. Ed.* **61**: e202201908.

80. Zhang, L., Cui, H., Zou, Z. et al. (2019). Directed evolution of a bacterial laccase (CueO) for enzymatic biofuel cells. *Angew. Chem. Int. Ed.* **58** (14): 4562–4565.

81. Pardo, I., Santiago, G., Gentili, P. et al. (2016). Re-designing the substrate binding pocket of laccase for enhanced oxidation of sinapic acid. *Catal. Sci. Technol.* **6**: 3900–3910.

82. Zhang, M., Hu, P., Zheng, Y.C. et al. (2020). Structure-guided engineering of *pseudomonas dacunhae*L-aspartate β-decarboxylase for L-homophenylalanine synthesis. *Chem. Commun.* **56** (89): 13876–13879.

83. Xu, X., Chen, J., Wang, Q. et al. (2016). Mutagenesis of key residues in the binding center of L-aspartate-ß-semialdehyde dehydrogenase from *Escherichia coli* enhances utilization of the cofactor NAD(H). *ChemBioChem* **17** (1): 56–64.

84. Sun, Z., Lonsdale, R., Kong, X.-D. et al. (2015). Reshaping an enzyme binding pocket for enhanced and inverted stereoselectivity: use of smallest amino acid alphabet in directed evolution. *Angew. Chem. Int. Ed.* **54**: 12410–12415.

85. Zheng, H.B. and Reetz, M.T. (2010). Manipulating the stereoselectivity of limonene epoxide hydrolase by directed evolution based on iterative saturation mutagenesis. *J. Am. Chem. Soc.* **132** (44): 15744–15751.

86. Wang, J., Wang, D., Wang, B. et al. (2012). Enhanced activity of rhizomucor miehei lipase by directed evolution with simultaneous evolution of the propeptide. *Appl. Microbiol. Biotechnol.* **96** (2): 443–450.

87. Reetz, M.T., Prasad, S., Carballeira, J.D. et al. (2010). Iterative saturation mutagenesis accelerates laboratory evolution of enzyme stereoselectivity: rigorous comparison with traditional methods. *J. Am. Chem. Soc.* **132** (26): 9144–9152.

88. Engström, K., Nyhlen, J., Sandström, A.G., and Bäckvall, J.E. (2010). Directed evolution of an enantioselective lipase with broad substrate scope for hydrolysis of alpha-substituted esters. *J. Am. Chem. Soc.* **132** (20): 7038–7042.

89. Wikmark, Y., Svedendahl Humble, M., and Bäckvall, J.-E. (2015). Combinatorial library based engineering of *Candida antarctica* lipase A for enantioselective transacylation of sec-alcohols in organic solvent. *Angew. Chem. Int. Ed.* **54** (14): 4284–4288.

90. Brundiek, H.B., Evitt, A.S., Kourist, R., and Bornscheuer, U.T. (2012). Creation of a lipase highly selective for trans fatty acids by protein engineering. *Angew. Chem. Int. Ed.* **51** (2): 412–414.

91. Kille, S., Acevedo-Rocha, C.G., Parra, L.P. et al. (2013). Reducing codon redundancy and screening effort of combinatorial protein libraries created by saturation mutagenesis. *ACS Synth. Biol.* **2**: 83–92.

92. (a) Zorn, K., Oroz-Guinea, I., Brundiek, H. et al. (2018). Alteration of chain length selectivity of *Candida antarctica* lipase A by semi-rational design for the enrichment of erucic and gondoic fatty acids. *Adv. Synth. Catal.* **360** (21): 4115–4131. (b) Oroz-Guinea, I., Zorn, K., and Bornscheuer, U.T. (2020). Enhancement of lipase CAL-a selectivity by protein engineering for the hydrolysis of erucic acid from crambe oil. *Eur. J. Lipid Sci. Technol.* **122** (1): 1900115.

93. Li, D., Chen, X., Chen, Z. et al. (2021). Directed evolution of lipase A from *Bacillus subtilis* for the preparation of enantiocomplementary *sec*-alcohols. *Green Syn. Catal.* **2** (3): 290–294.

94. Qin, B., Liang, P., Jia, X. et al. (2013). Directed evolution of *Candida antarctica* lipase B for kinetic resolution of profen esters. *Catal. Commun.* **38**: 1–5.

95. Li, D.-Y., Lou, Y.-J., Xu, J. et al. (2021). Electronic effect-guided rational design of *Candida antarctica* lipase B for kinetic resolution towards diarylmethanols. *Adv. Synth. Catal.* **363**: 1867–1872.

96. Colin, D.Y., Deprez-Beauclair, P., Silva, N. et al. (2010). Modification of pancreatic lipase properties by directed molecular evolution. *Prot. Eng. Des. Sel.* **23** (5): 365–373.

97. Panizza, P., Cesarini, S., Diaz, P., and Rodriguez Giordano, S. (2015). Saturation mutagenesis in selected amino acids to shift *Pseudomonas* sp. acidic lipase lip I.3 substrate specificity and activity. *Chem. Commun.* **51** (7): 1330–1333.

98. Li, X.-J., Zheng, R.-C., Ma, H.-Y. et al. (2014). Key residues responsible for enhancement of catalytic efficiency of *Thermomyces lanuginosus* lipase lip revealed by complementary protein engineering strategy. *J. Biotechnol.* **188**: 29–35.

99. Zheng, W., Yu, H., Fang, S. et al. (2021). Directed evolution of L-threonine aldolase for the diastereoselective synthesis of β-hydroxy-α-amino acids. *ACS Catal.* **11** (6): 3198–3205.

100. Wang, C., Zhang, K., Zhongjun, C. et al. (2015). Directed evolution and mutagenesis of lysine decarboxylase from *Hafnia alvei* AS1.1009 to improve its activity toward efficient cadaverine production. *Biotechnol. Bioprocess Eng.* **20** (3): 439–446.

101. Sun, S., Zhang, W., Mannervik, B., and Andersson, D.I. (2013). Evolution of broad spectrum β-lactam resistance in an engineered metallo-β-lactamase. *J. Biol. Chem.* **288** (4): 2314–2324.

102. Chen, Z., Ma, Y., He, M. et al. (2015). Semi-rational directed evolution of monoamine oxidase for kinetic resolution of rac-mexiletine. *Appl. Biochem. Biotechnol.* **176** (8): 2267–2278.

103. (a) Ghislieri, D., Green, A.P., Pontini, M. et al. (2013). Engineering an enantioselective amine oxidase for the synthesis of pharmaceutical building blocks and alkaloid natural products. *J. Am. Chem. Soc.* **135** (29): 10863–10869. (b) Ghislieri, D., Houghton, D., Green, A.P. et al. (2013). Monoamine oxidase (MAO-N) catalyzed deracemization of tetrahydro-β-carbolines: substrate dependent switch in enantioselectivity. *ACS Catal.* **3** (12): 2869–2872.

104. (a) Ren, X., Liu, N., Chandgude, A.L., and Fasan, R. (2020). An enzymatic platform for the highly enantioselective and Stereodivergent construction of cyclopropyl-δ-lactones. *Angew. Chem. Int. Ed.* **59** (48): 21634–21639. (b) Ren, X., Chandgude, A.L., and Fasan, R. (2020). Highly stereoselective synthesis of fused cyclopropane-γ-lactams via biocatalytic iron-catalyzed intramolecular cyclopropanation. *ACS Catal.* **10** (3): 2308–2313. (c) Chandgude, A.L., Ren, X., and Fasan, R. (2019). Stereodivergent intramolecular cyclopropanation enabled by engineered carbene transferases. *J. Am. Chem. Soc.* **141** (23): 9145–9150.

105. Xue, Y.-P., Shi, C.-C., Xu, Z. et al. (2015). Design of nitrilases with superior activity and enantioselectivity towards sterically hindered nitrile by protein engineering. *Adv. Synth. Catal.* **357** (8): 1741–1750.

106. Xu, Y., Liu, X.F., Chen, X.A., and Li, Y.Q. (2022). Directed evolution of a nonheme diiron *N*-oxygenase AzoC for improving its catalytic efficiency toward nitrogen heterocycle substrates. *Molecules* **27** (3): 868.

107. Demming, R.M., Hammer, S.C., Nestl, B.M. et al. (2019). Asymmetric enzymatic hydration of unactivated, aliphatic alkenes. *Angew. Chem. Int. Ed.* **58** (1): 173–177.

108. Ritter, C., Nett, N., Acevedo-Rocha, C.G. et al. (2015). Bioorthogonal enzymatic activation of caged compounds. *Angew. Chem. Int. Ed.* **54**: 13440–13443.

109. Roiban, G.-D., Agudo, R., and Reetz, M.T. (2014). Cytochrome P450 catalyzed oxidative hydroxylation of achiral organic compounds with simultaneous creation of two chirality centers in a single C–H activation step. *Angew. Chem. Int. Ed.* **53** (33): 8659–8663.

110. Chen, J., Kong, F., Ma, N. et al. (2019). Peroxide-driven hydroxylation of small alkanes catalyzed by an artificial P450BM3 peroxygenase system. *ACS Catal.* **9** (8): 7350–7355.

111. Li, F., Deng, H., and Renata, H. (2022). Remote B-ring oxidation of sclareol with an engineered P450 facilitates divergent access to complex terpenoids. *J. Am. Chem. Soc.* **144** (17): 7616–7621.

112. (a) Pham, S.Q., Pompidor, G., Liu, J. et al. (2012). Evolving P450pyr hydroxylase for highly enantioselective hydroxylation at non-activated carbon atom. *Chem. Commun.* **48** (38): 4618–4620. (b) Pham, S.Q., Gao, P., and Li, Z. (2013). Engineering of recombinant *E. coli* cells co-expressing P450pyrTM monooxygenase and glucose dehydrogenase for highly regio- and stereoselective hydroxylation of alicycles with cofactor recycling. *Biotechnol. Bioeng.* **110** (2): 363–373. (c) Review of directed evolution of P450-pyr:Yang, Y. and Li, Z. (2015). Evolving P450pyr monooxygenase for regio- and stereoselective hydroxylations. *Chimia* **69**: 136–141.

113. Yang, Y., Chi, Y.T., Toh, H.H., and Li, Z. (2015). Evolving P450pyr monooxygenase for highly regioselective terminal hydroxylation of *n*-butanol to 1,4-butanediol. *Chem. Commun.* **51** (5): 914–917.

114. Ba, L., Li, P., Zhang, H. et al. (2013). Semi-rational engineering of cytochrome P450sca-2 in a hybrid system for enhanced catalytic activity: insights into the important role of electron transfer. *Biotechnol. Bioeng.* **110** (11): 2815–2825.

115. Yang, Y., Liu, J., and Li, Z. (2014). Engineering of P450pyr hydroxylase for the highly regio- and enantioselective subterminal hydroxylation of alkanes. *Angew. Chem. Int. Ed.* **53** (12): 3120–3124.

116. Li, R.J., Xu, J.H., Chen, Q. et al. (2018). Enhancing the catalytic performance of a cyp116b monooxygenase by transdomain combination mutagenesis. *ChemCatChem* **10** (14): 2962–2968.

117. Iizaka, Y., Arai, R., Takahashi, A. et al. (2022). Engineering sequence and selectivity of late-stage C–H oxidation in the MycG iterative cytochrome P450. *J. Ind. Microbiol. Biotechnol.* **49** (1): kuab069.

118. Grobe, S., Badenhorst, C.P.S., Bayer, T. et al. (2021). Engineering regioselectivity of a P450 monooxygenase enables the synthesis of ursodeoxycholic acid via 7β-hydroxylation of lithocholic acid. *Angew. Chem. Int. Ed.* **60** (2): 753–757.

119. Rapp, L.R., Marques, S.M., Zukic, E. et al. (2021). Substrate anchoring and flexibility reduction in cyp153a m.aq leads to highly improved efficiency toward octanoic acid. *ACS Catal.* **11** (5): 3182–3189.

120. Bokel, A., Hutter, M.C., and Urlacher, V.B. (2021). Molecular evolution of a cytochrome P450 for the synthesis of potential antidepressant (2*R*,6*R*)-hydroxynorketamine. *Chem. Commun.* **57** (4): 520–523.

121. Bokel, A., Rühlmann, A., Hutter, M.C., and Urlacher, V.B. (2020). Enzyme-mediated two-step regio- and stereoselective synthesis of potential rapid-acting antidepressant (2*S*,6*S*)-hydroxynorketamine. *ACS Catal.* **10**: 4151–4159.

122. Deaguero, A.L., Blum, J.K., and Bommarius, A.S. (2012). Improving the diastereoselectivity of penicillin G acylase for ampicillin synthesis from racemic substrates. *Protein Eng. Des. Sel.* **25** (3): 135–144.

123. Toplak, A., de Oliveira, E.F.T., Schmidt, M. et al. (2021). From thiol-subtilisin to omniligase: design and structure of a broadly applicable peptide ligase. *Comput. Struct. Biotechnol. J.* **19**: 1277–1287.

124. Ye, L.J., Toh, H.H., Yang, Y. et al. (2015). Engineering of amine dehydrogenase for asymmetric reductive amination of ketone by evolving *Rhodococcus* phenylalanine dehydrogenase. *ACS Catal.* **5** (2): 1119–1122.

125. Abrahamson, M.J., Wong, J.W., and Bommarius, A.S. (2013). The evolution of an amine dehydrogenase biocatalyst for the asymmetric production of chiral amines. *Adv. Synth. Catal.* **355** (9): 1780–1786.

126. Naqvi, T., Warden, A.C., French, N. et al. (2014). A 5000-fold increase in the specificity of a bacterial phosphotriesterase for malathion through combinatorial active site mutagenesis. *PLoS One* **9** (4): e94177.

127. Emond, S., Petek, M., Kay, E.J. et al. (2020). Accessing unexplored regions of sequence space in directed enzyme evolution via insertion/deletion mutagenesis. *Nat. Commun.* **11** (1): 3469.
128. Hoque, M.A., Zhang, Y., Li, Z. et al. (2020). Remodeling enzyme active sites by stepwise loop insertion. *Methods Enzymol.* **643**: 111–127.
129. (a) Sundermann, U., Bravo-Rodriguez, K., Klopries, S. et al. (2013). Enzyme-directed mutasynthesis: a combined experimental and theoretical approach to substrate recognition of a polyketide synthase. *ACS Chem. Biol.* **8** (2): 443–450. (b) Kushnir, S., Sundermann, U., Yahiaoui, S. et al. (2012). Minimally invasive mutagenesis gives rise to a biosynthetic polyketide library. *Angew. Chem. Int. Ed.* **51** (42): 10664–10669.
130. Tournier, V., Topham, C.M., Gilles, A. et al. (2020). An engineered PET depolymerase to break down and recycle plastic bottles. *Nature* **580** (7802): 216–219.
131. Keys, T.G., Fuchs, H.L.S., Ehrit, J. et al. (2014). Engineering the product profile of a polysialyltransferase. *Nat. Chem. Biol.* **10** (6): 437–442.
132. Yang, J., Xiao, Y.Z., Li, R. et al. (2020). Repurposing a bacterial prolidase for organophosphorus hydrolysis: reshaped catalytic cavity switches substrate selectivity. *Biotechnol. Bioeng.* **117** (9): 2694–2702.
133. Zheng, H., Wang, X., Yomano, L.P. et al. (2013). Improving *Escherichia coli* FucO for furfural tolerance by saturation mutagenesis of individual acid positions. *Appl. Environ. Microbiol.* **79**: 3202–3208.
134. Guo, L.-T., Wang, Y.-S., Nakamura, A. et al. (2014). Polyspecific pyrrolysyl-tRNA synthetases from directed evolution. *Proc. Natl. Acad. Sci. U.S.A.* **111** (47): 16724–16729.
135. Kaya, E., Vrabel, M., Deiml, C. et al. (2012). A genetically encoded norbornene amino acid for the mild and selective modification of proteins in a copper-free click reaction. *Angew. Chem.Int. Ed.* **51** (18): 4466–4469.
136. Kumar, R., Karmilowicz, M.J., Burke, D. et al. (2021). Biocatalytic reductive amination from discovery to commercial manufacturing applied to abrocitinib JAK1 inhibitor. *Nat. Catal.* **4**: 775–782.
137. Ibach, J., Dietrich, L., Koopmans, K.R.M. et al. (2013). Identification of a T7 RNA polymerase variant that permits the enzymatic synthesis of fully 2′-*O*-methyl-modified RNA. *J. Biotechnol.* **167** (3): 287–295.
138. Exner, M.P., Kuenzl, T., To, T.M. et al. (2017). Design of *S*-allylcysteine in situ production and incorporation based on a novel pyrrolysyl-tRNA synthetase variant. *ChemBioChem* **18** (1): 85–90.
139. Frei, M.S., Tarnawski, M., Roberti, M.J. et al. (2022). Engineered HaloTag variants for fluorescence lifetime multiplexing. *Nat. Methods* **19** (1): 65–70.
140. a Püllmann, P., Knorrscheidt, A., Münch, J. et al. (2021). A modular two yeast species secretion system for the production and preparative application of unspecific peroxygenases. *Commun. Biol.* **4** (1): 562. (b) Püllmann, P. and Weissenborn, M.J. (2021). Improving the heterologous production of fungal peroxygenases through an episomal *Pichia pastoris* promoter and signal peptide shuffling system. *ACS Synth. Biol.* **10** (6): 1360–1372.
141. Dionne, U., Bourgault, É., Dubé, A.K. et al. (2021). Protein context shapes the specificity of SH3 domain-mediated interactions in vivo. *Nat. Commun.* **12** (1): 1597.
142. Bosshart, A., Hee, C.S., Bechtold, M. et al. (2015). Directed divergent evolution of a thermostable D-tagatose epimerase towards improved activity for two hexose substrates. *ChemBioChem* **16** (4): 592–601.
143. Deszcz, D., Affaticati, P., Ladkau, N. et al. (2015). Single active-site mutants are sufficient to enhance serine: pyruvate α-transaminase activity in an ω-transaminase. *FEBS J.* **282** (13): 2512–2526.

144. Wu, S., Liu, J., and Li, Z. (2017). Biocatalytic formal anti-markovnikov hydroamination and hydration of aryl alkenes. *ACS Catal.* **7**: 5225–5233.

145. Novick, S.J., Dellas, N., Garcia, R. et al. (2021). Engineering an amine transaminase for the efficient production of a chiral sacubitril precursor. *ACS Catal.* **11** (6): 3762–3770.

146. Ma, Y., Jiao, X., Wang, Z. et al. (2022). Engineering a transaminase for the efficient synthesis of a key intermediate for rimegepant. *Org. Process Res. Dev.* https://doi.org/10.1021/acs.oprd.1c00376.

147. Liu, Q., Xie, X., Tang, M. et al. (2021). One-pot asymmetric synthesis of an aminodiol intermediate of florfenicol using engineered transketolase and transaminase. *ACS Catal.* **11** (12): 7477–7488.

148. Ranoux, A., Karmee, S.K., Jin, J. et al. (2012). Enhancement of the substrate scope of transketolase. *ChemBioChem* **13** (13): 1921–1931.

149. Yi, D., Saravanan, T., Devamani, T. et al. (2015). A thermostable transketolase evolved for aliphatic aldehyde acceptors. *Chem. Commun.* **51** (3): 480–483.

150. Knorrscheidt, A., Soler, J., Hünecke, N. et al. (2021). Accessing chemo- and regioselective benzylic and aromatic oxidations by protein engineering of an unspecific peroxygenase. *ACS Catal.* **11**: 7327–7338.

151. Lu, Z., Li, X., Zhang, R. et al. (2019). Tunnel engineering to accelerate product release for better biomass-degrading abilities in lignocellulolytic enzymes. *Biotechnol. Biofuels* **12**: 275.

152. Knorrscheidt, A., Püllmann, P., Schell, E. et al. (2020). Identification of novel unspecific peroxygenase chimeras and unusual YfeX axial heme ligand by a versatile high-throughput GC-MS approach. *ChemCatChem* **12**: 4788–4795.

153. Hock, K.J., Knorrscheidt, A., Hommelsheim, R. et al. (2019). Tryptamine synthesis by iron porphyrin catalyzed C–H functionalization of indoles with diazoacetonitrile. *Angew. Chem. Int. Ed.* **58** (11): 3630–3634.

154. Eggbauer, B., Schrittwieser, J.H., Kerschbaumer, B. et al. (2022). Regioselective biocatalytic C4-prenylation of unprotected tryptophan derivatives. *ChemBioChem* **23**: e202200311.

155. Herrmann, S., Dippe, M., Pecher, P. et al. (2022). Engineered bacterial flavin-dependent monooxygenases for the regiospecific hydroxylation of polycyclic phenols. *ChemBioChem* **23** (6): e202100480.

156. Godehard, S.P., Badenhorst, C.P., Müller, H., and Bornscheuer, U.T. (2020). Protein engineering for enhanced acyltransferase activity, substrate scope, and selectivity of the *Mycobacterium smegmatis* acyltransferase MsAcT. *ACS Catal.* **10** (14): 7552–7562.

157. Godehard, S.P., Müller, H., Badenhorst, C. et al. (2021). Efficient acylation of sugars and oligosaccharides in aqueous environment using engineered acyltransferases. *ACS Catal.* **11**: 2831–2836.

158. García-Marquina, G., Núñez-Franco, R., Peccati, F. et al. (2022). Deconvoluting the directed evolution pathway of engineered acyltransferase LovD. *ChemCatChem* **14** (4): e202101349.

159. Staudt, A., Terholsen, H., Kaur, J. et al. (2021). Rational design for enhanced acyltransferase activity in water catalyzed by the *Pyrobaculum calidifontis* VA1 esterase. *Microorganisms* **9** (8): 1790.

160. Jakoblinnert, A., van den Wittenboer, A., Shivange, A.V. et al. (2013). Design of an activity and stability improved carbonyl reductase from *Candida parapsilosis*. *J. Biotechnol.* **165** (1): 52–62.

161. Ao, Y.-F., Hu, H.-J., Zhao, C.-X. et al. (2021). Reversal and amplification of the enantioselectivity of biocatalytic desymmetrization toward meso heterocyclic dicarboxamides enabled by rational engineering of amidase. *ACS Catal.* **11** (12): 6900–6907.

162. Femmer, C., Bechtold, M., and Panke, S. (2020). Semi-rational engineering of an amino acid racemase that is stabilized in aqueous/organic solvent mixtures. *Biotechnol. Bioeng.* **117** (9): 2683–2693.

163. Dindo, M., Pascarelli, S., Chiasserini, D. et al. (2022). Structural dynamics shape the fitness window of alanine:glyoxylate aminotransferase. *Protein Sci.* **31** (5): e4303.

164. Modi, T., Risso, V.A., Martinez-Rodriguez, S. et al. (2021). Hinge-shift mechanism as a protein design principle for the evolution of β-lactamases from substrate promiscuity to specificity. *Nat. Commun.* **12**: 1852.

165. Bayer, C.D., van Loo, B., and Hollfelder, F. (2017). Specificity effects of amino acid substitutions in promiscuous hydrolases: context-dependence of catalytic residue contributions to local fitness landscapes in nearby sequence space. *ChemBioChem* **18** (11): 1001–1015.

166. Muiños, F., Martínez-Jiménez, F., Pich, O. et al. (2021). In silico saturation mutagenesis of cancer genes. *Nature* **596** (7872): 428–432.

167. Park, A. and Park, S. (2022). Discovery and redesign of a family VIII carboxylesterase with high (*S*)-selectivity toward chiral *sec*-alcohols. *ACS Catal.* **12** (4): 2397–2402.

168. Yang, D., Su, W., Jiang, Y. et al. (2022). Biosynthesis of β-lactam nuclei in yeast. *Metab. Eng.* **72**: 56–65.

169. Park, S.-H., Seo, H., Seok, J. et al. (2021). Cβ-selective aldol addition of D-threonine aldolase by spatial constraint of aldehyde binding. *ACS Catal.* **11** (12): 6892–6899.

170. He, L., Tzarum, N., Lin, X. et al. (2020). Proof of concept for rational design of hepatitis C virus E2 core nanoparticle vaccines. *Sci. Adv.* **6** (16): eaaz6225.

171. Liu, Y.Y., Wu, M.D., Zhu, X. et al. (2021). Remarkable improvement in the regiocomplementarity of a *Glycine max* epoxide hydrolase by reshaping its substrate-binding pocket for the enantioconvergent preparation of (*R*)-hexane-1,2-diol. *Mol. Catal.* **514**: 111851.

172. Nobili, A., Tao, Y., Pavlidis, I.V. et al. (2015). Simultaneous use of in silico design and a correlated mutation network as a tool to efficiently guide enzyme engineering. *ChemBioChem* **16** (5): 805–810.

173. Teze, D., Zhao, J., Wiemann, M. et al. (2021). Rational enzyme design without structural knowledge: a sequence-based approach for efficient generation of transglycosylases. *Chem. Eur. J.* **27**: 10323.

174. Zhang, J., Tang, M., Chen, Y. et al. (2021). Catalytic flexibility of rice glycosyltransferase OsUGT91C1 for the production of palatable steviol glycosides. *Nat. Commun.* **12**: 7030.

175. Li, J., Qu, G., Shang, N. et al. (2021). Near-perfect control of the regioselective glucosylation enabled by rational design of glycosyltransferases. *Green Syn. Catal.* **2** (1): 45–53.

176. Caserta, G., Chino, M., Firpo, V. et al. (2018). Enhancement of peroxidase activity in artificial mimochrome VI catalysts through rational design. *ChemBioChem* **19** (17): 1823–1826.

177. Müller, J., Sowa, M.A., Fredrich, B. et al. (2015). Enhancing the acyltransferase activity of *Candida antarctica* lipase A by rational design. *ChemBioChem* **16** (12): 1791–1796.

178. (a) Xu, J., Cen, Y., Singh, W. et al. (2019). Stereodivergent protein engineering of a lipase To access all possible stereoisomers of chiral esters with two stereocenters. *J. Am. Chem. Soc.* **141** (19): 7934–7945. (b) Li, D., Wu, Q., and Reetz, M.T. (2020). Focused rational iterative site-specific mutagenesis (FRISM). *Methods Enzymol.* **643**: 225–242.

179. Tian, L., Zhou, J., Lv, Q. et al. (2021). Rational engineering of the *Plasmodium falciparum* l-lactate dehydrogenase loop involved in catalytic proton transfer to improve chiral 2-hydroxybutyric acid production. *Int. J. Biol. Macromol.* **179**: 71–79.

180. Huber, T.D., Clinger, J.A., Liu, Y. et al. (2020). Methionine adenosyltransferase engineering to enable bioorthogonal platforms for AdoMet-utilizing enzymes. *ACS Chem. Biol.* **15** (3): 695–705.

181. Schmidt, M., Toplak, A., Rozeboom, H.J. et al. (2018). Design of a substrate-tailored peptiligase variant for the efficient synthesis of thymosin-α_1. *Org. Biomol. Chem.* **16** (4): 609–618.

182. Liu, X., Kang, F., Hu, C. et al. (2018). A genetically encoded photosensitizer protein facilitates the rational design of a miniature photocatalytic CO_2-reducing enzyme. *Nat. Chem.* **10** (12): 1201–1206.

183. Roda, S., Fernandez-Lopez, L., Cañadas, R. et al. (2021). Computationally driven rational design of substrate promiscuity on serine ester hydrolases. *ACS Catal.* **11**: 3590–3601.

184. Wójcik, M., Szala, K., van Merkerk, R. et al. (2020). Engineering the specificity of *Streptococcus pyogenes* sortase A by loop grafting. *Proteins* **88** (11): 1394–1400.

185. Bastian, S.A., Hammer, S.C., Kreß, N. et al. (2017). Selectivity in the cyclization of citronellal introduced by squalene hopene cyclase variants. *ChemCatChem* **9** (23): 4364–4368.

186. Xia, Y., Li, X., Yang, L. et al. (2021). Development of thermostable sucrose phosphorylase by semi-rational design for efficient biosynthesis of alpha-D-glucosylglycerol. *Appl. Microbiol. Biotechnol.* **105** (19): 7309–7319.

187. Romero-Romero, S., Costas, M., Manzano, D.-A.S. et al. (2021). The stability landscape of de novo TIM barrels explored by a modular design approach. *J. Mol. Biol.* **433** (18): 167153.

188. Le-Huu, P., Rekow, D., Krüger, C. et al. (2018). Chemoenzymatic route to oxyfunctionalized cembranoids facilitated by substrate and protein engineering. *Chem. Eur. J.* **24** (46): 12010–12021.

189. Ma, N., Fang, W., Liu, C. et al. (2021). Switching an artificial P450 peroxygenase into peroxidase via mechanism-guided protein engineering. *ACS Catal.* **11** (14): 8449–8455.

190. Xu, J., Fan, J., Lou, Y. et al. (2021). Light-driven decarboxylative deuteration enabled by a divergently engineered photodecarboxylase. *Nat. Commun.* **12**: 3983.

191. Zeng, W., Li, X., Yang, Y. et al. (2022). Substrate-binding mode of a thermophilic PET hydrolase and engineering the enzyme to enhance the hydrolytic efficacy. *ACS Catal.* **12** (5): 3033–3040.

192. Reetz, M.T. (2011). Laboratory evolution of stereoselective enzymes: a prolific source of catalysts for asymmetric reactions. *Angew. Chem. Int. Ed.* **50** (1): 138–174.

193. (a) Reetz, M.T. (2022). Making enzymes suitable for organic chemistry by rational protein design. *ChemBioChem* e202200049. (b) Reetz, M.T. (2022). Witnessing the birth of directed evolution of stereoselective enzymes as catalysts in organic chemistry. *Adv. Synth. Catal.* https://doi.org/10.1002/adsc.2022004466. (c) Lovelock, S.L., Crawshaw, R., Basler, S. et al. (2022). The road to fully programmable protein catalysis. *Nature* **606** (7912): 49–58.

6
Protein Engineering of Enzyme Robustness Relevant to Organic and Pharmaceutical Chemistry and Applications in Biotechnology

6.1
Introductory Remarks

Increasing the robustness of enzymes under operating conditions for applications in organic chemistry and biotechnology has been the goal of protein engineers for almost four decades [1] (see also Chapter 1). The different kinds of forces stabilizing proteins and the influence of mutations have been discussed [2]. A discussion of whether the traditional thermodynamic hypothesis of folding should be questioned has recently appeared [3]. A concise review of current advances in the design and engineering strategies of industrial enzymes, including thermostabilization methods, has appeared [4]. Up-to-date reviews of progress in process engineering, which include immobilization techniques for enzyme stabilization, are available [5].

As in the case of enzyme selectivity, the two protein engineering approaches are rational design (Section 6.2) and directed evolution (Section 6.5), sometimes both being utilized in a given project. Both rational enzyme design of protein robustness, as summarized by V.G.H. Eijsink in 2004 [6], and the directed evolution option [7] rely on similar guidelines, which are as follows:

- Introduction of salt bridges
- Construction of H-bond networks
- Construction of disulfide bonds
- Phylogenetic analysis (consensus)
- Previous mutational data [8]
- Structural and mechanistic information
- Molecular dynamics (MD) computations

It has been found that certain enzymes in hyperthermophilic organisms have more salt bridges, mainly on the protein surface, relative to those in mesophilic organisms. Thus, these can be identified and introduced in homologous enzymes (consensus principle). Further insights can be gained in a monograph on extremophiles medicine and pharmaceutical industries [9]. Moreover, the loss of salt bridges generally results in a decrease in stability, although other parameters may be improved (tradeoff effect). The introduction of salt bridges [6]

Enzyme Engineering: Selective Catalysts for Applications in Biotechnology, Organic Chemistry, and Life Science, First Edition. Manfred T. Reetz, Zhoutong Sun, and Ge Qu.

is particularly important because they lead not just to enhanced thermostability and also to higher resistance to hostile organic solvents [10]. More perspectives regarding the stability-function tradeoff in proteins have been documented in a recent comprehensive review [11].

6.2 Rational Design of Enzyme Thermostability and Resistance to Hostile Organic Solvents

It is worthwhile to remind the reader of an older, yet highly important study reporting rational design of enzyme robustness in which many of the above bullet points were considered [12]. It set the stage for further refinements and novel developments. The moderately stable protease TLP-ste from *Bacillus stearothermophilus* was turned into a hyper-stable homolog by introducing only a limited number of point mutations (Figure 6.1). Briefly, amino acids were first replaced by considering equivalent residues of well-known robust proteases (consensus strategy). Designed disulfide bonds were also constructed, and the mutations were combined. The best mutant functioned well even at 100 °C (Figure 6.2). Most of the mutations were considered to induce stabilizing effects that are due to a reduction of entropy, i.e. such mutations as Gly → Ala or Ala→Pro and disulfide bonds cause rigidification.

Fortunately, solving the problem of insufficient enzyme thermostability and/or resistance to hostile organic solvents by rational design is usually less challenging than rationally enhancing or reversing stereoselectivity MTR625. Today, protein engineers essentially utilize the same guidelines as listed in

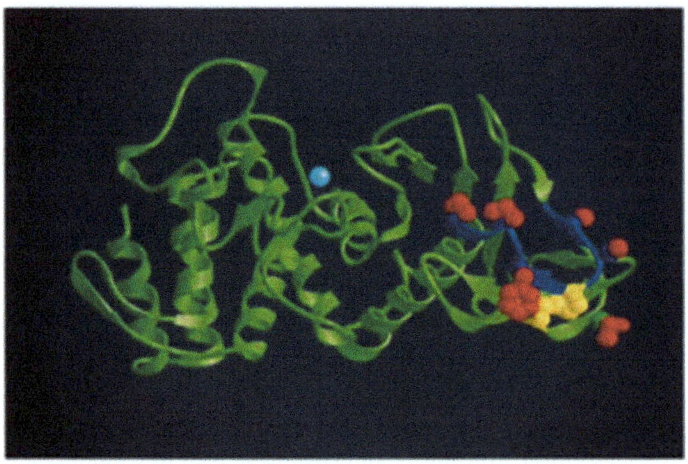

Figure 6.1 Overlay of the eightfold mutated TLP-ste variant. The substitutions (A4T, T56A, G58A, T63F, S65P, and A69P) are depicted in red, and the disulfide bridge between N60C and G8C is indicated in yellow. The zinc ion is shown in light blue. Source: Ref. [12]. Copyright (1998) National Academy of Sciences, U.S.A.

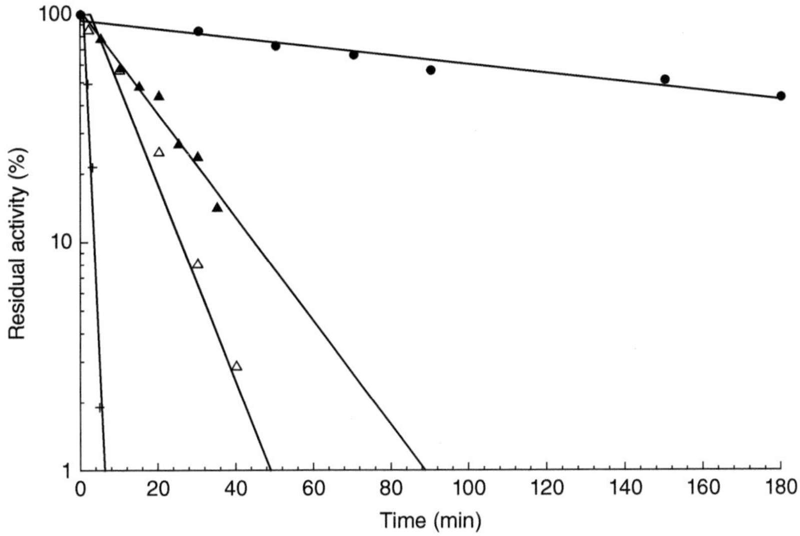

Figure 6.2 Residual activity and stability determined for TLP-ste at 80 °C (▲) and 90 °C (+), thermolysin at 90 °C (Δ), and the eightfold mutant at 100 °C (●). Source: Ref. [12]. Copyright (1998) National Academy of Sciences, U.S.A.

Section 6.1, except that a number of important refinements and improvements have since been reported. A prime advance is the RosettaDesign package used for a variety of purposes, not just for stability enhancement [13]. An early application of Rosetta concerns the thermostabilization of the homodimeric hydrolase yeast cytosine deaminase (yCD) having 153 residues, its natural function being the conversion of cytosine to uracil [14]. It was known to undergo irreversible unfolding at elevated temperatures, probably due to undesired aggregation. Target sequences threaded onto a template backbone were first evaluated by an energy function. A Monte Carlo procedure was then applied for searching sequence space. It was started with a random sequence in which single amino acid rotamers were replaced by a rotamer derived from the Dunbrack backbone-dependent rotamer library [15]. The Dunbrack algorithm enables a Bayesian statistical analysis of conformations of side chains, based on X-ray data of the Protein Data Bank. In the Rosetta study, the next step involved energy reevaluation. Low-energy sequences were identified and retained for further consideration. Residues participating in catalysis and those within 4 Å of the active site were "held fixed," while the remaining 65 residues were treated in reevaluations by introducing thereat all amino acids except cysteine. Following the computational predictions, site-directed mutagenesis was used to generate and test the mutants. Several predictions proved to be correct, e.g. a triple mutant with an increase in apparent melting temperature T_m of 10 °C and a 30-fold increase in half-life at 50 °C (Figure 6.3).

Based on the crystal structures of the double and triple mutants and the analysis of residue-by-residue packing, it was concluded that in the enzyme

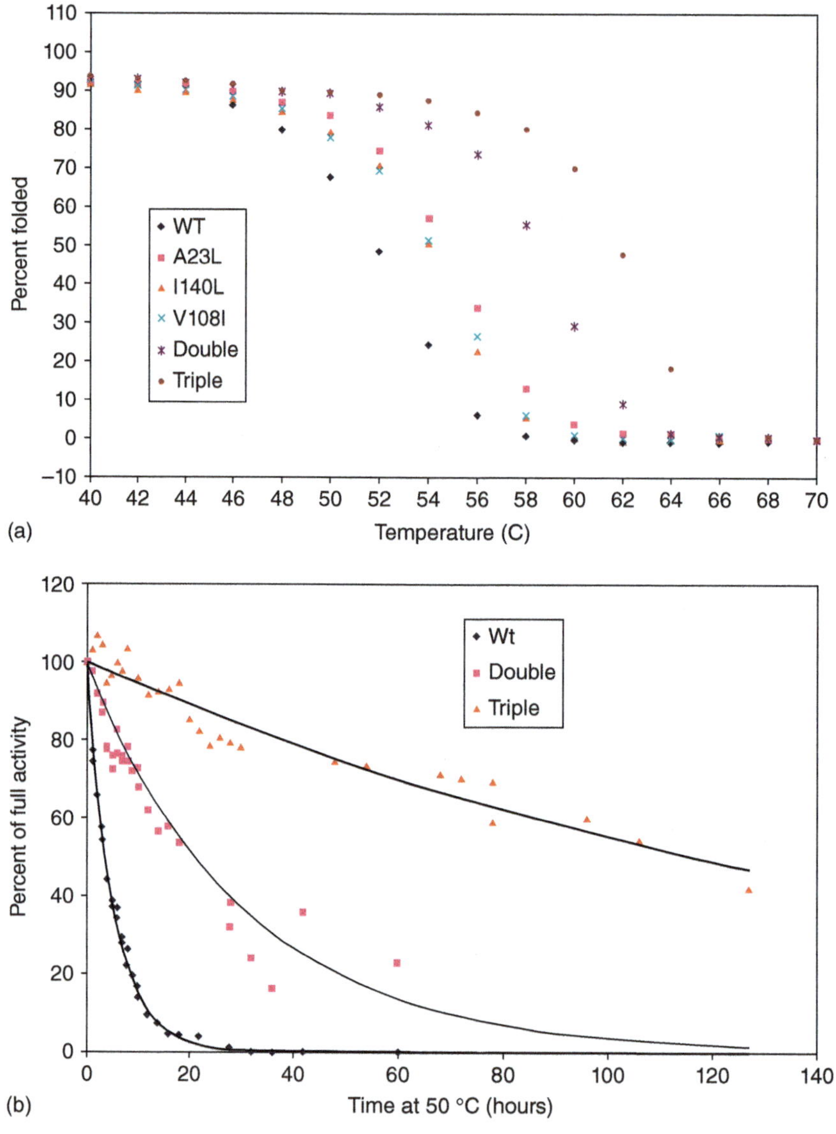

Figure 6.3 Temperature melt measuring (a) and activity half-life measurements (b). Source: Ref. [14]/American Association for the Advancement of Science.

core residues pack more closely, and that about 70 Å^2 additional buried surface results as a consequence of three mutations (Figure. 6.4). The structure of the active site remains as in wild-type (WT). Due to potential practical applications, it was important to study the mutational effects *in vivo*, and therefore a strain of *Eshcherichia coli* dependent on cytosine deaminase was engineered and transformed with the mutant yCD and WT. In this case, the stabilizing effects turned

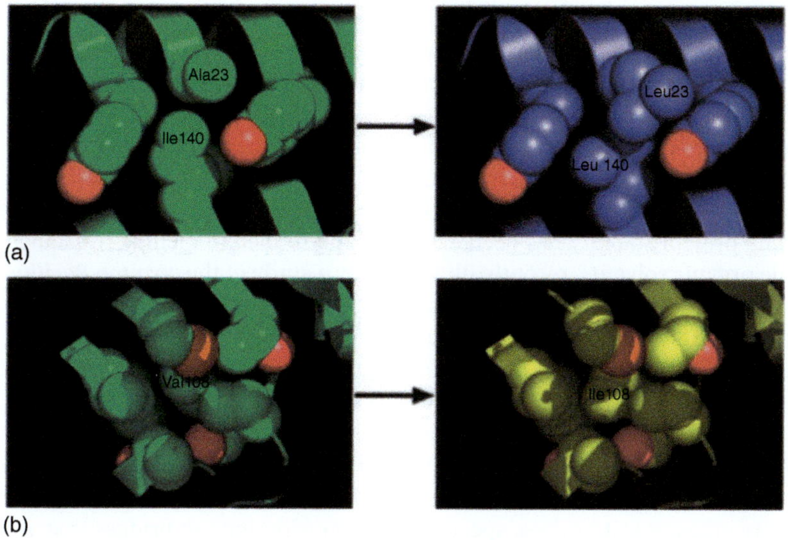

Figure 6.4 (a) van der Waals representation of sites A23 and I140 in WT yCD (left) and the substitutions A23L and I140L in mutant (right); (b) van der Waals radii representation of V108 in WT (left) and the substitution V108I in the triple mutant (right). Source: Ref. [14]/American Association for the Advancement of Science.

out to be modest, but the authors concluded that "there is no obvious reason why additional mutations predicted by the program could not be iteratively incorporated into the enzyme core, resulting in a panel of catalysts that display sequential increases in thermal stability."

RosettaDesign for protein thermostabilization has also been used in more recent studies [16a], sometimes in combination with rationally designed disulfide bonds [16b–e]. In one investigation, Rosetta was compared to the traditional consensus technique ("back to consensus") [17]. Accordingly, an *E. coli* transketolase served as the enzyme, and both approaches proved to be similarly successful. The best variant showed a threefold enhanced half-life at 60 °C and an increase in T_m of 5 °C, which are somewhat modest. The Rosetta program reached a prediction accuracy of 65%. In yet a different study with important implications regarding practical applications, a "Rosetta scan" was included [18]. Since it is more related to a semi-rational directed evolution approach coupled with fluorescence-activated cell sorting (FACS) (see Chapter 2), it will be highlighted in Section 6.4.

In another study featuring a different computational approach, the automated structure- and sequence-based design of a human acetylcholinesterase (hAChE) having enhanced stability and improved heterologous expression was reported, an enzyme that mediates synaptic transmission [19]. The computational strategy involved application of the consensus technique [20] and RosettaDesign [21], leading to a variant with 51 mutations. It has a 20 °C higher thermostability relative to WT, while also showing a 2000-fold higher expression level in *E. coli*.

The algorithm requires X-ray structural data and only a few dozen sequences of natural homologs and is available free of charge at http://pross.weizmann.ac.il. An analysis of the origin of improvements revealed increased core packing as well as modified surface polarity and backbone rigidity (Figure 6.5). The same approach proved to be successful in the directed evolution of human histone deacylase SIRT6, DNA methyltransferase Dnmt3a, and bacterial phosphotriesterase.

In other work, the notion of designed salt bridges and H-networks for enhancing protein stability has been extended and refined to include resistance to hostile organic solvents, recently illustrated by a study of the enzyme *Bacillus subtilis* lipase A (BSLA) [22]. In previous efforts, the authors had performed semi-rational NNK-based saturation mutagenesis with the creation of a mutant library BSLA-SSM, which was screened for variants showing improved resistance to 1,4-dioxane (DOX), dimethylsulfoxide (DMSO), and 2,2,2-trifluoroethanol (TFE) [23]. At each of the 181 residues of BSLA, mutations covering all 20 canonical amino acids were introduced (total of 3440 variants), assuming the absence of amino acid bias. In the new study, the data were utilized for identifying more efficient salt bridges, specifically by considering salt bridges that are formed in the presence of organic solvents but absent in water. For this purpose, salt bridges were replaced by two residues having the same charge. As shown by MD computations, this leads to local flexibility in the presence of organic solvents, tight attraction, and ultimately higher resistance to hostile organic solvents and even enhanced thermostability. The correlation between enhanced thermostability of mutants and higher resistance to denaturing processes in the presence of organic solvents is well-known in protein engineering [24, 25].

A number of predictions at various hotspots were made and their respective variants were evaluated in the laboratory. Information regarding the actual number of such mutants was not provided, which would have been important in a final assessment of this approach. To identify possible additive mutational effects [26] and to achieve further improvements, certain recombinations of mutations were considered at substitutions D34K, D64K, D144K, and K112E, 11 BSLA recombinants being prepared and tested for solvent resistance using crude culture supernatants (Figure 6.6). Relative to WT BSLA, all of them showed increased resistance to DOX, DMSO, and TFE. Especially double mutant D64K/D144K proved to be quite efficient, exhibiting a 6.6-fold, 5.3-fold, and 7.6-fold improved resistance against DOX, DMSO, and TFE, respectively. In this study, various other mutational combinations were discussed, and the origin of improvements was identified. For example, in D34K/D144K, two suboptimal salt bridges were replaced, leading to 3.0–4.2-fold higher resistance to the three hostile solvents [22].

The thermostability of some of the variants was also improved relative to WT BSLA, a result that is not overly surprising. Typical examples are shown in Figure 6.7 [22]. In summary, this study is a good example of modern rational enzyme design of a lipase with improved robustness. Further examples using other enzyme types would be needed for a final evaluation. From the viewpoint of laboratory effort, time, and costs, it would also be interesting to see how

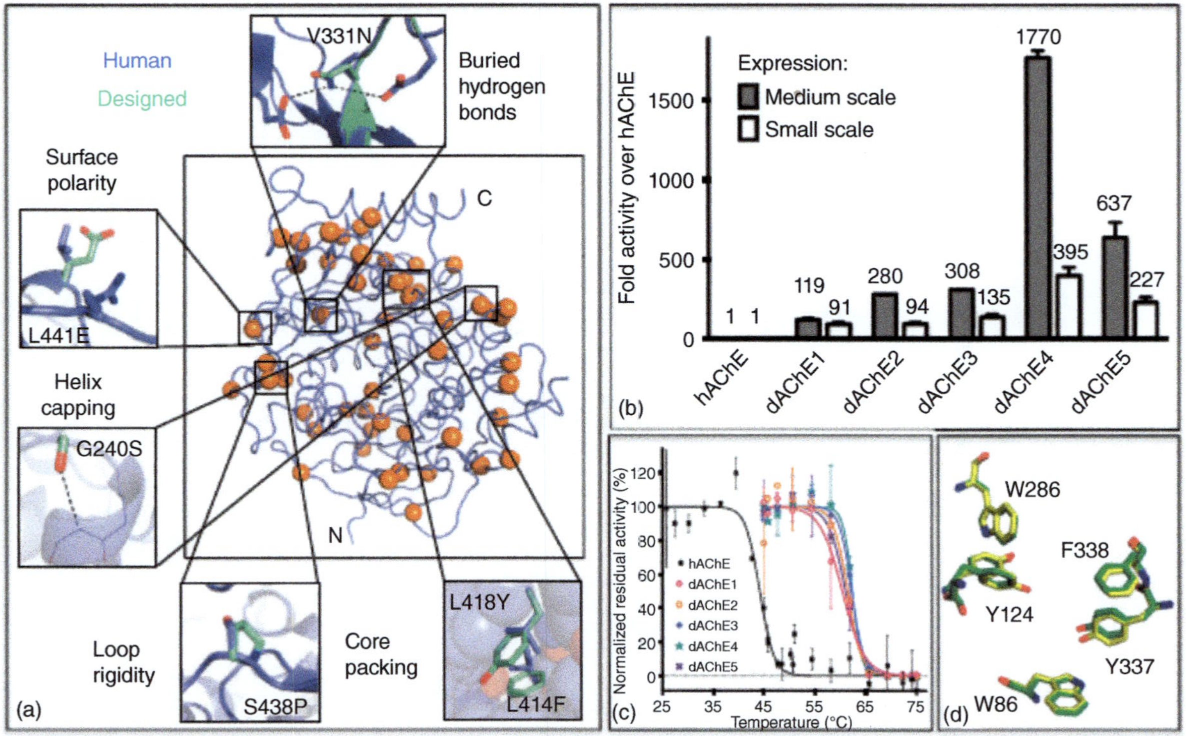

Figure 6.5 (a) Overlay of the 51 mutations in the designed variant. (b) Relative lysate activity assay of designed AChEs, and resistance to heat inactivation (c). Structural comparison of the active site between dAChE4 (PDB: 5HQ3, yellow) and hAChE (PDB: 4EY4, green). Source: Ref. [19]/With permission of Elsevier.

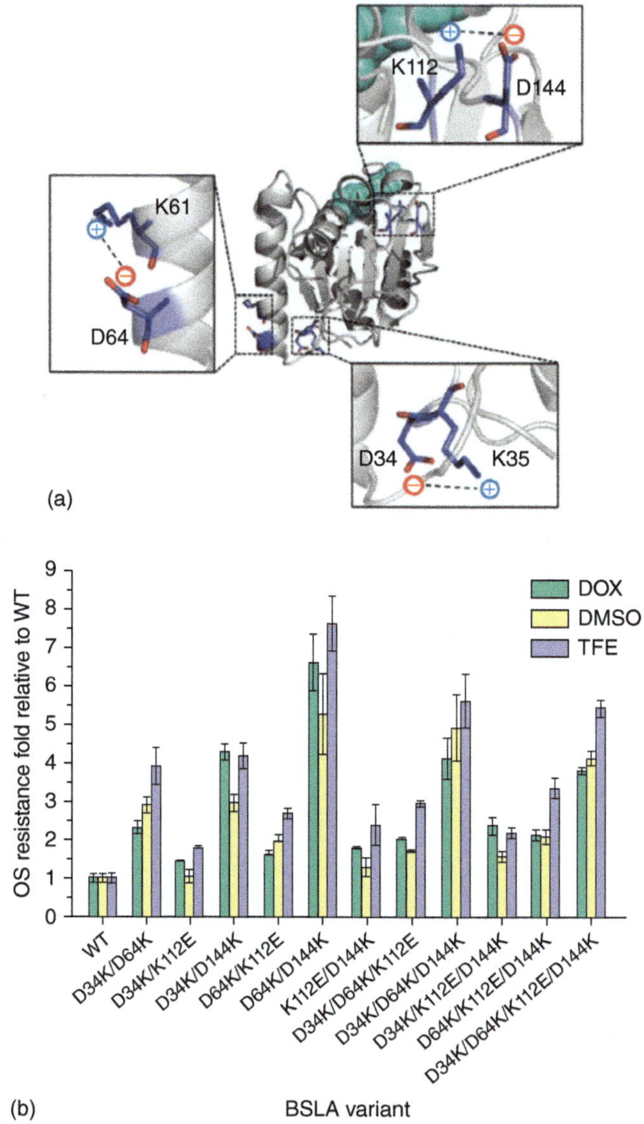

Figure 6.6 (a) Representations of the salt bridges in BSLA WT. (b) OSs resistance assay of BSLA mutants. Source: [22]/John Wiley & Sons.

semi-rational directed evolution approaches (Section 6.5) would work in the case of BSLA. Such comparisons have not been made to date.

Several other key studies on rational protein thermostabilization deserve mention [27]. In summary, a number of recent advances have been made in the rational design of thermostability and resistance to hostile organic solvents. The development of user-friendly algorithms for reliably predicting hotspots

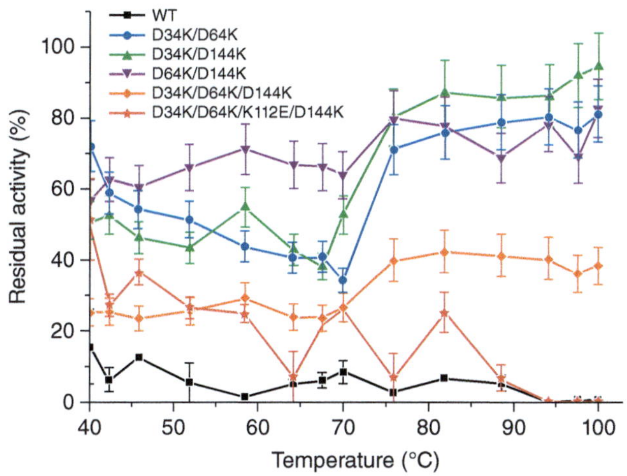

Figure 6.7 The thermal resistance assay of BSLA WT and its variants. Source: [22]/John Wiley & Sons.

was and still is an essential aspect of this ongoing research. Sometimes a given technique, e.g. Rosetta program, can be incorporated as one of the helping aids in the directed evolution approach to thermostabilization (Section 6.5).

6.3 Ancestral and Consensus Approaches and Their Structure-Guided Extensions

The ancestral technique for protein stabilization, first published in 1990 [28], assumes that early ancestors of today's organisms had higher thermostability than extant homologs. The method has not been applied very often, but interesting variations continue to be published [29]. In a study focusing on the B-subunit of DNA gyrase, the ancestral method was compared experimentally with the consensus strategy [29a]. The approach was also applied to the directed evolution of fungal Mesozoic laccases, in which thermostability and pH stability were improved [29c] A review on ancestral resurrection covering a variety of different applications, including the development of a vaccine against SARS-CoV-2 (but not of Covid-19), has recently appeared [29d].

The consensus approach for increasing the robustness of proteins is based on the alignment of orthologous proteins, leading to the identification of the most prevalent amino acid at a given position. The predicted mutations are then introduced into a small mutant library for screening [20a]. Early examples of using such multiple sequence alignments (MSAs) focused on the thermostabilization of a fungal phytase [30] and a ß-lactamase [31]. Interesting variations followed [32], including the use of Bayesian sequence-based algorithms applied to serine protease sequences in the quest, to identify stabilizing interactions in subtilisin E (increase in melting temperature by 13 °C) [33].

In another consensus study, the esterase from *Aspergillus fumigatus* was likewise stabilized [34]. A 24-fold longer half-life at 50 °C relative to the WT esterase was observed, which was related to the formation of new hydrogen bonds.

Low levels of sequence identity often exist, which means that consensus residues at a given position cannot be determined reliably. A step forward was the development of the structure-guided consensus approach [20, 35]. In the latter study, the challenging problem of thermostabilizing penicillin G acylase (PGA) was solved. It is a difficult endeavor because it involves a large heterodimeric enzyme (α subunit 23 kDa and ß subunit 63 kDa). The two subunits of eight known class IIa PGA sequences were separately aligned. This enabled the identification of 109 out of the 766 possible positions in *E. coli* PGA, which were different in *E. coli*. They occurred in more than 50% of the other sequences. In order to reduce the number of substitutions, the following structural criteria were applied [35]:

- Only mutations that are more than 10 Å away from the active site were considered.
- Mutations in a helix were considered only if they are not helix-destabilizing amino acids.
- Existing H-bonds or salt bridges were left intact.
- Stabilizing mutations previously reported in the literature were not considered.

This data-driven consensus procedure provided a dozen improved variants (out of the predicted 21) with little tradeoff in activity. Two of them displayed an almost threefold higher half-life at 50 ° C. It is important to note that none of the computational guides available at the time (FoldX, PROSA, or SCRATCH) predicted these variants [35]. The same group also championed the structure-guided consensus approach in the thermostabilization of glucose dehydrogenase [36].

In an interesting twist on the subject of consensus-based thermostabilization, design using binary polar/nonpolar patterning [37] was explored without phylogenetic bias [38]. The authors called attention to the fact that "because the sequences of natural proteins generally derive from a common ancestor, they tend to be heavily biased by evolutionary relationships." This is something to keep in mind [7]. Nevertheless, numerous studies utilizing the consensus approach for thermostabilization have proven to be successful [39–41].

6.4 Further Computationally Guided Methods for Protein Thermostabilization

Various computationally guided approaches to protein thermostabilization have been developed, sometimes in combination with structural and/or sequence data [42]. The respective algorithms predict a certain number of point mutations that are expected to be stabilizing, typically several dozen.

A small library comprising these single mutants is then generated and tested, of which only a fraction usually prove to be hit, which are then combined. The DeGrado-algorithm, which can be used for various applications including thermostabilization, is an early example [43]. Another computational package is the set of Rosetta algorithms that have been used successfully a number of times for increasing thermostability [44], see Section 6.2. An automated protocol as part of RosettaDesign has been developed, which can predict stabilizing point mutations that improve insufficiently packed protein cores [44c].

In the most recent computational approach, an algorithm dubbed Protein Engineering Supporter (Proteus) was developed, in which stabilizing mutational pairs in target 3-dimensional structures can be predicted [45]. It assumes that when two residues do not interact, exchanging them by an interacting pair will enable protein stabilization. Importantly, such an exchange is suggested only if the main-chain conformation of the pair is maintained and no steric burden is introduced. Several case studies were presented using enzymes of industrial interest, including *Bacillus polymyxa* β-glucosidase (biofuel production), lipase II from *Rhizopus niveus*, and a lysozyme enzyme. The stabilizing interactions were identified mostly as H-bonds, ionic interactions, and disulfide bonds. The workflow illustrated in the Proteus server is shown in Scheme 6.1. Several other approaches are highlighted in more detail in Sections 6.4.1–6.4.5.

6.4.1 SCHEMA Approach

In Chapter 3 [46, 47], the SCHEMA technique in directed evolution has already been described, which can be used for enhancing the robustness of proteins [47]. For example, the cellobiohydrolase (CBH) I from *Hypocrea jecorina* was subjected to a type of SCHEMA [48]. Unlike traditional SCHEMA recombination libraries based on swapping elements of sequence, elements of structure in the form of "blocks" are shuffled among homologous proteins. These elements need not involve contiguous polypeptide sequences. Structural blocks among *H. jecorina* CBH 1 and two thermostable homologs from *Talaromyces emersonii* and *Chaetomium thermophilum* were subjected to swapping, and a subset of CBHIs from a library of 5 000 000 possible chimeric sequences was analyzed. Central to NCR is the creation of a graph from non-native residue–residue contacts with nodes that correspond to residues and edges corresponding to non-native contact. In this approach, minimal cuts that partition the graph are identified, thereby minimizing the SCHEMA disruption. Several variants within two blocks were found to stabilize *H. jecorina* CBHI, the best one being characterized by one-point mutation, which results in the stabilization of this industrially important enzyme by 3 °C. Unfortunately, for an unknown reason, combining other stabilizing mutations failed to increase thermostability to a useful and practical level. The authors suggest that their approach can be improved by prioritizing point mutations on the basis of the consensus approach [48].

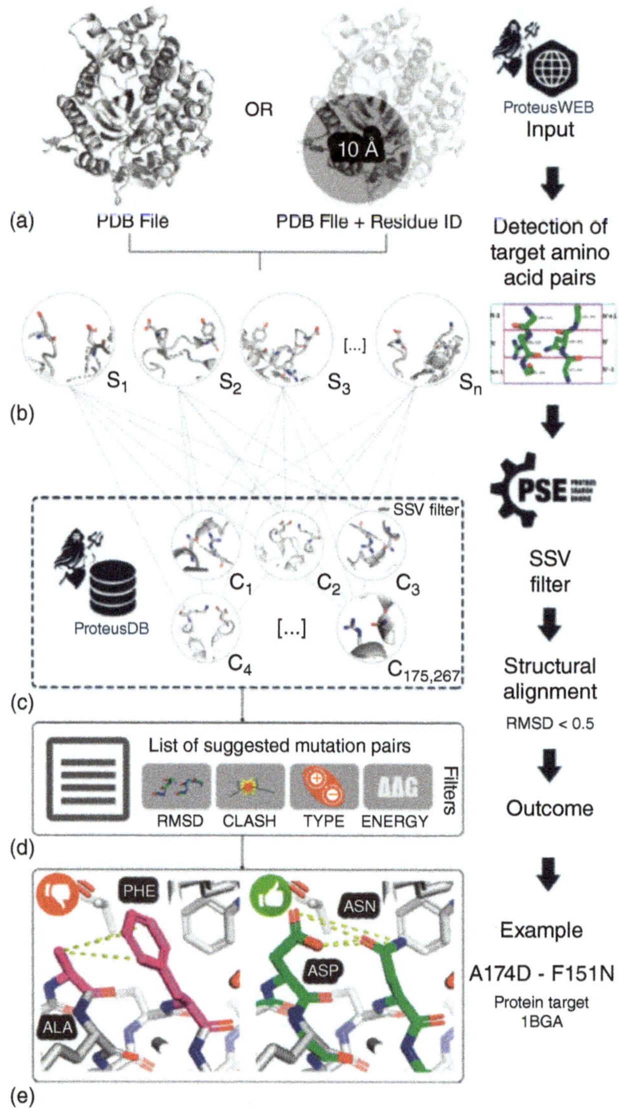

Scheme 6.1 The procedure of Proteus to predict beneficial mutations. (a) Using a PDB file or a particular region as input, and then detecting target amino acid pairs (b). The triad pairs with a good structural alignment (RMSD < 0.5 Å) with ProteusDB are selected (c) and gave the suggested substitutions. (d) The given mutation pair using a β-glucosidase (PDB: 1BGA) as an example (e). Source: Ref. [45]/Springer Nature/CC BY 4.0.

6.4.2
FRESCO Approach

Yet another approach to protein thermostabilization was called framework for rapid enzyme stabilization by computational libraries (FRESCO), as illustrated in Figure 6.8 [49a]. In the first step, three algorithms are used to identify stabilizing mutations, namely Rosetta ddg [49b], FoldX [50], and the newly developed dynamic disulfide discovery (DDD). These computational tools for choosing stabilizing mutations predict the respective change in the energy of folding ($\Delta\Delta G^{\text{Fold}}$). In steps 2 and 3, false positives are removed by filtering out unreasonable mutations and eliminating variants in which protein flexibility is increased. In step 4, experimental screening is performed prior to combining the most stabilizing mutations in the fifth step (Figure 6.8). A detailed experimental protocol with applications and additional citations [51] provides a good opportunity to learn the method [27b]. Only the basics are presented here.

The technique was originally applied to limonene epoxide hydrolase from *Rhodococcus erythropolis* DCL14 [49a]. Upon screening a library of 64 predicted

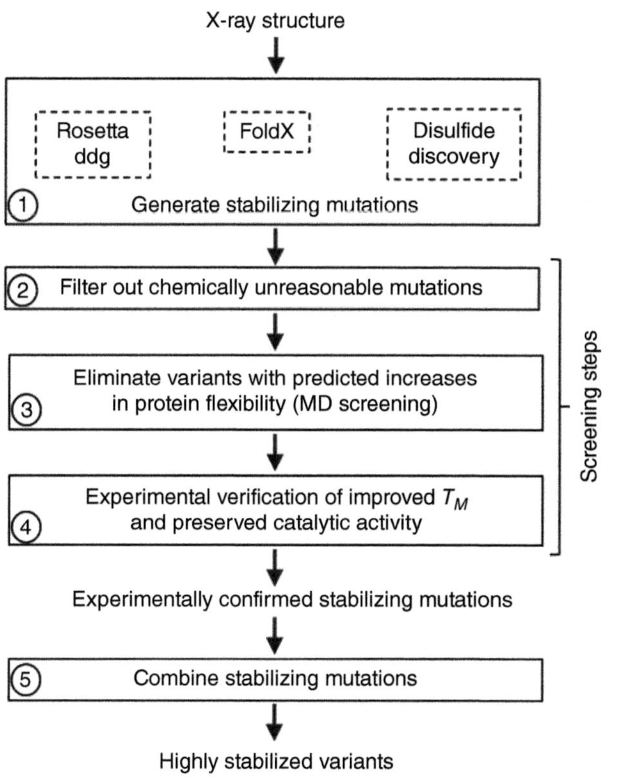

Figure 6.8 FRESCO strategy for protein thermostabilization. Source: Ref. [49]/Oxford University Press.

variants, 21 stabilizing mutations were found in flexible but also in rigid regions of the enzyme. Thus, the computational "error-rate" amounts to 43 false predictions. As a result of combining 10–12 of the experimentally stabilizing mutations, several multi-site mutants resulted showing an impressive increase in apparent melting temperature from WT 50 to 85 °C and a more than 250-fold longer half-life, enhanced catalytic activity and maintained regioselectivity also being observed. Control experiments showed that the orthogonal in silico screening is necessary for success. For example, when applying only FoldX, a significant number of predicted stabilizing mutations proved to be neutral or even destabilizing. According to the authors, the sole use of Rosetta ddg would have predicted only the introduction of aromatic amino acids on the protein surface, a known problem when attempting in silico design.

In a second study, the crystal structures of two of the most stable epoxide hydrolase variants allowed a detailed structural analysis of the source of thermostabilization [52]. Variants LEH-P and LEH-F1b are characterized by 8 and 12 mutations, respectively. Both carry the identical eight point mutations S15P, A19K, E45K, T76K, T85V, N92K, Y96F, and E124D, the latter variant also having four point mutations that allow the formation of two disulfide bonds (I5C-E84C and G89C-S91C). The crystal structure of LEH-F1b is shown in Figure 6.9.

All of the stabilizing point mutations occur on or near the surface of the enzyme. Many of the beneficial mutations lead to stronger H-bonds that

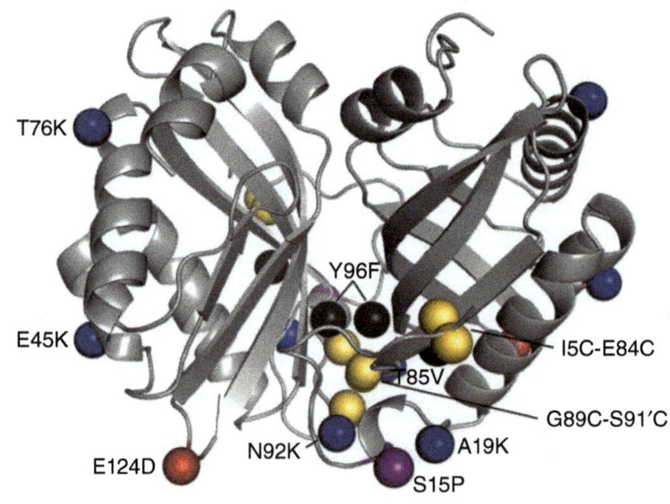

Figure 6.9 Positions of 12 stabilizing mutations as revealed by the crystal structure of the LEH-F1b and P dimers. Mutations introducing surface-located positively charged residues are shown in blue, surface-located negative charges are shown in red, and buried hydrophobic residues are shown in black. Proline residues in loops are in purple and disulfide bonds are in yellow. Mutations are indicated once per dimer. Source: Ref. [52]/John Wiley & Sons.

stabilize the local protein structure, the most important ones being localized inside or near the flexible N-terminus, and to a lesser degree in helices 3 and 4. The S15P mutation enhances the enzyme's melting temperature by 1.0 °C, whereas the formation of a disulfide bridge increases it by about 14–15 °C. A comparison of FRESCO with B-FIT and consensus was also made, although these methods were not directly applied to this epoxide hydrolase. It was shown that both FRESCO and B-FIT predict amino acid substitutions at positions 5, 15, and 92. However, FRESCO also led to mutations at positions that are not flexible and that would have been missed by B-FIT. FRESCO in turn misses some of the flexible positions predicted by B-FIT.

In another work, FRESCO was used for stabilizing the haloalkane dehalogenase from *Sphingomonas paucimobilis* [53]. A mere 150 predicted mutants were experimentally tested, of which 18 were found to be stabilizing. Thus, the vast majority of suggested mutants do not fulfill the predictions. Why the success rate of FRESCO was so poor, was not explained. A combined mutant, generated by selecting three stabilizing point mutations remote from the active site and one disulfide bond, led to an 11 °C increase in thermostability. The best variant was produced by choosing other combinations of stabilizing point mutations, including several that are close to the active site. This resulted in a 23 °C increase in apparent unfolding temperature over WT and a 200-fold lower rate of inactivation. Fast and optimal application of the three algorithms requires some experience.

6.4.3 FireProt Approach

In techniques such as FRESCO, the best-predicted point mutations are combined into a final variant, which may indeed result in improved protein stability, but also in a detrimental effect due to possible nonadditivity [26]. In contrast, it has been claimed that an alternative method called FireProt directly delivers variants characterized by multiple mutations [54]. The method combines two concepts, the energy- and evolution-based approaches (Figure 6.10). Among other steps, the performances of four prediction tools, FoldX [50], Rosetta [49b], ERIS [55], and CUPSAT [56], are evaluated using the ProTherm dataset. Additionally, the best multiple-point mutants derived from predictions of $\Delta\Delta G$ following mutation based on a set of crystal structures and evolutionary information obtained from multiple sequence alignment are combined. Intelligent pre- and post-processing filters need to be employed in order to maximize prediction reliability.

FireProt was applied to the haloalkane dehalogenase DhaA, which resulted in a notable increase in thermostability ($\Delta T_m = 24.6\,°C$), albeit at a 3.2-fold reduction in enzyme activity. A second enzyme, gamma-hexachlorocyclohexane dechlorinase (LinA) was also subjected to FireProt, which resulted in $\Delta T_m = 20.9\,°C$ [54]. Subsequently, it was used in the thermostabilization of a sucrose phosphorylase

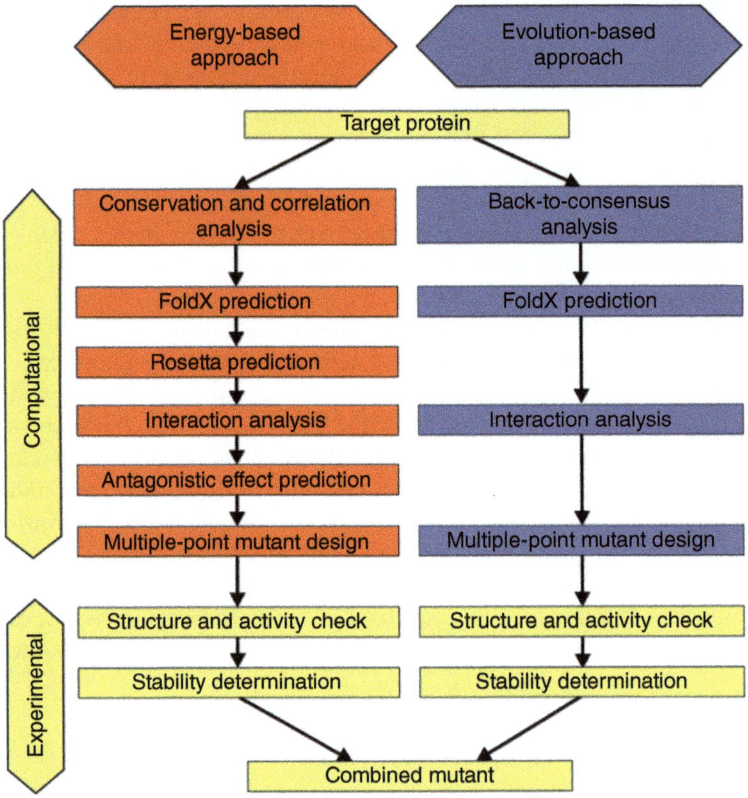

Figure 6.10 Workflow of the FireProt method. Individual steps involved in the energy- and evolution-based approaches. Source: Ref. [54]/PLOS/CC BY 4.0.

(SPase), which catalyzes transglycosylation [57]. Due to the low thermostability of this enzyme, SPase has not been exploited industrially. The authors developed a semi-rational design strategy that combined FireProt with structure–function analysis and molecular dynamics (MDs) computations (Figure 6.11). A single mutant T219L and a combination of mutations in the form of variant Mut4 (I31F/T219L/T263L/S360A) were obtained, showing notably enhanced thermostability. The half-lives at 50 °C of the two mutants increased twofold. On the practical side, the industrially important α-D-glucosylglycerol was obtained efficiently from sucrose and glycerol ($196\,g\,l^{-1}$) [57]. The utility of FireProt was recently reviewed [58]. It should be mentioned that SPase has been stabilized by immobilization on a solid support in a microstructured continuous flow format [59]. A product yield of α-D-glucosylglycerol (85%) was achieved as well as high productivity ($500\,mMh^{-1}$) and product titers up to 800 mM. Due to the different experimental setups, comparison with the FireProt approach is difficult.

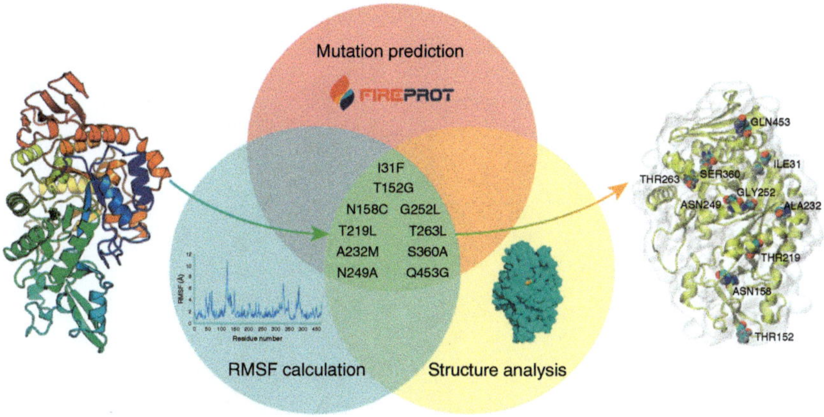

Figure 6.11 Prediction of the candidate residues for mutagenesis. Source: Ref. [57]/Springer Nature/CC BY 4.0.

6.4.4 Constrained Network Analysis (CNA) Approach

The so-called constrained network analysis (CNA) has been championed as a novel and effective computational guide to protein thermostabilization because it can identify weak spots in an enzyme [60]. It is based on a graph-theory rigidity analysis and therefore sets the stage for rational site-specific mutagenesis or saturation mutagenesis. The correlation of protein rigidity and thermodynamic stability is the basis of the method, as in other approaches. A protein is modeled as a constraint network in a "body-and-bar" representation followed by rigidity analysis leading to the identification of rigid clusters of atoms having no internal motion and flexible links in between. The rigidity analysis is performed by application of the pebble game algorithm. In one study [61], CNA nicely explained a number of mutations in the lipase A from *Bacillus subtilis*, which had been engineered earlier by other research groups [62–64]. The primary features of CNA are promising because it:

- Offers a refined modeling of thermal unfolding simulations, which include consideration of temperature dependence of hydrophobic tethers.
- Enables rigidity analyses on ensembles of network topologies on the basis of either structural ensembles or fuzzy noncovalent constraints.
- Provides a computed set of global and local indices needed in quantifying protein stability.

In a practical advance, VisualCNA was developed, which is a user-friendly PyMol plug helpful in rapidly establishing a setup for CNA runs and analyses thereof [65]. Figure 6.12 outlines the general scheme of VisualCNA [61]. In a recent CNA study, the endo-ß-glucanase EGLII from *Penicillium verruculosum*

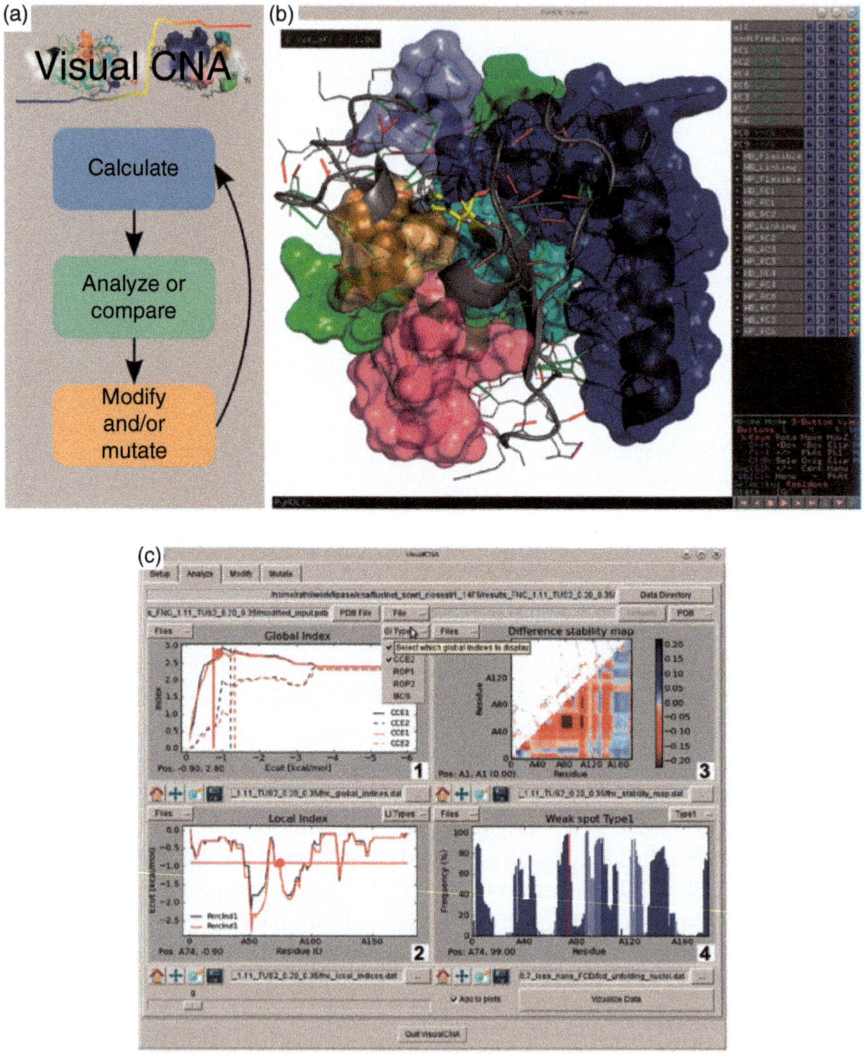

Figure 6.12 Schematic representation of VisualCNA. (a) Illustration of the technique's iterative work flow for optimization of protein thermostability. (b) PyMOL window showing the 3D protein structure at the melting point. Rigid clusters are shown as uniformly colored semi-transparent bodies. Constraints due to hydrogen bonds, salt bridges, and hydrophobic contacts are shown as red, magenta, and green sticks, respectively. A mutation is shown in yellow stick representation. Flexible regions are shown in gray. (c) The VisualCNA Analyze panel showing a comparison of multiple graphs from wild-type (black) and mutant (red) analyses. 1: Global indices with transition points indicated as vertical lines. 2: Local index with a red circle indicating the mutation, and a horizontal red line showing the unfolding state. 3: Difference stability map between wild-type and mutant. 4: Likelihood of a residue being a structural weak spot with the mutant shown in red. Source: Rathi et al. [65]/with permission of Oxford University Press.

served as the enzyme, which is important in the food industry [66]. A twofold improvement in thermostability was reported. It will be interesting to see how often this approach will be applied in future directed evolution studies, and whether it can be generalized to include other types of enzymes.

6.4.5 Alternative Approaches

Nowadays, essentially all studies on protein thermostabilization are guided by computational techniques, sometimes utilizing strategies similar to previous ones [7]. In an interesting contribution, a statistical computationally assisted design strategy (SCADS) was applied to the terpene toabacco 5-epi-aristolochene synthase (TEAS) [67]. The DeGrado-algorithm was applied [43], which enables predictions on the basis of side-chain interactions with the neighboring protein backbone. The activity of WT TEAS as a catalyst in the cyclization of farnesyl diphosphate (FPP) to 5-*epi*-aristolochene breaks down at temperatures above 40 °C, in contrast to a mutant having 12 point mutations, and which is active at 65 °C. Some problems with inclusion bodies were encountered. Focused saturation mutagenesis experiments at selected hot spots were not performed, which would probably lead to further improvements [67].

In an integrated approach for thermal stabilization of a mesophilic adenylate kinase, multiple stabilization techniques were applied, including bioinformatics- and structure-guided protocols [68]. Local structural entropy (LSE) played a major role in this endeavor, which is an empirical descriptor for describing conformational variability in short stretches of protein sequence computed on the basis of structural information available in the Protein Data Bank [69]. It was extended by including sequence-based techniques for stabilization [70]. Using previous mutational data [71], the integrated approach led to a variant showing a 25 °C increase in its thermal denaturation midpoint [68]. The crystal structures of three variants were obtained, which led to an insight into the origin of enhanced robustness. A different computational approach provided an adenylate kinase displaying a 21.5 °C increase in T_m [72]. In this study, 100 predicted variants were tested experimentally, the best one having mutations quite different from the ones generated by the integrated approach. This shows that very different strategies may lead to notably stabilized proteins, but the effects on a molecular level can be very different.

In what has been termed "a divide and combine approach" to thermal stabilization, the focus shifted to larger and non-fully cooperative proteins that are actually more abundant [73]. The key to success in such systems is to identify the regions of lower stability (Figure 6.13). A stepwise combination of structure-based and rationally designed mutations at such regions was employed in the thermostabilization of an engineered apoflavodoxin, the final variant displaying a 32 °C increase in melting temperature. Thus far, the underlying concept has not been used very often [74].

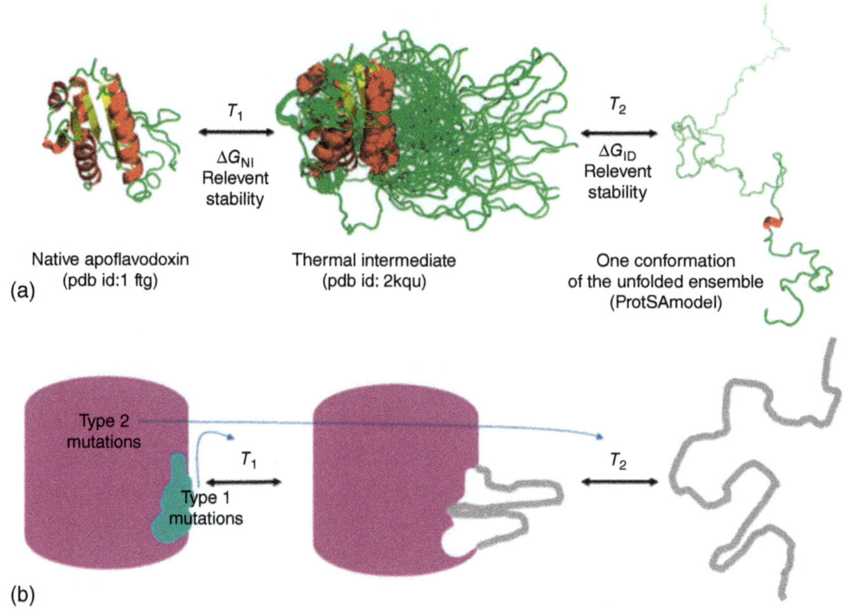

Figure 6.13 Divide and combine approach to protein thermostabilization featuring unfolding equilibria of a three-state protein. (a) Ribbon cartoons representing the conformation of apoflavodoxin in the three states populated in its thermal unfolding equilibrium. The native state is represented by the crystal structure of the WT protein (PDB ID: 1FTG), the intermediate state by the solution structure of the F98N variant (PBD ID: 2KQU), and the unfolded state by one of the 2000 conformations calculated for the unfolded ensemble using the ProtSA server. The low-temperature transition (T_1) signals the unfolding of the less stable region, leading to an equilibrium intermediate. The high-temperature transition (T_2) represents the unfolding of the intermediate, leading to the unfolded state. The free energy difference between the native and the intermediate conformation (ΔG_{NI}) is termed relevant stability of the protein, while that between the intermediate and the fully unfolded conformation (ΔG_{IU}) is termed residual stability of the protein. (b) Simplified scheme depicting a protein with two structural regions of different stability (less stable region in cyan, and more stable one in pink) and the likely effects of mutations on T_1 and T_2. Type 1 mutations, those introduced in the unstable region or at its interface with the more stable one, will mainly modify the relevant stability of the protein. Type 2 mutations, those introduced in the more stable region, will only modify the residual stability of the protein. Source: Ref. [73]/Springer Nature/CC BY 4.0.

Finally, protein thermostabilization has been achieved by systematically designing salt bridges [75]. Using this computational approach for enhancing the robustness of a ß-glycosidase, 10 556 surface salt bridges in 6493 X-ray structures were analyzed statistically. The model was coupled with B-factors, weighted contact number, relative solvent accessibility, and conservation screening. Electrostatic pairs at five positions were identified, and the combination of three of them led to an increase in T_m of 15.7 °C [75]. As in many other new approaches, it remains to be seen to what extent this technique can be generalized [7].

6.5 Directed Evolution of Enzyme Thermostability and Resistance to Hostile Organic Solvents

The utility of saturation mutagenesis for increasing enzyme robustness was first applied to enhance the oxidative stability of a protease in the presence of H_2O_2, the screening process, leading to two different mutants in which Ser or Ala replaced the oxidatively sensitive methionine, respectively, with dramatically enhanced oxidative stability [76]. As outlined in Chapter 1, the first case of non-focused random mutagenesis for increasing thermostability, using two mutagenesis/screening cycles, ushered in the era of directed evolution of thermostability and provided the basis for further advancements. In the 1990s numerous studies followed in which epPCR and/or saturation mutagenesis at hot spots as revealed by the random mutagenesis process were applied, while occasionally DNA shuffling was also tested [77a, b]. This development has continued to this day [77c–g]. As pointed out previously [7], the use of protein display systems such as yeast display for enhanced enzyme stability has been reviewed [77f].

Two other strategies also deserve mention. In one approach, a thermophilic error-prone strain from *Geobacillus kaustophilus* was engineered by deleting functional DNA repair genes, and then used in directed evolution of protein thermostabilization [77h]. This concept and related techniques [78] still need to be tested on a broad basis. In a different development, *B. subtilis* spore display of a laccase under extreme conditions of high concentrations of hostile organic solvents proved to be successful [79], but it is currently not clear how general this approach is [7].

It is difficult to compare the viability of the various strategies because different enzymes were used as well as different robustness indices. Many researchers reported melting temperatures (T_m), T_{50} values (temperature at which 50% of enzyme activity is lost after heating for a given period of time), or half-life at a defined temperature. Different assessments also result depending upon whether purified, partially purified enzymes, whole cells, or lysates are used in such measurements because robustness also depends upon the interaction of enzymes with other proteins and different biomolecules. Different kinds of strategies for stabilizing enzymes in organic solvents have been summarized, including protein engineering, chemical modification, immobilization, and the use of additives [40].

Importantly, structural and mechanistic insights were gained by studying the origin of thermal stability of thermophilic enzymes, which indicated higher rigidity due to mutations. [80] Therefore, a central hypothesis was to increase the rigidity of thermally sensitive enzymes by introducing structure-guided point mutations, which counteract the natural flexibility. In some of the early directed evolution studies at least a few of the discovered point mutations were interpreted on a molecular level. Such factors as newly formed salt bridges, H-bonding interactions, disulfide bridges, introduction of proline for reducing conformational flexibility, and/or interior hydrophobic packing effects were invoked [77, 81]. Unfortunately, many of the observed mutations could not be rationalized [7]. The likely possibility that some of the accumulated point mutations are actually superfluous was not considered. In 2005, comprehensive review articles appeared, showing that epPCR and/or DNA shuffling, which require no structural data, were the most often used mutagenesis methods during the period 1995–2007 [77c–e]. These reviews also stress the importance of kinetic versus thermodynamic stability. Since then a trend toward more "rational" strategies in laboratory evolution has emerged in which structural, sequence, and/or computational information is cleverly exploited. The sequence data-based consensus approach, first developed for stabilizing antibodies [20a], has been applied in the quest to enhance enzyme robustness for a long time [20b]. It is still used today, often augmented with structural data, which then constitutes a powerful strategy (see below). One of the central themes has been and continues to be the controlled rigidification of flexible sites or regions in a protein [41e].

An interesting early approach is the PROSIDE method, according to which designed large mutant libraries generated by random mutagenesis are evaluated by selection procedures based on phage or ribozome display systems [82]. In the absence of any guiding information, which is actually quite rare, epPCR and DNA shuffling are recommended even today. Several recent reviews cover all current approaches to protein engineering of enzyme robustness and consider the following properties [40, 41].

- Kinetic and thermodynamic thermostability.
- Resistance to hostile organic solvents including ionic liquids.
- Oxidative stability.
- Tolerance to different pH ranges.

An unusual technique for enhancing enzyme robustness does not require any changes in the sequence of the target enzyme, and therefore constitutes a completely different approach. Insertional fusion of enzymes to a thermophilic host protein is the underlying strategy [83]. Since evolvability is less difficult when using robust enzymes [84], it has been stressed that the idea of utilizing thermophilic enzymes for the directed evolution of other parameters such as selectivity and activity should be considered more often than in the past [85]. Deep learning in protein structural modeling and design is also noteworthy, of potential use in directed evolution and rational design [86].

Whereas progress in developing advanced methods for protein stabilization has been made in recent years, it is not evident which approach promises to be truly general and most effective [7]. Consequently, the situation is fundamentally different from evolving stereo- and regioselective enzymes in which saturation mutagenesis at sites lining the binding pocket has emerged as the method of choice (see Chapters 1–4). Several computational tools useful as guides when applying directed evolution for enhanced robustness have been developed (see Section 6.8). In the sections that follow, selected recent examples illustrating methodology development are critically analyzed. For the assessment of other case studies, the reader is referred to an excellent review [41a].

6.6 Application of epPCR and DNA Shuffling

Both epPCR and DNA shuffling continue to be used in protein stabilization, and indeed some degree of improvement can always be expected in laboratory evolution, irrespective of the mutagenesis strategy or method [7]. For example, a mutant of the insect α-carboxylesterase from *Lucilia cuprina*, which had been evolved earlier to accept organophosphate insecticides, was subjected to directed evolution to increase its thermostability [87]. Four rounds of epPCR were assessed according to the following simple procedure. About 100 000 random variants were plated onto agar plates and then replicated on filter paper, which was subsequently incubated at defined temperatures for one hour. Then the filter paper was immersed in a solution containing 2-naphthyl acetate. Colonies showing the highest activity were selected and pooled for the subsequent epPCR cycle. The final variant contained six new point mutations and displayed notable esterase activity after heat treatment at 54 °C for one hour, whereas WT showed no activity under the same conditions. Three of the mutations occur on the enzyme surface (Met364Leu, Lys530Glu, and Asp55Gly), the others are in the interior (Ile419Phe, Ala472Thr, and Ile505Thr). The structural basis for this notable improvement was partially elucidated with the realization that Ile419Phe fills space in a hydrophobic cavity [87]. The result of filling hydrophobic cavities with concomitant increase in thermal stability is a well-known mutational effect [88]. However, the interpretation of the other mutations proved to be difficult. Although an enormously high number of transformants had to be screened, the assay is fairly simple to perform.

The robustness of the feruloyl esterase from *Aspergillus niger* was also improved, an enzyme that is instrumental in the selective degradation of lignocellulose [89]. In an initial attempt using the PoPMuSiC algorithm [90] as a computational guide, a slightly improved variant with two point mutations was generated, but the degree of thermostabilization was far from meeting industrial standards. Much higher stability was needed to decrease the likelihood of microbial contamination, support the disorganization of the raw material (lignocellulose in biomass), and promote enzyme penetration. Using the previously generated double mutant as a template, epPCR was applied with the formation of a 40 000-membered library. About 10 000 colonies were screened by a UV/Vis-plate reader that monitors the formation of 2-chloro-4-nitrophenolate (425 nm) formed upon hydrolysis of 2-chloro-4-nitrophenyl ferrulate [89c]. A mutant with 12 additional mutations was identified showing 80% residual activity after heat treatment at 90 °C for 15 minutes. It was also demonstrated that ferulic acid was smoothly released from steam-exploded corn stalk as typical biomass (Figure 6.14).

In the successful attempt to enhance the robustness of the homodimeric haloperoxidase from *Streptomyces aureofaciens*, epPCR was also used [91]. A 1000-membered library was assayed in the halogenation of chlorodimedone. The best mutants show moderately enhanced thermostability, but

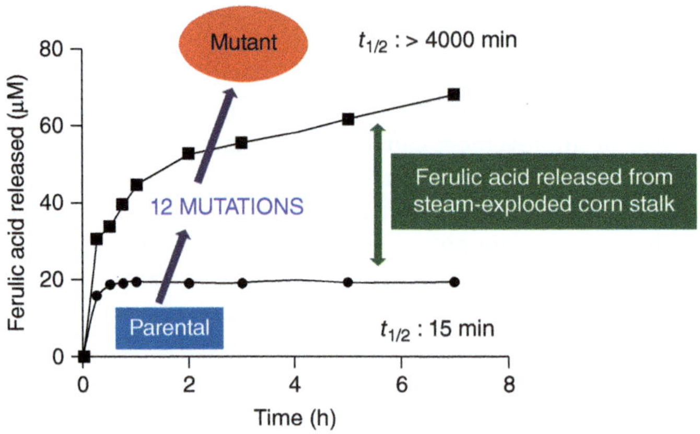

Figure 6.14 Catalytic performance of a variant of the feruloyl esterase from *Aspergillus niger* in the degradation of steam-exploded corn stalk as biomass. Source: Ref. [89]/With permission of Elsevier.

Table 6.1 Specific activity, T_{50}, and 1-propanol stability of WT and mutant BPO-A1 haloperoxidases.

BPO-A1[a)]	Specific activity (kU g^{-1})	T_{50} (°C)	Residual activity in 40% (v/v) 1-propanol (%)[b)]
Wild-type	56.4	80.6	25.9 ± 3.0
Mutant HT177	59.6	82.0	55.4 ± 4.1
Mutant R114H	50.3	81.9	63.2 ± 4.3
Mutant N146H	78.7	80.4	20.3 ± 5.3
Mutant HT507	101.7	84.5	15.2 ± 3.2
Mutant G106S	99.4	81.1	33.5 ± 2.4
Mutant V148I	117.5	83.7	7.60 ± 1.2

a) Mutants HT177 and HT507 were obtained by directed evolution. MutantsR114H, N146H, G106S, and V148I having single-amino-acid-substitutions were constructed by site-directed mutagenesis.
b) The data represent the averages of 2 independent experiments.
Source: Ref. [91]/With permission of Elsevier.

increased resistance to a variety of different solvents such as 1-butanol (Table 6.1). Mainly surface mutations occurred. It was deduced that the α-helix was stabilized and that the interaction between subunits contribute to the improvements.

Further studies based on random mutagenesis or DNA shuffling have appeared. To boost the robustness of the haloalkane dehalogenase from *Rhodococcus rhodochrous* toward heat and organic cosolvents, epPCR was first performed, leading to the identification of a stabilizing mutation at the access tunnel [92]. Therefore, the access tunnel was explored mutationally more closely by saturation mutagenesis, which provided the best mutant showing an increase in melting temperature of 19 °C and a notable resistance to DMSO as cosolvent. It would be interesting to apply iterative saturation mutagenesis (ISM) to this enzyme (see also Section 6.3 featuring the SM-based B-FIT approach).

In most directed evolution studies, only one property such as thermostability was targeted [7]. However, a particularly challenging task is to improve several properties (simultaneously), e.g. thermostability, alkaline stability, and activity, all needed for practical applications. The example featured here concerns the directed evolution of the xylanase from *B. subtilis*, in which thermostability, stability at pH 8, and activity were all improved [93]. In an earlier study, the authors had performed two rounds of epPCR, and assayed in each case about 12 000 transformants with a prescreen based on halo formation on solid agar containing Congo Red [93b]. This was followed by alternating DNA shuffling and further epPCR cycles while varying the "selection pressure." The workflow is shown in Figure 6.15.

The best variant has six point mutations with a temperature optimum at 80 °C under alkaline conditions (pH 8) and a threefold increase in specific activity [93b]. The total number of transformants screened in the overall process was not reported, but this work shows that with a well-designed strategy and a great deal of experimental work the difficult problem of mastering several enzyme properties for real (industrial) applications can be solved. Several other approaches to xylanase optimization (although not always addressing several parameters) have been reported, generally also relying on epPCR and DNA shuffling [94].

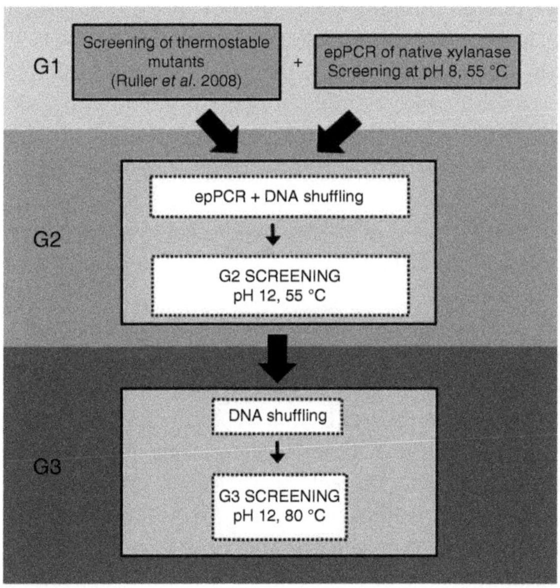

Figure 6.15 Workflow in the directed evolution of an improved xylanase. Source: Adapted from Ref. [93b].

To enhance the activity of the thermostable ß-1,3-1,4-glucanase from *Paecilomyces thermophilia* at acidic pH where it is needed in such industrial applications as beer brewing and animal feed, the combination of epPCR and DNA shuffling was applied. A variant characterized by three point mutations shifted the optimum pH from 7.0 to 5.0 with no tradeoff in thermostability, opening the way to practical applications.

The old idea of focusing saturation mutagenesis on the residues identified by epPCR for further thermostabilization [95a] or enhanced stereoselectivity [95b] is still used occasionally today [7]. Thermostability studies based on epPCR and/or DNA shuffling continue to appear [96, 97]. An interesting example pertains to the thermostabilization of aldehyde dehydrogenase by epPCR and saturation mutagenesis at such hot spots for application in synthetic cascade biomanufacturing [96b].

In a recent study, the lipase from *Pseudomonas fluorescens* was thermostabilized by application of epPCR [98].

6.7 Saturation Mutagenesis in the B-FIT Approach

The B-FIT approach to enzyme thermostabilization utilizes focused saturation mutagenesis, and if necessary with ISM [99]. The choice of randomization sites is made on the basis of B-factor values available from X-ray data. In a first step, the average B-factor values of all residues are obtained with the help of the computer aid B-FITTER, available free of charge from one of the authors (http://kofo.mpg.de/en/research/biocatalysis). It automatically lists the average B-factor values of all residues in a protein, starting from the highest to the lowest values. Residues exhibiting high average B-factors indicate positions of flexibility. Therefore, six to eight such residues are normally chosen for saturation mutagenesis. As in the case of CAST/ISM for enhancing or inverting stereoselectivity, statistical factors concerning library coverage and oversampling need to be considered (Chapters 3 and 4). Typically, residues exhibiting the highest average B-factors are grouped into 2- or 3-residue randomization sites, followed by saturation mutagenesis and possibly ISM.

B-FIT-based ISM was first applied to the lipase from *B. subtilis* (Lip A), leading to a variant characterized by five point mutations with significantly increased apparent thermostability [99]. Complete deconvolution allowed the construction of a fitness pathway landscape with 5! = 120 trajectories leading from WT Lip A to the best-evolved mutant. A theoretical analysis of the results revealed pronounced cooperative mutational effects occurring on the surface of the lipase in the form of a continuous H-bond network [100]. Biophysical and biochemical characterization including protein NMR spectroscopic studies, circular dichroism, X-ray structural analyses, and combining thermal inactivation profiles uncovered an interesting surface effect that results from the mutations [101]: Upon heat treatment, WT Lip A undergoes undesired irreversible aggregation with precipitation, whereas the "robust" variant aggregates at

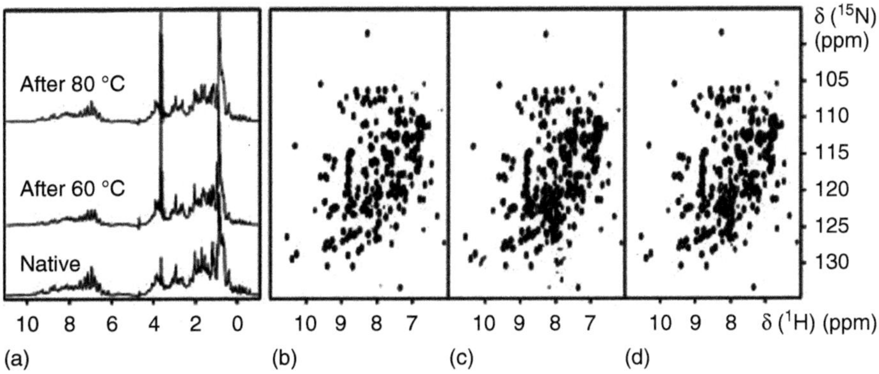

Figure 6.16 NMR spectra recorded for native and thermally treated ^{15}N-labeled Lipase A mutant XI. (a) 1D ^{1}H spectra. (b–d) 2D [^{15}N,^{1}H]-HSQC spectra of mutant XI Lipase A: (b) Native. (c) Recovered after 60 °C treatment. (d) Recovered after 80 °C treatment. Source: Ref. [101]/John Wiley & Sons.

higher temperatures to a significantly lower extent. Reduced aggregation and precipitation of the unfolding intermediates were shown to be responsible for activity retention at higher temperatures. The NMR spectroscopic investigation of the native and the thermally treated ^{15}N-labeled Lip A variant showed that it recovers almost all of the natural conformation after heat treatment and cooling to ambient temperature, in agreement with activity recovers. After heating and cooling the variant shows a conformation that is almost identical to the heat untreated enzyme, as identified by peaks in 1D^{1}H and 2D [^{15}N, ^{1}H]-HSQC spectra (Figure 6.16) [101].

It was also discovered that not 100% of the enzyme is recovered in its native fold. In a different study, it was demonstrated that the same evolved Lip A variant is also tolerant to hostile organic solvents such as acetonitrile, dimethylsulfoxide, and dimethylformamide [24]. The possible role of irreversible aggregation and precipitation propensity of thermally unfolded states of WT enzymes have not been addressed very often when performing directed evolution [24, 97j], a subject that needs further attention in the future. When reporting only T_{50} values, the true cause of enzyme improvement may not be fully evident, a caution that should be heeded in all protein engineering studies.

The thermostability of the epoxide hydrolase from *A. niger* was increased to a significant notable extent degree using B-FIT/ISM [102]. Eight of the 356 residues showing the highest average B-factor values as well as four crystallographically unresolved and possibly likewise flexible residues were considered. The 12 residues were then grouped into six 2-residue randomization sites A, B, C, D, E, and F for NDT-based saturation mutagenesis. Several but not all theoretically possible upward ISM pathways were explored, leading to the best variant with a 21 °C increase in the $T_{50}{}^{60}$ value, an 80-fold improvement in half-life at 60 °C, and a 44 kcal mol^{-1} improvement in inactivation energy. ISM exploration also provided seven other variants characterized by 10–14 °C increases in

T_{50}^{60} values, 20–30-fold increase in half-lives at 60 °C, and 15–20 kcal mol^{-1} elevations in inactivation energy. In several cases, a given library failed to contain improved variants, which means that a local minimum on the fitness landscape (dead end) was encountered. Such an event is not at all rare in the directed evolution in general, irrespective of the mutagenesis method. In such situations, a simple strategy to escape from these local minima was applied [103], namely by utilizing a non-improved or even inferior mutant in the library as a template for the subsequent ISM step. In a curiosity-driven experiment, a neutral variant and an inferior one were used in otherwise the same ISM scheme, and the inferior template led to better results (Figure 6.17) [102].

The cellulase from *Trichoderma reesei* served as the enzyme for economically saccharifying cellulosic biomass in the production of biofuels [104]. Using the B-FITTER computer aid, 10 residues with high B-factors were identified and grouped into 7 randomization sites. Amino acids with high B-Factors are spatially close to the N- or C-termini, but possible disulfide bridges or N-glycosylation sites were not considered. A total of 11 000 transformants were screened, which led

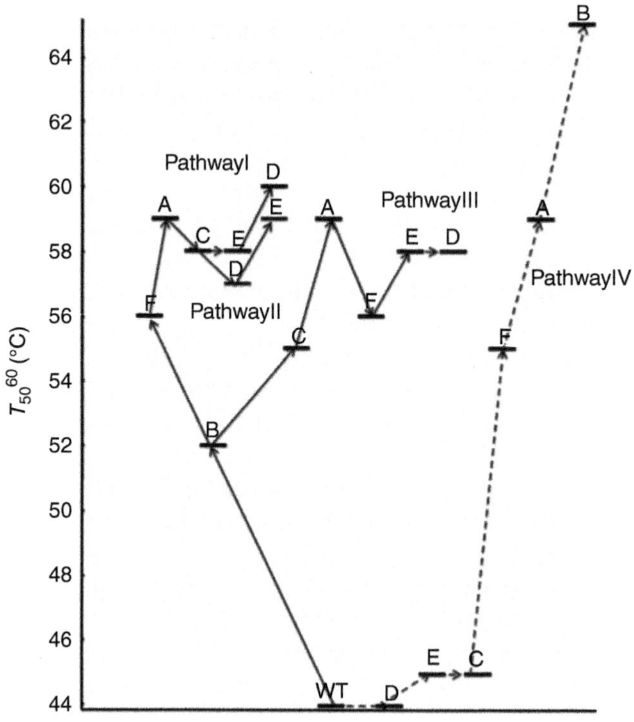

Figure 6.17 Results of limited ISM exploration starting from the best mutant, GUY-003 (site B), and the worst mutant, GUY-007, in the initial round of saturation mutagenesis at sites A–F. In all cases, NDT codon degeneracy was used except when performing saturation mutagenesis at site D, in which case NNG codon degeneracy was applied. Source: Ref. [102]/John Wiley & Sons.

Table 6.2 Summary of additional typical B-FIT studies of enzymes for enhanced thermostabilization and/or robustness.

Enzyme	E.C.	Organism	Structures	Comment	Reference
Lipase A	3.1.1.3	*Bacillus subtilis*	X-ray	10 sites were selected and NNK-based SM was applied	[62a]
Lipase	3.1.1.3	*Thermomyces lanuginosus*	X-ray	Also involved ISM technology	[112]
Lipase	3.1.1.3	*Rhizomucor miehei*	X-ray	Improved thermostability based on B-factor SM	[113]
Cutinase	3.1.1.74	*Glomerella cingulata*	X-ray	Improved thermostability using B-FIT	[114]
Choline oxidase	1.1.3.17	*Arthrobacter cholorphenolicus*	X-ray	Also improved thermostability using B-FIT	[115]
Phospholipase D	3.1.4.4	*Streptomyces antibioticus*	X-ray	Single-point mutants and then combinations of double and triple mutants	[116]
Epoxide hydrolase	3.3.2.3	*Aspergillus niger*	X-ray	Explored evolutionary pathways and enhanced stereoselectivity with maintained activity	[102, 117]
Type A feruloyl esterase	3.1.1.73	*Aspergillus usamii*	Homology	Combined with PoPMusiC algorithm	[99]
Endo-1,4-ß-galactanase	3.2.1.89	*Talaromyces stipitatus*	Homology	Supported by consensus technique and PoPMusiC	[118]
Endoglucanase I	3.2.1.4	*Trichoderma reesei*	X-ray	Increased the activity in parallel	[104]
ADP-glucose pyrophosphorylase	2.7.7.27	*Zea mays*	Homology	In combination with ISM	[107]
Xylanase	3.2.1.8	*Psychrobacter* sp. strain 2–17	Homology	Combined with focused epPCR	[119]
Alcohol dehydrogenase PedE	1.1.1.1	*Pseudomonas putida* KT2440	Homology	combination of consensus data and SM	[120]
3-Quinuclidinone reductase	1.1.1.B51	*Rhodotorula rubra*	X-ray	Improved thermostability using B-FIT	[121]
Uronate dehydrogenase	1.1.1.203	*Agrobacterium tumefaciens*	Homology	Consensus not possible; $\Delta T_{50}{}^{15} = 3.2\,°C$	[122]
Xylanase	3.2.1.8	*Aspergillus oryzae*	Homology	ISM; T_m up to 61 °C	[123]
Lipase PAL	3.1.1.3	*Pseudomonas aeruginosa*	X-ray	SM for *reducing* stability (see Section 3.3.11)	[124]
Lipase C	3.1.1.3	*P. aeruginosa*	Homology	Sevenfold increased thermal stability; SM was applied	[125]
γ-Lactamase	3.5.2.B2	*Microbacterium hydrocarbonoxydans*	X-ray	Eight flexible residues with high B-factor values were chosen for SM	[126]

to about 500 variants showing higher activity in the reaction of carboxymethyl cellulose following preincubation at 50 °C for 45 minutes. From this collection, 70 variants were sequenced, many of which showed notably enhanced robustness. Combining point mutations led to further improvements in some cases. The triple mutant G230A/D113S/D115T displayed a higher melting temperature by 3 °C and increased half-life at 60 °C ($t_{1/2}$ = 161 hours versus 74 hours of WT).

The high number of other B-FIT studies leading to the thermostabilization of enzymes does not surprise, after all, the choice of the focused randomization sites is rational and logical. Examples include *P. fluorescens* esterase [105], feruloyl esterase from *Aspergillus usamii* [99], *Penicillium expansum* [106], endo-1,4-ß-galactanase from *Talaromyces stipitatus* [101], ADP-glucose pyrophosphorylase [107], *Pseudomonas aeruginosa* lipase C [108], *Burgholderia* lipase [109], rhamnogalacturonan I lyase from *Bacillus licheniformis* [110], and *Bacillus Acidopullulyticus pullulanase* [111]. In the last two cited studies, B-FIT was compared with two other strategies based on the consensus approach (see Section 6.5) and the PoPMuSiC computational aid. B-FIT proved to be the superior strategy [110, 111]. Table 6.2 lists typical B-FIT studies including the most recent contributions.

A different development, which can be considered to be an extension of B-FIT, has been dubbed Active Center Stabilization (ACS) [127]. Accordingly, the active site is viewed as a fragile part of the enzyme as a whole, especially when considering the role of enzyme conformation, dynamics, and activity during the denaturing process [6, 128]. Consequently, it was postulated that rigidifying the active site by saturation mutagenesis at the respective residues of high B-factors should enhance protein robustness. However, in view of the insight that a certain degree of flexibility in the binding pocket is needed for maintaining high activity [129], too much rigidification should be avoided [130]. The flowchart of ACS is shown in Scheme 6.2.

Examples of ACS involve the thermostabilization of *Candida antarctica* B (CALB) [127a] and lipase I from *Candida rugose* (Lip I) having 534

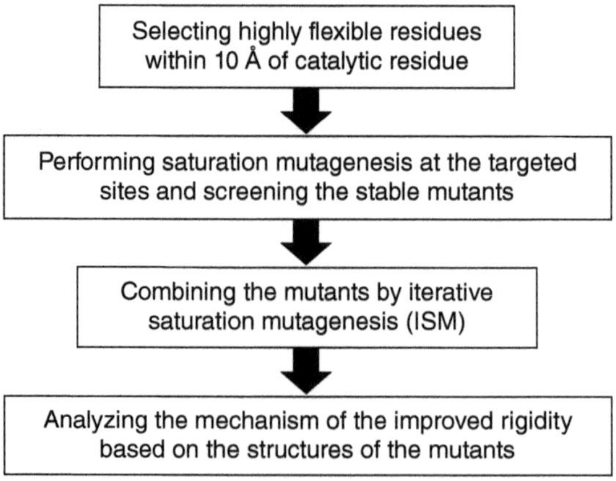

Scheme 6.2 The workflow of ACS strategy for improving enzyme stability. Source: Ref. [127b]/Springer Nature/CC BY 4.0.

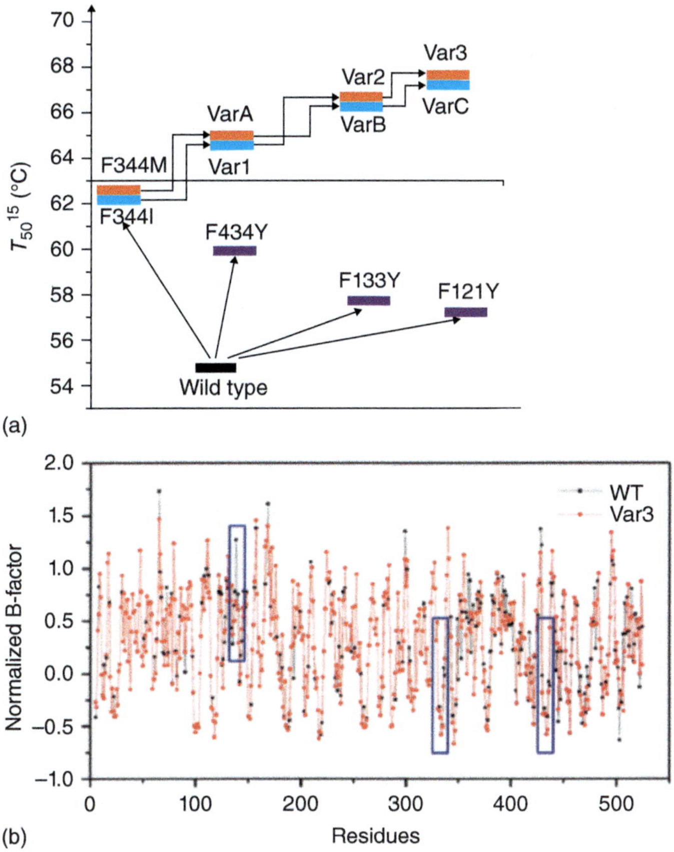

Figure 6.18 (a) Thermostability assay of Lip1 mutants. (b) Evaluation of the residues fluctuation in WT and variant based on normalized B-factors. Source: Refs. [127b, c]/Springer Nature/CC BY 4.0.

residues [127c]. The 18 residues surrounding the binding pocket of Lip I with highest B-factors were chosen for saturation mutagenesis (Figure 6.18). In five cases improved thermostability was found. Upon combining positive mutations sequentially, two different best variants were identified with increased $T_{50}{}^{15}$ values of 12–13 °C and a 40-fold enhancement of half-lives [127c]. It remains to be seen how general ACS is.

6.8
Iterative Saturation Mutagenesis (ISM) at Protein–Protein Interfacial Sites for Multimeric Enzymes

In a very different study, ISM was developed in which neither CAST nor B-FIT participate. Instead, randomization sites were chosen at a protein–protein interface of a dimeric enzyme [131] for introducing mutations that prevent dimer dissociation, because strengthening the inter-subunit interface would

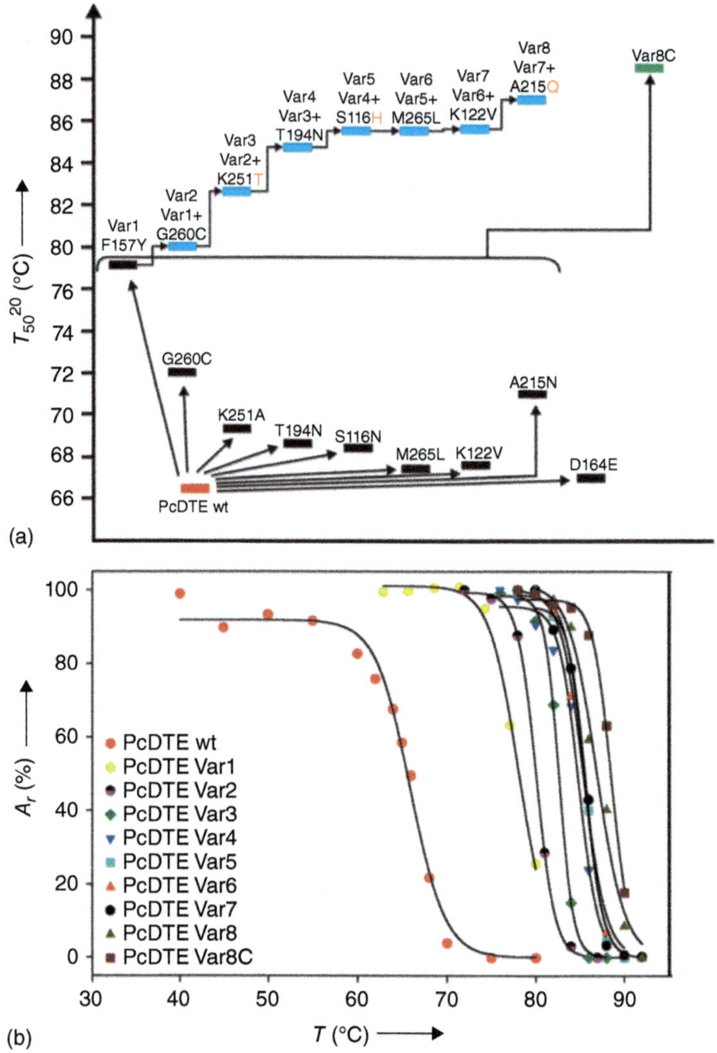

Figure 6.19 Thermostabilization of PcDTE following application of ISM [131]. (a) Thermostability, expressed as the T_{50}^{20} value, of all variants involved in this study: PcDTE wild-type (red bar), hits obtained in the first SM round (black bars), variants 2–8 obtained by ISM (blue bars), and variant 8C obtained by combination of the eight mutations from the first round (green bar). Mutation D164E was excluded in combinations as no improved variant could be identified during ISM. (b) Residual activity curves of WT PcDTE, variants 1–8, and variant 8C fitted to a second-order sigmoidal function. Source: Ref. [131]/John Wiley & Sons.

counteract disintegration [131]. The homodimeric D-tagatose 3-epimerase from *Pseudomonas chichorii* (PcDTE) served as the model enzyme. Initial application of B-FIT provided only a slightly improved variant, and the introduction of inter-subunit disulfide bonds failed completely. Therefore, the PcDTE crystal structure was analyzed for interfacial interactions using the software PDBePISA [132], giving rise to 44 residues as potential candidates for saturation mutagenesis. Three of these residues hardly contribute to the buried surface area and were therefore excluded from further consideration, as were 10 highly conserved residues, which were thought to fulfill other important functions. Each of the remaining 31 residues were subjected to NNK-based saturation mutagenesis. In nine libraries thermally improved hits were identified, the best variant being F157Y. Then ISM was applied, leading to an enormous increase in thermostability (Figure 6.19). Less than 4000 transformants had to be screened. This approach is likely to be useful in the future for improving the robustness of multimeric proteins [7].

6.9 Conclusions and Perspectives

As aforementioned, there are currently numerous rational and directed evolution studies on enhancing protein robustness from which the protein engineer can choose. The readers may therefore want to know what to apply best in a given new project. As an answer, we propose two general types of strategies (i) using the methodology that has been successfully deployed for proteins belonging to the same superfamily, and (ii) combining distinct approaches to improving robustness that generates optimal mutants for subsequent applications in organic and pharmaceutical chemistry as well as biotechnology while eliminating suboptimal ones. As a note of caution, the user must make sure that stability improvement is not achieved at the expense of selectivity. In general, genetic optimization of protein robustness is easier than protein engineering of stereo- and/or regioselective enzymes (see also Chapter 1).

References

1. (a) Lotti, M. and Secundo, F. (2015). Editorial: Protein stabilization – crossroad for protein-based processes and products. *Biotechnol. J.* **10** (3): 341–342. (b) Drauz, K., Gröger, H., and May, O. (ed.) (2012). *Enzyme Catalysis in Organic Synthesis*, 3e. Weinheim: Wiley-VCH. (c) Liese, A., Seelbach, K., and Wandrey, C. (2006). *Industrial Biotransformations*. Weinheim: Wiley-VCH Verlag GmbH.
2. Pace, C.N., Scholtz, J.M., and Grimsley, G.R. (2014). Forces stabilizing proteins. *FEBS Lett.* **588** (14): 2177–2184.
3. Sorokina, I., Mushegian, A., and Koonin, E.V. (2022). Is protein folding a thermodynamically unfavorable, active, energy-dependent process? *Int. J. Mol. Sci.* **23** (1): 521.
4. Dinmukhamed, T., Huang, Z., Liu, Y. et al. (2021). Current advances in design and engineering strategies of

industrial enzymes. *Systems Microbiol. Biomanuf.* **1**: 15–23.

5. (a) Sheldon, R.A. and Brady, D. (2021). Streamlining design, engineering, and applications of enzymes for sustainable biocatalysis. *ACS Sustainable Chem. Eng.* **9** (24): 8032–8052. (b) Rizwana, N., Ayob, S.N., Rahman, N. et al. (2020). Rational protein engineering of α-L-arabinofuranosidase from *Aspergillus niger* for improved catalytic hydrolysis efficiency on kenaf hemicellulose. *Process Biochem.* **102**: 349–359. (c) Chen, S., Li, Z., Gu, Z. et al. (2022). Immobilization of β-cyclodextrin glycosyltransferase on gelatin enhances β-cyclodextrin production. *Process Biochem.* **113**: 216–223.
6. Eijsink, V.G.H., Bjørk, A., Gåseidnes, S. et al. (2004). Rational engineering of enzyme stability. *J. Biotechnol.* **113** (1–3): 105–120.
7. Reetz, M.T. (2016). *Directed Evolution of Selective Enzymes Catalysts for Organic Chemistry and Biotechnology*. Weinheim: Wiley-VCH.
8. Guo, C., Ni, Y., Biewenga, L. et al. (2021). Using mutability landscapes to guide enzyme thermostabilization. *ChemBioChem* **22** (1): 170–175.
9. Patel, N.Y., Baria, D.M., and Raval, V. (2022). *Application of Extremophiles in Medicine and Pharmaceutical Industries*. Hershey, PA: IGI Global.
10. (a) Donald, J.E., Kulp, D.W., and DeGrado, W.F. (2011). Salt bridges: geometrically specific, designable interactions. *Proteins* **79** (3): 898–915. (b) Basu, S. and Biswas, P. (2018). Salt-bridge dynamics in intrinsically disordered proteins: a trade-off between electrostatic interactions and structural flexibility. *Biochim. Biophys. Acta, Proteins Proteomics* **1866** (5–6): 624–641. (c) Wu, J.P., Li, M., Zhou, Y. et al. (2015). Introducing a salt bridge into the lipase of *Stenotrophomonas maltophilia* results in a very large increase in thermal stability. *Biotechnol. Lett.* **37** (2): 403–407. (d) Folch, B., Rooman, M., and Dehouck, Y. (2008). Thermostability of salt bridges versus hydrophobic interactions in proteins probed by statistical potentials. *J. Chem. Inf. Model.* **48** (1): 119–127. (e) Missimer, J.H., Steinmetz, M.O., Baron, R. et al. (2007). Configurational entropy elucidates the role of salt-bridge networks in protein thermostability. *Protein Sci.* **16** (7): 1349–1359. (f) Meuzelaar, H., Tros, M., Huerta-Viga, A. et al. (2014). Solvent-exposed salt bridges influence the kinetics of α-helix folding and unfolding. *J. Phys. Chem. Lett.* **5** (5): 900–904. (g) Basu, S. and Mukharjee, D. (2017). Salt-bridge networks within globular and disordered proteins: characterizing trends for designable interactions. *J. Mol. Model.* **23** (7): 206.
11. Teufl, M., Zajc, C.U., and Traxlmayr, M.W. (2022). Engineering strategies to overcome the stability–function trade-off in proteins. *ACS Synth. Biol.* **11** (3): 1030–1039.
12. Van den Burg, B., Vriend, G., Veltman, O.R. et al. (1998). Engineering an enzyme to resist boiling. *Proc. Natl. Acad. Sci. U.S.A.* **95** (5): 2056–2060.
13. (a) Ovchinnikov, S., Kim, D.E., Wang, R.Y. et al. (2016). Improved de novo structure prediction in CASP11 by incorporating coevolution information into Rosetta. *Proteins* **84** (Suppl 1): 67–75. (b) Leman, J.K., Weitzner, B.D., Lewis, S.M. et al. (2020). Macromolecular modeling and design in Rosetta: recent methods and frameworks. *Nat. Methods* **17** (7): 665–680. (c) Schmitz, S., Ertelt, M., Merkl, R., and Meiler, J. (2021). Rosetta design with co-evolutionary information retains protein function. *PLoS Comput. Biol.* **17** (1): e1008568.
14. Korkegian, A., Black, M.E., Baker, D., and Stoddard, B.L. (2005). Computational thermostabilization of an enzyme. *Science* **308** (5723): 857–860.
15. Dunbrack, R.L. Jr., and Cohen, F.E. (1997). Bayesian statistical analysis of protein side-chain rotamer preferences. *Protein Sci.* **6** (8): 1661–1681.
16. (a) Iannuzzelli, J.A., Bacik, J.P., Moore, E.J. et al. (2022). Tuning enzyme thermostability via computationally guided covalent stapling and structural basis of enhanced stabilization. *Biochemistry*

61 (11): 1041–1054. (b) Li, G., Fang, X., Su, F. et al. (2018). Enhancing the thermostability of *Rhizomucor miehei* lipase with a limited screening library by rational-design point mutations and disulfide bonds. *Appl. Environ. Microbiol.* **84** (2): e02129–e02117. (c) Moore, E.J., Zorine, D., Hansen, W.A. et al. (2017). Enzyme stabilization via computationally guided protein stapling. *Proc. Natl. Acad. Sci. U.S.A.* **114** (47): 12472–12477. (d) Sammond, D.W., Kastelowitz, N., Donohoe, B.S. et al. (2018). An iterative computational design approach to increase the thermal endurance of a mesophilic enzyme. *Biotechnol. Biofuels* **11**: 189. (e) Nance, M.L., Labonte, J.W., Adolf-Bryfogle, J., and Gray, J.J. (2021). Development and evaluation of glycandock: a protein–glycoligand docking refinement algorithm in Rosetta. *J. Phys. Chem. B* **125** (25): 6807–6820.

17. Yu, H., Yan, Y., Zhang, C., and Dalby, P.A. (2017). Two strategies to engineer flexible loops for improved enzyme thermostability. *Sci. Rep.* **7**: 41212.
18. Harrington, L.B., Jha, R.K., Kern, T.L. et al. (2017). Rapid thermostabilization of *Bacillus thuringiensis* Serovar Konkukian 97-27 dehydroshikimate dehydratase through a structure-based enzyme design and whole cell activity assay. *ACS Synth. Biol.* **6** (1): 120–129.
19. Goldenzweig, A., Goldsmith, M., Hill, S.E. et al. (2016). Automated structure- and sequence-based design of proteins for high bacterial expression and stability. *Mol. Cell* **63** (2): 337–346. Erratum: *Mol. Cell*, 2018, **70** (2), 380.
20. (a) Steipe, B., Schiller, B., Plückthun, A., and Steinbacher, S. (1994). Sequence statistics reliably predict stabilizing mutations in a protein domain. *J. Mol. Biol.* **240** (3): 188–192. (b) Lehmann, M., Pasamontes, L., Lassen, S.F., and Wyss, M. (2000). The consensus concept for thermostability engineering of proteins. *Biochim. Biophys. Acta* **1543** (2): 408–415.
21. Whitehead, T.A., Chevalier, A., Song, Y. et al. (2012). Optimization of affinity, specificity and function of designed influenza inhibitors using deep sequencing. *Nat. Biotechnol.* **30** (6): 543–548.
22. Cui, H., Eltoukhy, L., Zhang, L. et al. (2021). Less unfavorable salt bridges on the enzyme surface result in more organic cosolvent resistance. *Angew. Chem. Int. Ed.* **60** (20): 11448–11456.
23. (a) Cui, H., Stadtmüller, T.H., Jiang, Q. et al. (2020). How to engineer organic solvent resistant enzymes: insights from combined molecular dynamics and directed evolution study. *ChemCatChem* **12**: 4073–4083. (b) Markel, U., Zhu, L., Victorine, F.M. et al. (2017). Are directed evolution approaches efficient in exploring nature's potential to stabilize a lipase in organic cosolvents? *Catalysts* **7** (5): 142. (c) Frauenkron-Machedjou, V.J., Fulton, A., Zhao, J. et al. (2018). Exploring the full natural diversity of single amino acid exchange reveals that 40-60% of BSLA positions improve organic solvents resistance. *Bioresour. Bioprocess.* **5** (1): 2. (d) Rathi, P.C., Fulton, A., Jaeger, K.E., and Gohlke, H. (2016). Application of rigidity theory to the thermostabilization of lipase A from *Bacillus subtilis*. *PLoS Comput. Biol.* **12** (3): e1004754. (e) Nutschel, C., Fulton, A., Zimmermann, O. et al. (2020). Systematically scrutinizing the impact of substitution sites on thermostability and detergent tolerance for *Bacillus subtilis* lipase A. *J. Chem. Inf. Model.* **60** (3): 1568–1584.
24. Reetz, M.T., Soni, P., Fernández, L. et al. (2010). Increasing the stability of an enzyme toward hostile organic solvents by directed evolution based on iterative saturation mutagenesis using the B-FIT method. *Chem. Commun.* **46** (45): 8657–8658.
25. Li, G., Qin, Y., Fontaine, N.T. et al. (2021). Machine learning enables selection of epistatic enzyme mutants for stability against unfolding and detrimental aggregation. *ChemBioChem* **22** (5): 904–914.
26. Reetz, M.T. (2013). The importance of additive and non-additive mutational effects in protein engineering. *Angew. Chem. Int. Ed.* **52** (10): 2658–2666.

27. (a) Modi, T., Risso, V.A., Martinez-Rodriguez, S. et al. (2021). Hinge-shift mechanism as a protein design principle for the evolution of β-lactamases from substrate promiscuity to specificity. *Nat. Commun.* **12** (1): 1852. (b) Wijma, H.J., Fürst, M.J.L.J., and Janssen, D.B. (2018). A computational library design protocol for rapid improvement of protein stability: FRESCO. *Methods Mol. Biol.* **1685**: 69–85. (c) Gupta, A. and Agrawal, S. (2022). Machine learning-based enzyme engineering of PETase for improved efficiency in degrading non-biodegradable plastic. *BioRxiv* **11**: 475766. (d) Scarabelli, G., Oloo, E.O., Maier, J.K.X., and Rodriguez-Granillo, A. (2022). Accurate prediction of protein thermodynamic stability changes upon residue mutation using free energy perturbation. *J. Mol. Biol.* **434** (2): 167375. (e) Jia, L.L., Sun, T.T., Wang, Y., and Shen, Y. (2021). A machine learning study on the thermostability prediction of (*R*)-ω-selective amine transaminase from *Aspergillus terreus*. *Biomed. Res. Int.* **2021**: 2593748.

28. Malcolm, B.A., Wilson, K.P., Matthews, B.W. et al. (1990). Ancestral lysozymes reconstructed, neutrality tested, and thermostability linked to hydrocarbon packing. *Nature* **345**: 86–89.

29. (a) Akanuma, S., Iwami, S., Yokoi, T. et al. (2011). Phylogeny-based design of a B-subunit of DNA gyrase and its ATPase domain using a small set of homologous amino acid sequences. *J. Mol. Biol.* **412** (2): 212–225. (b) Perez-Jimenez, R., Ingles-Prieto, A., Zhao, Z.M. et al. (2011). Single-molecule paleoenzymology probes the chemistry of resurrected enzymes. *Nat. Struct. Mol. Biol.* **18** (5): 592–596. (c) Gomez-Fernandez, B.J., Risso, V.A., Rueda, A. et al. (2020). Ancestral resurrection and directed evolution of fungal mesozoic laccases. *Appl. Environ. Microbiol.* **86** (14): e00778–e00720. Erratum: *Appl. Environ. Microbiol.*, 2020, **86** (17), e01732–20. (d) Selberg, A.G.A., Gaucher, E.A., and Liberles, D.A. (2021). Ancestral sequence reconstruction: from chemical paleogenetics to maximum likelihood algorithms and beyond. *J. Mol. Evol.* **89** (3): 157–164.

30. (a) Lehmann, M. and Wyss, M. (2001). Engineering proteins for thermostability: the use of sequence alignments versus rational design and directed evolution. *Curr. Opin. Biotechnol.* **12** (4): 371–375. (b) Lehmann, M., Loch, C., Middendorf, A. et al. (2002). The consensus concept for thermostability engineering of proteins: further proof of concept. *Protein Eng.* **15** (5): 403–411.

31. Amin, N., Liu, A.D., Ramer, S. et al. (2004). Construction of stabilized proteins by combinatorial consensus mutagenesis. *Protein Eng. Des. Sel.* **17** (11): 787–793.

32. (a) Binz, H.K., Stumpp, M.T., Forrer, P. et al. (2003). Designing repeat proteins: well-expressed, soluble and stable proteins from combinatorial libraries of consensus ankyrin repeat proteins. *J. Mol. Biol.* **332** (2): 489–503. (b) Watanabe, K., Ohkuri, T., Yokobori, S.-i., and Yamagishi, A. (2006). Designing thermostable proteins: ancestral mutants of 3-isopropylmalate dehydrogenase designed by using a phylogenetic tree. *J. Mol. Biol.* **355** (4): 664–674.

33. DiTursi, M.K., Kwon, S.-J., Reeder, P.J., and Dordick, J.S. (2006). Bioinformatics-driven, rational engineering of protein thermostability. *Protein Eng. Des. Sel.* **19** (11): 517–524.

34. Zhang, S., Wu, G., Feng, S., and Liu, Z. (2014). Improved thermostability of esterase from *Aspergillus fumigatus* by site-directed mutagenesis. *Enzyme Microb. Technol.* **64-65**: 11–16.

35. Polizzi, K.M., Chaparro-Riggers, J.F., Vazquez-Figueroa, E., and Bommarius, A.S. (2006). Structure-guided consensus approach to create a more thermostable penicillin G acylase. *Biotechnol. J.* **1** (5): 531–536.

36. Vázquez-Figueroa, E., Chaparro-Riggers, J., and Bommarius, A.S. (2007). Development of a thermostable glucose dehydrogenase by a structure-guided consensus concept. *ChemBioChem* **8** (18): 2295–2301.

37. Kamtekar, S., Schiffer, J., Xiong, H. et al. (1993). Protein design by binary patterning of polar and nonpolar amino acids. *Science* **262** (5140): 1680–1685.

38. (a) Besenmatter, W., Kast, P., and Hilvert, D. (2007). Relative tolerance of mesostable and thermostable protein homologs to extensive mutation. *Proteins* **66** (2): 500–506. (b) Jäckel, C., Bloom, J.D., Kast, P. et al. (2010). Consensus protein design without phylogenetic bias. *J. Mol. Biol.* **399** (4): 541–546.

39. (a) Huang, L., Ma, H.-M., Yu, H.-L., and Xu, J.-H. (2014). Altering the substrate specificity of reductase CgKR1 from *Candida glabrata* by protein engineering for bioreduction of aromatic α-keto esters. *Adv. Synth. Catal.* **356** (9): 1943–1948. (b) Anbar, M., Gul, O., Lamed, R. et al. (2012). Improved thermostability of clostridium thermocellum endoglucanase cel8a by using consensus-guided mutagenesis. *Appl. Environ. Microbiol.* **78** (9): 3458–3464. (c) Zhang, D., Zhu, F., Fan, W. et al. (2011). Gradually accumulating beneficial mutations to improve the thermostability of *N*-carbamoyl-D-amino acid amidohydrolase by step-wise evolution. *Appl. Microbiol. Biotechnol.* **90** (4): 1361–1371. (d) Trudeau, D.L., Lee, T.M., and Arnold, F.H. (2014). Engineered thermostable fungal cellulases exhibit efficient synergistic cellulose hydrolysis at elevated temperatures. *Biotechnol. Bioeng.* **111** (12): 2390–2397. (e) Komor, R.S., Romero, P.A., Xie, C.B., and Arnold, F.H. (2012). Highly thermostable fungal cellobiohydrolase I (Cel7A) engineered using predictive methods. *Protein Eng. Des. Sel.* **25** (12): 827–833. (f) Qian, H., Zhang, C., Lu, Z. et al. (2018). Consensus design for improved thermostability of lipoxygenase from *Anabaena* sp. *PCC 7120. BMC Biotechnol.* **18**: 57. (g) Cirri, E., Brier, S., Assal, R. et al. (2018). Consensus designs and thermal stability determinants of a human glutamate transporter. *Elife* **7**: e40110. (h) Okafor, C.D., Pathak, M.C., Fagan, C.E. et al. (2018). Structural and dynamics comparison of thermostability in ancient, modern, and consensus elongation factor Tus. *Structure* **26** (1): 118–129.e3.

40. Stepankova, V., Bidmanova, S., Koudelakova, T. et al. (2013). Strategies for stabilization of enzymes in organic solvents. *ACS Catal.* **3** (12): 2823–2836.

41. (a) Bommarius, A.S. and Paye, M.F. (2013). Stabilizing biocatalysts. *Chem. Soc. Rev.* **42** (15): 6534–6565. (b) Liszka, M.J., Clark, M.E., Schneider, E., and Clark, D.S. (2012). Nature versus nurture: developing enzymes that function under extreme conditions. *Annu. Rev. Chem. Biomol. Eng.* **3** (1): 77–102. (c) Suplatov, D., Voevodin, V., and Švedas, V. (2015). Robust enzyme design: bioinformatic tools for improved protein stability. *Biotechnol. J.* **10** (3): 344–355. (d) Socha, R.D. and Tokuriki, N. (2013). Modulating protein stability – directed evolution strategies for improved protein function. *FEBS J.* **280** (22): 5582–5595. e Yu, H. and Huang, H. (2014). Engineering proteins for thermostability through rigidifying flexible sites. *Biotechnol. Adv.* **32** (2): 308–315. f Naeem, M., Khalil, A.B., Tariq, Z., and Mahmoud, M. (2022). A review of advanced molecular engineering approaches to enhance the thermostability of enzyme breakers: from prospective of upstream oil and gas industry. *Int. J. Mol. Sci.* **23**: 1597.

42. (a) Pantazes, R.J., Grisewood, M.J., and Maranas, C.D. (2011). Recent advances in computational protein design. *Curr. Opin. Struct. Biol.* **21** (4): 467–472. (b) Wijma, H.J., Floor, R.J., and Janssen, D.B. (2013). Structure- and sequence-analysis inspired engineering of proteins for enhanced thermostability. *Curr. Opin. Struct. Biol.* **23** (4): 588–594. (c) Yang, L., Peplowski, L., Shen, Y., and Xia, Y. (2022). Enhancing thermostability and activity of sucrose phosphorylase for high-level production of 2-*O*-α-D-glucosylglycerol.

Syst. Microbiol. Biomanuf. https://doi.org/10.1007/s43393-022-00090-y.

43. (a) Bender, G.M., Lehmann, A., Zou, H. et al. (2007). De novo design of a single-chain diphenylporphyrin metalloprotein. *J. Am. Chem. Soc.* **129** (35): 10732–10740. (b) Nanda, V., Rosenblatt, M.M., Osyczka, A. et al. (2005). De novo design of a redox-active minimal rubredoxin mimic. *J. Am. Chem. Soc.* **127** (16): 5804–5805. (c) Calhoun, J.R., Kono, H., Lahr, S. et al. (2003). Computational design and characterization of a monomeric helical dinuclear metalloprotein. *J. Mol. Biol.* **334** (5): 1101–1115.
44. (a) Sheffler, W. and Baker, D. (2009). RosettaHoles: rapid assessment of protein core packing for structure prediction, refinement, design, and validation. *Protein Sci.* **18** (1): 229–239. (b) Sheffler, W. and Baker, D. (2010). RosettaHoles2: a volumetric packing measure for protein structure refinement and validation. *Protein Sci.* **19** (10): 1991–1995. (c) Borgo, B. and Havranek, J.J. (2012). Automated selection of stabilizing mutations in designed and natural proteins. *Proc. Natl. Acad. Sci. U.S.A.* **109** (5): 1494–1499. (d) Kiss, G., Celebi-Olcum, N., Moretti, R. et al. (2013). Computational enzyme design. *Angew. Chem. Int. Ed.* **52** (22): 5700–5725.
45. Barroso, J.R.M.S., Mariano, D., Dias, S.R. et al. (2020). Proteus: an algorithm for proposing stabilizing mutation pairs based on interactions observed in known protein 3D structures. *BMC Bioinf.* **21** (1): 275.
46. (a) Saab-Rincon, G., Li, Y., Meyer, M. et al. (2009). Protein engineering by structure-guided SCHEMA recombination. In: *Protein Engineering Handbook* (ed. S. Lutz and U.T. Bornscheuer), 481–492. Weinheim: Wiley-VCH Verlag GmbH. (b) Heinzelman, P., Snow, C.D., Smith, M.A. et al. (2009). SCHEMA recombination of a fungal cellulase uncovers a single mutation that contributes markedly to stability. *J. Biol. Chem.* **284**: 26229–26233. (c) Pantazes, R.J., Saraf, M.C., and Maranas, C.D. (2007). Optimal protein library design using recombination or point mutations based on sequence based scoring functions. *Protein Eng. Des. Sel.* **20**: 361–373.
47. (a) Romero, P.A., Stone, E., Lamb, C. et al. (2012). Schema-designed variants of human arginase I and II reveal sequence elements important to stability and catalysis. *ACS Synth. Biol.* **1** (6): 221–228. (b) Smith, M.A., Romero, P.A., Wu, T. et al. (2013). Chimeragenesis of distantly-related proteins by noncontiguous recombination. *Protein Sci.* **22** (2): 231–238. (c) Heinzelman, P., Komor, R., Kanaan, A. et al. (2010). Efficient screening of fungal cellobiohydrolase class I enzymes for thermostabilizing sequence blocks by SCHEMA structure-guided recombination. *Protein Eng. Des. Sel.* **23** (11): 871–880.
48. Smith, M.A., Bedbrook, C.N., Wu, T., and Arnold, F.H. (2013). *Hypocrea jecorina* cellobiohydrolase I stabilizing mutations identified using noncontiguous recombination. *ACS Synth. Biol.* **2** (12): 690–696.
49. (a) Wijma, H.J., Floor, R.J., Jekel, P.A. et al. (2014). Computationally designed libraries for rapid enzyme stabilization. *Protein Eng. Des. Sel.* **27** (2): 49–58. (b) Kellogg, E.H., Leaver-Fay, A., and Baker, D. (2011). Role of conformational sampling in computing mutation-induced changes in protein structure and stability. *Proteins* **79** (3): 830–838.
50. Guerois, R., Nielsen, J.E., and Serrano, L. (2002). Predicting changes in the stability of proteins and protein complexes: a study of more than 1000 mutations. *J. Mol. Biol.* **320** (2): 369–387.
51. Arabnejad, H., Lago, M.D., Jekel, P.A. et al. (2017). A robust cosolvent-compatible halohydrin dehalogenase by computational library design. *Protein Eng. Des. Sel.* **30** (3): 173–187.
52. Floor, R.J., Wijma, H.J., Jekel, P.A. et al. (2015). X-ray crystallographic validation of structure predictions used

in computational design for protein stabilization. *Proteins* **83** (5): 940–951.

53. Floor, R.J., Wijma, H.J., Colpa, D.I. et al. (2014). Computational library design for increasing haloalkane dehalogenase stability. *ChemBioChem* **15** (11): 1660–1672.

54. Bednar, D., Beerens, K., Sebestova, E. et al. (2015). FireProt: energy- and evolution-based computational design of thermostable multiple-point mutants. *PLoS Comput. Biol.* **11** (11): e1004556.

55. Yin, S., Ding, F., and Dokholyan, N.V. (2007). Eris: an automated estimator of protein stability. *Nat. Methods* **4** (6): 466–467.

56. Parthiban, V., Gromiha, M.M., and Schomburg, D. (2006). CUPSAT: prediction of protein stability upon point mutations. *Nucleic Acids Res.* **34** (11): W239–W242.

57. Xia, Y., Li, X., Yang, L. et al. (2021). Development of thermostable sucrose phosphorylase by semi-rational design for efficient biosynthesis of alpha-D-glucosylglycerol. *Appl. Microbiol. Biotechnol.* **105** (19): 7309–7319.

58. Markova, K., Chmelova, K., Marques, S.M. et al. (2020). Decoding the intricate network of molecular interactions of a hyperstable engineered biocatalyst. *Chem. Sci.* **11** (41): 11162–11178.

59. Bolivar, J.M., Luley-Goedl, C., Leitner, E. et al. (2017). Production of glucosyl glycerol by immobilized sucrose phosphorylase: options for enzyme fixation on a solid support and application in microscale flow format. *J. Biotechnol.* **257**: 131–138.

60. (a) Rathi, P.C., Radestock, S., and Gohlke, H. (2012). Thermostabilizing mutations preferentially occur at structural weak spots with a high mutation ratio. *J. Biotechnol.* **159** (3): 135–144. (b) Kruger, D.M., Rathi, P.C., Pfleger, C., and Gohlke, H. (2013). CNA web server: rigidity theory-based thermal unfolding simulations of proteins for linking structure, (thermo-)stability, and function. *Nucleic Acids Res.* **41**: W340–W348. (c) Pfleger, C., Rathi, P.C., Klein, D.L. et al. (2013). Constraint Network Analysis (CNA): a python software package for efficiently linking biomacromolecular structure, flexibility, (thermo-)stability, and function. *J. Chem. Inf. Model.* **53** (4): 1007–1015.

61. Rathi, P.C., Jaeger, K.E., and Gohlke, H. (2015). Structural rigidity and protein thermostability in variants of lipase A from *Bacillus subtilis*. *PLoS One* **10** (7): https://doi.org/10.1371/journal.pone.0130289.

62. (a) Reetz, M.T., Carballeira, J.D., and Vogel, A. (2006). Iterative saturation mutagenesis on the basis of b factors as a strategy for increasing protein thermostability. *Angew. Chem. Int. Ed.* **45** (46): 7745–7751. (b) Reetz, M.T. and Carballeira, J.D. (2007). Iterative saturation mutagenesis (ISM) for rapid directed evolution of functional enzymes. *Nat. Protoc.* **2** (4): 891–903.

63. (a) Augustyniak, W., Brzezinska, A.A., Pijning, T. et al. (2012). Biophysical characterization of mutants of *Bacillus subtilis* lipase evolved for thermostability: factors contributing to increased activity retention. *Protein Sci.* **21** (4): 487–497. (b) Augustyniak, W., Wienk, H., Boelens, R., and Reetz, M.T. (2013). ^{1}H, ^{13}C and ^{15}N resonance assignments of wild-type *Bacillus subtilis* lipase A and its mutant evolved towards thermostability. *Biomol. NMR Assign.* **7** (2): 249–252.

64. (a) Ahmad, S. and Rao, N.M. (2009). Thermally denatured state determines refolding in lipase: mutational analysis. *Protein Sci.* **18** (6): 1183–1196. (b) Srivastava, A. and Sinha, S. (2014). Thermostability of in vitro evolved *Bacillus subtilis* lipase A: a network and dynamics perspective. *PLoS One* **9**: e102856.

65. Rathi, P.C., Mulnaes, D., and Gohlke, H. (2015). VisualCNA: a GUI for interactive constraint network analysis and protein engineering for improving thermostability. *Bioinformatics* **31** (14): 2394–2396.

66. Contreras, F., Nutschel, C., Beust, L. et al. (2021). Can constraint network analysis guide the identification phase of KnowVolution? A case study on improved thermostability of an endo-β-glucanase. *Comput. Struct. Biotechnol. J.* **19**: 743–751.

67. Diaz, J.E., Lin, C.S., Kunishiro, K. et al. (2011). Computational design and selections for an engineered, thermostable terpene synthase. *Protein Sci.* **20** (9): 1597–1606.
68. Moon, S., Jung, D.K., Phillips, G.N. Jr., and Bae, E. (2014). An integrated approach for thermal stabilization of a mesophilic adenylate kinase. *Proteins* **82** (9): 1947–1959.
69. Chan, C.H., Liang, H.K., Hsiao, N.W. et al. (2004). Relationship between local structural entropy and protein thermostability. *Proteins* **57** (4): 684–691.
70. (a) Bae, E., Bannen, R.M., and Phillips, G.N. Jr., (2008). Bioinformatic method for protein thermal stabilization by structural entropy optimization. *Proc. Natl. Acad. Sci. U.S.A.* **105** (28): 9594–9597. (b) Bannen, R.M., Suresh, V., Phillips, G.N. Jr., et al. (2008). Optimal design of thermally stable proteins. *Bioinformatics* **24** (20): 2339–2343.
71. Bae, E. and Phillips, G.N. Jr., (2006). Roles of static and dynamic domains in stability and catalysis of adenylate kinase. *Proc. Natl. Acad. Sci. U.S.A.* **103** (7): 2132–2137.
72. Howell, S.C., Inampudi, K.K., Bean, D.P., and Wilson, C.J. (2014). Understanding thermal adaptation of enzymes through the multistate rational design and stability prediction of 100 adenylate kinases. *Structure* **22** (2): 218–229.
73. Lamazares, E., Clemente, I., Bueno, M. et al. (2015). Rational stabilization of complex proteins: a divide and combine approach. *Sci. Rep.* **5**: 9129.
74. Conde-Giménez, M. and Sancho, J. (2021). Unravelling the complex denaturant and thermal-induced unfolding equilibria of human phenylalanine hydroxylase. *Int. J. Mol. Sci.* **22** (12): 6539.
75. Lee, C.W., Wang, H.J., Hwang, J.K., and Tseng, C.P. (2014). Protein thermal stability enhancement by designing salt bridges: a combined computational and experimental study. *PLoS One* **9** (11): e112751.
76. Estell, D.A., Graycar, T.P., and Wells, J.A. (1985). Engineering an enzyme by site-directed mutagenesis to be resistant to chemical oxidation. *J. Biol. Chem.* **260** (11): 6518–6521.
77. (a) Arnold, F.H. (1998). Design by directed evolution. *Acc. Chem. Res.* **31**: 125–131. (b) Petrounia, I.P. and Arnold, F.H. (2000). Designed evolution of enzymatic properties. *Curr. Opin. Biotechnol.* **11** (4): 325–330. (c) Eijsink, V.G.H., Gåseidnes, S., Borchert, T.V., and van den Burg, B. (2005). Directed evolution of enzyme stability. *Biomol. Eng* **22** (1–3): 21–30. (d) Bommarius, A.S. and Broering, J.M. (2005). Established and novel tools to investigate biocatalyst stability. *Biocatal. Biotransform.* **23** (3–4): 125–139. (e) Polizzi, K.M., Bommarius, A.S., Broering, J.M., and Chaparro-Riggers, J.F. (2007). Stability of biocatalysts. *Curr. Opin. Chem. Biol.* **11** (2): 220–225. (f) Tokuriki, N. and Tawfik, D.S. (2009). Stability effects of mutations and protein evolvability. *Curr. Opin. Struct. Biol.* **19** (5): 596–604. (g) Traxlmayr, M.W. and Obinger, C. (2012). Directed evolution of proteins for increased stability and expression using yeast display. *Arch. Biochem. Biophys.* **526** (2): 174–180. (h) Suzuki, H., Kobayashi, J., Wada, K. et al. (2015). Thermoadaptation-directed enzyme evolution in an error-prone thermophile derived from *Geobacillus kaustophilus* HTA426. *Appl. Environ. Microbiol.* **81**: 149–158.
78. (a) Kobayashi, J., Furukawa, M., Ohshiro, T., and Suzuki, H. (2015). Thermoadaptation-directed evolution of chloramphenicol acetyltransferase in an error-prone thermophile using improved procedures. *Appl. Microbiol. Biotechnol.* **99** (13): 5563–5572. (b) Suzuki, H. (2018). Peculiarities and biotechnological potential of environmental adaptation by *Geobacillus* species. *Appl. Microbiol. Biotechnol.* **102** (24): 10425–10437. (c) Suzuki, H., Taketani, T., Tanabiki, M. et al. (2021). Frequent transposition of multiple insertion sequences in *Geobacillus*

kaustophilus HTA426. *Front. Microbiol.* **12**: 650461.

79. (a) Jia, H., Lee, F.S., and Farinas, E.T. (2014). Bacillus subtilis spore display of laccase for evolution under extreme conditions of high concentrations of organic solvent. *ACS Comb. Sci.* **16** (12): 665–669. (b) Farinas, E.T. (2021). Laccase and its mutant displayed on the *Bacillus subtilis* spore coat for oxidation of phenolic compounds in organic solvents. *Catalysts* **11**: 606. (c) Brissos, V., Ferreira, M., Grass, G., and Martins, O.L. (2015). Turning a hyperthermostable metallo-oxidase into a laccase by directed evolution. *ACS Catal.* **5** (8): 4932–4941.
80. (a) Szilágyi, A. and Závodszky, P. (2000). Structural differences between mesophilic, moderately thermophilic and extremely thermophilic protein subunits: results of a comprehensive survey. *Structure* **8** (5): 493–504. (b) Kumar, S., Tsai, C.-J., and Nussinov, R. (2000). Factors enhancing protein thermostability. *Protein Eng.* **13** (3): 179–191. (c) Purmonen, M., Valjakka, J., Takkinen, K. et al. (2007). Molecular dynamics studies on the thermostability of family 11 xylanases. *Protein Eng. Des. Sel.* **20** (11): 551–559.
81. (a) Oshima, T. (1994). Stabilization of proteins by evolutionary molecular engineering techniques. *Curr. Opin. Struct. Biol.* **4** (4): 623–628. (b) Ó'Fágáin, C. (2003). Enzyme stabilization—recent experimental progress. *Enzyme Microb. Technol.* **33** (2–3): 137–149.
82. (a) Sieber, V., Pluckthun, A., and Schmid, F.X. (1998). Selecting proteins with improved stability by a phage-based method. *Nat. Biotechnol.* **16** (10): 955–960. (b) Schmid, F.-X. (2011). Lessons about protein stability from in vitro selections. *ChemBioChem* **12** (10): 1501–1507. (c) Kristensen, P. and Winter, G. (1998). Proteolytic selection for protein folding using filamentous bacteriophages. *Fold Des.* **3** (5): 321–328.
83. Pierre, B., Labonte, J.W., Xiong, T. et al. (2015). Molecular determinants for protein stabilization by insertional fusion to a thermophilic host protein. *ChemBioChem* **16** (16): 2392–2402.
84. (a) Bloom, J.D., Labthavikul, S.T., Otey, C.R., and Arnold, F.H. (2006). Protein stability promotes evolvability. *Proc. Natl. Acad. Sci. U.S.A.* **103** (15): 5869–5874. (b) Tokuriki, N., Stricher, F., Serrano, L., and Tawfik, D.S. (2008). How protein stability and new functions trade off. *PLoS Comput. Biol.* **4** (2): e1000002. (c) Takano, K., Aoi, A., Koga, Y., and Kanaya, S. (2013). Evolvability of thermophilic proteins from archaea and bacteria. *Biochemistry* **52** (28): 4774–4780. (d) Bershtein, S., Goldin, K., and Tawfik, D.S. (2008). Intense neutral drifts yield robust and evolvable consensus proteins. *J. Mol. Biol.* **379** (5): 1029–1044.
85. Finch, A.J. and Kim, J.R. (2018). Thermophilic proteins as versatile scaffolds for protein engineering. *Microorganisms* **6** (4): 97.
86. (a) Gao, W., Mahajan, S.P., Sulam, J., and Gray, J.J. (2020). Deep learning in protein structural modeling and design. *Patterns* **1** (9): 100142. (b) Harmalkar, A. and Gray, J.J. (2021). Advances to tackle backbone flexibility in protein docking. *Curr. Opin. Struct. Biol.* **67**: 178–186.
87. Jackson, C.J., Liu, J.-W., Carr, P.D. et al. (2013). Structure and function of an insect α-carboxylesterase (α-esterase-7) associated with insecticide resistance. *Proc. Natl. Acad. Sci. U.S.A.* **110** (25): 10177–10182.
88. Ishikawa, K., Nakamura, H., Morikawa, K., and Kanaya, S. (1993). Stabilization of *Escherichia coli* ribonuclease HI by cavity-filling mutations within a hydrophobic core. *Biochemistry* **32** (24): 6171–6178.
89. (a) Zhang, S.-B., Pei, X.-Q., and Wu, Z.-L. (2012). Multiple amino acid substitutions significantly improve the thermostability of feruloyl esterase a from *Aspergillus niger*. *Bioresour. Technol.* **117**: 140–147. (b) Zhang, S.-B. and Wu, Z.-L. (2011). Identification of amino acid residues responsible for increased thermostability of feruloyl esterase A from *Aspergillus niger* using

the PoPMuSiC algorithm. *Bioresour. Technol.* **102** (2): 2093–2096. (c) Zhang, S.-B., Ma, X.-F., Pei, X.-Q. et al. (2012). A practical high-throughput screening system for feruloyl esterases: substrate design and evaluation. *J. Mol. Catal. B: Enzym.* **74** (1–2): 36–40.

90. Dehouck, Y., Grosfils, A., Folch, B. et al. (2009). Fast and accurate predictions of protein stability changes upon mutations using statistical potentials and neural networks: Popmusic-2.0. *Bioinformatics* **25** (19): 2537–2543.

91. Yamada, R., Higo, T., Yoshikawa, C. et al. (2014). Improvement of the stability and activity of the BPO-A1 haloperoxidase from *Streptomyces aureofaciens* by directed evolution. *J. Biotechnol.* **192**: 248–254.

92. Koudelakova, T., Chaloupkova, R., Brezovsky, J. et al. (2013). Engineering enzyme stability and resistance to an organic cosolvent by modification of residues in the access tunnel. *Angew. Chem. Int. Ed.* **52** (7): 1959–1963.

93. (a) Ruller, R., Alponti, J., Deliberto, L.A. et al. (2014). Concomitant adaptation of a GH11 xylanase by directed evolution to create an alkali-tolerant/thermophilic enzyme. *Protein Eng. Des. Sel.* **27** (8): 255–262. (b) Ruller, R., Deliberto, L., Ferreira, T.L., and Ward, R.J. (2008). Thermostable variants of the recombinant xylanase a from *Bacillus subtilis* produced by directed evolution show reduced heat capacity changes. *Proteins: Struct. Funct. Bioinform.* **70** (4): 1280–1293.

94. (a) Palackal, N., Brennan, Y., Callen, W.N. et al. (2004). An evolutionary route to xylanase process fitness. *Protein Sci.* **13** (2): 494–503. (b) McHunu, N.P., Singh, S., and Permaul, K. (2009). Expression of an alkali-tolerant fungal xylanase enhanced by directed evolution in *Pichia pastoris* and *Escherichia coli*. *J. Biotechnol.* **141** (1–2): 26–30. (c) Qaim, M., Subramanian, A., and Sadashivappa, P. (2009). Commercialized GM crops and yield. *Nat. Biotechnol.* **27** (9): 803–804. (d) Hokanson, C.A., Cappuccilli, G., Odineca, T. et al. (2011). Engineering highly thermostable xylanase variants using an enhanced combinatorial library method. *Protein Eng. Des. Sel.* **24** (8): 597–605. (e) Wang, Y., Feng, S., Zhan, T. et al. (2013). Improving catalytic efficiency of endo-β-1, 4-xylanase from *Geobacillus stearothermophilus* by directed evolution and H179 saturation mutagenesis. *J. Biotechnol.* **168** (4): 341–347. (f) Zheng, H., Liu, Y., Sun, M. et al. (2014). Improvement of alkali stability and thermostability of *Paenibacillus campinasensis* family-11 xylanase by directed evolution and site-directed mutagenesis. *J. Ind. Microbiol. Biotechnol.* **41** (1): 153–162. (g) Qian, C., Liu, N., Yan, X. et al. (2015). Engineering a high-performance, metagenomic-derived novel xylanase with improved soluble protein yield and thermostability. *Enzyme Microb. Technol.* **70**: 35–41.

95. (a) Miyazaki, K. and Arnold, F.H. (1999). Exploring nonnatural evolutionary pathways by saturation mutagenesis: rapid improvement of protein function. *J. Mol. Evol.* **49** (6): 716–720. (b) Liebeton, K., Zonta, A., Schimossek, K. et al. (2000). Directed evolution of an enantioselective lipase. *Chem. Biol.* **7** (9): 709–718.

96. (a) Dana, C.M., Saija, P., Kal, S.M. et al. (2012). Biased clique shuffling reveals stabilizing mutations in cellulase Cel7A. *Biotechnol. Bioeng.* **109** (11): 2710–2719. (b) Steffler, F., Guterl, J.-K., and Sieber, V. (2013). Improvement of thermostable aldehyde dehydrogenase by directed evolution for application in synthetic cascade biomanufacturing. *Enzyme Microb. Technol.* **53**: 307–314.

97. (a) Niederhauser, B., Siivonen, J., Määttä, J.A. et al. (2012). DNA family shuffling within the chicken avidin protein family – a shortcut to more powerful protein tools. *J. Biotechnol.* **157** (1): 38–49. (b) Taskinen, B., Airenne, T.T., Jänis, J. et al. (2014). A novel chimeric avidin with increased thermal stability using DNA shuffling. *PLoS One* **9** (3): e92058. (c) Wu, I. and Arnold, F.H. (2013). Engineered thermostable fungal Cel6A and

Cel7A cellobiohydrolases hydrolyze cellulose efficiently at elevated temperatures. *Biotechnol. Bioeng.* **110** (7): 1874–1883. (d) Liang, C., Gui, X., Zhou, C. et al. (2015). Improving the thermoactivity and thermostability of pectate lyase from *Bacillus pumilus* for ramie degumming. *Appl. Microbiol. Biotechnol.* **99** (6): 2673–2682. (e) Stephens, D.E., Khan, F.I., Singh, P. et al. (2014). Creation of thermostable and alkaline stable xylanase variants by DNA shuffling. *J. Biotechnol.* **187**: 139–146. (f) Buettner, K., Hertel, T., and Pietzsch, M. (2012). Increased thermostability of microbial transglutaminase by combination of several hot spots evolved by random and saturation mutagenesis. *Amino Acids* **42** (2–3): 987–996. (g) Gonzalez-Perez, D., Garcia-Ruiz, E., Ruiz-Dueñas, F.J. et al. (2014). Structural determinants of oxidative stabilization in an evolved versatile peroxidase. *ACS Catal.* **4** (11): 3891–3901. (h) Huang, L., Xu, J.-H., and Yu, H.-L. (2015). Significantly improved thermostability of a reductase CgSKR1 from *Candida glabrata* with a key mutation at Asp 138 for enhancing bioreduction of aromatic α-keto esters. *J. Biotechnol.* **203**: 54–61. (i) Parker, B.M., Taylor, I.N., Woodley, J.M. et al. (2011). Directed evolution of a thermostable L-aminoacylase biocatalyst. *J. Biotechnol.* **155** (4): 396–405. (j) Kamal, M.Z., Ahmad, S., Molugu, T.R. et al. (2011). In vitro evolved non-aggregating and thermostable lipase: structural and thermodynamic investigation. *J. Mol. Biol.* **413** (3): 726–741.

98. Guan, L., Gao, Y., Li, J. et al. (2020). Directed evolution of *Pseudomonas fluorescens* lipase variants with improved thermostability using error-prone PCR. *Front. Bioeng. Biotechnol.* **8**: 1034.

99. Yin, X., Li, J.-F., Wang, C.-J. et al. (2015). Improvement in the thermostability of a type A feruloyl esterase, AuFaeA, from *Aspergillus usamii* by iterative saturation mutagenesis. *Appl. Microbiol. Biotechnol.* **99** (23): 10047–10056.

100. Reetz, M.T., Soni, P., Acevedo, J.P., and Sanchis, J. (2009). Creation of an amino acid network of structurally coupled residues in the directed evolution of a thermostable enzyme. *Angew. Chem. Int. Ed.* **48** (44): 8268–8272.

101. Larsen, D., Nyffenegger, C., Swiniarska, M. et al. (2015). Thermostability enhancement of an endo-1,4-β-galactanase from *Talaromyces stipitatus* by site-directed mutagenesis. *Appl. Microbiol. Biotechnol.* **99** (10): 4245–4253.

102. Gumulya, Y. and Reetz, M.T. (2011). Enhancing the thermal robustness of an enzyme by directed evolution: least favorable starting points and inferior mutants can map superior evolutionary pathways. *ChemBioChem* **12** (16): 2502–2510.

103. Gumulya, Y., Sanchis, J., and Reetz, M.T. (2012). Many pathways in laboratory evolution can lead to improved enzymes: how to escape from local minima. *ChemBioChem* **13** (7): 1060–1066.

104. Chokhawala, H., Roche, C., Kim, T.-W. et al. (2015). Mutagenesis of *Trichoderma reesei* endoglucanase I: impact of expression host on activity and stability at elevated temperatures. *BMC Biotechnol.* **15** (1): 11.

105. Jochens, H., Aerts, D., and Bornscheuer, U.T. (2010). Thermostabilization of an esterase by alignment-guided focussed directed evolution. *Protein Eng. Des. Sel.* **23** (12): 903–909.

106. Frascari, D., Zannoni, A., Pinelli, D., and Nocentini, M. (2007). Chloroform aerobic cometabolism by butane-utilizing bacteria in bioaugmented and non-bioaugmented soil/groundwater microcosms. *Process Biochem.* **42** (8): 1218–1228.

107. Boehlein, S.K., Shaw, J.R., Stewart, J.D. et al. (2015). Enhancing the heat stability and kinetic parameters of the maize endosperm ADP-glucose pyrophosphorylase using iterative saturation mutagenesis. *Arch. Biochem. Biophys.* **568**: 28–37.

108. Cesarini, S., Bofill, C., Pastor, F.I.J. et al. (2012). A thermostable variant of *P. aeruginosa* cold-adapted Lip C obtained by rational design and saturation mutagenesis. *Process Biochem.* **47** (12): 2064–2071.

109. Liu, Y., Qiu, L., Huang, J. et al. (2015). Screening for mutants with thermostable lipase A from *Burgholderia* sp. ZYB002. *Acta Microbiol. Sin.* **55**: 748–754.

110. Silva, I., Larsen, D., Jers, C. et al. (2013). Enhancing RGI lyase thermostability by targeted single point mutations. *Appl. Microbiol. Biotechnol.* **97** (22): 9727–9735.

111. Chen, A., Li, Y., Nie, J. et al. (2015). Protein engineering of *Bacillus acidopullulyticus* pullulanase for enhanced thermostability using in silico data driven rational design methods. *Enzyme Microb. Technol.* **78**: 74–83.

112. Tian, K., Tai, K., Bjw, C., and Li, Z. (2017). Directed evolution of *Thermomyces lanuginosus* lipase to enhance methanol tolerance for efficient production of biodiesel from waste grease. *Bioresour. Technol.* **245**: 1491–1497.

113. Zhang, J.H., Lin, Y., Sun, Y.F. et al. (2012). High-throughput screening of B factor saturation mutated *Rhizomucor miehei* lipase thermostability based on synthetic reaction. *Enzyme Microb. Technol.* **50** (6–7): 325–330.

114. Hanapi, W.N.W., Iuan-Sheau, C., Mahadi, N.M. et al. (2015). Site-saturation mutagenesis of *Glomerella cingulata* cutinase gene for enhanced enzyme thermostability. *AIP Conf. Proc.* **1678** (1): 030021.

115. Heath, R.S., Birmingham, W.R., Thompson, M.P. et al. (2019). An engineered alcohol oxidase for the oxidation of primary alcohols. *ChemBioChem* **20** (2): 276–281.

116. Damnjanović, J., Takahashi, R., Suzuki, A. et al. (2012). Improving thermostability of phosphatidylinositol-synthesizing *Streptomyces* phospholipase D. *Protein Eng. Des. Sel.* **25** (8): 415–424.

117. Li, G., Zhang, H., Sun, Z. et al. (2016). Multiparameter optimization in directed evolution: engineering thermostability, enantioselectivity, and activity of an epoxide hydrolase. *ACS Catal.* **6** (6): 3679–3687.

118. Larsen, D.M., Nyffenegger, C., Swiniarska, M.M. et al. (2015). Thermostability enhancement of an endo-1,4-beta-galactanase from *Talaromyces stipitatus* by site-directed mutagenesis. *Appl. Microbiol. Biotechnol.* **99** (10): 4245–4253.

119. Acevedo, J.P., Reetz, M.T., Asenjo, J.A., and Parra, L.P. (2017). One-step combined focused epPCR and saturation mutagenesis for thermostability evolution of a new cold-active xylanase. *Enzyme Microb. Technol.* **100**: 60–70.

120. Wehrmann, M. and Klebensberger, J. (2018). Engineering thermal stability and solvent tolerance of the soluble quinoprotein PedE from *Pseudomonas putida* KT2440 with a heterologous whole-cell screening approach. *Microb. Biotechnol.* **11** (2): 399–408.

121. Watanabe, S., Ito, M., and Kigawa, T. (2021). DiRect: site-directed mutagenesis method for protein engineering by rational design. *Biochem. Biophys. Res. Commun.* **551**: 107–113.

122. Roth, T., Beer, B., Pick, A., and Sieber, V. (2017). Thermostabilization of the uronate dehydrogenase from *Agrobacterium tumefaciens* by semi-rational design. *AMB Express* **7** (1): 103.

123. Li, X.-Q., Wu, Q., Hu, D. et al. (2017). Improving the temperature characteristics and catalytic efficiency of a mesophilic xylanase from *Aspergillus oryzae*, AoXyn11A, by iterative mutagenesis based on *in silico* design. *AMB Express* **7** (1): 97.

124. Reetz, M.T., Soni, P., and Fernandez, L. (2009). Knowledge-guided laboratory evolution of protein thermolability. *Biotechnol. Bioeng.* **102** (6): 1712–1717.

125. Cesarini, S., Bofill, C., Pastor, F.I.J. et al. (2012). A thermostable variant of *P. aeruginosa* cold-adapted LipC obtained by rational design and saturation mutagenesis. *Process Biochem.* **47** (12): 2064–2071.

126. Gao, S.H., Zhu, S.Z., Huang, R. et al. (2018). Engineering the enantioselectivity and thermostability of a (+)-gamma-lactamase from microbacterium hydrocarbonoxydans for kinetic resolution of vince lactam (2-Azabicyclo 2.2.1 hept-5-en-3-one). *Appl. Environ. Microbiol.* **84** (1): e01780–e01717.

127. (a) Xie, Y., An, J., Yang, G. et al. (2014). Enhanced enzyme kinetic stability by increasing rigidity within the active site. *J. Biol. Chem.* **289** (11): 7994–8006. (b) Zhang, X.F., Yang, G.Y., Zhang, Y. et al. (2016). A general and efficient strategy for generating the stable enzymes. *Sci. Rep.* **6**: 33797. (c) Huang, J., Xie, D.F., and Feng, Y. (2017). Engineering thermostable (R)-selective amine transaminase from *Aspergillus terreus* through in silico design employing B-factor and folding free energy calculations. *Biochem. Biophys. Res. Commun.* **483** (1): 397–402.

128. Agarwal, P.K., Bernard, D.N., Bafna, K., and Doucet, N. (2020). Enzyme dynamics: looking beyond a single structure. *ChemCatChem* **12** (19): 4704–4720.

129. Li, G., Maria-Solano, M.A., Romero-Rivera, A. et al. (2017). Inducing high activity of a thermophilic enzyme at ambient temperatures by directed evolution. *Chem. Commun.* **53** (68): 9454–9457.

130. Sun, Z., Liu, Q., Qu, G. et al. (2019). Utility of B-factors in protein science: interpreting rigidity, flexibility, and internal motion and engineering thermostability. *Chem. Rev.* **119** (3): 1626–1665.

131. Bosshart, A., Panke, S., and Bechtold, M. (2013). Systematic optimization of interface interactions increases the thermostability of a multimeric enzyme. *Angew. Chem. Int. Ed.* **52** (37): 9673–9676.

132. Krissinel, E. and Henrick, K. (2007). Inference of macromolecular assemblies from crystalline state. *J. Mol. Biol.* **372** (3): 774–797.

7
Artificial Enzymes as Promiscuous Catalysts in Organic and Pharmaceutical Chemistry

7.1
Introductory Background Information

As will be shown in this chapter, artificial metal-free enzymes and artificial metalloenzymes continue to play an intriguing role in synthetic organic and pharmaceutical chemistry [1]. Artificial metalloenzymes deserve particular mention. Some are formed by the introduction of a transition metal complex into an appropriate protein scaffold. Others use natural metalloenzymes such as P450 monooxygenases, which are then tuned by directed evolution. A representative example of the former type involves a water-soluble and commercially available achiral phthalocyanine Cu(II)-complex (**1**), which is placed in serum albumins (Scheme 7.1, top-left) and then used as catalysts in enantioselective Diels–Alder reactions of azachalcones (**2**) with cyclopentadiene (**3**) (Scheme 7.1, bottom) resulting in adducts with 85–98% ee and typically endo:exo ratios of 95 : 5 when using human serum albumin (HSA) [2]. In designing this artificial metalloenzyme, previous cases of covalent and non-covalent introduction of metals into proteins (see below) and the known X-ray structure of HSA harboring protoporphyrin dimethyl ester were used as an inspiration and as a guide; the latter had revealed supramolecular bonding in the subdomain IB with weak axial coordination by Tyr161 [3]. In this way, a binding pocket for the cycloaddition reaction was created in HSA, which in nature is only a transport protein in human beings. Since selectivity proved to be exceedingly high [2], directed evolution of this promiscuous artificial metalloenzyme was not necessary. However, the reaction was quite slow, a disadvantage that has not been addressed to date [4]. It should be noted that for this particular Diels–Alder reaction, modern man-made catalysts have not been developed, i.e. the best-known transition metal catalysts or organocatalysts are likely to fail.

Numerous other examples of introducing transition metals into protein scaffolds have been reported, some of which are described later in this chapter. A systematization of techniques is presented in Scheme 7.2 [1, 2, 5b, c, 6]. The novelty in this scheme is not the different ways of introducing metals/ligands into protein scaffolds, which were known [5], but the important suggestion and implementation that such artificial metalloenzymes can be tuned by directed

Enzyme Engineering: Selective Catalysts for Applications in Biotechnology, Organic Chemistry, and Life Science, First Edition. Manfred T. Reetz, Zhoutong Sun, and Ge Qu.

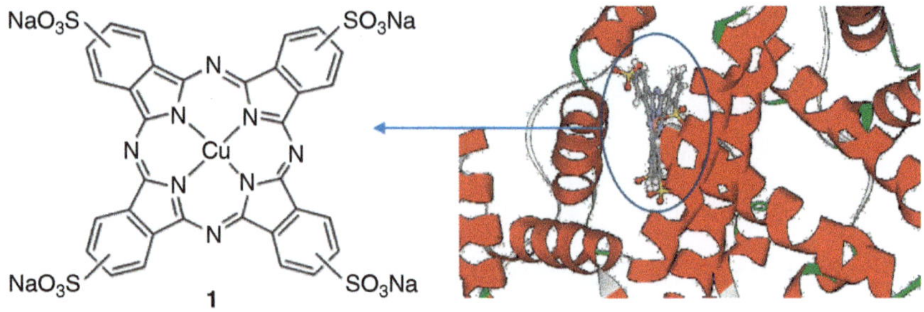

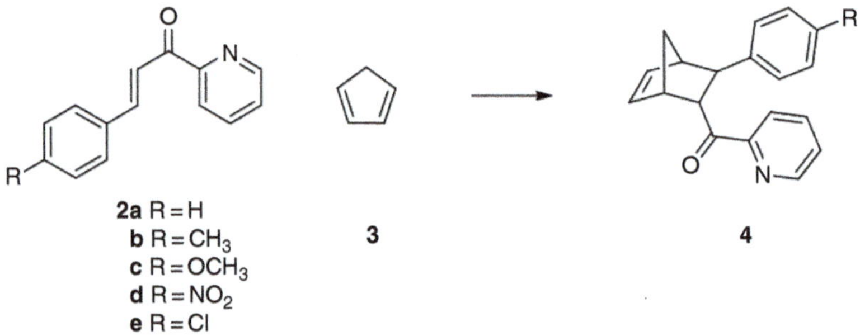

Scheme 7.1 Diels–Alder reaction of azachalcones **2** with **3** leading to endo-products **4** with 85–98% ee. Source: Adapted from Ref. [2].

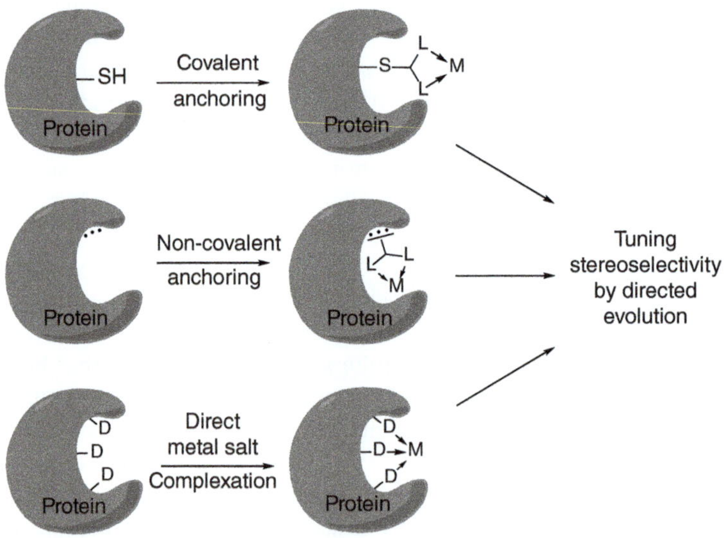

Scheme 7.2 Systematization for generating artificial metalloenzymes, originally dubbed "hybrid catalysts". L = synthetic ligand; M = transition metal; D = Donor atoms of side-chains of appropriate amino acids such as aspartate or cysteine which bind transition metals M directly. Source: Adapted from Refs. [1, 2, 5b, c, 6].

evolution for enhancing or reversing stereoselectivity [7]. This opened the door to an exciting new research field.

Over the decades, the term "promiscuous enzyme activity" has been used in different ways, some containing transition metals, others being metal-free [1]. For decades, biochemists and biotechnologists applied it whenever an enzyme was found to catalyze the reaction of a *natural* compound that was not the actual *natural substrate*, but later it was extended to include the transformation of unnatural compounds [8]. It also became clear that nature has evolved a given enzyme not just for a single purpose but sometimes to achieve several goals. Jensen postulated that broad specificity and promiscuous activities of enzymes constitute the starting points for the natural evolution of new and selective functions [9]. Much later, this phenomenon was mimicked in the laboratory by applying directed evolution to the human estrogen sulfotransferase with the aim of converting it from a generalist to a specialist [10]. The idea of enzyme ancestor reconstruction or resurrection, as proposed by Benner [11], has been implemented in laboratory experiments using directed evolution methods [12]. These and other insights are not only of central importance in evolutionary biology; they are also of potential practical interest in biotechnology [13].

In some rare cases, a given enzyme can catalyze a transformation that is usually mediated by another type of enzyme [1]. Examples include a decarboxylase that catalyzes the acyloin condensation [14a], and a carboxylic acid reductase that directs the intramolecular lactamization [14b]. Aside from the theoretical interest in this phenomenon, the practical benefits have not been sufficiently assessed [1]. This also applies to those studies in which such a switch was achieved by protein engineering, as in the conversion of an esterase into an epoxide hydrolase [15] (see Section 7.2), or in the conversion of a phosphotriesterase into an arylesterase [16]. These and other aspects of enzyme promiscuity have been addressed in numerous studies [17]. The ability of a protein or a chemically and/or genetically modified protein to catalyze a reaction type that does not occur in nature has also been described as being promiscuous [18]. Active-site redesign based on the introduction of a limited number of rationally chosen point mutations has proven to be successful in some studies, as summarized in an extensive review [19]. In addition to various review articles covering specific aspects [13], the general area of enzyme promiscuity has been summarized [20].

Some cases of this kind of promiscuity were discovered simply by screening a set of enzymes for a certain reaction type not known to be catalyzed by these biocatalysts, guided by mechanistic considerations and sometimes accompanied by designed point mutations. Seminal examples include lipases/esterases or proline polymers as catalysts in aldol and Michael reactions [21], epoxidation reactions [22], Mannich reactions [23], Markovnikov additions [24], and Baylis–Hillman reactions [25]. Progress in this particular research area has been reviewed [26].

As previously mentioned, promiscuous stereoselective catalytic behavior can be induced in a general manner by covalent or non-covalent (supramolecular) attachment of a ligand/transition metal moiety to a host protein, followed by directed evolution. The Whitesides system, comprising a biotinylated rhodium/diphosphine-complex which is conjugated non-covalently to avidin (or streptavidin, sav) is a seminal example [27]. The bound achiral Rh-complex was employed as the catalyst in asymmetric olefin hydrogenation of *N*-acyl acrylic acid (Scheme 7.3).

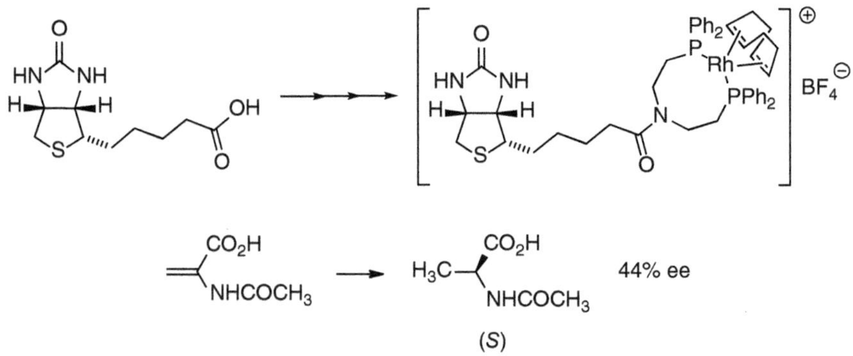

Scheme 7.3 Whitesides system comprises a biotinylated achiral diphosphine/Rh-complex non-covalently bound to avidin, which was used as the catalyst in the asymmetric olefin-hydrogenation of *N*-acyl acrylic acid [27]. Later, streptavidin was employed as the host protein. Source: Adapted from Refs. [5, 28].

Enantioselectivity proved to be moderate, but decades later this study opened the door to many interesting extensions, which included a variety of different transition metal catalyzed reaction types [5, 28]. Using wildtype (WT) streptavidin (or avidin), a single transition metal catalyst is produced, the catalytic profile of which is a matter of fortune. Fortunately, as first suggested in 2001–2002 [29] and further explained in Section 7.2, tuning by directed evolution offers exciting opportunities for developing truly efficient biocatalysts in organic chemistry, stereo- and/or regio-selectivity being the primary parameters of interest. However, this endeavor is not as easy as it may appear because several prerequisites have to be fulfilled:

- The host protein needs to be stable under operating conditions.
- The host protein needs to be expressed efficiently to provide sufficient quantities of protein in miniaturized and parallelized form.
- A simple and efficient protein purification in miniaturized form (microplates) has to be available.
- Bioconjugation should be regiospecific and essentially quantitative.

Following the original Whitesides publication, and prior to the application of directed evolution, many research groups focused in the 1980s and 1990s on devising other types of artificial metalloenzymes as catalysts in synthetic organic chemistry [6, 7, 30]. As detailed in review articles stressing various approaches, this intriguing research area is still inspiring chemists and biotechnologists today.

Alternative concepts for non-covalent bioconjugation with the introduction of transition metal catalysts into host proteins have been developed, which in principle can also be tuned by directed evolution [5b, c, 6]. For example, serum albumins binding non-covalently water-soluble sulfonylated Fe(III)- and Mn(III)-corroles are active catalysts in H_2O_2-mediated asymmetric sulfoxidation of prochiral thio-ethers (up to 74% ee) [31]. This work was based on the earlier finding that the sodium salts of di-, tri-, and tetra-sulfonic acid derivatives of porphyrins, phthalocyanines, and corroles bind strongly to serum albumins in a non-covalent manner [32].

Covalent bonding to the metal of porphyrin-type or structurally related transition metal complexes can also be envisioned in appropriate protein hosts, possibly supported by additional non-covalent interactions [1]. One of several possibilities is the Watanabe system, in which an achiral chromium(III) Schiff base catalyst is anchored to apo-myoglobin [33]. This artificial metalloenzyme was used as the catalyst in the asymmetric sulfoxidation of thioanisole (13% ee). Again, this is a case for directed evolution, especially if structural information could be obtained by X-ray crystallography [33]. Yet another potential case for directed evolution pertains to the Lu-system in which apo-myoblobin is used again but contains a designed covalent anchor between a manganese Schiff base complex and two cysteines introduced by site-directed mutagenesis [34]. This artificial metalloenzyme proved to be more active and enantioselective in the sulfoxidation of thioanisole (52% ee).

Covalent bioconjugation of proteins by way of C—S bond formation at a cysteine residue for a variety of purposes has been practiced for decades by S_N2 alkylation or Michael addition [1]. This traditional technique was applied a number of times in order to anchor achiral ligands/metal moieties (or organocatalysts) in protein hosts [5b, c, 6]. When aiming for this kind of artificial metalloenzyme, it is best to choose a host which is thermally robust and capable of tolerating the reaction conditions used in traditional aqueous transition metal catalysis [35]. An excellent protein host for this kind of endeavor is an enzyme called tHisF from the thermophilic organism *Thermotoga maritima*, which is involved in the biosynthesis of histidine [36]. It can be heated in the aqueous phase at 75 °C for one hour without any sign of denaturing. Moreover, expression in *Escherichia coli* is unusually efficient, and its X-ray structure has been analyzed [37]. tHisF has a barrel-like structure with one cysteine located fairly deep in the channel (Cys9). In order to enable smooth site-specific covalent anchoring, a double mutant Cys9Ala/Asp11Cys was created, with the reactive cysteine needed in bioconjugation being just below the rim of the barrel-like structure [6]. Appropriate Michael additions and S_N2 reactions were applied in order to anchor ligands, ligand/metal entities, and even organocatalysts (Scheme 7.4) [6]. A platform for potential directed evolution was developed, which includes miniaturization and parallelization of fermentation and mutant enzyme purification by simple heat treatment of 24-format deep-well plates.

This means that all foreign proteins are denatured and precipitated from solution, leaving behind tHisF for clean and selective bioconjugation. It would be interesting to apply directed evolution using a variety of different transition metal catalyzed transformations.

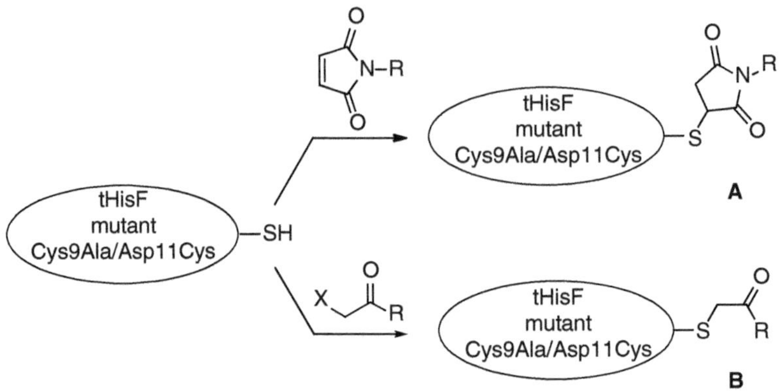

Scheme 7.4 Chemical modification of tHisF mutant Cys9Ala/Asp11Cys by means of Michael additions that lead to bioconjugates **A**, and S_N2-reactions that provide bioconjugates **B**. Source: Ref. [6]/John Wiley & Sons.

Developing artificial metalloenzymes by designing transition metal binding sites using His/Cys or other motifs inspired by natural proteins with the help of site-directed mutagenesis (Scheme 7.2) has a number of advantages, because it does not require a bioconjugation step. Synthetic ligands are not involved. One example of this approach uses the thermostable tHisF as the protein, in which a Cu(II) binding site was implemented by designing mutant Asp11/His50/His52 [38]. The apo-form was treated with $CuSO_4$, leading to the formation of the desired artificial metalloenzyme. It was characterized by electron paramagnetic resonance (EPR) spectroscopy, proving that two histidines are indeed involved in metal complexation. The catalyst was used in the same asymmetric Diels–Alder reaction as shown in Scheme 7.1, leading to an enantioselectivity of 46% ee and an endo/exo ratio of 13 : 1 [38]. Directed evolution of the catalyst in this asymmetric transformation or in other Cu(II)-catalyzed reactions has not been attempted to date.

As summarized in a review article [39], notable examples of computational design of artificial metalloenzymes and/or promiscuous enzymes in general based on the Rosetta algorithms have continued to appear in the literature. The multi-step computational procedure utilizes, inter alia, ab initio protein structure prediction, quantum mechanical (QM) energy refinement, and sequence design. Examples of the Rosetta-approach are designed proteins that catalyze the Kemp elimination, the Diels–Alder cycloaddition, and the Baylis–Hillman reaction, all metal-free promiscuous transformations. In the case of the Diels–Alder reaction shown in Scheme 7.5, the endo-(3*S*,4*R*)-configuration in adduct **7** was the desired outcome as stipulated by design [40]. From a collection of 84 computed and then *E. coli* expressed enzymes, two showed Diels–Alder activity, one of them leading to the predicted stereoselectivity. In the case of the Baylis–Hillman reaction, the results appear to be less successful, with no information regarding the enantioselectivity of the asymmetric transformation being reported [41]. Directed evolution was not attempted in this case, in contrast to the project focusing on the Kemp elimination (see Section 7.2).

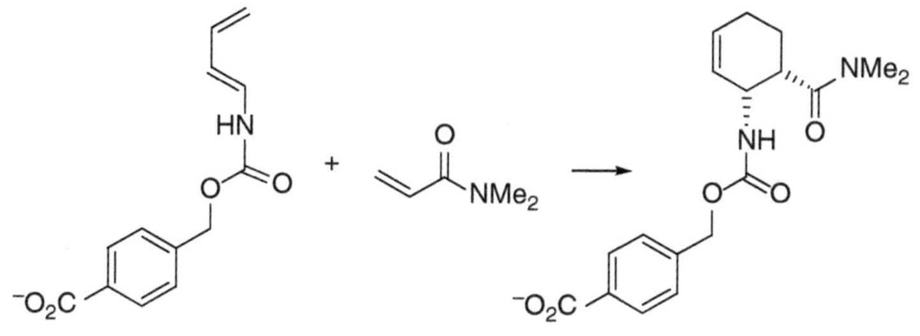

5 6 7

Scheme 7.5 Model Diels–Alder cycloaddition used in Rosetta-design. Source: Adapted from Ref. [40].

On a different note, promiscuous enzymes in unculturable bacterial communities promise to be a rich source of new biocatalysts, but their discovery by functional metagenomics has proven to be problematic. In order to solve this problem, the technique of microfluidic picoliter oil-in-water droplets (Chapter 2) has been applied [42]. Million-membered metagenome libraries can be assessed in this way. This advancement connects enzymes with distantly related sequences. It was found that most of the hits could not have been identified or predicted solely on the basis of sequence data. In yet another advance concerning promiscuous metalloenzymes, the Zn-dependent carbonic anhydrase kCAII was used to catalyze the highly enantioselective reduction of prochiral ketones, with organosilanes serving as the hydride source [43]. Protein engineering was not needed (99% ee, (*S*)-product), but that would be necessary if the (*R*)-configurated alcohols were the desired products.

7.2 Applying Protein Engineering for Tuning the Catalytic Profile of Promiscuous Enzymes

In some studies, the term "promiscuous behavior" was used when engineering an expanded substrate scope, which may be a misleading designation. Today, the majority of researchers who perform directed evolution for the purpose of manipulating substrate acceptance generally do not use this terminology. The same applies to the present monograph.

Saturation mutagenesis and Iterative Saturation Mutagenesis (ISM) at residues lining the active site as part of the Combinatorial Active-Site Saturation Test

(CAST), as described in Chapters 3 and 4, are reliable tools for reshaping the binding pocket of enzymes in the quest to manipulate activity, stereo- and/or regio-selectivity. This was not anticipated when the concept of directed evolution of artificial metalloenzymes (hybrid catalysts) was first proposed [29]. At the time, Error-prone polymerase chain reaction (epPCR) and DNA shuffling were used for tuning the stereoselectivity of enzymes by directed evolution [44]. The first example of directed evolution of an artificial metalloenzyme as a catalyst in a stereoselective transformation made use of the Whitesides system. In this seminal study, the ester of *N*-acyl acrylic acid was used, not the acid itself, because its reduction product is easier to extract from the aqueous phase (Scheme 7.6) [5].

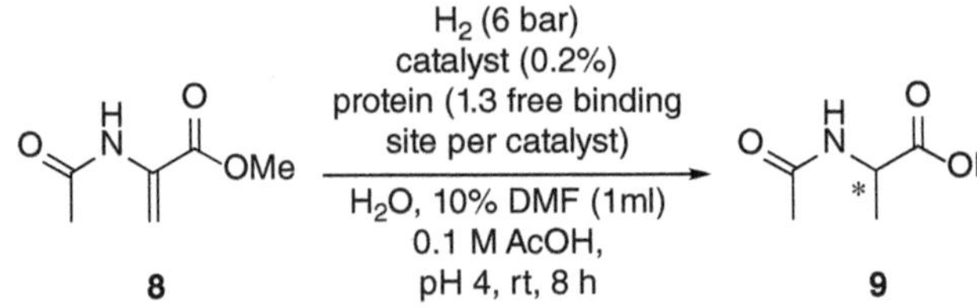

Scheme 7.6 Model reaction used in the directed evolution of the promiscuous Whitesides system. Source: Ref. [5]/Royal Society of Chemistry.

Several problems prevented the rapid realization of the envisioned laboratory evolution. The major difficulty was the fact that all the available expression systems of streptavidin were not very efficient at the time. It was not possible to produce enough protein in the conventional small wells of 96-format micro-titer plates. Therefore, expression, bioconjugation, and reaction were performed in 500 ml Erlenmeyer flasks. As a consequence, only very small mutant libraries were made and screened [5]. The biotinylated Rh-diphosphine complex was modeled in streptavidin, to identify potential randomization residues for saturation mutagenesis, which revealed several CAST residues about 4–5 Å from the Rh-center (Asn49, Leu110, Ser112, and Leu124) as well as second sphere CAST sites Glu51, Tyr54, Trp79, Asn81, Arg84, Asn85, and His87 (Figure 7.1). WT streptavidin containing the ligand which binds Rh was found to deliver an ee-value of only 23% in favor of (*R*)-**9**.

The outcome of a restricted ISM process provided the results shown in Scheme 7.7. It can be seen that ISM leads stepwise to clearly improved enantioselectivity (65% ee) and that even reversal of enantioselectivity is possible. As noted before [1], this study constitutes proof-of-principle regarding the use of directed evolution as a means to tune the stereoselectivity of promiscuous metalloenzymes, but it is far from being practical [5]. It is likely that today's "normal-sized" saturation mutagenesis libraries amounting to typically 500–2000 transformants would provide much higher stereoselectivity. Other olefinic substrates could also be studied.

As previously mentioned [1], Ward and coworkers optimized the expression system of streptavidin (Sav) and nicely extended the Whitesides system to include

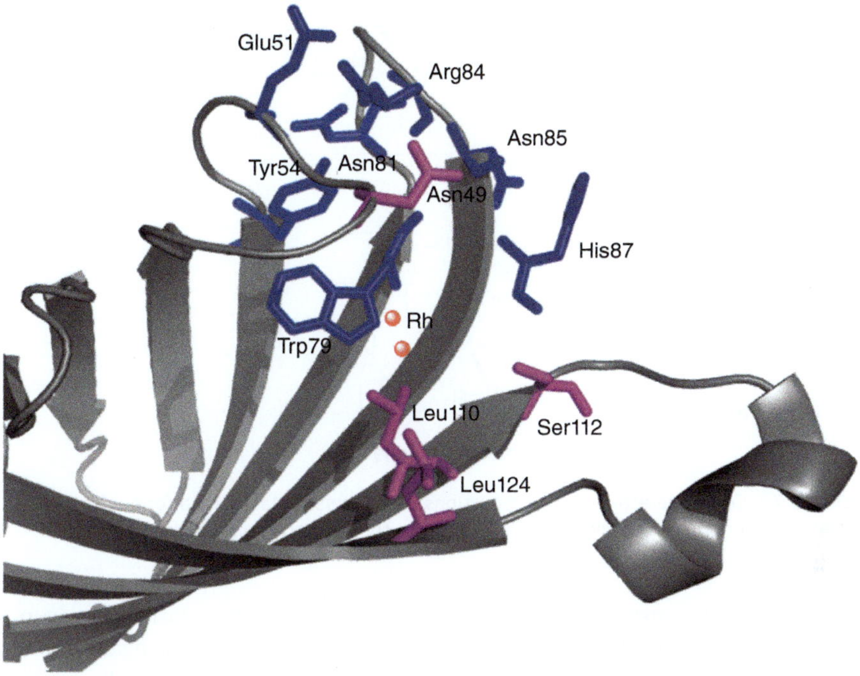

Figure 7.1 Model of the biotinylated diphosphine-Rh-complex in streptavidin obtained by docking. Source: Ref. [5a]/Royal Society of Chemistry.

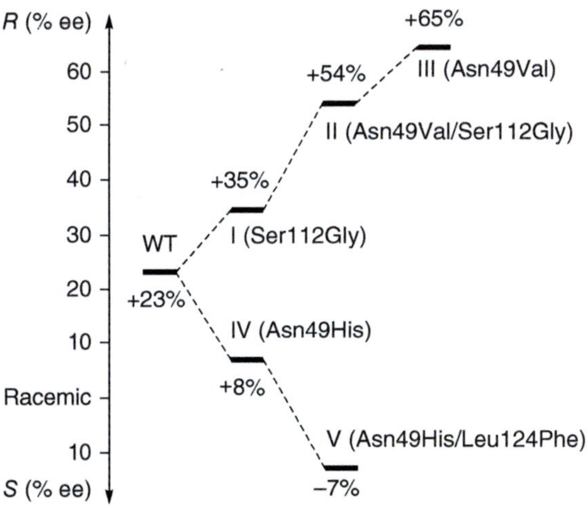

Scheme 7.7 Directed evolution of stereoselectivity of a promiscuous enzyme based on the Whitesides system, Iterative Saturation Mutagenesis (ISM) at first and second sphere CAST residues being employed as the genetic tool in the Rh-catalyzed hydrogenation. Source: Ref. [5]/Royal Society of Chemistry.

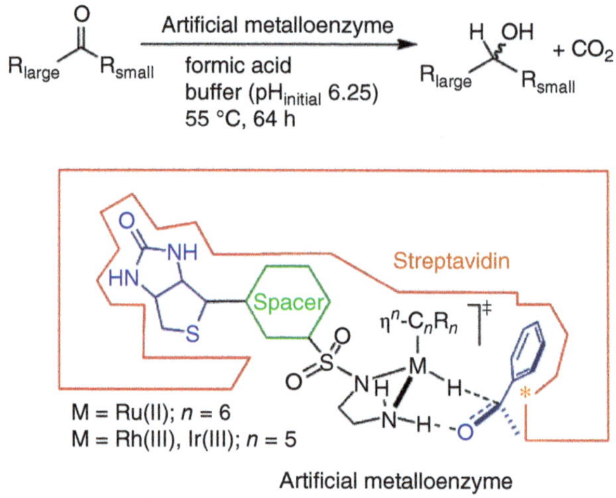

Scheme 7.8 Enantioselective ketone reductions using the streptavidin platform with formic acid serving as reducing equivalents of the chosen transition metals. Source: Ref. [45]/American Chemical Society.

many different transition metal catalyzed stereoselective transformations, generally by optimizing the spacer length between biotin and the ligand/metal entity and applying CAST-type saturation mutagenesis [45]. Up to 98–99% ee was obtained in ketone reductions (Scheme 7.8).

In another study focusing on Noyori-type transfer hydrogenation of prochiral ketones, the initial CAST libraries did not provide sufficiently improved variants, and therefore it was logical to invoke ISM [46] (Figure 7.2). Several achiral Ru-complexes were biotinylated with variation in the spacer length and tested as catalysts in the asymmetric reduction of seven different prochiral ketones. Thanks to ISM, high enantioselectivities have evolved.

Two different piano stool Ru-complexes were employed in the ISM experiments, leading to the structural results summarized in Figure 7.3 [46]. One of the fascinating results of this study is the crystal structure of a mutant, which reveals that the metal (Ru) itself constitutes a center of chirality. Another application of the Whitesides streptavidin system is regioselective Rh(III)-catalyzed CH-activation, which has been successfully enhanced by site-specific mutagenesis employed [47]. Key reviews describing this system with comparisons to other scaffolds have appeared [4, 20c, e, g], including an eye-opening perspective on the possibility of exploiting earth-abundant metals [48].

The streptavidin scaffold played a key role in a fascinating effort to control the formation of nitrogen heterocycles from alkynes via dual gold (Au) catalysis (Scheme 7.9) [49]. Based on docking computations, a limited number of rational enzyme experiments were first carried out, leading to minimal promiscuous hydroaminase activity.

In order to control regioselectivity in terms of Markovnikov/anti-Markovnikov preference and to enhance activity, ISM was again implemented successfully

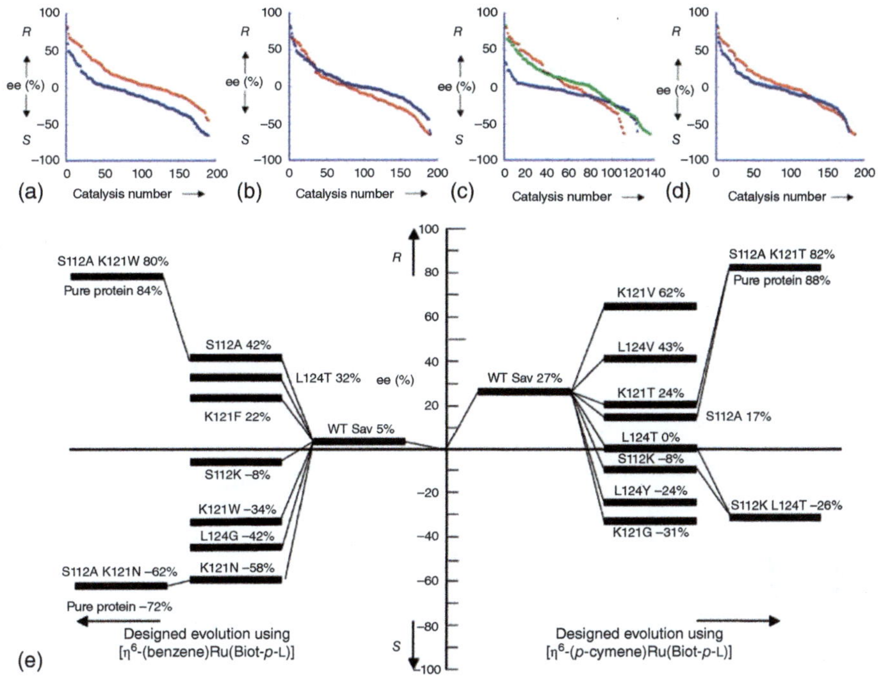

Figure 7.2 Iterative Saturation Mutagenesis (ISM) in the directed evolution of the streptavidin-system for enantioselective ketone reductions via transfer hydrogenation. The nature of the capping η6-arene: benzene (blue) and *p*-cymene (red) (a), the products 1-(4-bromophenyl)ethan-1-ol (red) or 4-phenylbutan-2-ol (blue) (b), and the variants S112A (red), S112K (blue), S112 (green) (c). (d) The positions of K121X (red) or L124X (blue). (e) The evolutionary trajectory of the best *R*- and *S*-selective hybrid catalysts in the conversion of 4-phenyl-2-butanone. Source: Ref. [46]/John Wiley & Sons.

(Scheme 7.10) [49]. A review on further applications of the Whitesides artificial protein scaffold has appeared [20b].

A completely different type of protein scaffold suitable for introducing transition metals with the formation of complexes, which in principle can catalyze many different types of promiscuous transformations, utilizes the multidrug resistance regulator LmrR (Scheme 7.11) [50]. The homodimeric protein has a hydrophobic inner cavity, which can serve as a binding pocket for many potential nonpolar substrates of practical interest.

Following the concept of bioconjugation at cysteine residues of proteins (Scheme 7.2) [30], appropriate cysteine introduction in LmrR was ensured by mutations M89C and N19C [52]. Various bidentate ligands were then covalently attached to either LmrR mutant M89C or N19C, followed by treatment with $Cu(NO_3)_2$, which provided new artificial metalloenzymes. Stereoselective promiscuous transformations such as Friedel–Crafts alkylations in water,

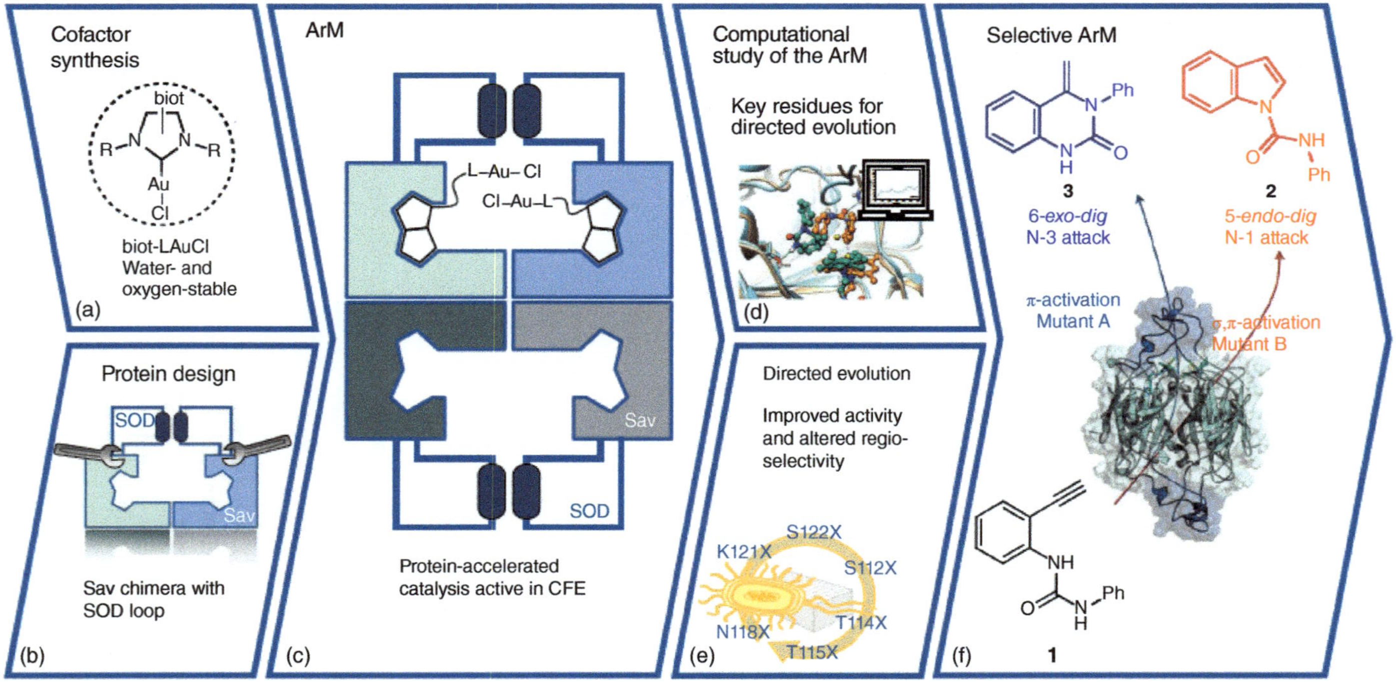

Scheme 7.9 Protein engineering of a HAMase based on dual gold activation of alkynes. Chemo-genetic optimization of the catalytic properties by combining a biotinylated (biot) cofactor (a) and a tailored chimeric protein (b) to assemble an ArM paired with two neighboring gold cofactors (c). Computer-aided structural modeling (d) to identify hot spot amino acids (e) for mutagenesis, with the aim of transforming the alkyne substrate via either σ,π-activation or π-activation (f). Source: Ref. [49]/Springer Nature.

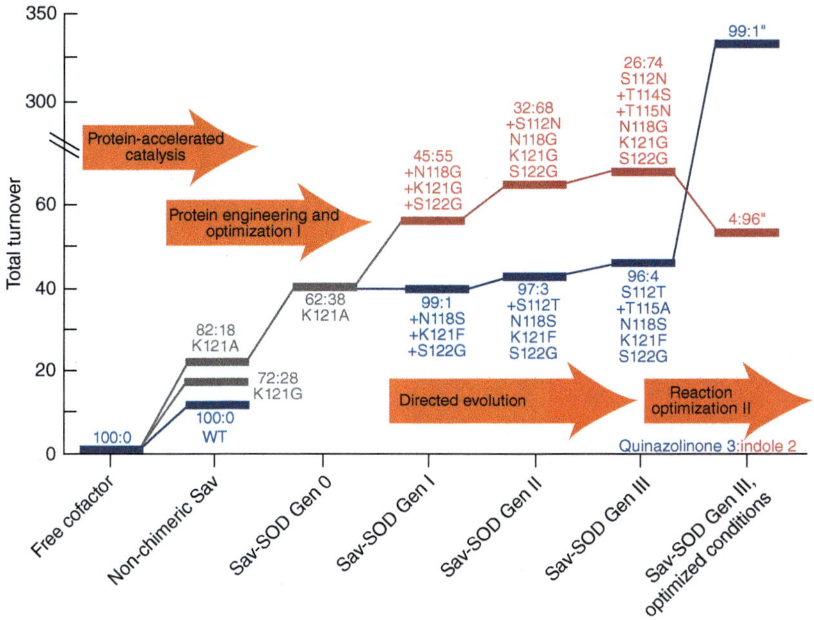

Scheme 7.10 Directed evolution of a HAMase in an iterative fashion. Source: Ref. [49]/Springer Nature.

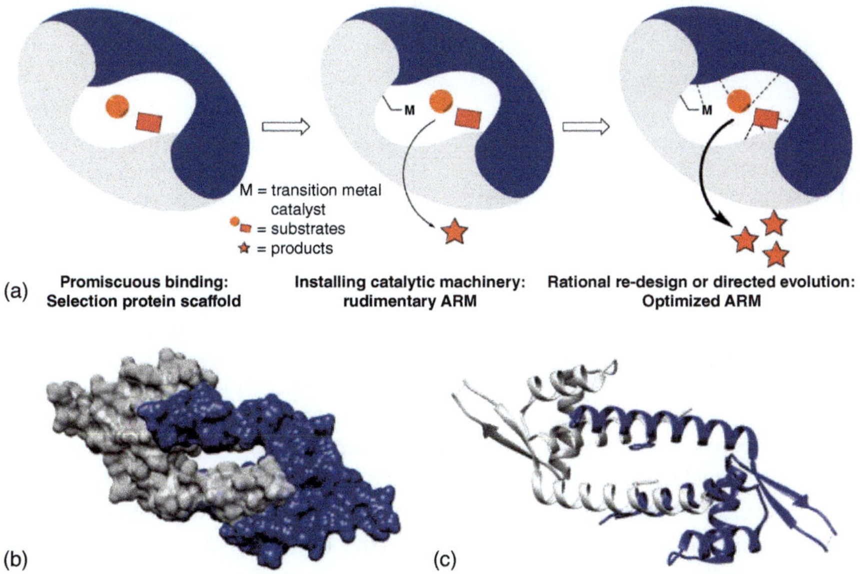

Scheme 7.11 Representation of a unique artificial metalloenzyme (ARM). (a) Design concept; (b) Space-filling representation. Source: Ref. [50]; (c) Ribbon-representation of the X-ray structure of LmrR (PDB: 3F8C). Source: Ref. [51].

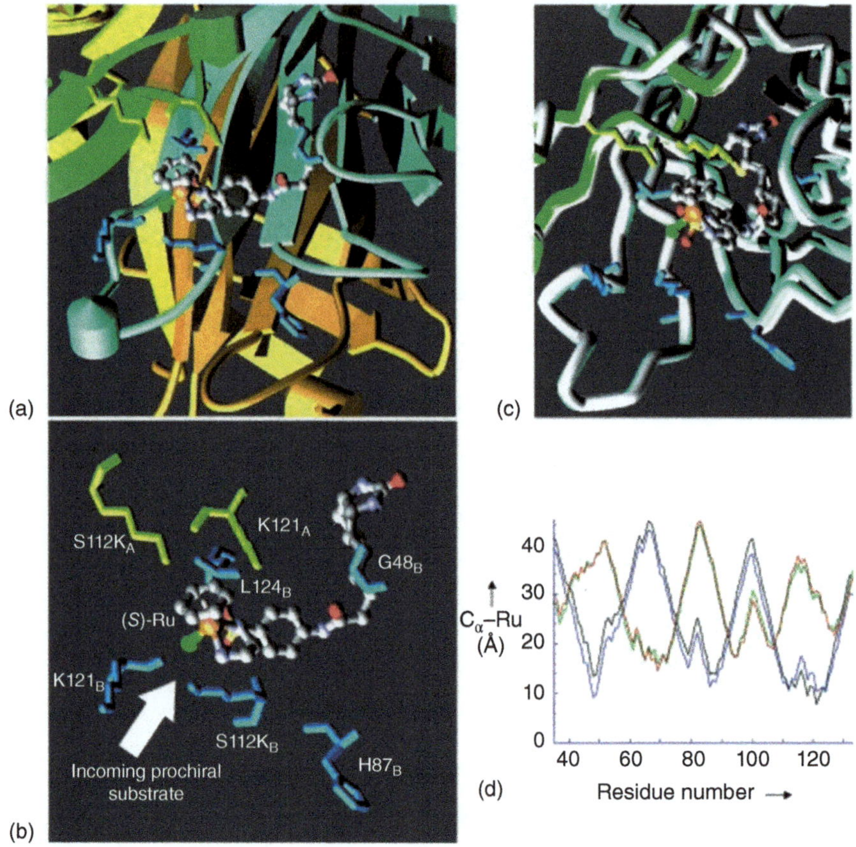

Figure 7.3 X-ray crystal structure of [η^6-(benzene)RuCl(Biot-*p*-L)]⊂S112K Sav. (a) Close-up view (only monomer B (blue) occupied by the biotinylated catalyst (ball-and-stick representation); monomers A (green), C (orange), and D (yellow)). (b) Highlight of amino acid sidechain residues displaying short contacts with Ru. The absolute configuration of ruthenium is S. (c) Superimposition of the structure of [η^6-(benzene)RuCl(Biot-*p*-L)]⊂S112K Sav with the structure of biotin⊂core streptavidin (PDB reference code 1STP, only monomers A and B displayed for clarity; biotin: white stick, core streptavidin: white tube). (d) Ru–Ca distances extracted from the X-ray structure of [η^6-(benzene)RuCl(Biot-*p*-L)]⊂S112K Sav; monomers: A black, B blue, C green, and D red. Source: Ref. [46]/John Wiley & Sons.

enantioselective hydration reactions, and stereoeselective Diels–Alder cycloadditions of azachalcones were realized (Scheme 7.12) [52]. Whenever WT resulted in low stereoselectivity, rational enzyme design and directed evolution were applied. In an interesting variation of the LmrR concept, metal-binding unnatural amino acids were introduced genetically during the biosynthesis of LmrR *in vivo* by using an expanded genetic code [53], followed by treatment with Cu(II) salts and similar transformations as before [54]. The results derived from the initial approach proved to be superior in activity and selectivity. Since

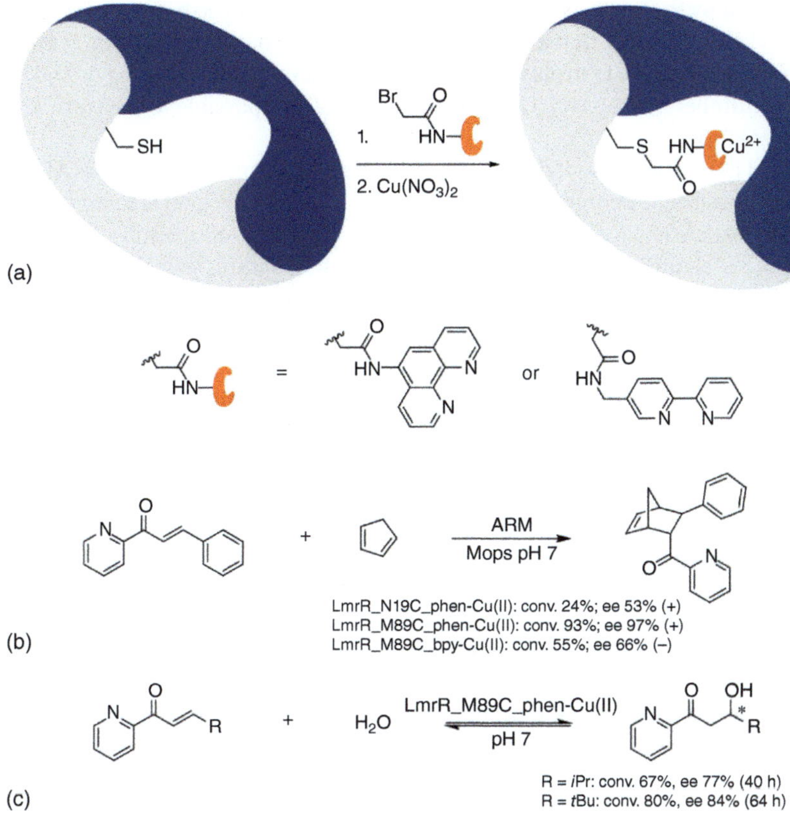

Scheme 7.12 New artificial metalloenzymes based on the LmrR scaffold in which chemical conjugation by reaction of bromoacetamide-functionalized phenanthroline and bipyridine ligands enables bidentate complexation of Cu(II). (a) Chemical bioconjugation procedure; (b) Stereoselective Cu(II)-catalyzed Diels–Alder cycloadditions; (c) Cu(II)-catalyzed enantioselective hydration of azachalcones. Source: Adapted from Ref. [52].

the appearance of the review article [50], further developments and applications of the LmrR concept have been reported [55].

In other work with a focus on metal-free systems, the esterase from *Pseudomonas fluorescens* (PFE) was converted into an epoxide hydrolase [15]. Its mechanism was compared to that of the epoxide hydrolase from *Agrobacterium radiobacter* (EchA), which formed the basis for subsequent directed evolution experiments [15]. The sequence data of esterases and epoxide hydrolases were consulted, as was the structural comparison of six epoxide hydrolases. Several rationally designed mutants were generated, e.g. by switching serine to aspartate or by introducing two tyrosines, but these showed no activity in the hydrolysis of styrene epoxide. Since rational design failed, the whole loop in a PFE mutant containing mutations L29P, F93H, S94D, F125Y, V139Y, and V195Y was replaced by the corresponding element in EchA by a PCR-based procedure. This provided a variant that could be expressed as a soluble protein in *E. coli*, showing low epoxide hydrolase activity (initial activity of 9 mU mg^{-1} and a turnover number of 0.01 s^{-1}). Unfortunately, it was not possible to measure K_m nor V_{max} values.

As listed in Chapter 5, switching cofactor dependency from NADPH to NADH, or reverse, has been accomplished several times. Usually, the goal is to exchange the more expensive NADPH for NADH. In one study concerning cofactor switching, a change in enantioselectivity in sulfoxidation was unexpectedly observed upon introducing mutations by site-directed mutagenesis and saturation mutagenesis, the result being termed "cofactor promiscuity" [56]. The flavoprotein monooxygenase from *Stenotrophomonas maltophilia* (SMFMO) is unusual because it can utilize either the non-phosphorylated cofactor NADH or NADPH for the reduction of the FAD coenzyme, the specificity being 1.5 : 1 in favor of NADH. Following protein engineering, the specificity switched to 1 : 3.5 in favor of NADPH. Seven different prochiral thio-ethers were tested using the evolved mutant, resulting either in a reduction or in a reversal of enantioselectivity. Although practical applications did not result from this study, it shows that switching cofactor specificity can influence streoselectivity in a manner that is currently not well understood [1, 56].

Protein engineering has also been applied successfully in the attempt to increase the activity of a promiscuous enzyme called Kemp eliminase, which has been produced by computational design using the Rosetta algorithm [57]. The Kemp elimination [58] is a synthetically trivial base-catalyzed transformation leading to ring-opened products of benzisoxazoles. Nevertheless, much can be learned from designing protein-mediated catalysts [1]. A review describes the standards set by the various strategies in the race to produce the most active Kemp eliminase, with the specific transformation **10** → **11** serving as the standard model reaction (Scheme 7.13) [59].

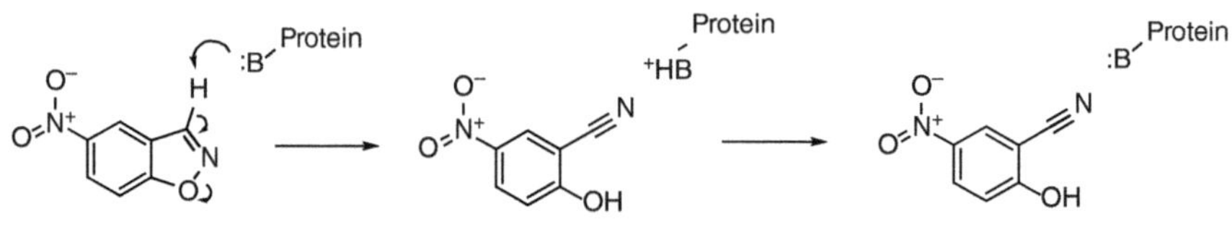

10 **11**

Scheme 7.13 Protein-catalyzed Kemp elimination **10** → **11**.

One of the best original Rosetta design motifs predicted a TIM barrel scaffold of HisF with glutamine serving as the base (Figure 7.4) [57]. Following the detection of very low activity, directed evolution was applied. After seven rounds of epPCR and DNA shuffling, a mutant characterized by 8-point mutations was evolved, showing a 200-fold increase in catalytic efficiency (k_{cat}/K_m = 2600 M^{-1} s^{-1}) [57a, b].

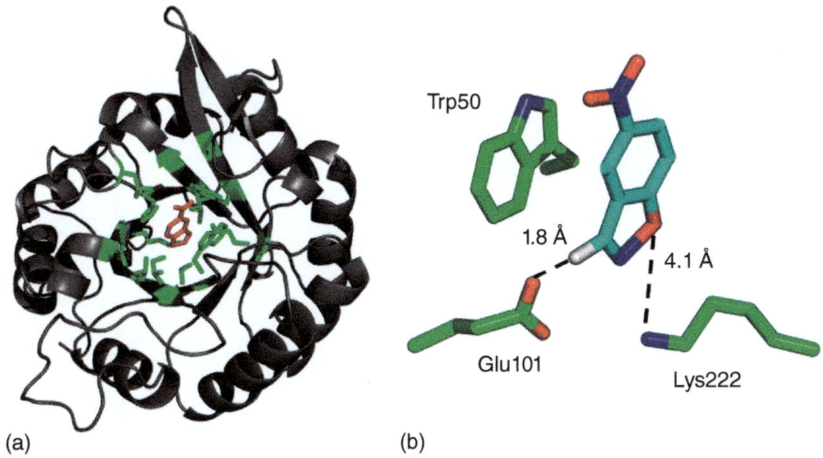

Figure 7.4 (a) The KE07 design, showing the TIM barrel scaffold of HisF (PDB accession code 1THF), the modeled 5-nitrobenzisoxazole substrate (red), and the 13 residues that were replaced to create the designed Kemp eliminase active site (green). (b) Details of the designed KE07's active site. Shown are the 5-nitrobenzisoxazole substrate (cyan), the catalytic base (Glu101), the general acid/H-bond donor (Lys222), and the stacking residue (Trp50). Source: Ref. [57b]/With permission of Elsevier.

Considering the simplicity of Kemp-type elimination, with common bases being highly active in the absence of any protein, the catalytic efficiency achieved by this eliminase is far from optimal despite extensive design and protein engineering. Therefore, a second attempt was undertaken to further improve activity [57c]. One of the earlier designed Kemp eliminases based on the TIM barrel scaffold of the deoxyribose phosphate aldolase of *E. coli*, KE70, was subjected to nine rounds of directed evolution, resulting in the best presently known protein-based catalyst ($k_{cat}/K_m = 5 \times 10^4\ M^{-1}\ s^{-1}$). This extensive experimental effort (Table 7.1) was flanked by computational design, including ensemble generation, ß-strand perturbations, pK_a modulation, and loop redesign. The best mutant was characterized structurally and biochemically, which led to the conclusion that (i) the active-site cavity was reshaped to achieve tighter substrate binding, (ii) the electrostatics around the catalytic dyad His–Asp were optimized, and (iii) the active-site dyad was stabilized in a conformation which is optimal for catalysis [57c]. Thus, these are valuable mechanistic lessons learned as a consequence of such an undertaking, which underscore the primal characteristics of natural enzyme catalyzed reactions as involving precisely positioned substrates as defined by the Jencks hypothesis [60] and electrostatic stabilization of the transition state according to the Warshel hypothesis [61]. The results are indeed impressive, because catalytic efficiency begins to approach that of enzyme-catalyzed natural transformations, which are known to be in the range $k_{cat}/K_m = 10^5–10^8\ M^{-1}\ s^{-1}$. Nevertheless, there is still room for further rate enhancement.

Table 7.1 Summary of the directed evolution of the Kemp eliminase KE70.

Round	Random mutagenesis	Recombination	Mutation spiking by incorporating synthetic oligonucleotides via gene reassembly (ISOR)	Fold improvement measured with crude lysates[a)]
1	2 ± 1 random mutations per gene	—	—	≤6-fold relative to designed KE70; best variant is R1 8/9C
2	—	Shuffling of the 15 best variants from Round 1	—	≤1.5-fold relative to R1 8/9C; best variant is R2 7/12F
3	—	—	Shuffling of the 15 best variants from Round 2, with incorporation of designed mutations: Library 1 (design categories 1a +2) – Met16Ile/Leu/Val/Phe, Leu18Ile/Leu/Val/Phe, Trp72Ser/Cys/His/Leu, Gly101Glu/Gln/Ala/Ser, Ser138Ala, His166Tyr/Asp/Asn/Ala/Ser; Library 2 (design category 4a) – insertions after the residues Thr20 (Gly/Ser), Asn22 (Gly/Ser), Thr171 (Asn/Ala/Pro/Gly/Ser), Val204 (Ala/Pro/Gly/Ser), and Ser239 (Asn/Ala/Pro/Gly/Ser)	≤3-fold relative to R2 7/12F; best variants are R3 2/6D (Library 1) and R3 9/3B (Library 2)
4	—	Shuffling of the 18 best variants from Round 3, both from Libraries 1 and 2	—	≤3-fold relative to R3 9/3B; best variants are R4 4/1B and R4 4/5B
5	—	—	Shuffling of the 12 best variants from Round 4, with incorporation of designed mutations: design category 1b – Ser74Ala/Gly, Phe77Tyr, Leu136Trp, Ala178Ser, Lys173Asn/Thr, Ala231Ser, Ala238Ser, Ser239Thr/Asn/His/Arg; design category 4a – Ala21Asn/Gln/Arg, Asn22Gln/Arg	≤3-fold relative to R4 4/1B; best variant is R5 7/4A

6	—	Shuffling of the 10 best variants from Round 5	—	≤1.2-fold relative to R5 7/4A; best variants are R6 6/10A and R6 4/8B
7	4 best variants from Round 5 and 4 variants from Round 6; 2 ± 1 mutations per gene	—	—	≤1.2-fold relative to R6 6/10A; best variant is R7 7/1C
8	—	—	Shuffling of the 14 best variants of Round 7, with the incorporation of designed mutations: design category 2 – Met16Ala/Val, Leu18Ile/Val/Leu/Phe; design category 4b – Ala238Met, Ser240Gly, Leu241Ala	≤1.2-fold relative to R7 4/2E and 3/2B; best variants are R8 12/12B and R8 15/11E
9	—	—	—	≤1.2-fold relative to R8 12/12B and R8 15/11E

a) The activity improvement measured in crude lysates is not corrected for protein expression and is therefore only a preliminary measure for an increase in protein activity.

Source: Ref. [57c]/With permission of Elsevier.

Following these reports, a considerably shorter approach was undertaken, which also involved directed evolution, but was accompanied by fewer mutagenesis efforts [62]. In this case, a non-catalytic calmodulin scaffold was employed, a fairly small (16 706 Da, 148 amino acid) regulatory binding protein (CaM). In a simple and fast computational design based on the interplay of protein folding and functional group tuning, CaM was converted into a Kemp eliminase called AlleyCat. It was truncated in the process (74 amino acids), only the C-terminal domain of CaM being utilized. It showed (low) activity in the reaction of the standard substrate **10** → **11** (Scheme 7.13). A single point mutation was shown to be essential [62a]. In order to boost activity, directed evolution was subsequently applied [62b]. First, the N-terminal domain was re-introduced in order to enhance protein expression efficiency in *E. coli*. This protein was shown to have the same activity as the originally designed AlleyCat. Then eight positions lining the binding pocket were subjected individually to saturation mutagenesis using NNK codon degeneracy encoding all 20 canonical amino acids. Several variants displayed more than a sixfold increase in enzymatic efficiency. This was followed by one round of DNA shuffling and two subsequent epPCR cycles. The N-terminal domain was then removed, affording the catalyst AlleyCat7. A total of only seven rounds of mutagenesis and screening of less than 6000 transformants were required, leading to a catalytic efficiency of $k_{cat}/K_m = 1283\,M^{-1}\,s^{-1}$ [62b]. The simple introduction of glutamine at the correct position was shown to be responsible for activity. Interestingly, it was shown that the Kemp eliminase AlleyCat 7 can be cycled through the on and off states, which does not result in any activity loss or product inhibition [62b]. This approach shows that a fairly simple strategy can be successful, but the final result in terms of catalytic efficiency does not match the best Kemp eliminases generated by a more elaborate and labor-intensive procedure [57c]. It can be concluded that more research is necessary, which entails less effort but higher catalytic efficiency and additional insights.

An unsurpassed record concerning activity of a Kemp eliminase was set by applying design and extensive directed evolution using a xylanase as scaffold [63a]. Starting with the mutant HG3 designed and evolved previously, 17 further rounds of mutagenesis were performed using epPCR, DNA shuffling, saturation mutagenesis at hot spots identified by epPCR, and ISM. This provided variant HG3-17, showing a catalytic efficiency of $k_{cat}/K_m = 230\,000\,M^{-1}\,s^{-1}$ [63a]. On the basis of X-ray structural analysis, it was concluded that precise positioning of the catalytic machinery with respect to the substrate is essential for approaching typical natural enzyme activity. Attaining the activity characteristic of enzymes catalyzing natural transformations has yet to be achieved [1]. In a very different approach, a redox-mediated P450-BM3 based Kemp eliminase was evolved, but its activity did not reach the previous best record [64]. Nevertheless, upon performing the known mutation C400S at the Fe-binding position, it showed a sixfold improvement in catalytic efficiency relative to WT.

Several other studies describing the use of directed evolution in the quest to increase promiscuous enzyme activity have appeared [65], in addition to the report describing directed evolution for the functional analysis of multi-specific proteins [63b]. Although involving "only" substrate acceptance (which is treated in Chapters 4 and 5), a study describing the directed evolution of a nucleotidyltransferase deserves mention here. In an attempt to expand the nucleotide and sugar 1-phosphate promiscuity of the nucleotidyltransferase RmlA, a combination of epPCR and saturation mutagenesis was employed [65a]. Based on structural data, 22 residues lining the binding pocket (CAST sites) were selected for individual NNK-based randomization. The final variants were shown to accept non-native substrates like pyrimidine and purine-based nucleotides as well as non-native D- and L-sugars with α- and ß-isomers, which has obvious practical implications [65a]. Another example is the engineering of cytochrome c from *Rhodothermus marinus* (RMC). The WT RMC showed low activity toward Si–H carbene insertion (~30 turnovers) with 97% ee [65b]. Inspection of the crystal structure of RMC identified the coordinatively labile residue M100 as the initial hotspot for site-saturation mutagenesis, followed by two rounds of ISM on the vicinal residues V75 and M103. As a result, a triple mutant V75T/M100D/M103E was created, which performed over 2500 turnovers in four hours with >99% ee [65b]. More recently, a prolyl oligopeptidase from *Pyrococcus furiosus* was engineered to perform as an artificial metalloenzyme, which showed good cyclopropanase activity by means of epPCR [65c].

7.3 Applying Protein Engineering to P450 Monooxygenases for Manipulating Activity and Stereoselectivity of Promiscuous Transformations

The topic highlighted here occupies a separate section, focusing first on the works of Frances H. Arnold (see Chapter 1) on a number of intriguing P450 monooxygenase mediated promiscuous transformations [20]. We begin with a seminal study, which appeared more than 30 years ago authored by Dawson, Breslow, and coworkers, who had reported that P450 enzymes show promiscuous reactivity by catalyzing inter- and intra-molecular insertion of nitrenes into nonactivated C—H bonds [66]. The source of the *N*-sulfonyl-nitrenes were compounds of the type $ArS(O)_2N{=}I{-}C_6H_5$, which were transferred onto Fe(II) as metal nitrenes. The respective Fe-bound nitrenes underwent C–H activating amidations (Scheme 7.14), but unfortunately, this report remained unnoticed for several decades. At that time no one thought of studying stereoselectivity, or just as importantly, the Fe-*carbenoid* analogs for a wide range of carbene reactions.

This report set the stage for systematic studies of promiscuous P450 monooxygenases, which in turn provided intriguing opportunities. A different cytochrome P450 monooxygenase was chosen by the Arnold group, specifically P450-BM3, which was genetically modified so that catalysis of several different reaction types became possible, which traditionally belonged to the realm of conventional transition metal catalysis. The first of a series of studies in this area focused on Fe-catalyzed carbene transfer reactions using ethyl diazo acetate as the carbene

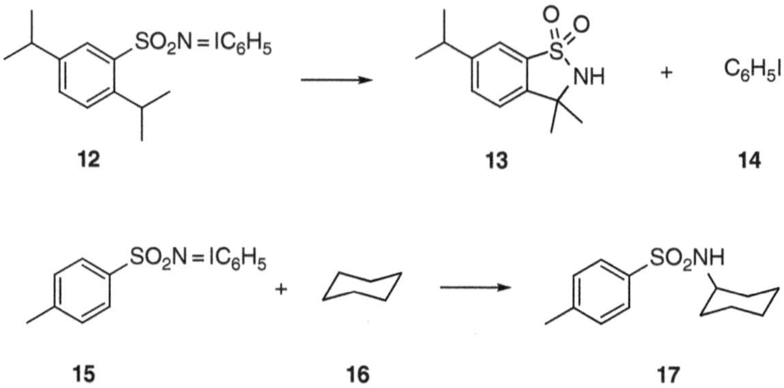

Scheme 7.14 P450-catalyzed insertion of nitrenes into nonactivated C—H bonds. Source: Ref. [66]/American Chemical Society.

source and styrene as the olefinic substrate, with the formation of the respective cyclopropane derivative (Scheme 7.15, top) [67]. Interestingly, the identical transformation, but using a chiral Cu-complex, was the first transition metal catalyzed reaction in synthetic organic chemistry, which opened a new research era [68]!

Prior to the Arnold study, it had been shown that numerous porphyrin-based transition metal complexes are active catalysts in a variety of different transition metal catalyzed reactions in the absence of any proteins [69]. For example, an achiral porphyrin-Fe(II) complex was known to catalyze the cyclopropanation of styrene using ethyl diazo acetate with the formation of all four stereoisomers, provided an oxygen-free system was applied [70]. In the presence of air, Fe(II) is oxidized to Fe(III), which was shown to be inactive. Moreover, in the P450-catalyzed metabolism of 1,3-benzodioxole, the intermediacy of a 1,3-benzodioxole *carbenoid* at the heme-Fe center has already been postulated [71]. In the new approach (Scheme 7.15), air was not excluded, but the reducing agent $Na_2S_2O_4$ was added under aerobic conditions. In spite of some mechanistic uncertainty [72], WT P450-BM3 proved to be active but not very stereoselective, yet this observation was interpreted as an indication of a Fe-carbenoid intermediate. Therefore, a library of P450-BM3 variants, evolved earlier for other purposes, was screened, leading to improved enantio- and diastereo-selectivity. Final optimization was accomplished by saturation mutagenesis at some of the CAST sites in the binding pocket. The final variant showed high diastereo-selectivity (cis:trans = 92 : 8) and high enantioselectivity (97% ee) in favor of the (*S*,*R*)-adduct. Several other styrene-type substrates were also subjected to cyclopropanation, but these reactions occurred with moderate to poor diastereo- and enantio-selectivity [67].

The researchers also chose another strategy by optimizing the axial ligand at heme-Fe, i.e. via the exchange of cysteine by other amino acids, guided by the known electronic influences of ligands in simple porphyrin-Fe catalysis [20]. The amino acids that were found to impart very different promiscuous catalytic

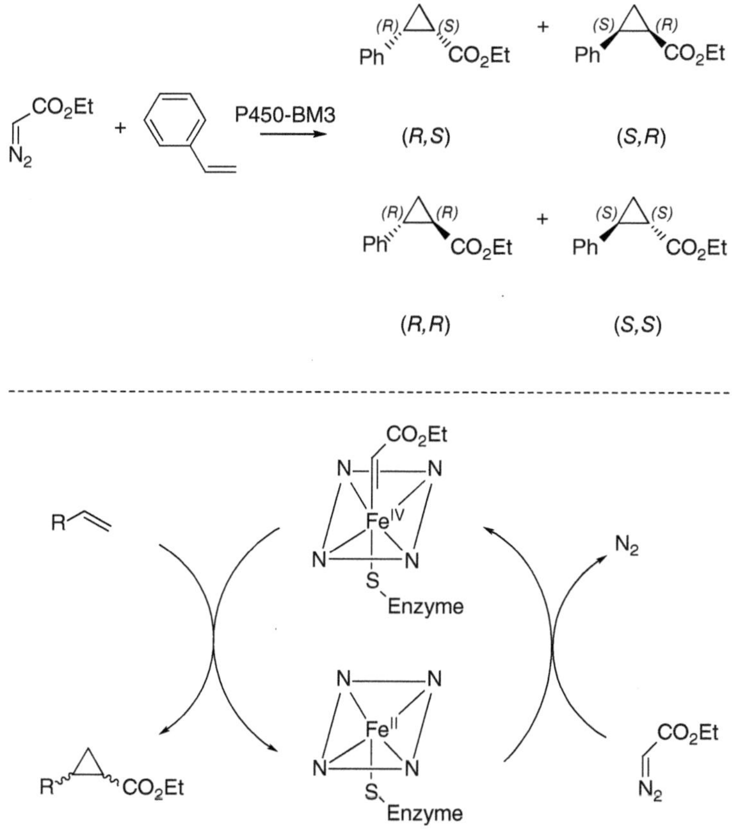

Scheme 7.15 Promiscuous reactivity of P450-BM3. Source: Adapted from Ref. [67].

properties to P450-BM3 are serine or histidine, which is an important discovery. In the cyclopropanation reaction, activity was enhanced dramatically, probably because these mutations allow ready reduction of the heme-Fe to Fe(II) [73]. This also enabled whole cell catalysis of carbene transfer at a high rate, which began to compete with the most active Rh-catalysts. Cyclopropanation was also applied to the formal synthesis of the therapeutic drug levomilnacipran by using a P450-BM3 variant with only five amino acid substitutions [74]. Carbene insertion into aryl N—H bonds using variants of P450-BM3 was also demonstrated [74].

P450-catalyzed carbene additions to terminal and internal acetylenes with the formation of chiral cyclopropenes were reported in further studies, and in one investigation, a second carbene addition was exploited in order to access highly strained bicyclobutane-products, likewise with essentially complete stereoselectivity (Scheme 7.16) [75]. The combination of testing known in-house P450-BM3 mutants, utilizing previous variants for other purposes, rational introduction of point mutations, and especially focused saturation mutagenesis based on the

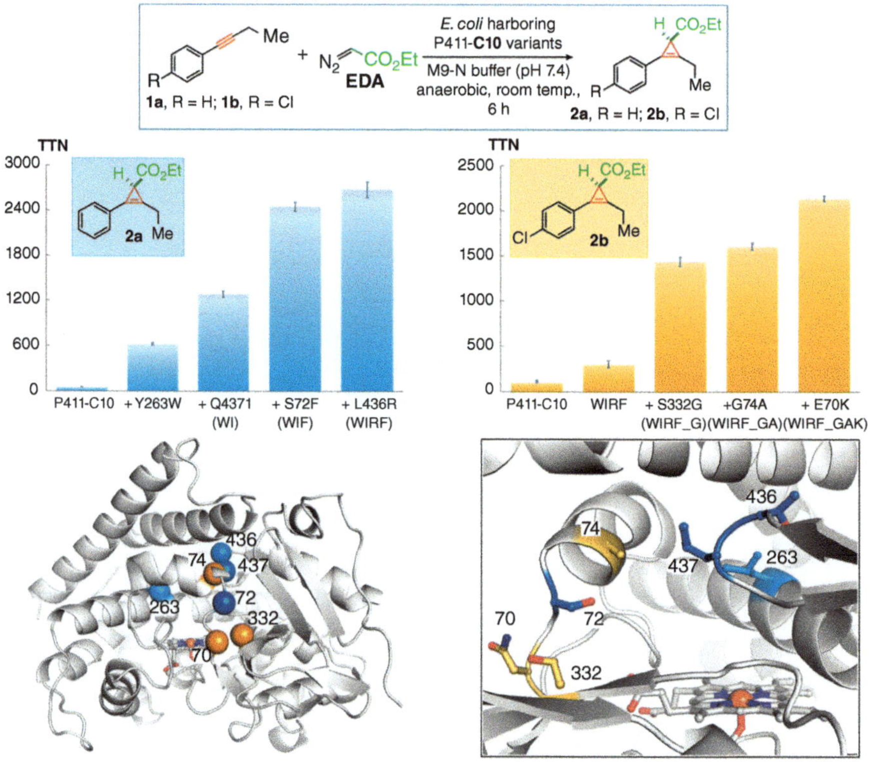

Scheme 7.16 Protein engineering of P411-C10 in the synthesis of internal cyclopropene. Source: Ref. [75a]/American Chemical Society.

22c-trick (see Chapter 3), enabled high selectivity. Such transformations are known reactions in synthetic organic chemistry utilizing man-made transition metal complexes, but generally with lower substrate scope and selectivity.

Inspired by the already mentioned early report on P450-catalyzed nitrene insertions via intermediate Fe-nitrenoids (Scheme 7.15) [66], it appeared logical to test the particular P450-BM3 system for nitrene transfer reactions [76]. The Arnold group hoped that C–H activating amidation would occur rather than aziridine formation, which would be analogous to the Fe-carbenoid mediated cyclopropanation reactions. Using sulfonyl azides as the nitrene precursors, this was indeed observed. Upon optimizing the structure of the aryl group in the nitrene precursor and testing a variety of P450-BM3 mutants as catalysts, smooth intramolecular amidation was observed with enantioselectivities up to 87–89% ee [76]. The point mutations Thr268Ala and Cys400Ser were shown to be particularly important, the latter occurring as the axial ligand at heme-Fe. In a follow-up study, saturation mutagenesis was performed again using the 22c-trick for saturation mutagenesis in order to switch CH-activating amidation from the α- to the less activated ß-position with the formation of six-membered heterocycles [77]. It should be mentioned that the Arnold group also published a startling paper on CH-amidation enabling the stereoselective synthesis of lactams, which was later retracted [78]. Bond strength, inductive effects, steric accessibility, and ring strain were originally invoked, which shows that theoretical analyses can be completely offline. Recently, Arnold has published several reviews of the general area of biocatalytic carbenoid and nitrenoid reactions [79].

Other groups have also utilized P450-BM3 or other heme-dependent proteins in the quest to study promiscuous biocatalytic transformations. For example, different P450-BM3 mutants, originally evolved by systematic saturation mutagenesis at sites lining the binding pocket of P450-BM3 as part of extensive CASTing, provided even better results when opted for α-activating amidation [80]. In a study directed toward switching the regioselectivity of intramolecular C–H activating amidation, six-membered instead of the usual five-membered cyclic sulfonamides were observed as products using a P450-BM3 variant characterized by a dozen point mutations [77].

The Fe-heme-dependent protein myoglobin has also been exploited as a scaffold for promiscuous transformations using the natural Fe but also other transition metals [81]. The WT myoglobin from whale sperm was chosen, which showed 86% diastereoselectivity favoring the *E*-diastereomer (trans configuration), but as a racemate. Relying on the known crystal structure of this myoglobin, residues F43, H64, and V68 "above" the heme-plane as well as H93 "below" it were identified, all four sites serving as hotspots at which rational enzyme design (see Chapter 3) was considered in the model cyclopropanation of styrene with ethyl diazo acetate (Scheme 7.17). The structural data allowed speculation as to possible amino acid exchanges, and a small set of small and larger amino acids were introduced at the hotspots. Some single mutants showed improvements in enantioselectivity, which led the researchers to form various combinatorial double mutants, especially H64V/V68A, leading to complete trans-preference in the model reaction. The myoglobin of whale sperm is a small

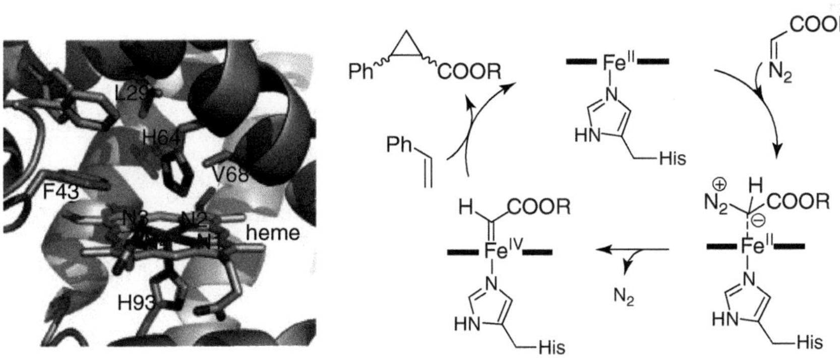

(a) (b)

Scheme 7.17 (a) Site-directed mutagenesis of sperm whale myoglobin on the key residues F43, H64, and V68. (b) Proposed mechanism of styrene cyclopropanation catalyzed by myoglobin. Source: Ref. [81a]/John Wiley & Sons.

protein (17 kDa), which as a scaffold for promiscuous transformations has some advantages over P450-BM3.

Particularly impressive results were reported by replacing the natural Fe in heme-dependent proteins by other metals such as iridium (Ir), which provided highly active and selective promiscuous biocatalysts (Scheme 7.18) [82]. Methods for exchanging Fe in cofactors with other transition metals were a strategy that had been used for a long time, mainly for mechanistic studies [83]. The concept was first applied using myoglobin [82a] and in rapid succession to the thermostable CYP119 from *Sulfolobus solataricus* [84], which proved to be the better scaffold [82b]. Based on rationally chosen mutations for a limited directed evolution campaign comprising about 200 variants, extraordinarily active and selective mutants were evolved for several different promiscuous transformations (Scheme 7.18) [82b]. Earlier, residues T213 and T214 had

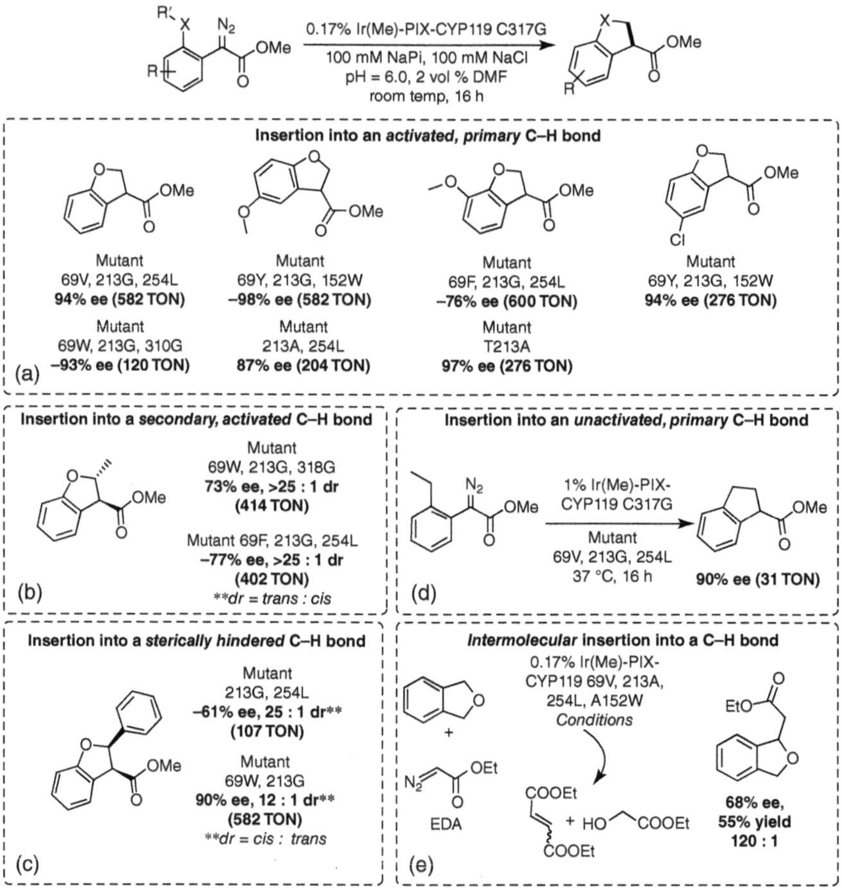

Scheme 7.18 Intra- and inter-molecular promiscuous reactions of mutants of Ir(Me)-PIX CYP119. Me = methyl; PIX = acronym for promiscuous metal porphyrins. Source: [82b]/American Association for the Advancement of Science.

already been identified as hotspots, which regulate the active site structure, spin state, and natural activity [84]. In the study of Ir-CYP119, steric and electronic factors based on structural information were considered, including amino acid exchange T213G, leading to activities that match those of natural enzymes!

In summary, the concept of exchanging the natural Fe with other transition metals in porphyrins, especially via introduction of Ir, opens a new chapter in the field of promiscuous biocatalytic reactions. Thus far, the results appear to be superior those obtained using Fe-based heme proteins. Recently, several exciting new advancements were reported [85]. A noteworthy example is the use of a combination of semi-rational mutagenesis and Focused Rational Iterative Site-specific Mutagenesis (FRISM) [85a]. In this particular application, Ir(Me)-CYP119 was mutationally manipulated in order to enforce a drastic enhancement of activity (Scheme 7.19). Steps toward increasing the

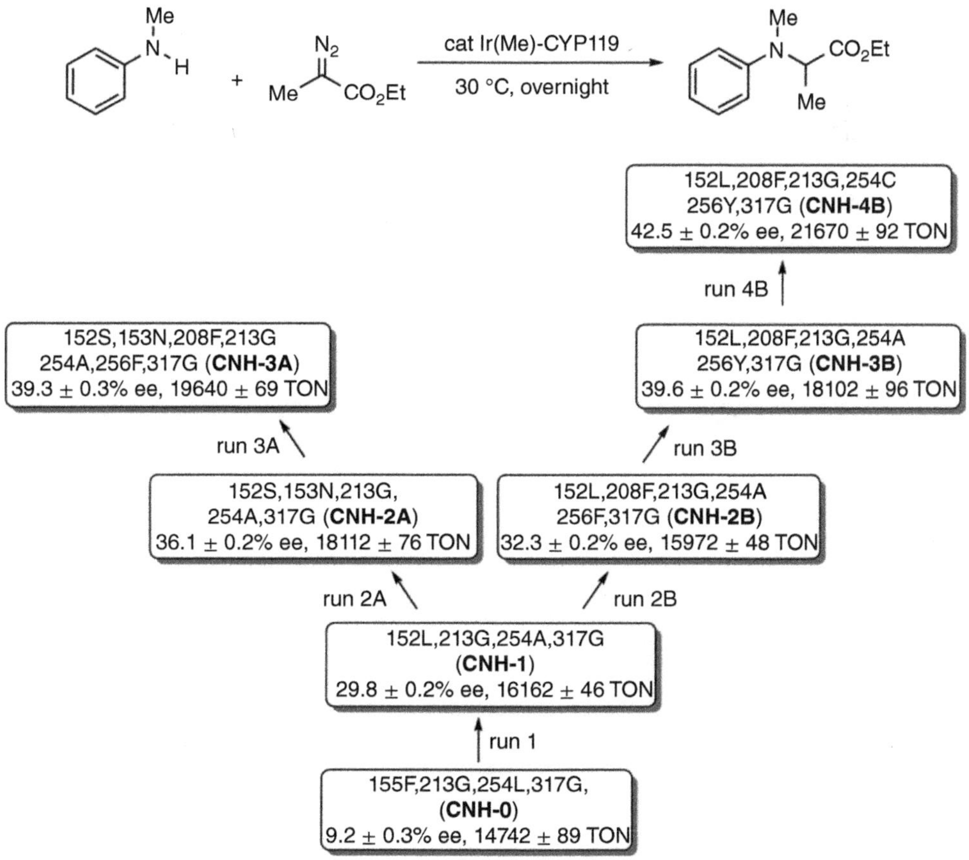

Scheme 7.19 Rational enzyme design based on a combination of semi-rational design and Focused Rational Iterative Site-specific Mutagenesis (FRISM) for enhancing activity of Ir(Me)-CYP119 as the catalyst for carbene insertions into N—H bonds in whole cells. Source: Ref. [85a]/John Wiley & Sons.

enantioselectivity were also undertaken. In another noteworthy contribution, the unnatural biosynthesis of terpenoids was enabled by the application of ISM [85c]. Further studies pertain to cooperative asymmetric reactions by combining photocatalytic or chemocatalytic cis/trans olefin-isomerization and ene-reductases [86]. These are examples of combining chemo-and biocatalysis in one system [87].

Along a different line, artificial metalloenzymes were devised for olefin metathesis in water [88]. In one study, a Grubbs–Hoveyda type Ru-catalyst was anchored covalently according to Scheme 7.4**A** to the ß-barrel structured protein nitrobindin, this artificial metalloenzyme mediating olefin metathesis in aqueous medium at an unusually high catalytic rate (Figure 7.5) [88a]. Some mutational effects were observed. In a second study, noncanonical amino acids employing the expanded genetic code were incorporated site-specifically into the ribosomal lasso peptide capistruin, which enabled the anchoring of the Grubbs–Hoveyda Ru-catalyst [88b]. The field of protein-based olefin metathesis transformations has since been reviewed [89]. These systems display olefin metathesis activity, but thus far they have not solved the problem of cis/trans selectivity in a general way. For example, the exceedingly challenging problem of cis-selectivity when employing different kinds of olefinic starting materials remains unsolved, in contrast to novel developments of protein-free catalysts reported recently by the Grubbs group [90].

Finally, a unique approach to the construction of promiscuous metalloenzymes has been proposed based on the utilization of designed short self-assembling (amyloid-type) peptides that bind hemin [91]. Such biocatalysts were first applied in cyclopropanation reactions, resulting in modest enantioselectivity (Scheme 7.20). However, protein engineering can now be applied. What also speaks for this platform is the possibility to design the same peptides with amino acids of opposite absolute configuration, which leads to opposite enantiomers in product formation. Moreover, unnatural amino acids accessible by the expanded genetic code can also be introduced in the designed amyloid-like peptides.

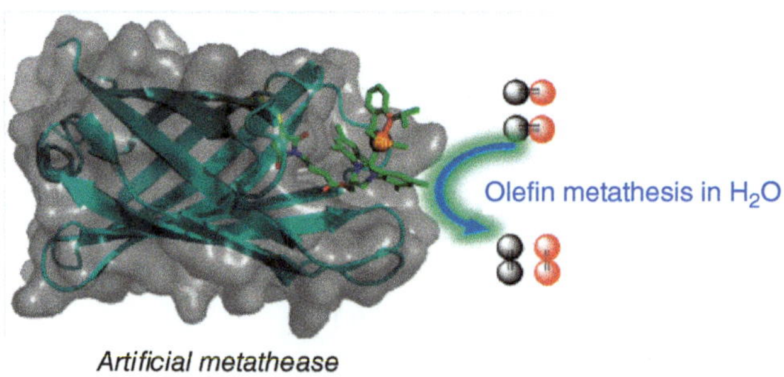

Figure 7.5 An artificial olefin metathease based on anchoring a Grubbs–Hoveyda Ru-catalyst covalently to the protein nitrobindin. Source: Ref. [88a]/American Chemical Society.

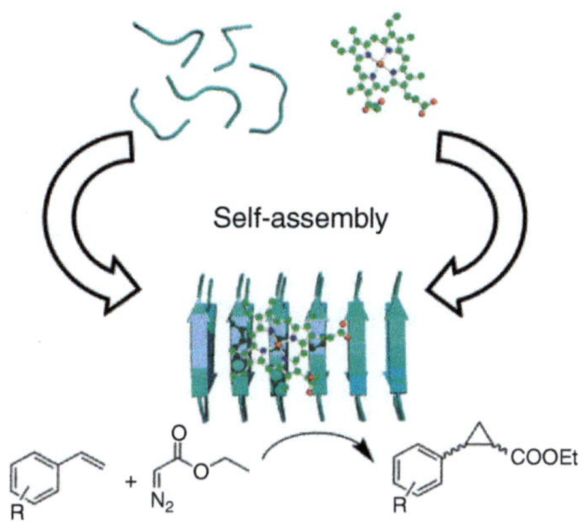

Scheme 7.20 Representation of hemin association with catalytic self-assembling short peptides of the amyloid-type, making cyclopropanation reactions possible. Source: Ref. [91]/John Wiley & Sons.

7.4 Conclusions and Perspectives

Without any doubt, the initial reports on the directed evolution of artificial metalloenzymes opened the door to a new research field [30]. Discovering promiscuous enzymes in nature or creating such biocatalysts by designing artificial metalloenzymes continues to this day [1]. However, more basic and applied research is necessary to prove that these studies are not provisional hype. In most of the cases studied thus far, activity was found to be low. As shown in Section 7.2, the application of directed evolution provides the potential to remedy the situation. Indeed, examples have been reported in which notable enhancement of activity was accomplished, sometimes accompanied by excellent stereo- and/or regio-selectivity. On a (self)-critical note, it needs to be stated that such cases are the exception [88a]. Moreover, many more improvements are necessary before real (industrial) applications become a reality in any of the approaches delineated in this chapter. This pertains especially to artificial metalloenzymes, because these biocatalysts have to be compared to the growing number of efficient man-made transition metal catalysts characterized by much lower molecular weights and higher activities [1]. As an example, directed evolution-based mutants of a biotinylated Rh-diphosphine complex anchored to streptavidin led to enhanced enantioselectivity at each evolutionary stage, but activity was not improved (Scheme 7.7) [5]. Thus, in the area of artificial metalloenzymes, the primary challenge revolves around the basic question of how to easily enhance the intrinsic activity of a given transition metal. In addition, the following challenges and questions should be considered in this field:

- How to make the metal–ligand binding site beyond the inner cavity, inter alia, on the surface.

- How to balance the energetics when protein folding with artificial ligand in order to keep the desired coordination geometry.
- How to balance the stabilization and conformational dynamics to ensure the catalytic promiscuity, when designing and/or engineering promiscuous enzymes.
- How to choose an appropriate scaffold as a starting point, since not all protein scaffolds are evolvable.
- Why is it sometimes necessary to add a reducing agent, e.g. $Na_2S_2O_4$, to the reaction system, and in similar cases not?

References

1. Reetz, M.T. (2016). *Directed Evolution of Selective Enzymes Catalysts for Organic Chemistry and Biotechnology*. Weinheim: Wiley-VCH.
2. Reetz, M.T. and Jiao, N. (2006). Copper–phthalocyanine conjugates of serum albumins as enantioselective catalysts in Diels–Alder reactions. *Angew. Chem. Int. Ed.* **45** (15): 2416–2419.
3. Zunszain, P.A., Ghuman, J., Komatsu, T. et al. (2003). Crystal structural analysis of human serum albumin complexed with hemin and fatty acid. *BMC Struct. Biol.* **3** (1): 1–9.
4. Reetz, M.T. (2019). Directed evolution of artificial metalloenzymes: a universal means to tune the selectivity of transition metal catalysts? *Acc. Chem. Res.* **52** (2): 336–344.
5. (a) Reetz, M.T., Peyralans, J.J.P., Maichele, A. et al. (2006). Directed evolution of hybrid enzymes: evolving enantioselectivity of an achiral Rh-complex anchored to a protein. *Chem. Commun.* **41**: 4318–4320. (b) Reetz, M.T. (2012). Artificial metalloenzymes as catalysts in stereoselective Diels–Alder reactions. *Chem. Rec.* **12** (4): 391–406. (c) Reetz, M.T. (2009). *Bio-inspired Catalysts*, vol. **25** (ed. T. Ward), 63–92. Springer Berlin Heidelberg.
6. Reetz, M.T., Rentzsch, M., Pletsch, A. et al. (2008). A robust protein host for anchoring chelating ligands and organocatalysts. *ChemBioChem* **9** (4): 552–564.
7. (a) Ueno, T., Abe, S., Yokoi, N., and Watanabe, Y. (2007). Coordination design of artificial metalloproteins utilizing protein vacant space. *Coord. Chem. Rev.* **251** (21–24): 2717–2731. (b) Lu, Y. (2005). Design and engineering of metalloproteins containing unnatural amino acids or non-native metal-containing cofactors. *Curr. Opin. Chem. Biol.* **9** (2): 118–126. (c) Köhler, V., Wilson, Y.M., Dürrenberger, M. et al. (2013). Synthetic cascades are enabled by combining biocatalysts with artificial metalloenzymes. *Nat. Chem.* **5** (2): 93–99. (d) Bos, J. and Roelfes, G. (2014). Artificial metalloenzymes for enantioselective catalysis. *Curr. Opin. Chem. Biol.* **19**: 135–143. (e) Petrik, I.D., Liu, J., and Lu, Y. (2014). Metalloenzyme design and engineering through strategic modifications of native protein scaffolds. *Curr. Opin. Chem. Biol.* **19**: 67–75.
8. (a) O'Brien, P.J. and Herschlag, D. (1999). Catalytic promiscuity and the evolution of new enzymatic activities. *Chem. Biol.* **6** (4): R91–R105. (b) Copley, S.D. (2003). Enzymes with extra talents: moonlighting functions and catalytic promiscuity. *Curr. Opin. Biol.* **7** (2): 265–272. (c) Khersonsky, O., Roodveldt, C., and Tawfik, D.S. (2006). Enzyme promiscuity: evolutionary and mechanistic aspects. *Curr. Opin. Chem. Biol.* **10** (5): 498–508.
9. Jensen, R.A. (1976). Enzyme recruitment in evolution of new function. *Annu. Rev. Microbiol.* **30** (1): 409–425.
10. Amar, D., Berger, I., Amara, N. et al. (2012). The transition of human estrogen sulfotransferase from generalist

to specialist using directed enzyme evolution. *J. Mol. Biol.* **416** (1): 21–32.

11. Benner, S.A. (2002). The past as the key to the present: resurrection of ancient proteins from eosinophils. *Proc. Natl. Acad. Sci. U.S.A.* **99** (8): 4760–4761.
12. Bar-Rogovsky, H., Stern, A., Penn, O. et al. (2015). Assessing the prediction fidelity of ancestral reconstruction by a library approach. *Protein Eng. Des. Sel.* **28** (11): 507–518.
13. (a) Tawfik, D.S. (2010). Enzyme promiscuity: a mechanistic and evolutionary perspective. *Annu. Rev. Biochem.* **79** (1): 471–505. (b) Nobeli, I., Favia, A.D., and Thornton, J.M. (2009). Protein promiscuity and its implications for biotechnology. *Nat. Biotechnol.* **27** (2): 157–167. (c) Risso, V.A., Gavira, J.A., Mejia-Carmona, D.F. et al. (2013). Hyperstability and substrate promiscuity in laboratory resurrections of precambrian β-lactamases. *J. Am. Chem. Soc.* **135** (8): 2899–2902. (d) Babtie, A., Tokuriki, N., and Hollfelder, F. (2010). What makes an enzyme promiscuous? *Curr. Opin. Chem. Biol.* **14** (2): 200–207. (e) Atkins, W.M. (2015). Biological messiness vs. biological genius: mechanistic aspects and roles of protein promiscuity. *J. Steroid Biochem. Mol. Biol.* **151**: 3–11. (f) Merkl, R. and Sterner, R. (2016). Ancestral protein reconstruction: techniques and applications. *Biol. Chem.* **397** (1): 1–21. (g) Trudeau, D.L. and Tawfik, D.S. (2019). Protein engineers turned evolutionists—the quest for the optimal starting point. *Curr. Opin. Struct. Biol.* **60**: 46–52. (h) Spence, M.A., Kaczmarski, J.A., Saunders, J.W., and Jackson, C.J. (2021). Ancestral sequence reconstruction for protein engineers. *Curr. Opin. Struct. Biol.* **69**: 131–141.
14. (a) Ward, O.P. and Singh, A. (2000). Enzymatic asymmetric synthesis by decarboxylases. *Curr. Opin. Biotechnol.* **11** (6): 520–526. (b) Qin, Z., Zhang, X., Sang, X. et al. (2022). Carboxylic acid reductases enable intramolecular lactamization reactions. *Green Syn. Catal.* https://doi.org/10.1016/j.gresc.2022.05.009.
15. Jochens, H., Stiba, K., Savile, C. et al. (2009). Converting an esterase into an epoxide hydrolase. *Angew. Chem. Int. Ed.* **48** (19): 3532–3535.
16. (a) Roodveldt, C. and Tawfik, D.S. (2005). Directed evolution of phosphotriesterase from *Pseudomonas diminuta* for heterologous expression in *Escherichia coli* results in stabilization of the metal-free state. *Protein Eng. Des. Sel.* **18** (1): 51–58. (b) Tokuriki, N., Jackson, C.J., Afriat-Jurnou, L. et al. (2012). Diminishing returns and tradeoffs constrain the laboratory optimization of an enzyme. *Nat. Commun.* **3**: 1257.
17. (a) Pandya, C., Farelli, J.D., Dunaway-Mariano, D., and Allen, K.N. (2014). Enzyme promiscuity: engine of evolutionary innovation. *J. Biol. Chem.* **289** (44): 30229–30236. (b) Hiblot, J., Gotthard, G., Elias, M., and Chabriere, E. (2013). Differential active site loop conformations mediate promiscuous activities in the lactonase SsoPox. *PLoS One* **8** (9): e75272. (c) Brizendine, A.M., Odokonyero, D., McMillan, A.W. et al. (2014). Promiscuity of *Exiguobacterium* sp. At1b *o*-succinylbenzoate synthase illustrates evolutionary transitions in the OSBS family. *Biochem. Biophys. Res. Commun.* **450** (1): 679–684. (d) Noda-García, L., Juárez-Vázquez, A.L., Ávila-Arcos, M.C. et al. (2015). Insights into the evolution of enzyme substrate promiscuity after the discovery of (βα)8 isomerase evolutionary intermediates from a diverse metagenome. *BMC Evol. Biol.* **15** (1): 1–14.
18. (a) Bornscheuer, U.T. and Kazlauskas, R.J. (2004). Catalytic promiscuity in biocatalysis: using old enzymes to form new bonds and follow new pathways. *Angew. Chem. Int. Ed.* **43** (45): 6032–6040. (b) Berglund, P. and Park, S. (2005). Strategies for altering enzyme reaction specificity for applied biocatalysis. *Curr. Org. Chem.* **9**: 325–336. (c) Hult, K. and Berglund, P. (2007). Enzyme promiscuity: mechanism and applications. *Trends Biotechnol.* **25** (5): 231–238.
19. Toscano, M.D., Woycechowsky, K.J., and Hilvert, D. (2007). Minimalist active-site

redesign: teaching old enzymes new tricks. *Angew. Chem. Int. Ed.* **46** (18): 3212–3236.

20. Selected recent reviews of artificial metalloenzymes:(a) Arnold, F.H. (2019). Innovation by evolution: bringing new chemistry to life (nobel lecture). *Angew. Chem. Int. Ed.* **58** (41): 14420–14426. (b) Liang, A.D., Serrano-Plana, J., Peterson, R.L., and Ward, T.R. (2019). Artificial metalloenzymes based on the biotin–streptavidin technology: enzymatic cascades and directed evolution. *Acc. Chem. Res.* **52** (3): 585–595. (c) Chen, K. and Arnold, F.H. (2020). Engineering new catalytic activities in enzymes. *Nat. Catal.* **3**: 203–213. (d) Ren, X. and Fasan, R. (2021). Engineered and artificial metalloenzymes for selective C–H functionalization. *Curr. Opin. Green Sustainable Chem.* **31**: 100494. (e) Lewis, J.C. (2019). Beyond the second coordination sphere: engineering dirhodium artificial metalloenzymes to enable protein control of transition metal catalysis. *Acc. Chem. Res.* **52** (3): 576–584. (f) Yu, Y., Liu, X., and Wang, J. (2019). Expansion of redox chemistry in designer metalloenzymes. *Acc. Chem. Res.* **52** (3): 557–565. (g) Dong, J., Liu, Y., and Cui, Y. (2021). Artificial metal-peptide assemblies: bioinspired assembly of peptides and metals through space and across length scales. *J. Am. Chem. Soc.* **143** (42): 17316–17336. (h) Wittwer, M., Markel, U., Schiffels, J. et al. (2021). Engineering and emerging applications of artificial metalloenzymes with whole cells. *Nat. Catal.* **4**: 814–827. (i) Leveson-Gower, R.B., Mayer, C., and Roelfes, G. (2019). The importance of catalytic promiscuity for enzyme design and evolution. *Nat. Rev. Chem.* **3**: 687–705. (j) Chalkley, M.J., Mann, S.I., and DeGrado, W.F. (2022). De novo metalloprotein design. *Nat. Rev. Chem.* **6**: 31–50. (k) Ebensperger, P. and Jessen-Trefzer, C. (2022). Artificial metalloenzymes in a nutshell: the quartet for efficient catalysis. *Biol. Chem.* **403** (4): 403–412. (l) Bloomer, B.J., Clark, D.S., and Hartwig, J.F. (2022). Progress, challenges, and opportunities with artificial metalloenzymes in biosynthesis. *Biochemistry* https://doi.org/10.1021/acs.biochem.1c00829. (m) Maity, B., Taher, M., Mazumdar, S., and Ueno, T. (2022). Artificial metalloenzymes based on protein assembly. *Coord. Chem. Rev.* **469**: 214593.

21. (a) Branneby, C., Carlqvist, P., Magnusson, A. et al. (2003). Carbon–carbon bonds by hydrolytic enzymes. *J. Am. Chem. Soc.* **125** (4): 874–875. (b) Li, C., Feng, X.-W., Wang, N. et al. (2008). Biocatalytic promiscuity: the first lipase-catalysed asymmetric aldol reaction. *Green Chem.* **10** (6): 616–618. (c) Torre, O., Gotor-Fernández, V., Alfonso, I. et al. (2005). Study of the chemoselectivity in the aminolysis reaction of methyl acrylate catalysed by lipase B from *Candida Antarctica*. *Adv. Synth. Catal.* **347** (7–8): 1007–1014. (d) Cai, Y., Wu, Q., Xiao, Y.-M. et al. (2006). Hydrolase-catalyzed Michael addition of imidazoles to acrylic monomers in organic medium. *J. Biotechnol.* **121** (3): 330–337. (e) Svedendahl, M., Hult, K., and Berglund, P. (2005). Fast carbon–carbon bond formation by a promiscuous lipase. *J. Am. Chem. Soc.* **127** (51): 17988–17989. (f) Qian, C., Xu, J.-M., Wu, Q. et al. (2007). Promiscuous acylase-catalyzed aza-Michael additions of aromatic N-heterocycles in organic solvent. *Tetrahedron Lett.* **48** (35): 6100–6104. (g) Strohmeier, G.A., Sović, T., Steinkellner, G. et al. (2009). Investigation of lipase-catalyzed Michael-type carbon–carbon bond formations. *Tetrahedron* **65** (29–30): 5663–5668. (h) Jiang, L., Wang, B., Li, R.-R. et al. (2014). Catalytic promiscuity of *Escherichia coli* BioH esterase: application in the synthesis of 3,4-dihydropyran derivatives. *Process Biochem.* **49** (7): 1135–1138. (i) Steffen-Munsberg, F., Vickers, C., Thontowi, A. et al. (2013). Revealing the structural basis of promiscuous amine transaminase activity. *ChemCatChem* **5**: 154–157. (j) Zhang, N., Sun, Z., and Wu, C. (2022). Artificial enzymes combining proteins with proline polymers

for asymmetric aldol reactions in water. *ACS Catal.* **12** (8): 4777–4783.

22. Svedendahl, M., Carlqvist, P., Branneby, C. et al. (2008). Direct epoxidation in *Candida antarctica* lipase B studied by experiment and theory. *ChemBioChem* **9** (15): 2443–2451.
23. Li, K., He, T., Li, C. et al. (2009). Lipase-catalysed direct Mannich reaction in water: utilization of biocatalytic promiscuity for C–C bond formation in a "one-pot" synthesis. *Green Chem.* **11** (6): 777–779.
24. Lou, F.-W., Liu, B.-K., Wu, Q. et al. (2008). *Candida antarctica* lipase B (Cal-B)-catalyzed carbon–sulfur bond addition and controllable selectivity in organic media. *Adv. Synth. Catal.* **350** (13): 1959–1962.
25. (a) Reetz, M.T., Mondière, R., and Carballeira, J.D. (2007). Enzyme promiscuity: first protein-catalyzed Morita–Baylis–Hillman reaction. *Tetrahedron Lett.* **48** (10): 1679–1681. (b) Jiang, L. and Yu, H.-W. (2014). An example of enzymatic promiscuity: the Baylis–Hillman reaction catalyzed by a biotin esterase (BIOH) from *Escherichia coli*. *Biotechnol. Lett.* **36** (1): 99–103.
26. Humble, M.S. and Berglund, P. (2011). Biocatalytic promiscuity. *Eur. J. Org. Chem.* **2011** (19): 3391–3401.
27. Wilson, M.E. and Whitesides, G.M. (1978). Conversion of a protein to a homogeneous asymmetric hydrogenation catalyst by site-specific modification with a diphosphine-rhodium(I) moiety. *J. Am. Chem. Soc.* **100** (1): 306–307.
28. (a) Thomas, C.M. and Ward, T.R. (2005). Artificial metalloenzymes: proteins as hosts for enantioselective catalysis. *Chem. Soc. Rev.* **34** (4): 337–346. (b) Dürrenberger, M. and Ward, T.R. (2014). Recent achievements in the design and engineering of artificial metalloenzymes. *Curr. Opin. Chem. Biol.* **19**: 99–106. (c) Dundas, C., Demonte, D., and Park, S. (2013). Streptavidin–biotin technology: improvements and innovations in chemical and biological applications. *Appl. Microbiol. Biotechnol.* **97** (21): 9343–9353.
29. Early papers describing for the first time the concept of directed evolution of hybrid catalysts, i.e., artificial metalloenzymes for stereoselective transformations: (a) Reetz, M.T. (2001). Optimization of Synthetic Catalysts by Means of Directed Evolution. Patent WO 92/18645. (b) Reetz, M.T., Rentzsch, M., Pletsch, A., and Maywald, M. (2002). Towards the directed evolution of hybrid catalysts. *Chimia* **56** (12): 721–723. (c) Reetz, M.T. (2002). Directed evolution of selective enzymes and hybrid catalysts. *Tetrahedron* **58** (32): 6595–6602.
30. Prior studies of artificial metalloenzymes and other hybrid catalysts without any protein engineering:(a) Qi, D., Tann, C.-M., Haring, D., and Distefano, M.D. (2001). Generation of new enzymes via covalent modification of existing proteins. *Chem. Rev.* **101** (10): 3081–3112. (b) Polgar, L. and Bender, M.L. (1966). A new enzyme containing a synthetically formed active site. Thiol-subtilisin1. *J. Am. Chem. Soc.* **88** (13): 3153–3154. (c) Schultz, P.G. (1988). The interplay between chemistry and biology in the design of enzymatic catalysts. *Science* **240**: 426–433. (d) Khumtaveeporn, K., DeSantis, G., and Jones, J.B. (1999). Expanded structural and stereospecificity in peptide synthesis with chemically modified mutants of subtilisin. *Tetrahedron: Asymmetry* **10** (13): 2563–2572. (e) Smith, H.B. and Hartman, F.C. (1988). Restoration of activity to catalytically deficient mutants of ribulosebisphosphate carboxylase/oxygenase by aminoethylation. *J. Biol. Chem.* **263** (10): 4921–4925. (f) Nicholas, K.M., Wentworth, P., Harwig, C.W. et al. (2002). A cofactor approach to copper-dependent catalytic antibodies. *Proc. Natl. Acad. Sci. U.S.A.* **99** (5): 2648–2653. (g) Hamachi, I. and Shinkai, S. (1999). Chemical modification of the structures and functions of proteins by the cofactor reconstitution method. *Eur. J. Org. Chem.* **1999** (3): 539–549. (h) Lu, Y. and Valentine, J.S. (1997). Engineering metal-binding sites in proteins. *Curr. Opin. Struct. Biol.* **7**

(4): 495–500. (i) Lu, Y., Berry, S.M., and Pfister, T.D. (2001). Engineering novel metalloproteins: design of metal-binding sites into native protein scaffolds. *Chem. Rev.* **101** (10): 3047–3080. (j) Kaiser, E.T. (1988). Catalytic activity of enzymes altered at their active sites. *Angew. Chem. Int. Ed. Engl.* **27** (7): 913–922. (k) Choma, C.T., Lear, J.D., Nelson, M.J. et al. (1994). Design of a heme-binding four-helix bundle. *J. Am. Chem. Soc.* **116** (3): 856–865.

31. Mahammed, A. and Gross, Z. (2005). Albumin-conjugated corrole metal complexes: extremely simple yet very efficient biomimetic oxidation systems. *J. Am. Chem. Soc.* **127** (9): 2883–2887.
32. Mahammed, A., Gray, H.B., Weaver, J.J. et al. (2004). Amphiphilic corroles bind tightly to human serum albumin. *Bioconjugate Chem.* **15** (4): 738–746.
33. Ohashi, M., Koshiyama, T., Ueno, T. et al. (2003). Preparation of artificial metalloenzymes by insertion of chromium(III) Schiff base complexes into apomyoglobin mutants. *Angew. Chem. Int. Ed.* **42** (9): 1005–1008.
34. (a) Carey, J.R., Ma, S.K., Pfister, T.D. et al. (2004). A site-selective dual anchoring strategy for artificial metalloprotein design. *J. Am. Chem. Soc.* **126** (35): 10812–10813. (b) Lu, Y., Yeung, N., Sieracki, N., and Marshall, N.M. (2009). Design of functional metalloproteins. *Nature* **460** (7257): 855–862.
35. Cornils, B. and Herrmann, W.A. (2006). *Aqueous-Phase Organometallic Catalysis.* Weinheim: Wiley-VCH.
36. (a) Douangamath, A., Walker, M., Beismann-Driemeyer, S. et al. (2002). Structural evidence for ammonia tunneling across the (beta alpha)(8) barrel of the imidazole glycerol phosphate synthase bienzyme complex. *Structure* **10** (2): 185–193. (b) Beismann-Driemeyer, S. and Sterner, R. (2001). Imidazole glycerol phosphate synthase from *Thermotoga maritima*: quaternary structure, steady-state kinetics, and reaction mechanism of the bienzyme complex. *J. Biol. Chem.* **276** (23): 20387–20396.
37. Lang, D., Thoma, R., Henn-Sax, M. et al. (2000). Structural evidence for evolution of the β/α barrel scaffold by gene duplication and fusion. *Science* **289** (5484): 1546–1550.
38. Podtetenieff, J., Taglieber, A., Bill, E. et al. (2010). An artificial metalloenzyme: creation of a designed copper binding site in a thermostable protein. *Angew. Chem. Int. Ed.* **49** (30): 5151–5155.
39. Kiss, G., Çelebi-Ölçüm, N., Moretti, R. et al. (2013). Computational enzyme design. *Angew. Chem. Int. Ed.* **52** (22): 5700–5725.
40. Siegel, J.B., Zanghellini, A., Lovick, H.M. et al. (2010). Computational design of an enzyme catalyst for a stereoselective bimolecular Diels–Alder reaction. *Science* **329** (5989): 309–313.
41. Bjelic, S., Nivón, L.G., Çelebi-Ölçüm, N. et al. (2013). Computational design of enone-binding proteins with catalytic activity for the Morita–Baylis–Hillman reaction. *ACS Chem. Biol.* **8** (4): 749–757.
42. Colin, P.-Y., Kintses, B., Gielen, F. et al. (2015). Ultrahigh-throughput discovery of promiscuous enzymes by picodroplet functional metagenomics. *Nat. Commun.* **6**: https://doi.org/10.1038/ncomms10008.
43. Ji, P., Park, J., Gu, Y. et al. (2021). Abiotic reduction of ketones with silanes catalysed by carbonic anhydrase through an enzymatic zinc hydride. *Nat. Chem.* **13** (4): 312–318.
44. Reetz, M.T. (2004). Controlling the enantioselectivity of enzymes by directed evolution: practical and theoretical ramifications. *Proc. Natl. Acad. Sci. U.S.A.* **101** (16): 5716–5722.
45. Letondor, C., Pordea, A., Humbert, N. et al. (2006). Artificial transfer hydrogenases based on the biotin–(strept)avidin technology: fine tuning the selectivity by saturation mutagenesis of the host protein. *J. Am. Chem. Soc.* **128** (25): 8320–8328.
46. Creus, M., Pordea, A., Rossel, T. et al. (2008). X-ray structure and designed evolution of an artificial transfer hydrogenase. *Angew. Chem. Int. Ed.* **47** (8): 1400–1404.
47. Hyster, T.K., Knörr, L., Ward, T.R., and Rovis, T. (2012). Biotinylated Rh(III) complexes in engineered streptavidin for

accelerated asymmetric C–H activation. *Science* **338** (6106): 500–503.

48. Bullock, R.M., Chen, J.G., Gagliardi, L. et al. (2020). Using nature's blueprint to expand catalysis with earth-abundant metals. *Science* **369** (6505): eabc3183.
49. Christoffel, F., Igareta, N.V., Pellizzoni, M.M. et al. (2021). Design and evolution of chimeric streptavidin for protein-enabled dual gold catalysis. *Nat. Catal.* **4**: 643–653.
50. Roelfes, G. (2019). LmrR: a privileged scaffold for artificial metalloenzymes. *Acc. Chem. Res.* **52** (3): 545–556.
51. Madoori, P.K., Agustiandari, H., Driessen, A.J., and Thunnissen, A.M. (2009). Structure of the transcriptional regulator LmrR and its mechanism of multidrug recognition. *EMBO J.* **28** (2): 156–166.
52. Bos, J., García-Herraiz, A., and Roelfes, G. (2013). An enantioselective artificial metallo-hydratase. *Chem. Sci.* **4**: 3578–3582.
53. Xie, J., Liu, W., and Schultz, P.G. (2007). A genetically encoded bidentate, metal-binding amino acid. *Angew. Chem. Int. Ed.* **46** (48): 9239–9242.
54. (a) Drienovská, I., Rioz-Martínez, A., Draksharapu, A., and Roelfes, G. (2015). Novel artificial metalloenzymes by *in vivo* incorporation of metal-binding unnatural amino acids. *Chem. Sci.* **6** (1): 770–776. (b) Bos, J., Browne, W.R., Driessen, A.J., and Roelfes, G. (2015). Supramolecular assembly of artificial metalloenzymes based on the dimeric protein LmrR as promiscuous scaffold. *J. Am. Chem. Soc.* **137** (31): 9796–9799.
55. (a) Chordia, S., Narasimhan, S., Lucini Paioni, A. et al. (2021). In vivo assembly of artificial metalloenzymes and application in whole-cell biocatalysis. *Angew. Chem. Int. Ed.* **60** (11): 5913–5920. (b) Leveson-Gower, R.B., Zhou, Z., Drienovská, I., and Roelfes, G. (2021). Unlocking iminium catalysis in artificial enzymes to create a Friedel–Crafts alkylase. *ACS Catal.* **11** (12): 6763–6770. (c) Zhou, Z. and Roelfes, G. (2021). Synergistic catalysis of tandem Michael addition/enantioselective protonation reactions by an artificial enzyme. *ACS Catal.* **11** (15): 9366–9369. (d) Roelfes, G. (2021). Repurposed and artificial heme enzymes for cyclopropanation reactions. *J. Inorg. Biochem.* **222**: 111523.
56. Jensen, C.N., Ali, S.T., Allen, M.J., and Grogan, G. (2013). Mutations of an NADP(H)-dependent flavoprotein monooxygenase that influence cofactor promiscuity and enantioselectivity. *FEBS Open Biol.* **3**: 473–478.
57. (a) Rothlisberger, D., Khersonsky, O., Wollacott, A.M. et al. (2008). Kemp elimination catalysts by computational enzyme design. *Nature* **453** (7192): 190–195. (b) Khersonsky, O., Röthlisberger, D., Dym, O. et al. (2010). Evolutionary optimization of computationally designed enzymes: Kemp eliminases of the KE07 series. *J. Mol. Biol.* **396** (4): 1025–1042. (c) Khersonsky, O., Röthlisberger, D., Wollacott, A.M. et al. (2011). Optimization of the in-silico-designed Kemp eliminase KE70 by computational design and directed evolution. *J. Mol. Biol.* **407** (3): 391–412. (d) Kries, H., Bloch, J.S., Bunzel, H.A. et al. (2020). Contribution of oxyanion stabilization to Kemp eliminase efficiency. *ACS Catal.* **10** (8): 4460–4464.
58. Casey, M.L., Kemp, D.S., Paul, K.G., and Cox, D.D. (1973). Physical organic chemistry of benzisoxazoles. I. Mechanism of the base-catalyzed decomposition of benzisoxazoles. *J. Org. Chem.* **38** (13): 2294–2301.
59. Korendovych, I.V. and DeGrado, W.F. (2014). Catalytic efficiency of designed catalytic proteins. *Curr. Opin. Struct. Biol.* **27**: 113–121.
60. Jencks, W.P. (1975). Binding energy, specificity, and enzyme catalysis: the Circe effect. *Adv. Enzymol. Relat. Areas Mol. Biol.* **43**: 219–410.
61. Warshel, A. (1998). Electrostatic origin of the catalytic power of enzymes and the role of preorganized active sites. *J. Biol. Chem.* **273** (42): 27035–27038.
62. (a) Korendovych, I.V., Kulp, D.W., Wu, Y. et al. (2011). Design of a switchable eliminase. *Proc. Natl. Acad. Sci. U.S.A.* **108** (17): 6823–6827. (b) Moroz, O.V., Moroz, Y.S., Wu, Y. et al. (2013).

A single mutation in a regulatory protein produces evolvable allosterically regulated catalyst of nonnatural reaction. *Angew. Chem. Int. Ed.* **52** (24): 6246–6249.

63. (a) Blomberg, R., Kries, H., Pinkas, D.M. et al. (2013). Precision is essential for efficient catalysis in an evolved Kemp eliminase. *Nature* **503**: 418–421. (b) Levin, M., Amar, D., and Aharoni, A. (2013). Employing directed evolution for the functional analysis of multi-specific proteins. *Bioorg. Med. Chem.* **21** (12): 3511–3516.
64. Li, A., Wang, B., Ilie, A. et al. (2017). A redox-mediated Kemp eliminase. *Nat. Commun.* **8**: 14876.
65. (a) Moretti, R., Chang, A., Peltier-Pain, P. et al. (2011). Expanding the nucleotide and sugar 1-phosphate promiscuity of nucleotidyltransferase RmLa via directed evolution. *J. Biol. Chem.* **286** (15): 13235–13243. (b) Kan, S.B.J., Lewis, R.D., Chen, K., and Arnold, F.H. (2016). Directed evolution of cytochrome c for carbon–silicon bond formation: bringing silicon to life. *Science* **354**: 1048–1051. (c) Yang, H., Swartz, A.M., Park, H.J. et al. (2018). Evolving artificial metalloenzymes via random mutagenesis. *Nat. Chem.* **10**: 318–324.
66. Svastits, E.W., Dawson, J.H., Breslow, R., and Gellman, S.H. (1985). Functionalized nitrogen atom transfer catalyzed by cytochrome P-450. *J. Am. Chem. Soc.* **107** (22): 6427–6428.
67. Coelho, P.S., Brustad, E.M., Kannan, A., and Arnold, F.H. (2013). Olefin cyclopropanation via carbene transfer catalyzed by engineered cytochrome P450 enzymes. *Science* **339** (6117): 307–310.
68. (a) Nozaki, H., Moriuti, S., Takaya, H., and Noyori, R. (1966). Asymmetric induction in carbenoid reaction by means of a dissymmetric copper chelate. *Tetrahedron Lett.* **7** (44): 5239–5244. (b) Noyori, R. (2012). Facts are the enemy of truth—reflections on serendipitous discovery and unforeseen developments in asymmetric catalysis. *Angew. Chem. Int. Ed.* **52** (1): 79–92.
69. Kadish, K.M., Smith, K.M., and Guilard, R. (ed.) (2003). *The Porphyrin Handbook*, vol. **1–20**. New York: Academic Press.
70. (a) Wolf, J.R., Hamaker, C.G., Djukic, J.-P. et al. (1995). Shape and stereoselective cyclopropanation of alkenes catalyzed by iron porphyrins. *J. Am. Chem. Soc.* **117** (36): 9194–9199. (b) Maxwell, J.L., O'Malley, S., Brown, K.C., and Kodadek, T. (1992). Shape-selective and asymmetric cyclopropanation of alkenes catalyzed by rhodium porphyrins. *Organometallics* **11**: 645–652.
71. (a) Philpot, R.M. and Hodgson, E. (1972). The effect of piperonyl butoxide concentration on the formation of cytochrome P-450 difference spectra in hepatic microsomes from mice. *Mol. Pharmacol.* **8** (2): 204–214. (b) Mansuy, D. (1980). New iron-porphyrin complexes with metal-carbon bond – biological implications. *Pure Appl. Chem.* **52**: 681–690.
72. Roiban, G.-D. and Reetz, M.T. (2013). Enzyme promiscuity: using a P450 enzyme as a carbene transfer catalyst. *Angew. Chem. Int. Ed.* **52** (21): 5439–5440.
73. Coelho, P.S., Wang, Z.J., Ener, M.E. et al. (2013). A serine-substituted P450 catalyzes highly efficient carbene transfer to olefins in vivo. *Nat. Chem. Biol.* **9** (8): 485–487.
74. Wang, Z.J., Renata, H., Peck, N.E. et al. (2014). Improved cyclopropanation activity of histidine-ligated cytochrome P450 enables enantioselective formal synthesis of levomilnacipran. *Angew. Chem. Int. Ed. Engl.* **53** (26): 6810–6813.
75. (a) Chen, K. and Arnold, F.H. (2020). Engineering cytochrome P450s for enantioselective cyclopropenation of internal alkynes. *J. Am. Chem. Soc.* **142** (15): 6891–6895. (b) Chen, K., Huang, X., Kan, S.B.J. et al. (2018). Enzymatic construction of highly strained carbocycles. *Science* **360** (6384): 71–75.
76. McIntosh, J.A., Coelho, P.S., Farwell, C.C. et al. (2013). Enantioselective intramolecular C–H amination catalyzed by engineered cytochrome P450

enzymes in vivo and in vitro. *Angew. Chem. Int. Ed.* **52** (35): 9309–9312.

77. Hyster, T.K., Farwell, C.C., Buller, A.R. et al. (2014). Enzyme-controlled nitrogen-atom transfer enables regiodivergent C–H amination. *J. Am. Chem. Soc.* **136** (44): 15505–15508.

78. (a) Cho, I., Jia, Z.J., and Arnold, F.H. (2019). Site-selective enzymatic C–H amidation for synthesis of diverse lactams. *Science* **364** (6440): 575–578. (b) Cho, I., Jia, Z.J., and Arnold, F.H. (2020). Retraction. *Science* **367** (6474): 155.

79. (a) Liu, Z. and Arnold, F.H. (2021). New-to-nature chemistry from old protein machinery: carbene and nitrene transferases. *Curr. Opin. Biotechnol.* **69**: 43–51.

80. Singh, R., Bordeaux, M., and Fasan, R. (2014). P450-catalyzed intramolecular sp^3 C–H amination with arylsulfonyl azide substrates. *ACS Catal.* **4** (2): 546–552.

81. (a) Bordeaux, M., Tyagi, V., and Fasan, R. (2015). Highly diastereoselective and enantioselective olefin cyclopropanation using engineered myoglobin-based catalysts. *Angew. Chem. Int. Ed.* **54** (6): 1744–1748. (b) Bordeaux, M., Singh, R., and Fasan, R. (2014). Intramolecular C (sp^3)–H amination of arylsulfonyl azides with engineered and artificial myoglobin-based catalysts. *Bioorg. Med. Chem.* **22** (20): 5697–5704. (c) Sreenilayam, G., Moore, E.J., Steck, V., and Fasan, R. (2017). Metal substitution modulates the reactivity and extends the reaction scope of myoglobin carbene transfer catalysts. *Adv. Synth. Catal.* **359** (12): 2076–2089.

82. (a) Key, H.M., Dydio, P., Clark, D.S., and Hartwig, J.F. (2016). Abiological catalysis by artificial haem proteins containing noble metals in place of iron. *Nature* **534** (7608): 534–537. (b) Dydio, P., Key, H.M., Nazarenko, A. et al. (2016). An artificial metalloenzyme with the kinetics of native enzymes. *Science* **354** (6308): 102–106.

83. Fruk, L., Kuo, C.H., Torres, E., and Niemeyer, C.M. (2009). Apoenzyme reconstitution as a chemical tool for structural enzymology and biotechnology. *Angew. Chem. Int. Ed.* **48** (9): 1550–1574.

84. Koo, L.S., Tschirret-Guth, R.A., Straub, W.E. et al. (2000). The active site of the thermophilic CYP119 from *Sulfolobus solfataricus*. *J. Biol. Chem.* **275** (19): 14112–14123.

85. (a) Gu, Y., Bloomer, B.J., Liu, Z. et al. (2022). Directed evolution of artificial metalloenzymes in whole cells. *Angew. Chem. Int. Ed.* **61** (5): e202110519. (b) Gu, Y., Natoli, S.N., Liu, Z. et al. (2019). Site-selective functionalization of (sp^3) C–H bonds catalyzed by artificial metalloenzymes containing an iridium-porphyrin cofactor. *Angew. Chem. Int. Ed.* **58** (39): 13954–13960. (c) Huang, J., Liu, Z., Bloomer, B.J. et al. (2021). Unnatural biosynthesis by an engineered microorganism with heterologously expressed natural enzymes and an artificial metalloenzyme. *Nat. Chem.* **13** (12): 1186–1191.

86. Litman, Z.C., Wang, Y., Zhao, H., and Hartwig, J.F. (2018). Cooperative asymmetric reactions combining photocatalysis and enzymatic catalysis. *Nature* **560** (7718): 355–359.

87. Rudroff, F., Mihovilovic, M.D., Gröger, H. et al. (2018). Opportunities and challenges for combining chemo- and biocatalysis. *Nat. Catal.* **1**: 12–22.

88. (a) Sauer, D.F., Himiyama, T., Tachikawa, K. et al. (2015). A highly active biohybrid catalyst for olefin metathesis in water: impact of a hydrophobic cavity in a ß-barrel protein. *ACS Catal.* **5**: 7519–7522. (b) Toma, R.S.A., Kuthning, A., Exner, M.P. et al. (2015). Site-directed and global incorporation of orthogonal and isostructural noncanonical amino acids into the ribosomal lasso peptide capistruin. *ChemBioChem* **16**: 503–509.

89. (a) Sauer, D.F., Gotzen, S., and Okuda, J. (2016). Metatheases: artificial metalloproteins for olefin metathesis. *Org. Biomol. Chem.* **14** (39): 9174–9183. (b) Markel, U., Sauer, D.F., Schiffels, J. et al. (2019). Towards the evolution of artificial metalloenzymes – a protein Engineer's perspective. *Angew. Chem. Int. Ed.* **58** (14): 4454–4464.

90. Xu, Y., Gan, Q., Samkian, A.E. et al. (2022). Bulky cyclometalated ruthenium nitrates for challenging Z-selective metathesis: efficient one-step access to α-oxygenated Z-olefins from acrylates and allyl alcohols. *Angew. Chem. Int. Ed.* **61** (4): e202113089.

91. Zozulia, O. and Korendovych, I.V. (2020). Semi-rationally designed short peptides self-assemble and bind hemin to promote cyclopropanation. *Angew. Chem. Int. Ed.* **59** (21): 8108–8112.

8
Learning Lessons from Protein Engineering

8.1
Introductory Remarks

The introduction of point mutations into proteins with the aim of clarifying basic enzyme mechanisms has been practiced for a long time [1, 2]. For example, active site residues of suspected catalytic triads can be identified by this technique. In other studies, two point mutations, dubbed X and Y, were introduced separately, likewise for mechanistic purposes. Their individual contributions to the catalytic profile (activity) were measured and subsequently compared to the mutational effect of the respective combined double mutant [3]. In most cases, additive effects regarding mutation X and mutation Y were found according to the Fersht equation (Eq. (8.1)), in which the "additional" free energy term $\Delta\Delta G_{\mathrm{I}}$ is zero.

$$\Delta\Delta G_{(\mathrm{X,Y})} = \Delta\Delta G_{(\mathrm{X})} + \Delta\Delta G_{(\mathrm{Y})} + \Delta\Delta G_{(\mathrm{I})} \tag{8.1}$$

This means that the two mutations, being independent, do not interact with one another on a molecular level, as shown by a survey of several studies at the time [3c] (Figure 8.1). Further examples of classical mutational additivity were reported later [4].

A different situation arises when applying protein engineering to manipulate the catalytic profile of an enzyme for practical purposes, especially when aiming to enhance or invert stereo- and/or regioselectivity (see Chapters 4 and 5). In such cases, it is rewarding to invest further research efforts to unveil the origin of mutational effects on a molecular level, which in turn helps researchers in future protein engineering studies [5, 6]. The term "Learning from Directed Evolution" was coined as early as 2004 [7] and continues to play an important role in terms of basic research and its influence on practical applications [6]. It is important to realize that two different types of lessons have been learned [6]:

- Unraveling the reasons for altered catalytic profiles with the generation of knowledge exposing as yet unknown intricacies of enzyme mechanisms (see Section 8.2).

Enzyme Engineering: Selective Catalysts for Applications in Biotechnology, Organic Chemistry, and Life Science, First Edition. Manfred T. Reetz, Zhoutong Sun, and Ge Qu.

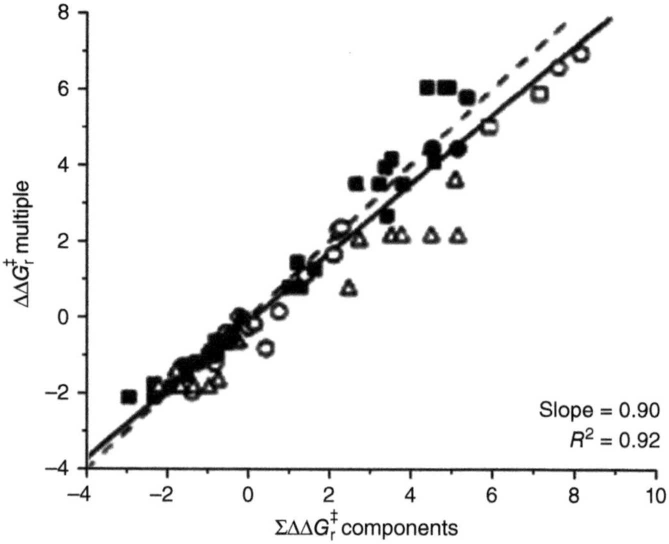

Figure 8.1 Changes in transition-state stabilization energies for the multiple mutants versus the sum of the component mutants. The data represent mutants from subtilisin, tyrosyl-tRNA synthetase, trypsin, dihydrofolate reductase (DHFR), and glutathione reductase. The dashed line has a slope of 1 representing perfect additivity, and the solid line corresponds to the best fit of the data. Source: Ref. [3c]/American Chemical Society.

- Discovering whether the effect of mutations in proteins is classically additive or nonadditive (cooperative or antagonistic) by way of mutational deconvolution and construction of fitness pathway landscapes (Section 8.2).

In this chapter, attention is also paid to the concept of *unexplored fleeting chiral intermediates* that occur in different types of enzymes, but which have been largely ignored in enzymology, biochemistry, and biotechnology [8] (Section 8.3). It has started to influence directed evolution and interpretations of mutations. Finally, protein engineering has been used to address fundamental questions in the field of evolutionary biology when asking how many upward evolutionary pathways exist connecting a WT enzyme and a multi-mutational variant in natural environments [9, 10]. However, since the authors of this monograph believe that evolution in nature is generally different from evolution in the laboratory [5, 6, 11], caution in making far-reaching conclusions is called for.

8.2 Additive Versus Nonadditive Mutational Effects in Fitness Landscapes Revealed by Partial or Complete Deconvolution

As noted in Section 8.1, mutational effects can be mathematically additive or more than additive, i.e. nonadditive in the form of synergism or antagonism.

Rare exceptions to additivity were noted early on. It was suggested that such odd phenomena are induced when the side chains of the two amino acids in the double mutant are spatially in close vicinity to one another, but the molecular basis was not explored [3]. In an investigation of mutational effects in dihydrofolate, a visionary conclusion was made: "... with an enhanced knowledge of the molecular origin of nonadditive effects, it may be possible to optimize an approach to improve the enzyme's efficiency by coupling mutations" [3j, k]. As shown in the following discourse, this challenging goal has been reached.

With the current level of theory, it is difficult to predict which type of mutational effect (epistasis) will result in such experiments, if any. In contrast to the classical double mutant free-energy cycles [3], directed evolution usually involves multiple rounds of mutagenesis/expression/screening with the accumulation of several point mutations. In the absence of deconvolution experiments, only the catalytic effect of the first mutation(s) is accessible experimentally in addition to the effect of all mutations acting in concert in the final variant. While some researchers insist that this is the way to proceed in research, this attitude actually prevents important lessons from being learned.

Indeed, partial or ideally full deconvolution of a multi-mutational variant constitutes a unique way to learn from directed evolution [12]. An eye-opening example concerns the hydrolytic kinetic resolution of *rac*-**1** with preferential formation of (*S*)-**2** catalyzed by the lipase from *Pseudomonas aeruginosa* (PAL) (Scheme 8.1) [13], this transformation being the model reaction of a long series of studies in which different mutagenesis strategies were compared (see Chapter 4) [5]. At an intermediate stage of methodology development, a variant characterized by 6-point mutations, introduced in several cycles by error-prone polymerase chain reaction (epPCR), saturation mutagenesis, and DNA shuffling

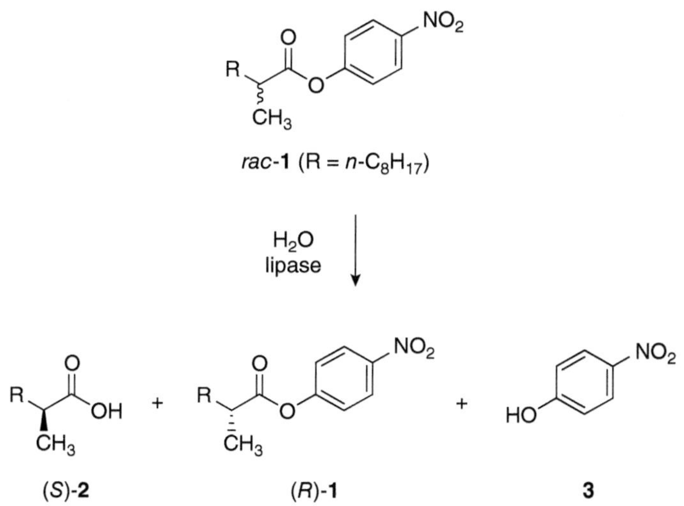

Scheme 8.1 Model reaction catalyzed by the lipase PAL. Source: Adapted from Refs. [13, 14].

led to a selectivity factor of $E = 51$, at the time a record [14a]. However, a quantum mechanics/molecular mechanics (QM/MM) study predicted that 4 of the 6-point mutations are superfluous, and indeed the remaining double mutant proved to be even better ($E = 63$) [7b]. A reasonable model was proposed on the basis of a relay effect originating from a remote position and extending to a residue next to the binding pocket.

Subsequently, iterative saturation mutagenesis (ISM) was applied to the same model system, leading to a dramatically improved variant, while screening considerably less transformants [13]. A 3-site combinatorial active-site saturation test (CAST)/ISM scheme composed of 2-residue sites A, B, and C was designed, the best pathway B → A leading to the final triple mutant composed of mutation Leu162Asn obtained by saturation mutagenesis at site B (Leu159/Leu162) and of Met16Ala/Leu17Phe by subsequent saturation mutagenesis at site A (Met16/Leu17) (Figure 8.2). The final best variant 1B2 (Leu162Asn/Met16Ala/Leu17Phe) showed an unprecedented selectivity factor of $E = 594$ in favor of (*S*)-**2** [13]. In addition to enhanced stereoselectivity, the reaction rate of the preferred enantiomer (*S*)-**1** was increased

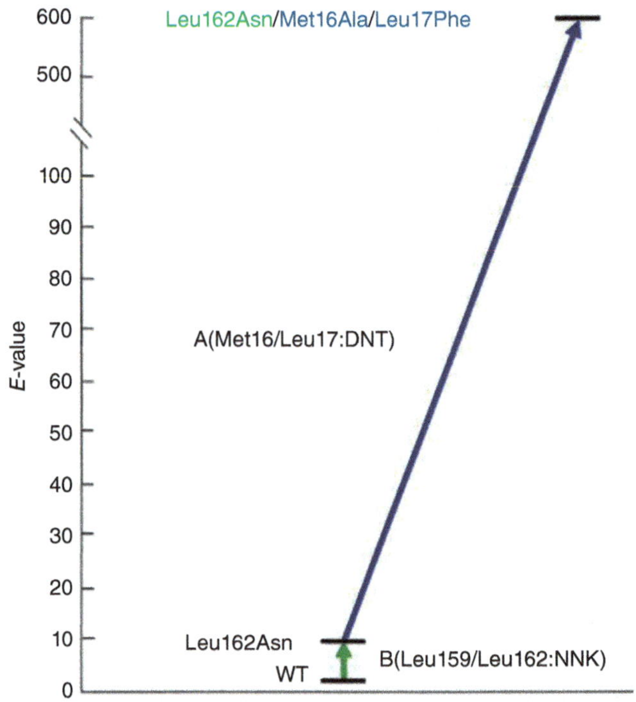

Figure 8.2 Best ISM pathway B → A leading to the triple mutant 1B2 (Leu162Asn/Met16Ala/Leu17Phe), which displays a selectivity factor of $E = 594$ in the hydrolytic kinetic resolution of *rac*-**1** with preferential formation of (*S*)-**2**. Source: Adapted from Ref. [13].

notably: WT PAL ($k_{cat} = 37 \times 10^{-3}$ s^{-1}; $k_{cat}/K_m = 43.5$ s^{-1}M^{-1}) versus variant 1B2 ($k_{cat} = 1374 \times 10^{-3}$s^{-1}; $k_{cat}/K_m = 4041$ s^{-1}M^{-1}).

The effect of the first mutation (Leu162Asn) alone enhances enantioselectivity only slightly to $E = 8$ (*S*). *The influence of the second set of mutations alone is not accessible from the original data unless deconvolution is performed.* Upon preparing and testing the respective double mutant Met16Ala/Leu17Phe as part of partial deconvolution, it was discovered that this variant by itself improves enantioselectivity relative to WT PAL hardly at all, the selectivity factor amounting to only $E = 2.6$ (*S*)! If additivity were to be operating, then the selectivity factor would be expected to be $E = \approx 22$. Since it is much larger ($E = 594$), a notable cooperative nonadditive effect has to be involved, amounting to a significant energy contribution of ≈2 kcal mol^{-1} [13]. Relative to WT PAL, the calculated difference in stabilization energy of the two enantiomers amounts to about 3 kcal mol^{-1}. Complete deconvolution by generating and testing separately the two single mutants Met16Ala and Leu17Phe was not investigated in this study.

Upon deconvoluting the final mutants in other CAST- or ISM-based directed evolution studies, positive epistatic effects were also uncovered [15], as summarized in a short review of this emerging phenomenon [16]. The occurrence of strongly cooperative mutational effects in ISM seems to be the underlying factor responsible for the efficacy of this approach to protein engineering [5]. In the present case, a second lesson was learned upon unveiling the reason for the positive epistatic effect on a molecular level. The mechanism of PAL involves the catalytic triad Asp229/His251/Ser82, which enables rate- and stereoselectivity-determining nucleophilic addition of activated Ser82 to the carbonyl function of esters with formation of short-lived oxyanion intermediates, followed by rapid product formation (Scheme 8.2). It is the typical lipase mechanism [17].

Using the crystal structure of WT PAL [18] as the starting structure, docking and molecular dynamics (MD) computations were performed [13]. The 3-point mutations were then studied by a docking program. Substrates (*R*)- and (*S*)-**1** were introduced in the PAL binding pocket as the respective oxyanions covalently bound to catalytically active Ser82. For comparison, Figure 8.3 shows the favored (*S*)-substrate bound in WT PAL and in the best mutant 1B2, respectively.

Several important features stand out that provide significant insight. Firstly, in WT PAL, the bulky side chain of Leu162 clashes with the *n*-octyl moiety of the ester, in contrast to the situation in variant 1B2 in which the position of Asn162 provides sufficient space for the long alkyl chain of (*S*)-**1**. It can be argued that the sidechain of asparagine is not that much smaller than that of leucine, but in the present case, another factor is involved. In 1B2, Asn162 forms a hydrogen bond to Ser158, thereby positioning the asparagine side chain further away from the bound substrate, which avoids steric clashes. This also explains the increase in activity when going from WT to variant 1B2.

Secondly, the substitution Met16→Ala16 provides more space in a sterically congested part of WT PAL, which allows more side chain flexibility of

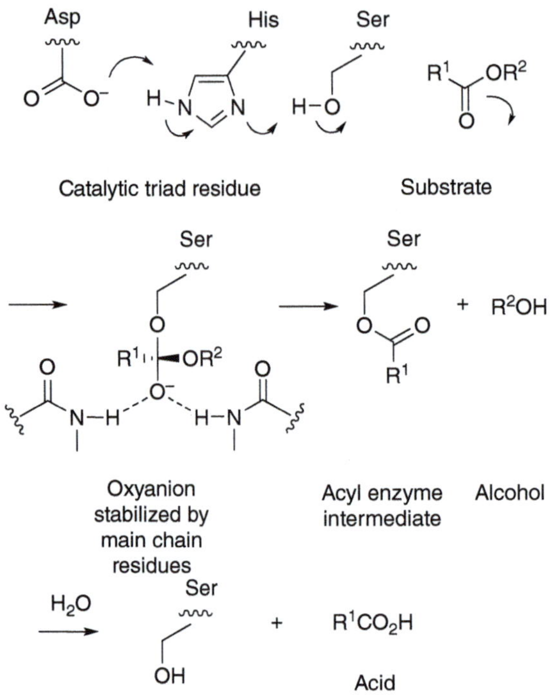

Scheme 8.2 Mechanism of lipase-catalyzed hydrolysis of esters.

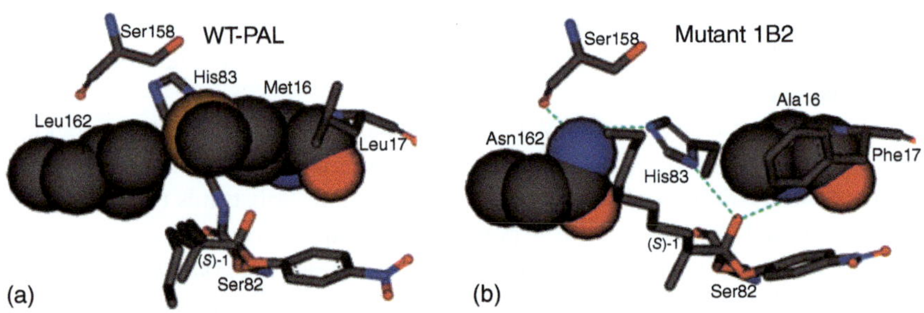

Figure 8.3 Comparison of the oxyanions with bound (*S*)-substrate at the catalytically active Ser82 of WT PAL (a) versus best variant 1B2 (b). Source: Adapted from Ref. [13].

normally uninvolved His83. The new pose enables additional stabilization of the short-lived oxyanion by way of hydrogen bond formation (Figure 8.3b). It should be noted that Ser158 and His83 are not part of the catalytically active triad Asp229/His251/Ser82, which means that in WT PAL they are "innocent" residues. In variant 1B2, however, they participate in a H-bond network that involves Asn162, Ser158, His83, and the O-atom of the oxyanion (Figure 8.3b) [13].

Thirdly, the substitution Leu17 → Phe17 enables π-stacking between the phenyl sidechain of phenylalanine and the *p*-nitrophenyl moiety of the ester substrate, which is not possible in WT PAL (Figure 8.3). This model explains the origin of the observed strong cooperative effect brought about by the ISM mutational substitutions. Mutation Leu162Asn alone exerts little influence, nor does Met16Ala/Leu17Phe by itself, but in concert more than additivity results. Finally, upon modeling the disfavored substrate (*R*)-**1** into variant 1B2, the respective methyl group at the stereogenic center points "upward," thereby preventing His83 from additionally stabilizing the oxyanion. This explains why the (*S*)-substrate is favored. In summary, much has been learned in this study. However, the configuration of the chiral short-lived oxyanion, an unexplored fleeting chiral intermediate [8], was not considered, which may have provided further mechanistic lessons.

The number of studies focusing on stereoselectivity of enzymes in which deconvolution experiments were performed is increasing, leading to new insights [14, 15]. In one study, an ISM-evolved quadruple mutant of a Baeyer–Villiger monooxygenase obtained in two ISM steps with accumulation of two new mutations in each cycle, was used as the catalyst in asymmetric sulfoxidation of methyl tolyl thioether (95% ee in favor of the (*R*)-sulfoxide starting from WT showing 90% ee of opposite (*S*)-enantioselectivity) [15a]. Deconvolution with formation of the respective four single mutants led to a surprising result. All of the four single mutants showed (*S*)-selectivity, which constitutes a counterintuitive finding:

$$\text{Mutation 1 } (\boldsymbol{S}) + \text{mutation 2 } (\boldsymbol{S}) + \text{mutation 3 } (\boldsymbol{S}) + \text{mutation 4 } (\boldsymbol{S}) \rightarrow \text{combined mutations } (\boldsymbol{R})$$

A common technique in protein engineering is to combine separately generated point mutations to improve a catalytic property (Chapters 4 and 5). However, it has been observed that this procedure does not always work for reasons that are not well understood [19]. By going the reverse way, i.e. deconvoluting multi-mutational variants, some insight as delineated here has been gained. Using enantioselectivity as the catalytic parameter, additive and nonadditive mutational effects as revealed by deconvolution experiments can be systematized (Scheme 8.3). This illustration features the case of an initial set of mutations A followed a second set of mutations B, which accumulated in a hypothetical directed evolution experiment, irrespective of the mutagenesis method. Deconvolution of the two sets with generation of B alone can in principle signal classical additivity (Scheme 8.3a). In this case, A and B do not interact with one another, and both favor the same direction of enantioselectivity, e.g. (*R*). Several kinds of nonadditivity are possible. In one type, deconvolution reveals that the contribution of B is less than expected, but the sense of enantioselectivity is the same as displayed by A (Scheme 8.3b). A second type of nonadditivity is uncovered upon deconvolution with formation of B alone, which favors the opposite enantiomer (Scheme 8.3c).

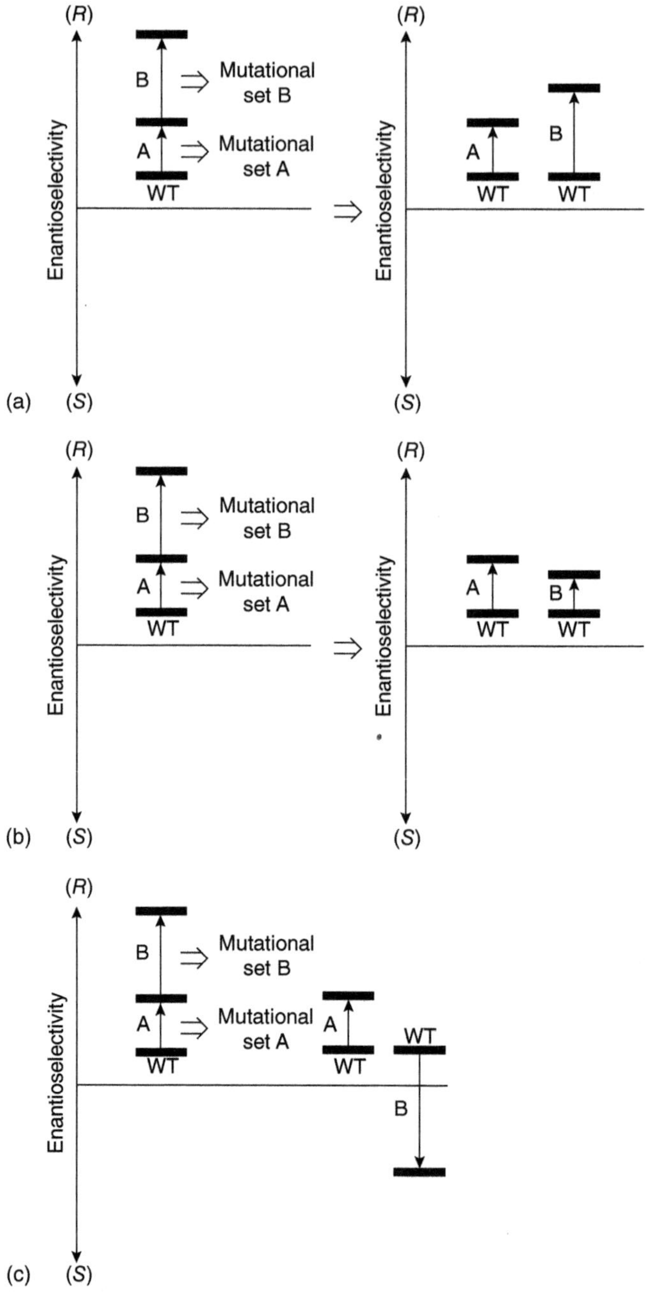

Scheme 8.3 Systematization of additive and nonadditive mutational effects in protein engineering, in this scheme using two sets of mutations A and B, illustrated by employing enantioselectivity as the catalytic parameter. (a) Classical additive mutational effect; (b) nonadditive mutational effect in which set B shows lower than expected enantioselectivity but in the same direction; (c) nonadditive effect in which mutational set B shows reversed enantioselectivity. Source: Refs. [5, 6]/Royal Society of Chemistry.

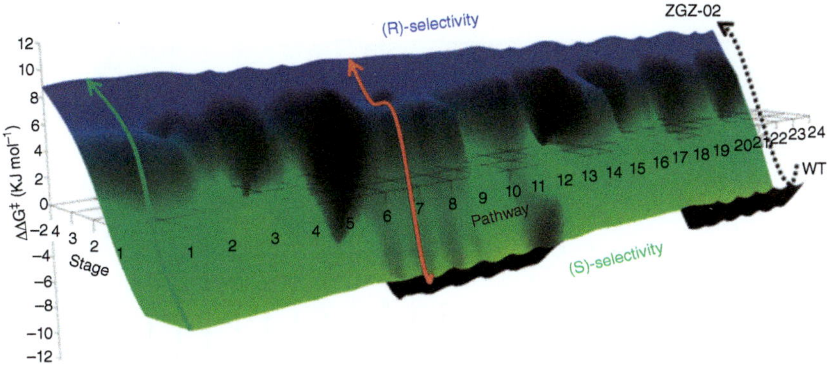

Figure 8.4 Fitness pathway landscape showing the 24 pathways leading from WT PAMO (bottom) to best (*R*)-selective variant ZGZ-2, a typical trajectory lacking local minima (green pathway) and one having local minima (red) being featured. Source: Ref. [15a]/American Chemical Society.

The situation becomes even more complex when deconvoluting both sets of mutations A and B individually with formation of the respective single mutants that may show different types of nonadditivity or classical additivity. Complete deconvolution does not only dissect a multi-mutational variant into the respective single mutants, it also calls for the generation of all theoretically possible combinations of point mutations (double, triple mutants, etc.). When these are prepared by site-specific mutagenesis and used as catalysts in an enantioselective transformation, it is possible to construct a complete fitness pathway landscape that features the mapping of all theoretically possible pathways leading from WT to the final best mutant. This type of fitness pathway landscape was constructed in the directed evolution project of the abovementioned Baeyer–Villiger monooxygenase as catalyst in the asymmetric sulfoxidation of methyl tolyl thioether [15a]. The final reversed (*R*)-selective quadruple mutant was fully deconvoluted, making the construction of 4! = 24 evolutionary pathways possible (Figure 8.4) [15a]. Out of the 24 trajectories, 6 proved to lack any local minima, while 18 displayed such local "valleys," meaning that at those points in the upward climb the respective mutant library contained no improved variants with improved enantioselectivity (Figure 8.5). This does not imply that the original CAST/ISM procedure was inefficient, the opposite is true as demonstrated by the experimental results. Deconvolution simply allows for important hitherto unnoticed insights [6].

In an earlier study in which the identical procedure was applied to an enantioselective epoxide hydrolase, a 5-step ISM pathway was deconvoluted that enabled the construction of the respective fitness landscape characterized by 5! = 120 pathways [20]. A total of 55 trajectories proved to be characterized by the absence of any local minima, which means that in all respective mutant libraries at least one variant was found that displays enhanced enantioselectivity.

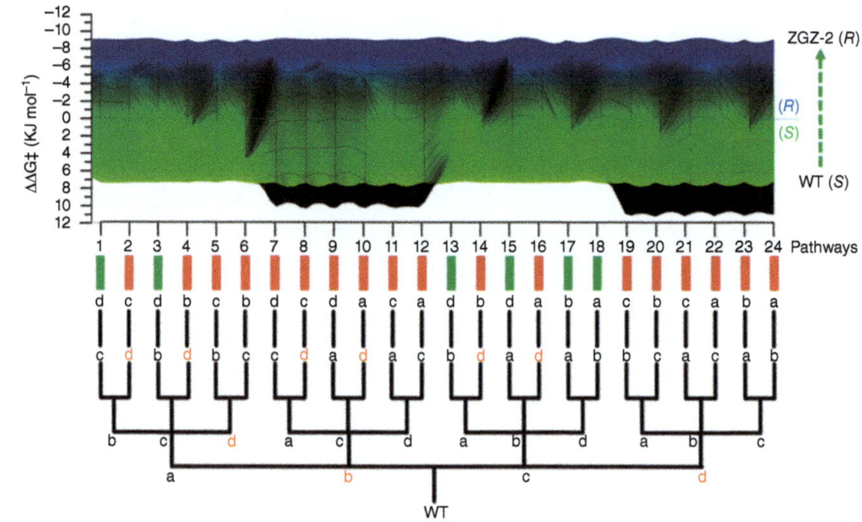

Figure 8.5 Fitness pathway landscape in the frontal view of Figure 8.4 of all 24 trajectories leading from WT PAMO to variant ZGZ-2 characterized by 4-point mutations [15a]. Green notations indicate energetically favored pathways, whereas red notations represent disfavored trajectories having local minima. Letters in red in the dendrogram denote a local minimum after the introduction of this mutation. Source: Ref. [15a]/American Chemical Society.

In the other 65 pathways, at least one library was encountered that failed to harbor an improved variant. In such cases an inferior set of mutations was necessarily utilized, thereby mapping trajectories that terminate with the same final variant, as all other pathways. When comparing the fitness pathway landscapes of the Baeyer–Villiger [15a] and the epoxide hydrolase [20] studies, it can be seen that the respective number of pathways characterized by the absence of local minima differs considerably: 6 out of 24 (25%) versus 55 out of 120 (46%), respectively. There is no reason to expect similar percentages. Each enzyme system, and particularly the mutagenesis method and strategy, will lead to different results. *It needs to be emphasized that this type of "constrained" fitness pathway landscape, which maps all trajectories connecting WT with a given previously evolved best mutant, is different from exploring all theoretically possible trajectories of a defined ISM system ("unconstrained" fitness pathway landscape)* as delineated in Chapter 4 [9]. To date, this is the only study in which all pathways of an (unconstrained) ISM system were explored experimentally. It led to a counterintuitive lesson regarding a simple way to escape from local minima, namely to choose an *inferior* mutant as the template in the subsequent ISM step [9].

Although both constrained and unconstrained fitness pathway landscapes were constructed and analyzed to learn how to increase the efficacy of directed evolution of enzymes for use in organic chemistry and biotechnology

[9, 15a, c, 16, 20], the results also touch on a very different research field, namely evolutionary biology [5, 6]. Using a mutant of a ß-lactamase characterized by 5-point mutations evolved previously, an experimental selection-based platform was devised on which a "constrained" fitness pathway landscape featuring 5! = 120 pathways were constructed [10]. Four of the mutations were based on Stemmer's first report on DNA shuffling [21]. The purpose of the study was to address the question of how many pathways in natural evolution lead from a starting gene to an evolved mutant gene. It turned out that of the 120 trajectories connecting WT with the quintuple mutant, 102 were characterized by local minima. This led the authors to conclude that "Darwinian evolution can follow only very few mutational pathways to fitter proteins" [10]. They described these 102 pathways as being "inaccessible to Darwinian selection" and stated that "many of the remaining trajectories have negligible probabilities of realization" [10].

The reader is advised to study the original publication for more details regarding the experimental platform [10]. The question arises whether the general conclusion of this study [10] contradicts the results of the subsequent studies featured above [9, 15a, c, 16, 20]. Would the application of ISM or some other molecular biological method for evolving a ß-lactamase variant with five (different) point mutations lead to a higher number of pathways to fitter proteins, and therefore to a different conclusion? It needs to be pointed out that the experimental platforms used in the ß-lactamase study [10] and in the subsequent studies [9, 15a, c, 16, 20] are quite different and therefore not directly comparable, the first utilizing *selection based on bacterial survival and the others employing screening based on analytical assays*, respectively. Nevertheless, as already stressed, care must be taken when exploiting the results of laboratory (directed) evolution for drawing general conclusions regarding the nature of Darwinian evolution [5, 6].

8.3
Unexplored Chiral Fleeting Intermediates and Their Role in Protein Engineering

Biochemists and enzymologists have invested decades of work in elucidating the gross mechanism of essentially all enzyme types, intriguing details often being part of the analyses. Recently, it was pointed out that an essential aspect has been forgotten, namely the role of "fleeting chiral intermediates" that occur in the mechanism of a surprisingly high number of enzyme types [8]. Not knowing whether such short-lived species possess the (*R*)- or the (*S*)-configuration prevents mechanistic intricacies from being identified and explained on a molecular basis. Fleeting chiral intermediates may occur in biocatalytic transformations of chiral substrates with formation of chiral products. Strangely enough, this is also the case when starting materials and products are achiral (Scheme 8.4).

As a classical example, the mechanism of lipases involves short-lived tetrahedral oxyanions, which are chiral even when all compounds in the transformation are achiral (Scheme 8.4b). In most biochemistry textbooks and journal articles,

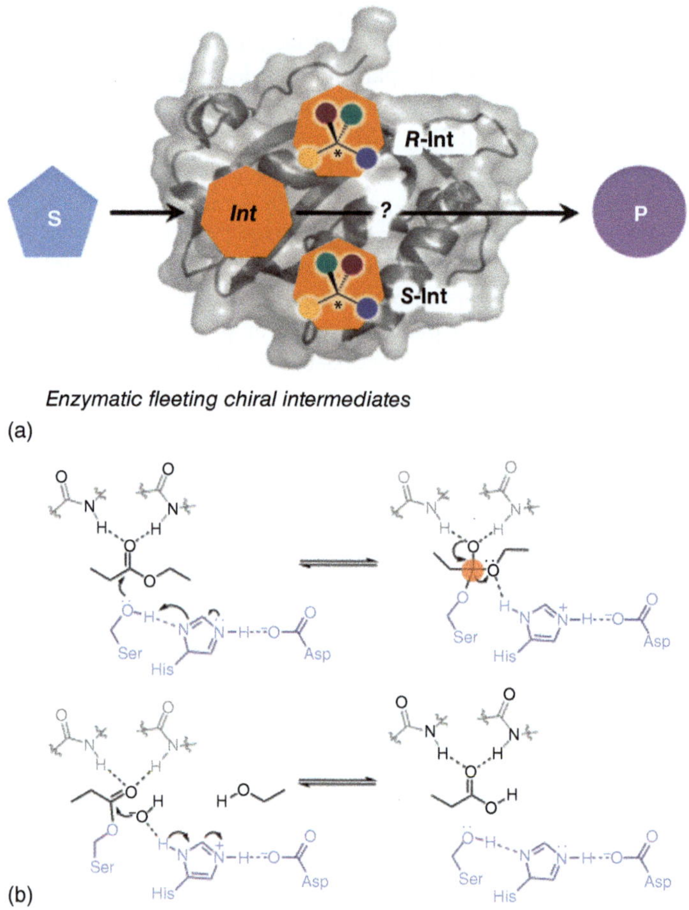

Scheme 8.4 Concept of fleeting chiral intermediates. (a) General representation of the concept (S: substrate; P: product; Int: intermediate having either the (*R*)- or (*S*)-configuration); (b) Generally accepted mechanism of lipase-catalyzed reactions characterized by the traditional catalytic triad Asp – His – Ser, here achiral propionic acid ethyl ester serving as the substrate. The orange-marked C-atom pinpoints the unexplored chirality of the short-lived intermediate. Source: Ref. [8]/American Chemical Society.

the fact that the oxyanions in lipase-catalyzed reactions are chiral is not mentioned, nor are QM/MM computations of the respective (*R*)- and (*S*)-configurated fleeting chiral intermediates presented. The same applies to the mechanistic discussions of esterases, proteases, epoxide hydrolases, transaminases, and Baeyer–Villiger monooxygenases (BVMOs) [8]. Thus, important or even crucial mechanistic details remain unexplored. The first example in which this issue was *not* neglected concerns the regioselectivity of a BVMO, highlighted in Section 8.4.5.

8.4 Case Studies Featuring Mechanistic, Structural, and/or Computational Analyses of the Source of Evolved Stereo- and/or Regioselectivity

8.4.1 Esterase

While various esterases continue to be subjected to directed evolution along with mechanistic explanations (Chapter 5), a study of the enantioselective esterase from *Archaeoglobus fulgidus* (AFEST) is highlighted here because it is noteworthy both in terms of methodology development and drawing insights [22]. A dual-channel microfluidic droplet screening system was first constructed for detecting (*R*)- and (*S*)-selectivity, respectively, based on the use of pseudo-enantiomeric iboprufen-related substrates in kinetic resolution, fluorescence-activated cell sorting (FACS) allowing for super high-throughput screening of 10^7 transformants (Figure 8.6). This approach is reminiscent of a study of the ester from *Pseudomonas aeruginosa* (EstA) in which hydrolysis of pseudo-enantiomeric esters triggers green or red fluorescence signaling (*R*)- or (*S*)-selectivity, respectively, as screened by FACS [23] (see Section 2.3). The new study using AFEST as the enzyme has the important advantage in which the products are not pinned to the cell surface, their release requiring further steps which makes the overall procedure cumbersome [23]. Today, devices are commercially available that allow the reactions to proceed in micro-droplets harboring single cells, but they are mainly used for the development of biologics [24].

The best mutants were used as catalysts in the hydrolytic kinetic resolution of (*R*)- and (*S*)-iboprufen *p*-nitrophenylesters. As anticipated, variant 6A8 resulted in inversion of enantioselectivity needed for pharmaceutical application (E =>100 (*S*)), and variant 4F12 improved non-desired (*R*)-selectivity to E = 74, compared to WT AFEST (E = 7.4 (*R*)). However, when testing other profen substrates, different trends were observed. The authors then proceeded to explain the molecular mechanism responsible for at least some of the experimental results (Figure 8.7). Residues S160, D255, and H285 form the catalytic triad, and G88, G89, and A166 shape the oxyanion hole by undergoing three H-bond interactions with the carbonyl group of both (*R*)- and (*S*)-substrate, as revealed by the docking poses. Differences were noted. The (*R*)-substrate was shown to be in a folded conformation at the beginning of the binding pocket, while the (*S*)-substrate occurs in a linear conformation deeply inside the binding pocket (Figure 8.7a). The introduction of mutation G89C into variant IE9 causes steric interaction with the methyl group of the (*R*)-substrate, thereby pushing it outwards (Figure 8.7b). The introduction of G89P into IE9 results in severe steric clash with the phenyl group of the (*S*)-substrate, causing it to shift in a different way (Figure 8.7c). Significant differences in the distance between the catalytically active serine (S160) and the respective carbonyl groups were not

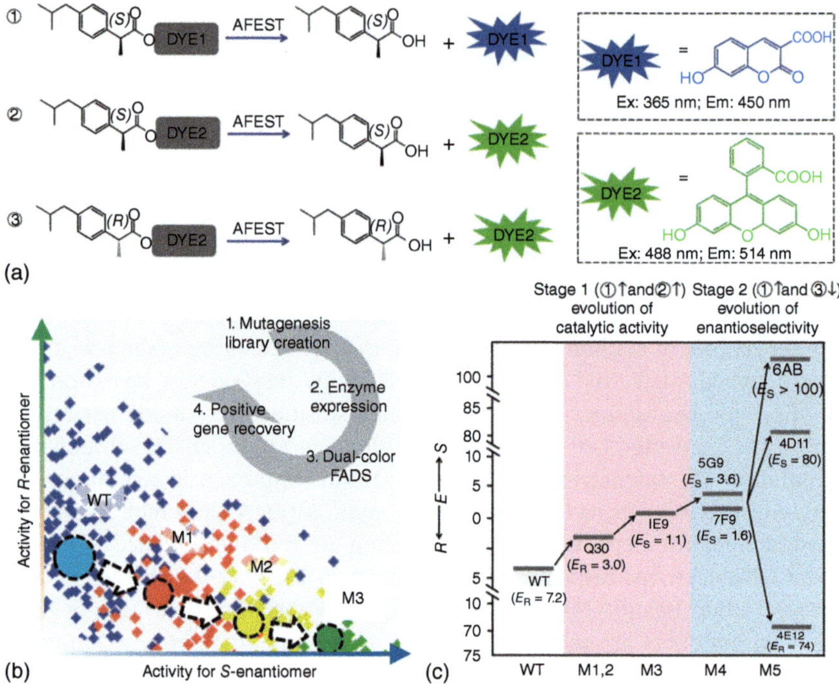

Figure 8.6 Directed evolution of the esterase from *Archaeoglobus fulgidus* (AFEST) as a catalyst in the enantioselective hydrolytic kinetic resolution using a FACS-based dual-channel microfluidic droplet screening platform. (a) Fluorescence-labeled chiral substrates; (b) Activity of (*R*)- versus (*S*)-selective mutants as evolution proceeds; (c) Combination of random mutagenesis and Iterative Saturation Mutagenesis (ISM) for evolving (*R*)- and (*S*)-selective variants. Source: Ref. [22]/Springer Nature/CC BY 4.0.

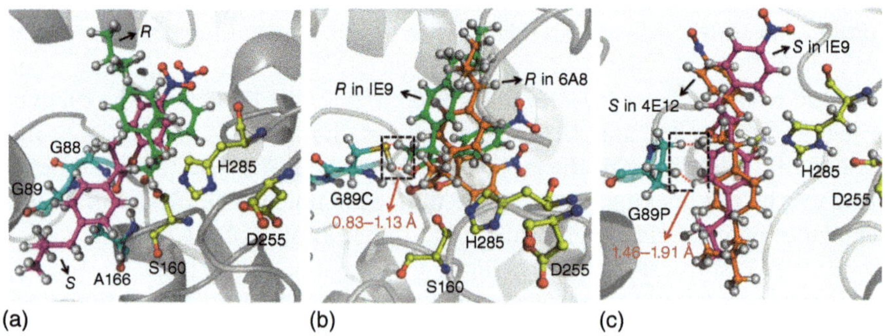

Figure 8.7 Structural basis for reversed enantioselectivity of AFEST. (a) Docking poses of (*R*)/(*S*)-ibuprofen *p*-nitrophenyl ester; (b) Mutation G89C results in steric clash with (*R*)-substrate; (c) Mutation G89P causes shift of the (*S*)-substrate. Source: Ref. [22]/Springer Nature/CC BY 4.0.

observed. Nevertheless, in some cases, a more stable tetrahedral transition state was postulated as part of a reasonable model.

In summary, the theoretical analyses in this study, based on MD simulations, provide some interesting insights. More work is required for advancing further mechanistic lessons. This would include the QM/MM-based analysis of the fleeting chiral intermediates (see Section 8.4.5), which could generate an understanding of the experimentally observed switch from (*R*)- to (*S*)-selectivity. On the practical side regarding potential pharmaceutical applications, it would be important to test the mutants not just with *p*-nitrophenyl esters, but with economically and ecologically viable substrates such as the corresponding ethyl esters.

8.4.2 Epoxide Hydrolase

An informative mechanistic and structural study that includes the crystal structure of a stereoselective variant produced earlier by directed evolution concerns the epoxide hydrolase from *Aspergillus niger* (ANEH) as the catalyst in the hydrolytic kinetic resolution of *rac*-**4** with preferential formation of (*S*)-**5** (Scheme 8.5) [25]. WT ANEH is characterized by poor (*S*)-selectivity ($E = 4.6$). In the prior directed evolution study, six CAST sites had been chosen for ISM, A (comprising amino acid positions 193/195/196), B (215/217/219), C (329/330), D (349/350), E (317/318), and F (244/245/249). An arbitrarily chosen pathway B → C → D → F → E provided the best variant LW202 showing a selectivity factor of $E = 115$ in favor of (*S*)-**5** [26]. Due to the very high degree of enantioselectivity, the upward climb was terminated without visiting site A (see also Chapter 3) [26]. This variant has 9-point mutations L215F/A217N/R219S/L249Y/T317W/T318V/M329P/L330Y/C350V, which accumulated along the ISM pathway WT → B(variant LW081) → C(LW086) → D(LW123) → F(LW44) → E(LW202).

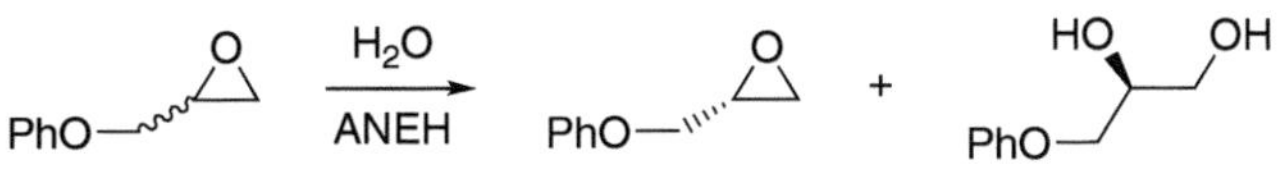

rac-**4** (*R*)-**4** (*S*)-**5**

Scheme 8.5 Hydrolytic kinetic resolution of *rac*-**1** catalyzed by the epoxide hydrolase from *Aspergillus niger* (ANEH). Source: Adapted from Refs. [25, 26].

Before these studies were initiated, the crystal structure of WT ANEH had been solved [27] and the basic mechanism elucidated [28]. It involves binding of the substrate and activation by H-bonds to the epoxide O-atom originating from Tyr251 and Tyr314, and catalytically active Asp192 then inducing an S_N2 reaction in the rate-determining initial step followed by fast hydrolysis of the short-lived acyl-enzyme intermediate (Scheme 8.6). Notice that the nucleophilic addition of water to the ester function (Step 1) leads to a chiral tetrahedral species (fleeting chiral intermediate), which is not shown.

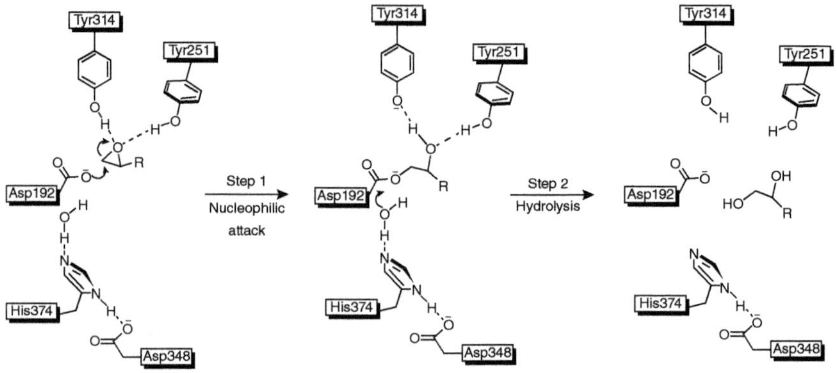

Scheme 8.6 Mechanism of the epoxide hydrolase ANEH. Source: Adapted from Refs. [25, 28].

With the aim to unravel the source of enhanced stereoselectivity of the best variant LW202, kinetic analyses based on the Michaelis–Menten equation were first carried out using in separate experiments enantiomerically pure (*R*)- and (*S*)-**4**, respectively. A nearly ideal behavior of a kinetic resolution was demonstrated, the reaction of the disfavored (*R*)-enantiomer being essentially shut down (Figure 8.8). The data also allows for a more exact determination of the selectivity factor, which is even higher ($E = 195$) than the original estimation based on the standard Sih equation [25]. Moreover, the relative values of k_{cat}/K_m for the two enantiomers also reflect pronounced (*S*)-selectivity. The K_m-value of the reaction

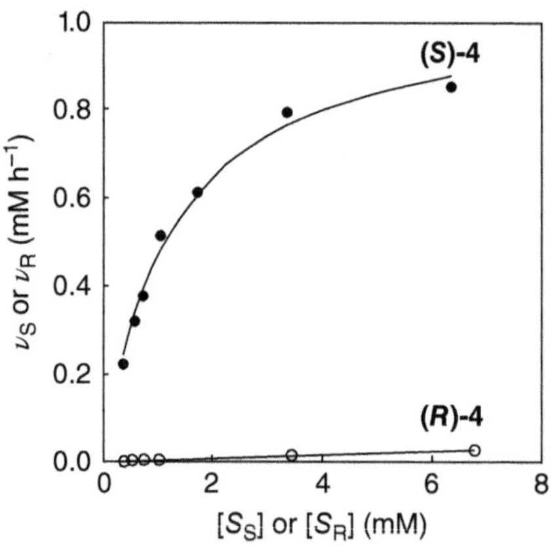

Figure 8.8 Kinetic analysis of variant LW202 as catalyst in separate reactions of (*R*)- and (*S*)-**4**, where v_R and v_S are the initial rates of hydrolysis of (*R*)- and (*S*)-**4** at different substrate concentrations [S_R] or [S_S]. Source: Ref. [25]/American Chemical Society.

of (*R*)-**4** is considerably higher than that of (*S*)-**4**, while the situation reverses when comparing the k_{cat}-values (factor of 4.7 in favor of (*S*)-**5**) [25, 26].

To identify the factors which cause enhanced (*S*)-enantioselectivity at every stage of the 5-step evolutionary process, WT → LW081 → LW086 → LW123 → LW44 → LW202, extensive MD simulations were performed using (*R*)- and (*S*)-**4** as substrates separately [25]. The distance, *d*, between the attacking O-atom of Asp192 and the epoxide C-atom undergoing S_N2 reaction was defined as the crucial parameter (Scheme 8.7) [25]. It was assumed that a sufficiently small *d*-value would correspond fairly well to a pose resembling near-attack conformation [29]. In other words, productive binding can be expected if *d* is relatively short in the range of ≈3.5 Å. Large values were foreseen in the reaction of the disfavored enantiomer (*R*)-**4**. Indeed, a striking correlation (R^2 = 0.86) was observed between the experimental *E*-values and the differences in the computed distance, Δd_{R-S}, for the two enantiomeric substrates (Table 8.1) [25]. It can be seen that in the case of disfavored (*R*)-**4** this difference increases as the evolutionary process proceeds. In the final variant LW202, d_R amounts to 5.4 Å, a clear indication that (*R*)-**4** is disfavored. This model is in full agreement with the kinetics (Scheme 8.7). It means that variant LW202 binds (*R*)-**1** in an unproductive mode, thereby essentially shutting down the reaction, quite different from complexed (*S*)-**4**. In contrast, in the binding pocket of WT ANEH, both enantiomers are bound in productive poses. The reasons for the different binding modes in LW202 were elucidated by molecular dynamics (MD) simulations and docking computations.

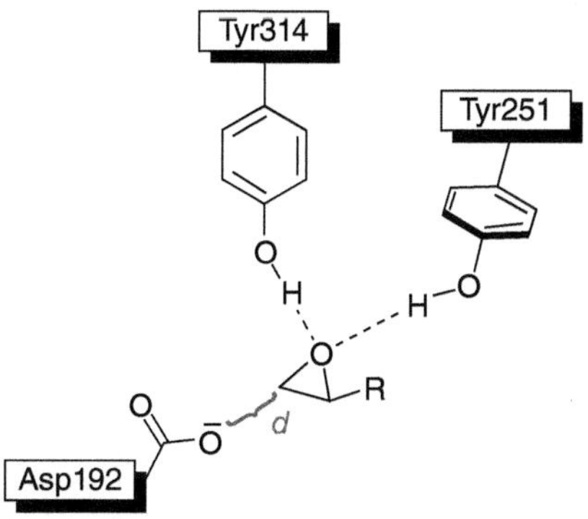

Scheme 8.7 Definition of the distance *d* in the rate-determining step of the ANEH-catalyzed reaction of *rac*-**4**. Source: Ref. [25]/American Chemical Society.

The determination of two crystal structures, that of WT ANEH harboring the inhibitor valpromide (2-propyl-pentanoic acid amide) and that of apo (unbound) variant LW2002 is an important asset of this study [25]. A comparison with apo WT ANEH [27] was also made. Whereas the gross features of all structures are almost identical (essentially same fold), clear differences in the shape of the binding pocket of LW202 relative to apo or bound WT became visible.

Table 8.1 Results of MD calculations.

Mutant	d_R	d_S	Δd_{R-S}	*E* (expl)
WT	4.3	3.5	0.8	4.6
LW081	4.8	4.0	0.8	14
LW086	4.9	4.0	0.9	21
LW123	5.1	4.0	1.1	24
LW44	5.1	3.9	1.2	35
LW202	5.4	3.8	1.6	115

Source: Ref. [25]/American Chemical Society.

The structures were used for docking the favored (*S*)-**4** and disfavored (*R*)-**4** in the respective binding pockets in a manner that ensures smooth attack by nucleophilic Asp192 (Figure 8.9). The preferred (*S*)-enantiomer fits well into the WT (Figure 8.9a) and into best variant LW202 (Figure 8.9c) without any steric clashes, while maintaining activation by Tyr251/Tyr314 as well as optimal positioning for nucleophilic attack by Asp192. In sharp contrast, the disfavored (*R*)-enantiomer fits well into the WT binding pocket (Figure 8.9b), but not into

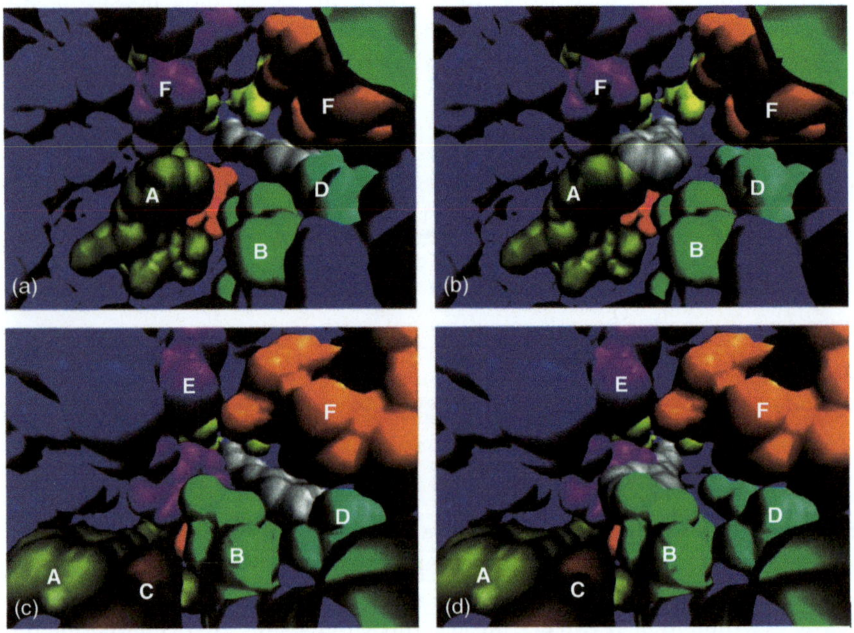

Figure 8.9 Interpretation of crystal structures of WT ANEH and evolved variants by manually docking (*R*)- and (*S*)-**4** into binding pockets, A, B, C, D, E, and F representing the originally designed randomization sites in the ISM process. (a) Favored (*S*)-**4** in WT ANEH binding pocket; (b) disfavored (*R*)-**4** in binding pocket of WT ANEH; (c) favored (*S*)-**4** in variant LW202; (d) disfavored (*R*)-**4** in variant LW201. Source: Ref. [25]/American Chemical Society.

LW202 because in this "forced" pose severe steric clashes occur between the substrate and the sidechains of mutated residues, especially at sites B and E (Figure 8.9d). Thus, productive binding is strongly prevented. This interpretation of the crystal structures is in line with the original MD computations performed in the absence of the LW202 X-ray structure and with the results of the kinetic study [25].

In conclusion, important mechanistic lessons can be learned from directed evolution, provided sufficient efforts are invested in interpreting altered catalytic profiles. This study not only uncovered the source of enhanced enantioselectivity on a molecular level. It also contributed to a deeper understanding of the mechanistic intricacies of this enzyme, and probably of other structurally and mechanistically related epoxide hydrolases [25]. Further computational progress in understanding epoxide hydrolases has been made since [12, 30]. The prediction of epoxide variants with considerably higher enantioselectivity in the model reaction based on the use of machine learning (ML) and deep learning (DL) has also been reported [31] (see Section 3.9.2). However, the role of fleeting chiral intermediates [8] that occur in epoxide hydrolases was not explored at the time, nor in subsequent directed evolution studies to date [12, 30].

8.4.3 Ene-reductase of the Old Yellow Enzyme (OYE)

Ene-reductases (enoate reductases) are valuable biocatalysts for the enantioselective reduction of a wide range of activated olefins bearing electron-withdrawing groups, e.g. α,ß-unsaturated ketones, esters, nitriles, etc. [32]. They generally belong to the class of old yellow enzymes (OYE). However, many substrates of interest to synthetic organic or pharmaceutical chemists show poor or the wrong enantioselectivity, in which case directed evolution has been applied successfully [5, 33] (see also Table 5.1).

Strangely enough, QM/MM studies of OYEs have not been reported until recently [34]. Nevertheless, many theoretical studies concerning the mechanism have provided some interesting lessons. The mechanism of OYEs involves the traditional binding mode in which an asparagine and a histidine (or two histidines) form H-bonds to the carbonyl moiety of the substrate, the reduced flavine FMNH then delivering a hydride to one π-face of the activated ß-position, while tyrosine spends a proton on the opposite π-face in an overall trans-specific manner (Scheme 8.8a). In the case of a prochiral substrate such as 3-substituted cyclohexene ones, the addition of two hydrogens leads to an enantiomer of defined absolute configuration. In the so-called flipped binding mode (Scheme 8.8b), the enantiomer of opposite absolute configuration results, which has been achieved by directed evolution [33a, b, 35]. When performing OYE-catalyzed reductions, it is necessary to employ an NADPH regeneration system, such as glucose dehydrogenase/glucose.

In a series of illuminating protein engineering studies of the prototypical ene-reductase OYE1 from *Saccharomyces pastorianus* as a catalyst in the

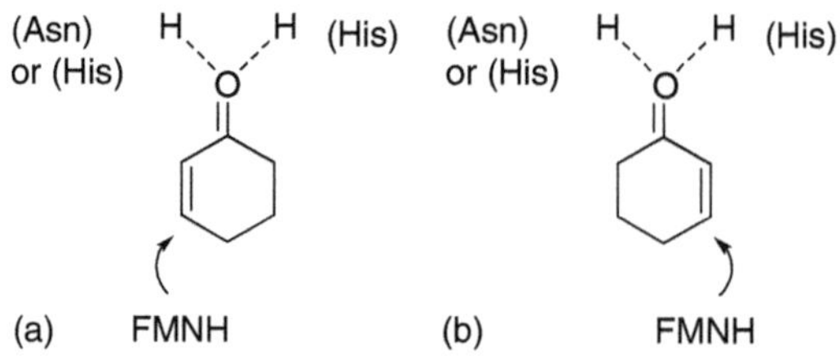

Scheme 8.8 Binding modes in the active site of ene-reductases. (a) Traditional (normal) binding mode. (b) Flipped binding mode. Source: Ref. [5]/John Wiley & Sons.

reduction of structurally different prochiral substrates, crystal structures of the evolved variants coupled with computational analyses uncovered the source of altered stereoselectivity [35]. In a preliminary investigation, it was postulated on the basis of the X-ray structure of WT OYE1 [36] that substitutions at position Trp116 could strongly influence the degree and direction of enantioselectivity. Indeed, upon screening an NNK saturation mutagenesis library at this position, improved variants were found for the reduction of such substrates as 2-methylcyclohexenone and (*R*)- and (*S*)-carvone [37]. Later this library was screened for a wide variety of other substrates, the mutants being characterized by crystal structures [35].

In the initial study, the OYE1 catalyzed reduction of Baylis–Hillman adducts was investigated; it was shown, inter alia, that in the reaction of 2-hydroxymethylcyclopentenone variant Trp116Ile causes reversal of enantioselectivity (91% ee (*S*) versus 60% ee (*R*) of WT) (Table 8.2) [35a].

Variant Trp116Ile was soaked with 2-hydroxymethylcyclopentenone, leading to crystals harboring the substrate, which diffracted at 1.7–1.4 Å. Surprisingly, two different binding poses were identified in the crystal, mode 1 with the reface of the π-system facing the $FMNH_2$ hydride source and leading to the minor (*R*)-product, and mode 2 in the flipped si-orientation leading to the observed (*S*)-product (Figure 8.10). Analysis of the two structures revealed different H-bond interactions of the hydroxymethyl group of the substrate with the respective environment in the two modes. Moreover, mode 1 was identified as an inferior Michaelis complex for catalysis, because the angle formed by FMN N_{10}-N_5-substrate ß-carbon is 78°, outside of the usual 96°–117° range observed previously in smooth hydride transfer [38]. This led to a model in which mode 2 is the productive pose. The results teach us that crystal structures are useful, but alone insufficient for understanding the source of enantioselectivity. Careful interpretations are necessary, at least with MD/docking computations, which were not reported in this study [35a].

In the second study, the source of enhanced and reversed diastereoselectivity of OYE1-catalyzed reduction of (*R*)- and (*S*)-carvone, observed earlier [37], was elucidated [35b]. The X-ray crystal structures of several key Trp116 variants and of WT OYE1 harboring (*R*)- or (*S*)-carvone revealed subtle changes that appear to control the orientation of substrate binding and thus determine diastereoselectivity. In line with the hydrophobic character of carvone (and of

Table 8.2 Stereoselective reduction of Baylis–Hillman adducts catalyzed by variants of OYE1 produced by saturation mutagenesis at position 116.

Protein	Substrate					
	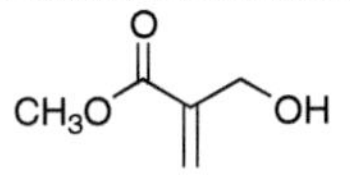		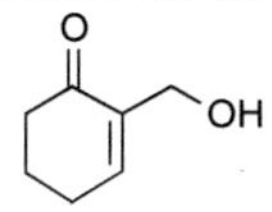		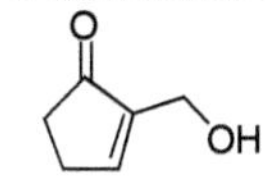	
	% conv	% ee[a)]	% conv	% ee[b)]	% conv	% ee[c)]
S. pastorianus OYE1 mutants (116 residue)						
Trp (wt)	19	>98 (*R*)	≤5	N.D.	51	60 (*R*)
Ala	9	90 (*R*)	84	>98 (*S*)	>98	72 (*S*)
Val	52	86 (*R*)	84	>98 (*S*)	97	92 (*S*)
Tyr	68	76 (*R*)	>98	>98 (*S*)	>98	87 (*S*)
Phe	37	70 (*R*)	>98	>98 (*S*)	98	>98 (*S*)
Ser	13	46 (*R*)	84	>98 (*S*)	87	>98 (*S*)
Ile	50	9 (*R*)	>98	>98 (*S*)	>98	91 (*S*)
Arg	≤5	N.D.	≤5	N.D.	≤5	N.D.
Pro	≤5	N.D.	14	>98 (*S*)	16	77 (*S*)
Thr	≤5	N.D.	28	>98 (*S*)	44	>98 (*S*)
Cys	≤5	N.D.	31	>98 (*S*)	47	77 (*S*)
Lys	≤5	N.D.	60	>98 (*S*)	75	76 (*S*)
Glu	≤5	N.D.	93	90 (*S*)	96	88 (*S*)
Asp	≤5	N.D.	>98	91 (*S*)	95	77 (*S*)
Gly	14	16 (*S*)	98	>98 (*S*)	>98	86 (*S*)
Leu	>98	20 (*S*)	>98	>98 (*S*)	>98	57 (*S*)
Asn	>98	41 (*S*)	>98	>98 (*S*)	>98	89 (*S*)
Met	15	64 (*S*)	>98	>98 (*S*)	>98	86 (*S*)
His	67	97 (*S*)	>98	>98 (*S*)	>98	77 (*S*)
Gln	78	>98 (*S*)	>98	>98 (*S*)	>98	89 (*S*)
P. stipitis OYE 2.6	>98	>98 (*S*)	>98	>99 (*S*)	>98	76 (*S*)

a) Absolute configurations were assigned by retention time comparison to an authentic sample of the *R* reduction product.
b) Absolute configurations were assigned by retention time comparisons to an authentic sample of *S* reduction product.
c) Absolute configurations were assigned by retention time comparisons to an authentic sample of *S* reduction product.
Source: Ref. [35a]/American Chemical Society.

most other substrates), substrate binding appears to be controlled largely by steric factors, and indeed the binding pocket of the ene-reductase is defined primarily by hydrophobic amino acids (Thr37, Met39, Phe74, Tyr82, Ala85, and Leu118), where substrates having large substituents can be accommodated, but access is blocked by tryptophan at position 116 (Figure 8.11). Substitutions at this position with incorporation of polar amino acids generally reduce activity sharply, with few exceptions. In the reaction of (*S*)-carvone, appropriate mutants allow the substrate to enter and also to determine either normal (e.g. Trp116Leu) or flipped binding modes (e.g. Trp116Ala). In the case of (*R*)-carvone, all variants

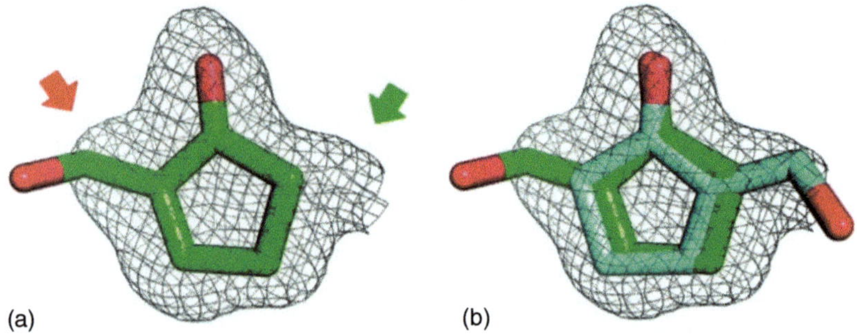

Figure 8.10 Location of substrate 2-hydroxymethylcyclopentenone in OYE1 mutant Trp116Ile within the observed electron density (0.4σ contour level). (a) Attempted poor fit by a single substrate orientation. Red and green arrows indicate regions of negative and positive electron density peaks, respectively, in the difference density map (not shown). (b) Successful fit by two substrate populations. C-atoms in binding mode 1 are pictured in green, those in binding mode are shown in light blue. Source: Ref. [35a]/American Chemical Society.

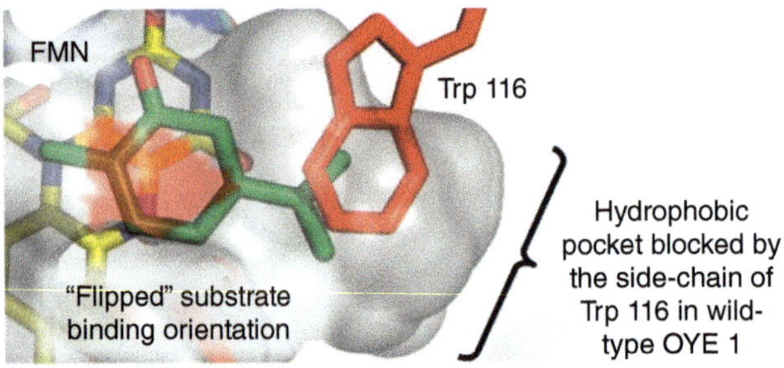

Figure 8.11 Schematic representation of the role of OYE1 variants characterized by mutations at position 116. Source: Ref. [35b]/American Chemical Society.

lead to the normal binding mode, with the exception of mutants Trp116Ala and vTrp116Val. This is surprising since variants having leucine or isoleucine at position 116 bind (*R*)-carvone only in the normal orientation. Subtle effects play a crucial role, but these are not easy to identify. Nevertheless, reasonable models based on the stereochemical and X-ray structural results were proposed.

The third study focuses on the origin of enhanced and reversed enantioselectivity in the bioreduction of (*Z*)-ß-aryl-ß-cyanoacrylates catalyzed by OYE1 mutants [35c]. The original saturation mutagenesis library at residue Trp116 was used once more. Both (*R*)- and (*S*)-products were observed, depending upon the amino acid substitution at position 116 of the variant. The degree of enantioselectivity varied considerably depending upon the substitution pattern of the aryl substituent, and of the particular mutation. Normal but also flipped binding modes were deduced. Soaking experiments with the aim of visualizing the binding mode of this class of substrates in the OYE1 variants were successful only in the case of the Trp116Ala variant. Two crystal structures were analyzed harboring (*Z*)-ß-phenyl-ß-cyano methyl acrylate and the *p*-fluoro analog, respectively. Both showed a nonproductive pose, which would be expected to provide the disfavored enantiomeric products. Even though such intermediates are not involved in the formation of the observed enantiomers, the structural data is nevertheless useful in interpreting the stereochemical results [5]. It was concluded that small amino acids at position 116 induce the substrate to adopt a classical orientation with formation of (*S*)-configurated products, while larger amino acids such as leucine result in a flipped pose leading to (*R*)-products [35c].

In conclusion, the combination of stereochemical results, X-ray structural data of OYE1 variants harboring prochiral substrates, and detailed analyses of the data have resulted in sound models, which explain the observed activity and enantioselectivity. The categorization of non-flipped and flipped substrate orientations in pre-transition states is an insightful lesson. A QM/MM investigation of OYE1 would be needed for further insights.

In a different approach, laboratory evolution of the ene-reductase from *Candida macedoniens* AKU4588 (CmOYE) was performed by introducing mutations in the substrate-recognition loop near the catalytically active site [39]. The results show that such a focus could also be successful in protein engineering of other OYEs. When comparing the crystal structure of WT CmOYE (apo) with the WT harboring the inhibitor *p*-hydroxybenzaldehye (*p*-HBA), the high flexibility of loop 6 became apparent (Figure 8.12), which is crucial in determining substrate acceptance and stereoselectivity [39]. In the apo form, the loop acts as a lid that closes the active site.

The WT enzyme was used in the reduction of ketoisophorone, which produced solely the desired (*R*)-product, but unfortunately, the reaction proved to be too slow for industrial applications [39]. The authors employed mainly

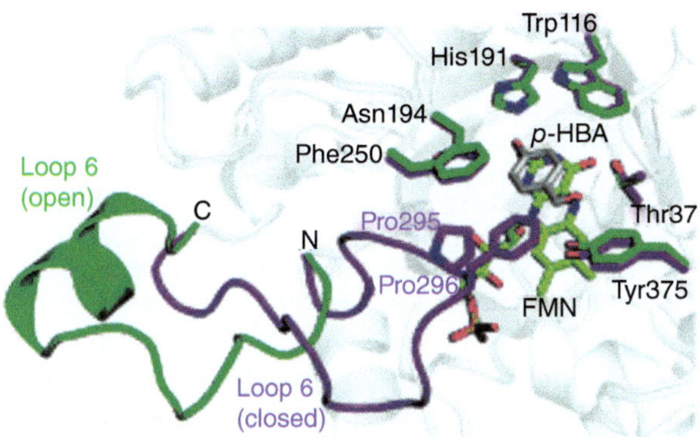

Figure 8.12 Superposition of CmOYE structures in the absence (green) and presence (magenta) of *p*-HBA in the catalytic pockets. The structures shown in green and magenta represent open and closed forms of CmOYE (loop 6), respectively. Amino acid residues in the catalytic sites, FMN (yellow), and *p*-HBA (gray) are shown as stick models. Source: Ref. [39]/John Wiley & Sons.

structure-guided rational enzyme design with the generation of small sets of variants, and in one case CAST-like saturation mutagenesis at a single residue Phe296. Mutations in loop 6 led to a twofold increase in activity while maintaining complete enantioselectivity, single mutant Pro295Gly being a prominent example. This variant was also effective in the stereoselective reduction of (4*S*)-phorenol with formation of the industrially desired (4*R*,6*R*)-actinol (Scheme 8.9), a 12-fold increase in activity being observed. It was postulated that Pro295 collides sterically with the dimethyl group of ketoisophorone and (4*S*)-phorenol, which means that it acts as substrate gatekeeper [39]. Mutation to Pro295Gly leads, inter alia, to greater flexibility and provides more space for substrate binding, an intriguing proposition that appears to be in line with the data. Thus, this substitution increases conformational flexibility of loop 6 that lids the catalytic site, thereby altering the substrate specificity. To corroborate this conclusion, MD/docking computations would have to be performed [5]. Since their first study, the authors have provided several X-ray structures of CmOYE and the other ene-reductase from *Pichia* sp. *AKU4542* (PsOYE), which show that the active site Phe246 (PsOYE)/Phe250 (CmOYE) has different positions. It causes the catalytic pocket of PsOYE to be wider than that of CmOYE, leading to 1.7-fold and 3.3-fold higher catalytic activities toward ketoisophorone and (4*S*)-phorenol, respectively [40].

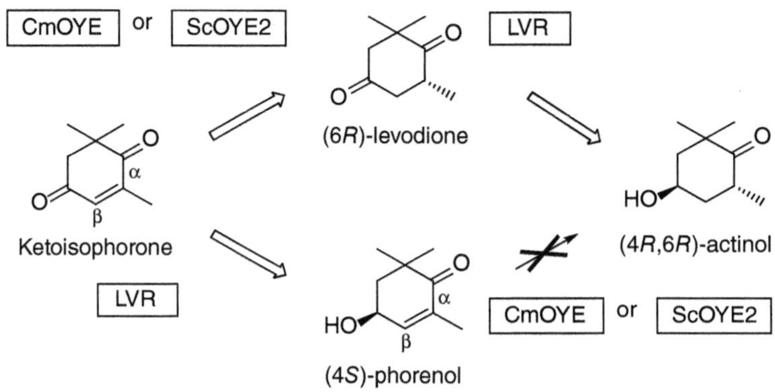

Scheme 8.9 Two-step biocatalytic conversion of ketoisophorone to (4*R*,6*R*)-actinol. Biocatalytic synthesis of (4*R*,6*R*)-actinol from ketoisophorone is performed by CmOYE (or ScOYE2) and LVR. CmOYE and ScOYE2 show less catalytic activity in the reduction of (4*S*)-phorenol than in the other reactions. Source: Ref. [39]/John Wiley & Sons.

The first and thus far only QM/MM study of an OYE (wildtype YqjM from *Bacillus subtilis*) appeared that extends the mechanistic picture of this class of enzymes [34]. This OYE has been used extensively in stereoselective reductions and is believed to follow a mechanism similar to other members of this class of enzymes except that two histidines (His164 and His167) bind and activate the substrate (instead of an Asn/His pair) [41]. Using 2-cyclohexenone as the substrate, two-dimensional QM reaction pathways at the B3LYP-D/OPLS2005 level

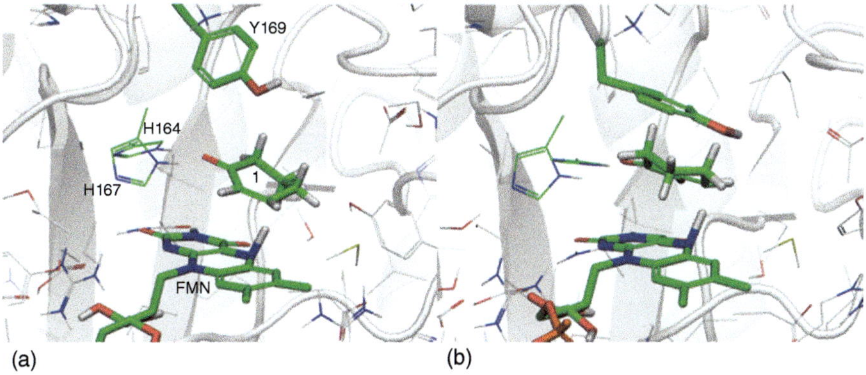

Figure 8.13 Computed transition state geometries of the lowest energy pathways for hydride transfer: (a) in the normal pose; (b) in the flipped pose. Source: Ref. [34]/American Chemical Society.

suggest that the hydride adds in a distinct rate-determining step followed by protonation of the stabilized enolate at the C-atom (not O-atom) in the second step (Figure 8.13). Thus, *hydride and proton motions are not concerted as often postulated.*

The computed activation barriers ($\Delta E^{\ddagger}$) and reaction energies (ΔE) for hydride transfer from N5 of the reduced flavin FMNH to the ß-C-atom of 2-cyclohexenone are summarized in Table 8.3. The identification of weak substrate binding, polarization, and activation of the carbonyl O-atom as well as the structural information of the transition state provide detailed models of the catalytic machinery of OYEs in general. Details of the origin of the proton

Table 8.3 Computed activation barriers and reaction energies[a)] for rate-determining hydride transfer from FMNH to 2-cyclohexenone in kcal mol^{-1}.

Substrate orientation	Profile	$\Delta E^{\ddagger}$	ΔE
Normal	1	15.7	7.3
	2	16.9	7.0
	3	18.0	9.3
Flipped	1[b)]	32.5	22.6
	2[b)]	32.9	24.5
	3[b)]	33.2	19.1
	4[c)]	22.5	16.4
	5[c)]	25.5	20.2

a) Calculated at the B3LYP-D3:cc-pVTZ(-f)/OPLS2005//B3LYP:6-31G(d)/OPLS2005 level.
b) Denotes starting structures that were obtained from the MD simulation prior to the observed conformation rearrangement.
c) Denotes starting structures that were obtained from the MD simulation following the observed conformation rearrangement.
Source: Ref. [34]/American Chemical Society.

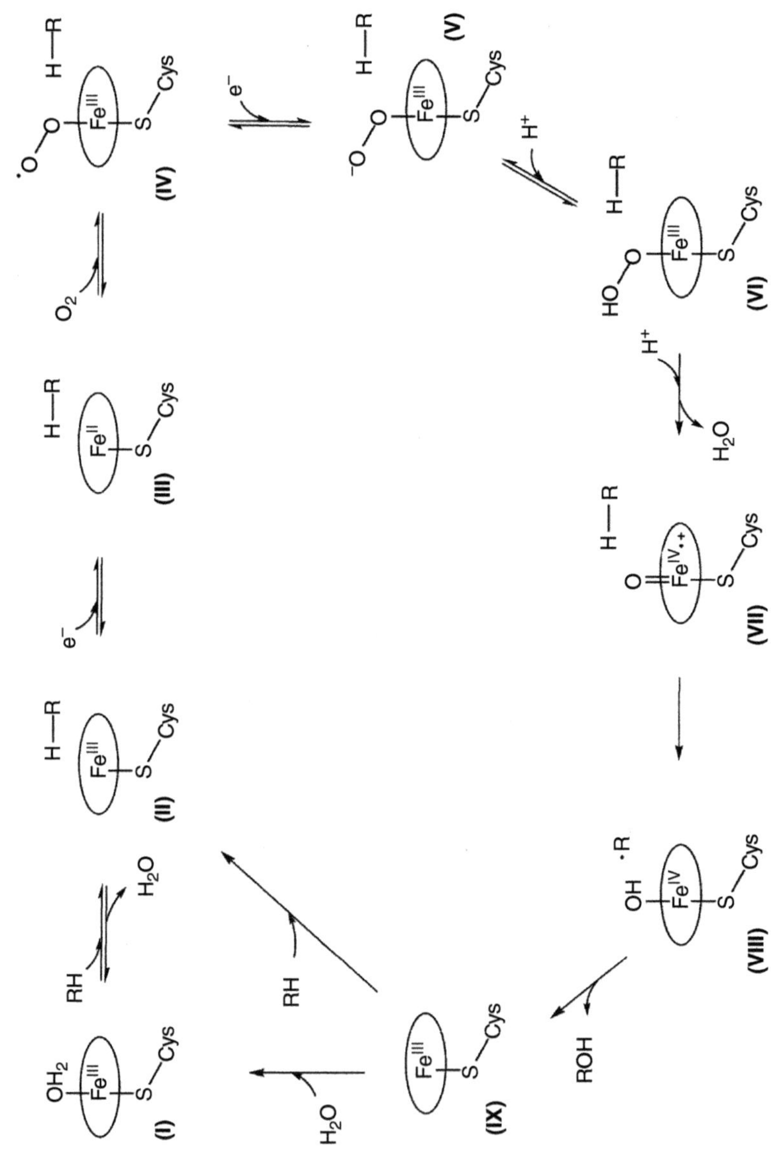

Scheme 8.10 Mechanism of P450 monooxygenases. Intermediate **VII** is catalytically active Compound **I**. Source: Ref. [43]/Royal Society of Chemistry.

adding to the α-C-atom of the reacting cyclohexanone are also included; it does not appear to be Tyr169 as traditionally assumed. The insights generated by this theoretical study may prove to be useful in designing optimal mutagenesis experiments in future directed evolution studies [5].

8.4.4 Cytochrome P450 Monooxygenase

Cytochrome P450 monooxygenases (CYPs) are heme-Fe-dependent enzymes, which catalyze the oxidative hydroxylation R–H → R–OH of a wide range of natural and unnatural organic compounds [42]. Olefin epoxidation has also been observed in some cases. The mechanism of oxidative hydroxylation involves the intermediacy of a catalytically active high-spin heme-Fe=O intermediate (so-called "Compound **I**") as shown in Scheme 8.10, which induces in the rate-determining step, the abstraction of an H-atom from the substrate R–H with formation of the short-lived radical R· followed by rapid C—O bond formation.

This high-energy process has been exploited in industrially relevant regio- and stereoselective transformations for a long time, as in steroid hydroxylation [44]. During the last decades, several other highly selective transformations have been reported, but unfortunately, the vast majority of tested substrates fail to react with high regio- and stereoselectivity. Moreover, many alkanes are not even accepted for various reasons [5]. Due to the very large binding pockets of CYPs, particularly small substrates, such as alkanes of the type propane, ethane, or methane are not oxidized because they appear in poses that, on average, are too far away from the catalytically active high-spin heme-Fe=O intermediate (Scheme 8.10) [45]. Many, but not all of these problems have been solved by directed evolution; for reviews of protein engineering of CYPs, see [46].

It is clear that CYPs are quite different from most other enzyme types, which means that different interpretations for explaining catalytic profiles are necessary [5]. Emil Fischer's lock-and-key postulate emphasizing shape complementarity [47] was extended by Linus Pauling's general enzyme theory based on the hypothesis that the transition states of enzyme-catalyzed transformations are stabilized by intimate interactions with the protein environment [48]. CYP-catalyzed oxidative hydroxylations may not correspond completely to the Pauling hypothesis, because the energy of a radical process may not be reduced to a notable extent by such interactions [42e]. Therefore, other explanations for substrate acceptance, regio- and stereoselectivity were developed, directed evolution playing an important role in this endeavor. As a result of many protein studies flanked by theoretical analyses, it has become clear that it is the protein environment in the binding pocket that "holds" the substrate above the catalytically active high-spin Compound **I** (Scheme 8.11) in a pose in which at least one of the CH-entities points toward the O-atom of heme-Fe=O, the angle of approach ideally being 130° as computed by QM [49a]. In the case of enantio- or diastereoselectivity, the hydrogen of the substrate that spends the largest amount of time in a sufficiently close position reacts preferentially and therefore defines the stereochemical outcome. This may well mean that the weakest C—H bond in a molecule is *not* the one that is oxidized, as protein engineers thought for a long time. Nevertheless, the particular C—H bond energy should not be neglected. Whenever the intermediate radical R is tetrahedral and chiral, the effect of its absolute configuration, (*R*) or (*S*), remains to be investigated. Such radicals are fleeting chiral intermediates [8]. Apart from this information gap, some important insights regarding the source of regio- and stereoselectivity were gained by MD/docking simulations [50]. An important QM study revealed the energetically optimal pose of a substrate so that H-atom abstraction can proceed optimally, the ideal O–H–C angle being about 130° [49].

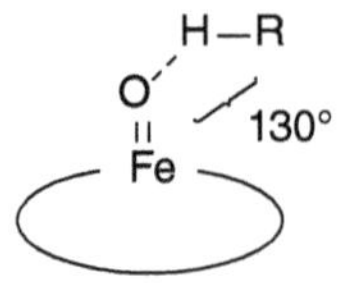

Scheme 8.11 Ideal pose of a substrate for smooth oxidative hydroxylation initiated by H-atom abstraction and formation of an intermediate short-lived radical R· which undergoes rapid C—O bond formation. Source: Adapted from Ref. [49].

Several CYPs play a central role in protein engineering, especially self-sufficient P450-BM3 from *Bacillus megaterium,* which consists of a heme-Fe domain and a diflavin reductase domain [51]. Fatty acids are the natural substrates. It has been characterized by X-ray crystallography [52] and used in much directed evolution and rational enzyme design studies [28, 46, 50]. In one report, it was noted that the C12 cycloalkane, cyclododecane, is not accepted by WT P450-BM3 [53]. Learning from an earlier rational enzyme design regarding hot spots and amino acid exchanges [54], a small collection of 24 mutants was tested, generated by utilizing five hydrophobic amino acids (Ala, Val, Phe, Leu, and Ile) at positions 87 and 328 directly above the heme group. The combined double mutant Phe87Ala/Ala328Val proved to be active. It had been known for some time that mutation Phe87Ala is instrumental in expanding the substrate scope of P450-BM3 because Phe87 appears to shield heme-Fe═O to some extent. To learn more about the effect of the double mutant as the catalyst in cyclododecane oxidation, MD/docking computations were performed (Figure 8.14) [53].

The MD simulation revealed that in the case of the double mutant Phe87Ala/Ala328Val, the pose of the substrate remains stable, the compound being in close proximity to the catalytically active heme-Fe═O (Compound **I**) with the C–H···O–Fe distance of 2.95 Å. It can also be seen that the small sidechain of Ala in the double mutant creates space that is occupied by part of cyclododecane (Figure 8.14a). The situation in the case of inactive WT P450-BM3 is quite different. The bulky sidechain of Phe87 occupies this space, thereby blocking an optimal substrate pose (Figure 8.14). However, this effect alone does

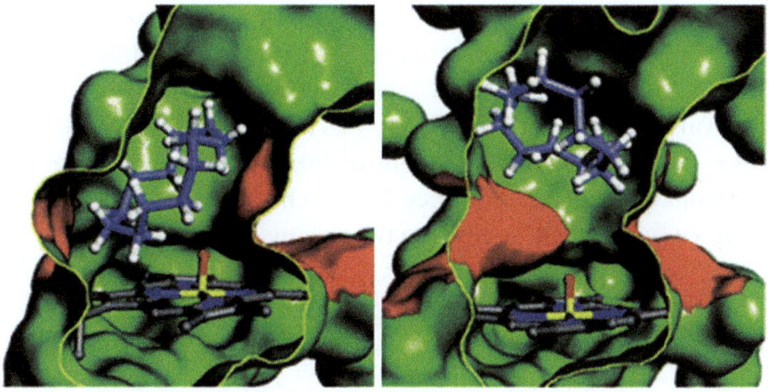

Figure 8.14 The substrate binding cavity of P450 BM3 harboring cyclododecane. (a) Double mutant F87A/A328V; (b) WT P450-BM3, both after 3 ns of unrestraint MD simulation. The mutated positions are depicted in red. Positions 87 (left) and 328 (right) stabilize the substrate in the active site cavity. The activated oxygen of the heme-Fe=O is shown in orange. Source: Ref. [53]/Royal Society of Chemistry.

not explain high activity toward the substrate, because single mutant Phe87Ala leads to a mere 4% conversion under the same experimental conditions. The other single mutant Ala328Val fails completely, which shows that the combined action of the two mutations is necessary for smooth oxidation, indicating a pronounced positive epistatic effect (more than additivity) (Figure 8.14a) [53]. MD/docking data of the single mutant Ala328Val was not reported but would provide additional insight into the details of substrate acceptance.

More challenging are P450-catalyzed transformations in which stereoselectivity needs to be engineered for applications in organic and pharmaceutical chemistry [5]. Indeed, P450-BM3 has been used in directed evolution for manipulating regio- *and* stereoselectivity in one and the same optimization process [5]. Prominent examples are the oxidative hydroxylation of steroids [50a], of 1-cyclohexene carboxylic acid ester with formation of the (*R*)- and (*S*)-enantiomers [50b], of ketones affording chiral acyloins [50c], and of 1-tetralones leading to hydroxylation at the 4-position [50d], of appropriate substrates with the creation of two chirality centers [50e], and of caged substrates in bioorthogonal processes [50f]. In all cases, ISM at CAST sites lining the binding pocket was applied. MD/docking calculations showed that in the binding pockets of the P450-BM3 mutants the substrates occupy an energetically preferred pose in which one of the two stereotopic C–H entities of a methylene group (CH_2) points to the high-spin heme-Fe=O. In contrast, when performing such computations using WT P450-BM3 or the "standard" mutant Phe87Ala, poor regio- as well as diastereo- and enantioselectivity are indicated. The optimal juxtapositioning of the substrates is a crucial lesson.

One of many examples of this kind of analysis was reported in the ISM-based study of P450-BM3 as a catalyst in the oxidative hydroxylation of testosterone (**6**) and other steroids (Scheme 8.12) [50a]. WT P450-BM3 fails to accept this substrate, while variant Phe87Ala is active but delivers a 50 : 50 mixture of regioisomers **7** and **8** in addition to small amounts of other alcohols. The triple mutant Arg47Ile/Thr49Ile/Phe87Ala leads to 94% overall selectivity in favor of the 2ß-product **7**, while variant Arg47Tyr/Thr49Phe/Val78Leu/Ala82Met/Phe87Ala is 96% selective for the 15ß-product **8**. A total of 9000 transformants had to be screened by automated high-performance liquid chromatography (HPLC), much more than typically 2000 mutants at the time using simpler substrates.

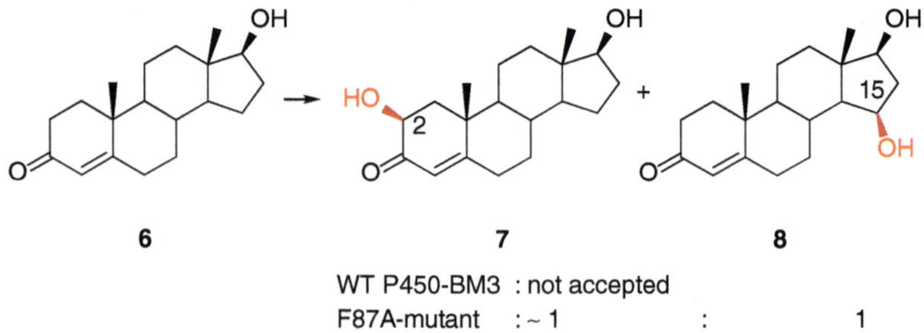

6 7 8

Scheme 8.12 P450-BM3 catalyzed oxidative hydroxylation of testosterone. Source: Ref. [50a]/Springer Nature.

Extensive MD/docking simulations using substrate **6** and mutant Phe87Ala identified *two poses of essentially equal energy*, one predicting 2ß-selectivity and the other 15ß-selectivity, in line with the experimental results [50a]. In sharp contrast, the analogous computations using the two ISM-evolved variants indicate only one reactive pose in each case. Figure 8.15 shows that in the 2ß-selective variant, the pro-ß H-atom of the methylene group at the 2C-position points toward the heme-Fe=O, while in the case of the 15ß-variant it is the pro-ß H-atom at the 15C-position, which is closest to the catalytic center [50a]. Similar analyses have been performed in the successful attempt to unveil the source of regio- and stereoselectivity in reactions of smaller unnatural substrates leading to (*R*)- and (*S*)-enantiomers [50].

A more rigorous theoretical treatment followed a decade later, in which the 2ß-selective mutant R47I/T49I/F87A was subjected to *complete deconvolution with construction of a multiparameter fitness pathway landscape* [55]. Up to this point (2021), fitness pathway landscapes of multi-mutational variants involved only a *single catalytic parameter* such as enantioselectivity. In the new study, a unique fitness landscape was constructed experimentally in which activity, diastereoselectivity, and enantioselectivity were considered simultaneously for the first time. Extensive QM and MD computations as well as a movie revealed that cooperative and deleterious epistatic effects (more than additive) are modulated by long-range interactions in loops, helices, and ß-strands that control the substrate access tunnel (Scheme 8.13). Thus, hitherto unsuspected conformational dynamics were uncovered in CYP catalysis [55]. The model supports the original hypothesis that proper positioning of the steroid for

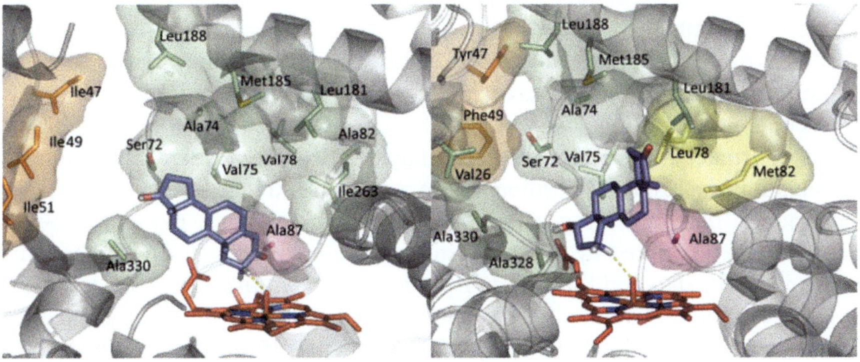

Figure 8.15 Computed pose of testosterone (**6**) explaining 2ß-selectivity (mutant R47I/T49I/F87A) and the respective pose leading to 15ß-selectivity (mutant R47Y/T49F/V78L/A82M/F87A). Source: Adapted from Refs. [46a, 50a].

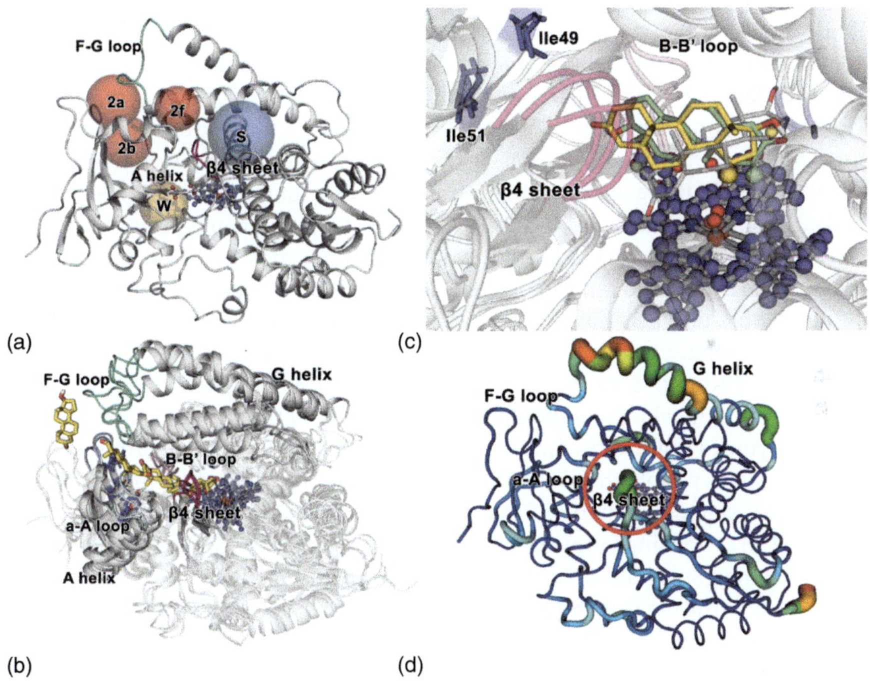

Scheme 8.13 Secondary structural elements determining regioselectivity and activity of 2ß-selective P450-BM3 mutant. (a) The channels observed in the WT crystal structure (PDB: 1FAG) and in the mutants of our MD simulations are shown as spheres (red and blue colors). (b) The representative snapshots derived from MD trajectories. (c) Substrate rotation from pose15 to pose2 in the active site of mutant III. (d) Principal component analysis of mutant III. The red circle indicates the β4 sheet. Source: Ref. [55]/Springer Nature/CC BY 4.0.

rapid, regio- and diastereoselective hydroxylation is required, and adds some interesting details.

At the time when the first steroid study based on semi-rational directed evolution appeared [50a], a great deal of excitement in the chemistry and biotechnology community was visible. However, representatives of the international pharmaceutical field were somewhat skeptical, because hydroxylation of steroids at the C2 and C15 positions was not needed and therefore of no practical importance. In contrast, especially positions C16 and C7 were reported to be of acute interest. Inspired by this criticism, protein engineering of P450-BM3 for achieving both 16α- *and* 16ß-selectivity *in a planned manner* was first attempted (MTR604). The technique of Mutability Landscaping [56] was first applied to identify optimal residues at which ISM was then performed in a subsequent step using appropriately reduced amino acid alphabets and mutant F87A as template [57]. Essentially complete 16α- and 16ß-selectivities were evolved, each transformation with unusually high activity (total turnover number (TTN) = 8660 and 9044, respectively). Only 3000 transformants had to be screened. Details of the directed evolution campaign can be found in the original publication [57]. As part of the mechanistic work, enzyme kinetics, electron spin resonance (ESR) experiments, and extensive MD and docking computations followed, leading to the model displayed in Scheme 8.14 and in two movies which reveal some dynamic features [57]. It can be seen that the computed poses of testosterone in the binding pockets of the two diastereoselective mutants are consistent with the experimental results. An even more rigorous theoretical analysis, as in the analysis of the 2ß-selective mutant [55], could provide further lessons. On the practical side, the evolved mutants proved to be active and selective in reactions to four other steroids in addition to testosterone (**6**) [57].

Remarkable progress has been made in learning from directed evolution of active and stereoselective P450 monooxygenases. In summary, semi-rational directed evolution of P450-BM3 and other CYPs based on CAST and ISM, and flanked by MD/docking computations, has thrown light on the mechanism of these enzymes [5, 46]. The function of the immediate protein environment at the active site is to hold the substrate in a juxtaposition that leads to the observed regio- and stereoselectivity of mutants generated. Standard or preferably advanced MD/docking software is generally successful [50], but some studies do not include such theoretical analyses. In the case of directed evolution of P450-BM3, as the catalyst in the oxidative hydroxylation of methylcyclohexane, a variant was produced leading to the simultaneous creation of two centers of chirality with high regio- as well as diastereo- and enantioselectivity [50e]. Amazingly, this substrate contains no functional groups that could participate in binding, and still, all three types of selectivities are high. Interestingly, standard MD/docking computations were not successful in predicting the experimental outcome. This shows that more refined theoretical approaches need to be tested (or developed!), a painful lesson. Perhaps more than one molecule of such small hydrophobic substrates occurs in the binding pocket of CYPs [58]. After all, in the crystal structure of one of the P450-BM3 mutants, it was discovered that

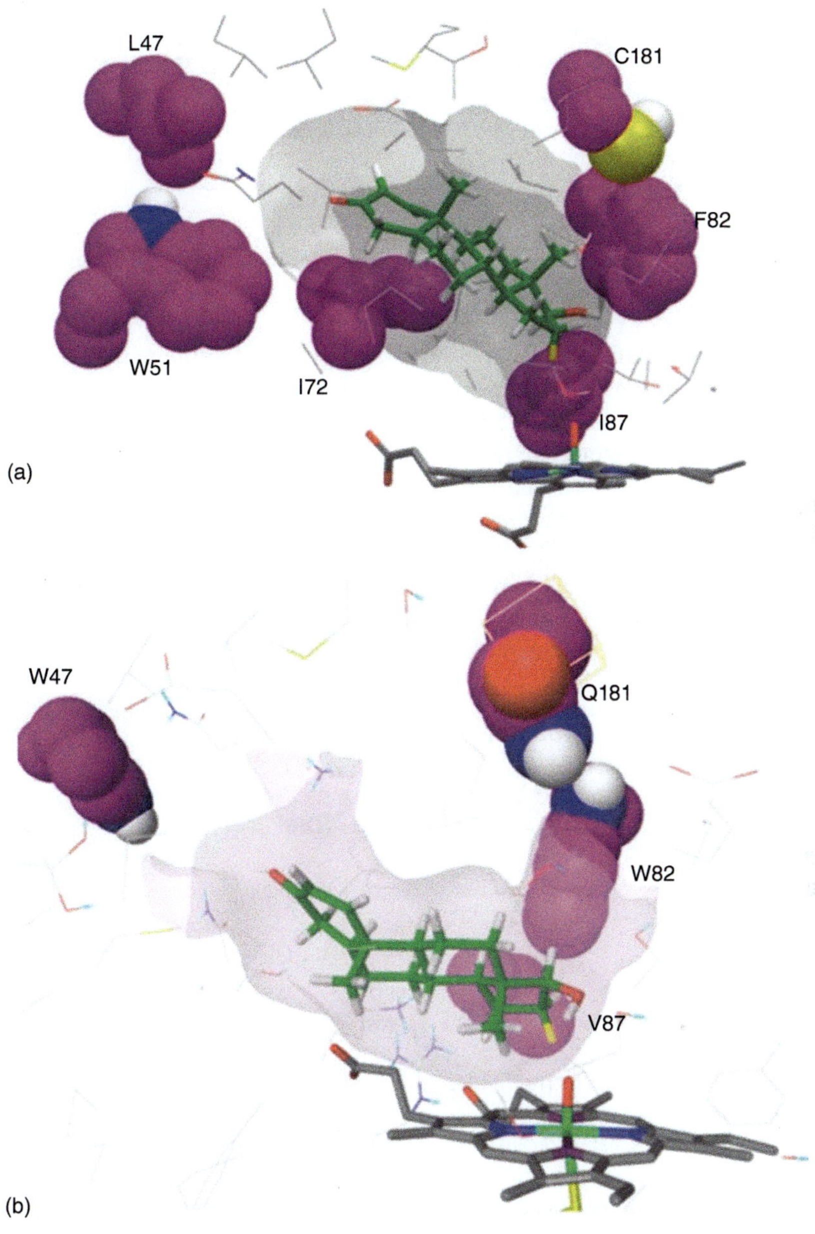

Scheme 8.14 Computed preferred poses of testosterone in binding pockets of P450-BM3 mutants leading to complete 16α-selectivity (a) and 16ß-selectivity (b), respectively. Source: Ref. [57]/American Chemical Society.

the reshaped binding pocket harbors two steroid molecules [57]! Consequently, a dozen or so much smaller alkanes should fit into the binding pockets of CYPs. Another revealing study of WT P450 BM3 as the catalyst in fatty acid hydroxylation appeared subsequently, in which MD and QM/MM computations were shown to predict substrate-induced gating as well as regio- and enantioselectivity [51c]. The generality of this approach still needs to be demonstrated by further investigations.

8.4.5 Analysis of Baeyer–Villiger Monooxygenase with Consideration of Fleeting Chiral Intermediates

Baeyer–Villiger oxidations of ketones have been practiced for more than a century using chemical reagents such as H_2O_2 or alkyl hydroperoxides ROOH [59]. It was found empirically that in reactions of unsymmetrical ketones $R^1(C{=}O)R^2$, one of the R-groups migrates preferentially, usually the "larger" one. Decades of theoretical work has shown that it is the one that stabilizes the incipient positive charge at one of the O-atoms as the O—O-bond breaks heterolytically at the stage of the Criegee intermediate, and that an anti-periplanar conformation must be maintained for maximum sigma-orbital overlap (Scheme 8.15) [61]. BV reactions can also be performed using BVMOs, in which case the same trends regarding preferential group migration are observed [62].

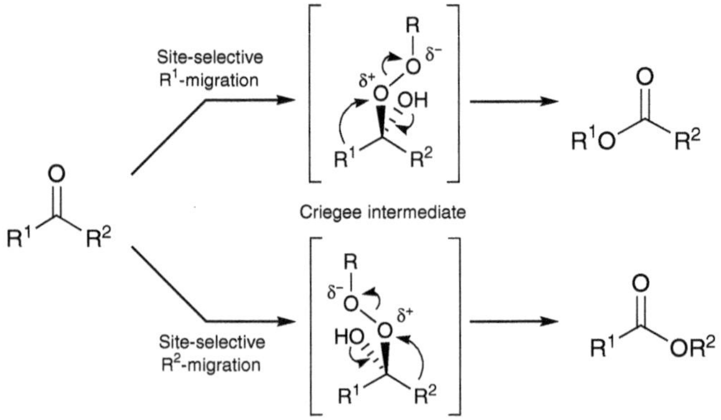

Scheme 8.15 Mechanism of Baeyer–Villiger reactions using synthetic reagents such as ROOH or BVMOs. Source: Ref. [60]/American Chemical Society.

Directed evolution has been applied previously to various enzyme types for manipulating regioselectivity (also called site-selectivity), summarized in a review [63] (see also Chapter 5). In the case of BVMOs, inverting normal reactivity with formation of abnormal products is a particular challenge. This difficult goal was finally reached by applying a directed evolution campaign, which included the use of CAST/ISM [60]. The thermostable BVMO from

Thermocrispum municipale DSM 44069 (TmCHMO) served as the enzyme and 4-phenyl-2-butanone (**9**) as the model substrate (Scheme 8.16). The final variant LGY3-D-E1 is a quadruple mutant (L145G/F434G/T435F/L437T) that reverses regioselectivity completely from WT TmCHMO (99 : 1 in favor of the normal product **11**) to 2 : 98 in favor of the abnormal product **10** in which the normally disfavored methyl group has migrated [60]. This seminal study included the QM-based calculation of the (*R*)-configurated Criegee intermediate, which seemed to fit best. A reasonable model for interpreting the experimental results was suggested. However, the transition state energy of the breakdown of the (*S*)-configurated Criegee intermediate was not computed, nor were those of the fleeting chiral intermediates at all stages of a given evolutionary upward climb connecting WT TmCHMO with the final best mutant.

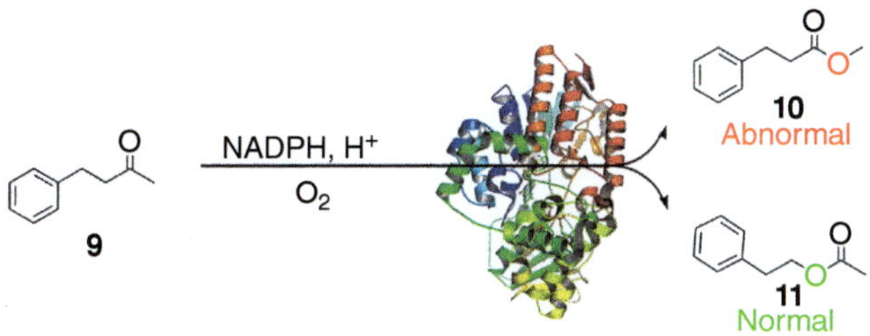

Scheme 8.16 Model TmCHMO-catalyzed reaction used in two BVMO studies. Source: Adapted from Refs. [60, 64].

In a follow-up study [64], this was achieved by performing complete deconvolution of the quadruple variant with construction of a fitness pathway landscape, in this case comprising 4! = 24 upward pathways based on the respective four single, six double, and four triple mutants (Figure 8.16). As in previous deconvolution exercises in ISM studies [16], pronounced cooperative (more than additive) mutational effects were uncovered in all 24 trajectories connecting WT TmCHMO with the best mutant. For details and mechanistic explanations, the reader is advised to read the original paper. Here, the focus is on the question of the role of chirality of the Criegee intermediates, which indeed was answered by performing extensive QM computations.

The role of the absolute configuration of the chiral Criegee intermediate was studied by extensive QM computations, only some typical results being summarized here in Table 8.4. This includes the energy of the respective Criegee intermediate and the transition state energies of the subsequent transformation. It can be seen that in WT TmCHMO the (*R*)-configuration is clearly preferred, but at later stages of the evolutionary process, a switch to the (*S*)-configuration occurs.

Part of the interpretations on a molecular level is presented here with reference to Figure 8.17, details can be found in the original publication [64]. For example, triple mutant L437T/L145G/F434G contains the additional exchange of phenylalanine to glycine at residue 434, which creates more space for better

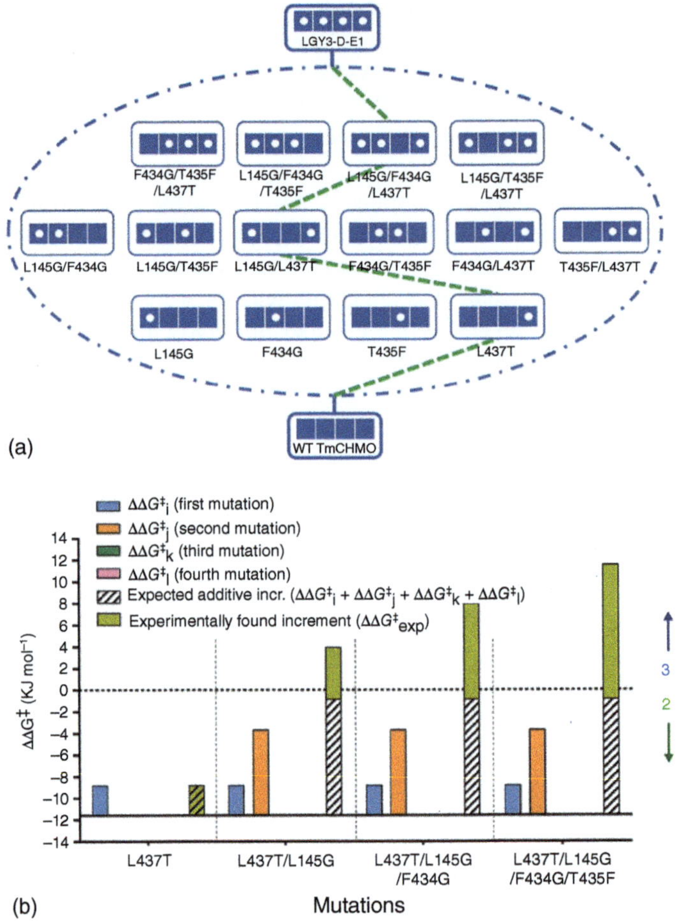

Figure 8.16 Results of deconvoluting the best TmCHMO mutant LGY3-D-E1 (L145G/F434G/T435F/L437T). (a) The 14 mutants necessary for deconvolution, made by site-specific mutagenesis, the green lines symbolizing an arbitrarily chosen pathway for illustrative purpose; (b) representation of positive epistatic mutational effects of the arbitrarily chosen upward pathways. Source: Ref. [64]/American Chemical Society.

accommodating the phenyl ring of substrate **9**. The QM-computed energy of the transition state to the abnormal product **10** ($TS_{M3\text{-}R\text{-}abnormal}$) is 0.2 kcal mol^{-1} lower than that of the transition state energy leading to the normal product ($TS_{M3\text{-}S\text{-}normal}$) (Figure 8.17). Moreover, the triple mutation allows F507 to move closer to the Criegee intermediate, keeping the phenyl ring of the Criegee intermediate in a specific orientation that disfavors the migration of the normally preferred phenylethyl. In the case of the quadruple mutant, the optimized structures of the transition states are similar to those of the triple mutant, which means that the additional mutation T435 has a small effect on regioselectivity. Indeed, the experimentally measured ratio of the two products is very similar,

Table 8.4 Calculated energies (kcal mol^{-1}) of the optimized structures for the Criegee intermediates (Int_{normal} and $Int_{abnormal}$) and respective transition states (TS_{normal} and $TS_{abnormal}$) along the reactions of WT enzyme TmCHMO and mutants.

	Int_{normal}		$Int_{abnormal}$		TS_{normal}		$TS_{abnormal}$	
	R	*S*	*R*	*S*	*R*	*S*	*R*	*S*
WT	**0.0**	6.8	3.2	7.5	**9.3**	21.0	**17.5**	20.4
L437T	1.3	2.7	**0.0**	9.5	**11.6**	19.1	**13.5**	22.7
L145G/L437T	1.2	2.4	**0.0**	9.3	**13.0**	18.5	**13.2**	22.4
L145G/F434G/L437T	2.9	**0.0**	1.7	5.6	15.2	**14.9**	**14.7**	20.0
L145G/F434G/T435F/L437T	3.2	**0.0**	2.6	5.3	16.8	**16.1**	**15.8**	20.0

Note: The subscription "normal" and "abnormal" indicate that the structures are from the pathways leading to the formation of the normal (**11**) and abnormal (**10**) products, respectively. "*R*" and "*S*" are the absolute configuration of the Criegee intermediates and the corresponding transition states. The relative energies of the lowest-energy Criegee intermediates are set to zero. The intermediates and transition states from the lowest energy pathway leading to different products are highlighted in bold.
Source: Ref. [64]/American Chemical Society.

although differences in activity are evident. The configuration of the Criegee intermediate is also the same as in the case of the triple mutant.

A final theoretical study to be highlighted here concerns the alcohol dehydrogenase (ADH) from *Ralstonia* sp. (RasADH) as the catalyst in the reductive desymmetrization of bulky 1,3-cyclodiketones needed for the production of pharmaceutically relevant steroidal drugs (Scheme 8.17) [65]. Stereoselective mutants were first evolved by CAST/ISM using ethyl secodione as the model substrate with formation of a product having the desired (13*R*,17*S*)-configuration. Subsequent synthetic steps leading to levonorgestrel and gestodene had already been described in the literature (Scheme 8.17b).

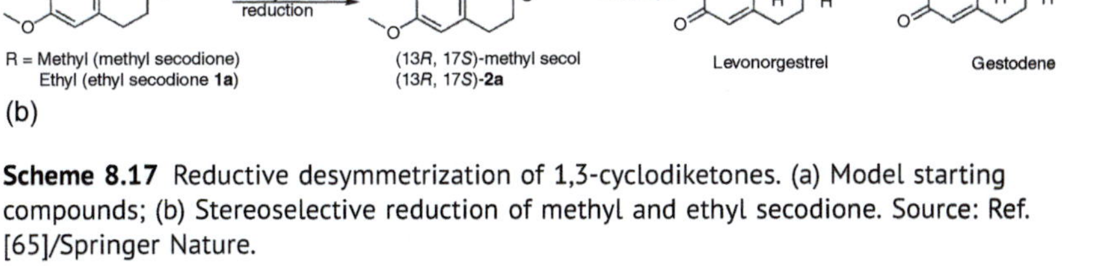

Scheme 8.17 Reductive desymmetrization of 1,3-cyclodiketones. (a) Model starting compounds; (b) Stereoselective reduction of methyl and ethyl secodione. Source: Ref. [65]/Springer Nature.

The CAST/ISM procedure was guided by docking ethyl secodione into the binding pocket the RasADH crystal structure [66], which led to crucial insights.

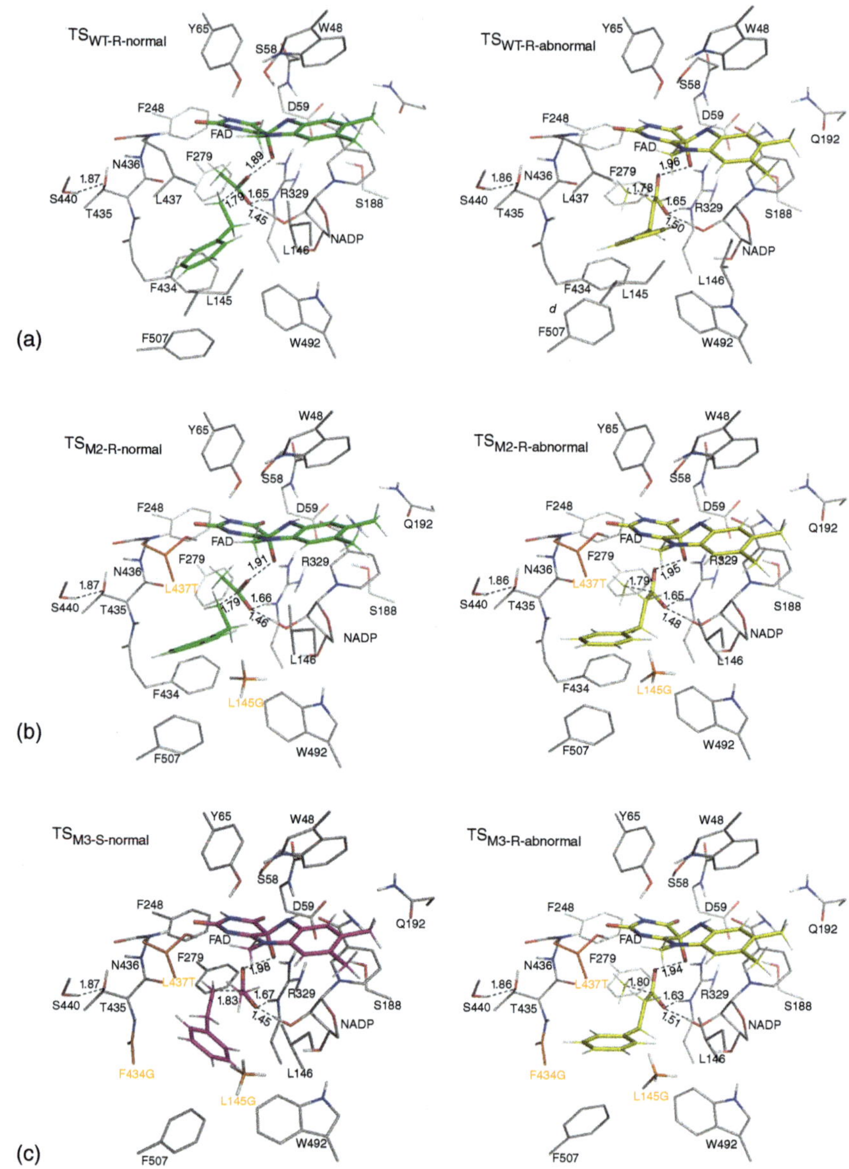

Figure 8.17 Computationally-optimized structures of the transition states involved in the formation of the normal product 11 and the abnormal product 10. (a) WT TmCHMO; (b) the double mutant L437T/L145G (M2); (c) the triple mutant L437T/L145G/F434G (M3). In the structures leading to the formation of the normal product, the substrate and the cofactor FAD are highlighted in green (Criegee intermediate: (*R*)-configurated) or purple (Criegee intermediate (*S*)-configurated), and for the structure leading to the formation of the abnormal product, **1** and FAD are shown in yellow (Criegee intermediate (*R*)-configurated). Selected distances are given in Å. Source: Ref. [64]/American Chemical Society.

It was found that S137 stabilizes and polarizes the carbonyl function of the substrate, while Y150 donates a proton to the O-atom of the carbonyl moiety. K154 was interpreted as enhancing the acidity of K150 and also interacts with the nicotinamide ribose group. The α6-helix, comprising residues 195–209, is at the entrance to the binding pocket, which is fairly close to the protein surface. Most importantly, F205 was found to cause steric interaction with the bulky tetrahydronaphthenyl ring, and L144 is on the other side of the ring, likewise preventing the substrate from entering the binding pocket (low activity). Residues I187, I188, Q91, Q93, H147, I187, and Y150 are all close to the cyclopentanedione ring. It became clear that L201 is close to F205, and that both are hydrophobic and cause steric hindrance effects. Therefore, in the first round of ISM, L201, and F205 were chosen for NNK-based saturation mutagenesis in separate experiments, although a reduced amino acid alphabet may have been the better option. Mutant F205A with alanine as the hydrophobic and smaller amino acid proved to be the best mutant, which subsequently served as the template for the second mutagenesis round with simultaneous randomization at double site I187/I188, resulting in variant 3B3 characterized by 3-point mutations I187S/I188L/F205A [65].

Then five saturation mutagenesis experiments were performed at sites I91/Q93, L144, H147, Y150, and Q191, providing 3B3 (I191V). To check whether valine is really optimal at residue 191, saturation mutagenesis was carried out at this position, and indeed, the best variant F12 emerged (I187S/I188L/F205A/I91V/Q191N), showing >99.5% stereoselectivity and a 183-fold enhancement of specific activity, in addition to leaving the remaining carbonyl function unreacted. The essential ISM steps are shown in

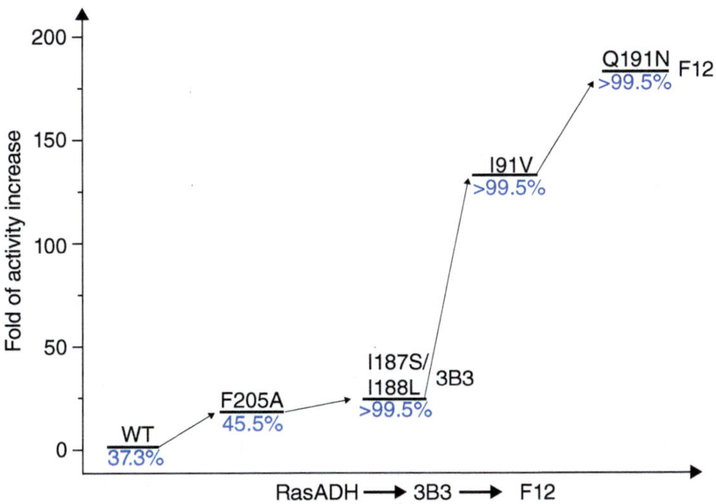

Scheme 8.18 ISM campaign of RasADH. Enzyme activity of WT is taken as 1.0, and the percentage of the desired (13*R*,17*S*)-product is shown in blue. Source: Adapted from Ref. [65].

Table 8.5 Catalytic profiles of WT RasADH and of selected single point mutations and combinations thereof in a type of deconvolution process.

Entry	Enzymes	Ratio of 13*R*, 17*S*%[a)]	Specific activity ($U\,mg^{-1}$)	Enhanced fold enzyme activity	Ratio of diol products (%)[a)]
1	RasADH wild type	37.3	0.008	1.0	18.8
2	F205A	45.5	0.142	17.7	6.0
3	3B3 (I187S/ I188L/F205A)	> 99.5	0.165	20.6	0.4
4	I187S/I188L/ F205A/I91V	> 99.5	1.073	131.4	< 0.1
5	F12 (I91V/I187S/I188L/ Q191N/F205A)	> 99.5	1.464	183.0	< 0.1
6	I187S	—	—	—	—
7	I188L	30.5	0.153	19.1	22.8
8	I187S/I188L	49.1	0.003	0.8	11.4
9	I187S/F205A	98.9	0.005	0.6	1.4
10	I188L/F205A	99.2	0.004	0.5	1.0

a) All of the four ketol products were considered 100%, and the ratio of ketol products was determined by the peak area as in the high-performance liquid chromatography (HPLC) plots.
Source: Ref. [65]/Springer Nature.

Scheme 8.18 [65]. Some of the point mutations were also prepared separately and combined, resulting in information in a type of deconvolution process (Table 8.5). For example, mutations I187S and I188L alone have quite different catalytic effects, yet when each is combined with F205A, the catalytic profile shifts toward improvement in both cases (Table 8.5, entry 3). As the authors point out, hot spot residues 205, 187, and 188 exert substantial synergistic effects (more than additive), an important lesson. Complete deconvolution of the final variant would have provided even more revealing information, possibly also demonstrating cooperative mutational effects.

The second type of lesson learned in this study concerns the structure- and MD-based investigation of the source of enhanced activity and stereoselectivity. Crystal structures of mutants 3B3/I91V and F12 were obtained housing both NADPH and the desymmetrization product (R = ethyl) with the desired (13*R*,17*S*)-configuration. The overall structures of WT RasADH and mutants with NADPH are very similar as shown by alignment (Figure 8.18). The active site center of variant F12 reveals that the amino acid S187 does not interact with the desymmetrization product, rather, it forms an H-bond with the phosphate moiety of NADPH. Importantly, F205A provides more space for accommodating the 6-methoxy-1,2,3,4-tetrahydronaphthalenyl moiety in the binding pocket, while I91V does the same for the ethyl group of the cyclopentane ring. Steric clashes are thus prevented. These and other effects determine activity and stereoselectivity.

Despite these insights, the structural analyses did not reveal any details of the role of the α6-helix which contains F205A. Therefore, five independent 250 ns

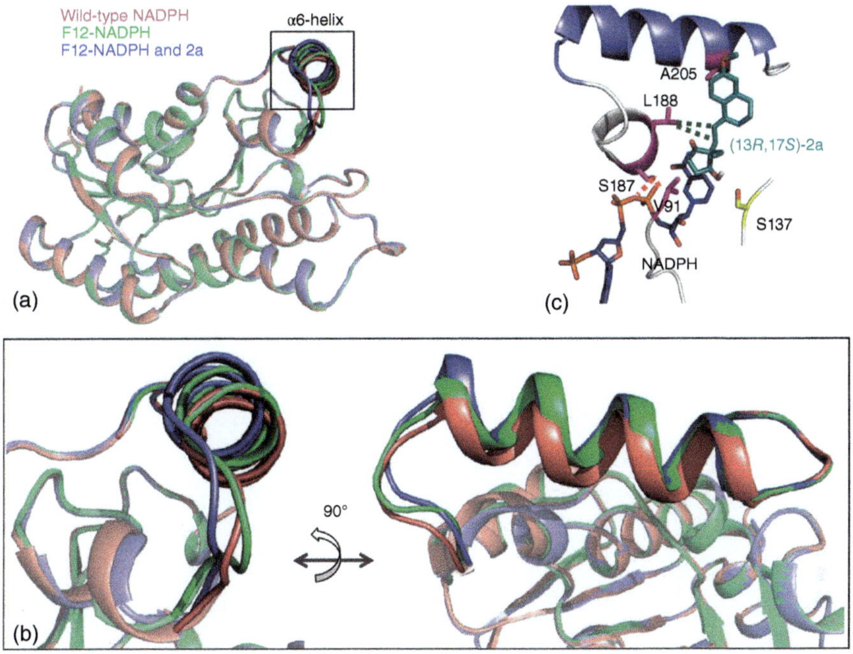

Figure 8.18 X-ray structures of WT RasADH and variant F12. (a) Overlay of WT RasADH/NADPH and variant F12/NADPH with or without housing the desymmetrization product; (b) Magnification of the enzyme lid; (c) Details of active site view of F12/NADPH with desymmetrization product. WT/NADPH, red; F12/NADPH, green; Chain B: F12/NADPH, blue. Source: Ref. [65]/Springer Nature.

MD computations were first performed for the systems having bound NADPH, which allowed conformational population analyses (Figure 8.19) [65]. The most general conclusion concerns the finding that F12 hardly samples the closed lid conformation in contrast to WT, which suggests a dynamic enzyme behavior.

To study the source of stereoselectivity more closely, a typical conformation from each minima A–E was chosen for calculating the active site volume and binding affinities of the substrate. This meant four conformations of WT that correspond to A–C and E were considered, and three conformations (C–E) for variant F12. The predictions were then validated by five independent 250 ns MD simulations. The dihedral angle formed between NADPH and the substrate was employed as an indicator of the desired (13*R*,17*S*)-enantiopreference (Figure 8.20). The lid conformation of variant F12 is strongly influenced by the respective mutations, which induces a reshaping of the binding pocket. Extra space is evolved for accommodating the substrate. On the practical side, variant F12 was also tested successfully with other structurally related bulky starting compounds, quite different from WT RasADH [65]. In summary, this is a deep-seated theoretical study of an enzyme that provides significant lessons of the type outlined in the introduction of the chapter.

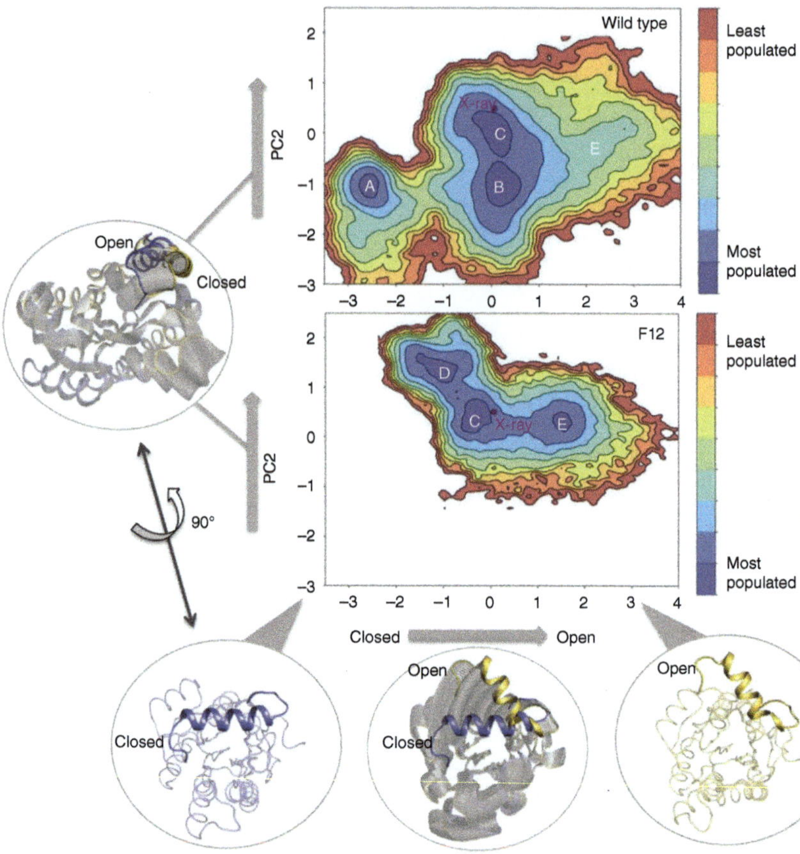

Figure 8.19 Conformational population analyses of the two most important components PC1 and PC2 that are based on Cα contacts for WT RasADH and the F12 mutant. Residues 195–209 are shown in purple, which depict conformations exploring the closed states of the α6-helix (lid), while the open states are shown in yellow having high PC1 values. The start of the MD simulations is projected on the conformational landscape, shown in violet. Source: Ref. [65]/Springer Nature.

8.5
Conclusions and Suggestions for Further Theoretical Work

As noted in Section 8.1, two types of lessons have been learned in protein engineering of enzymes used as catalysts in organic and pharmaceutical chemistry and in biotechnology: Uncovering the origin of enhanced or inverted stereo- and/or regioselectivity, and finding out whether mutations are additive or more than additive, as revealed by deconvolution of multi-mutational variants with construction of fitness landscapes. These lessons will continue to be learned, provided that researchers realize that such efforts are truly rewarding. The insights that are gained can be expected to help future campaigns in directed

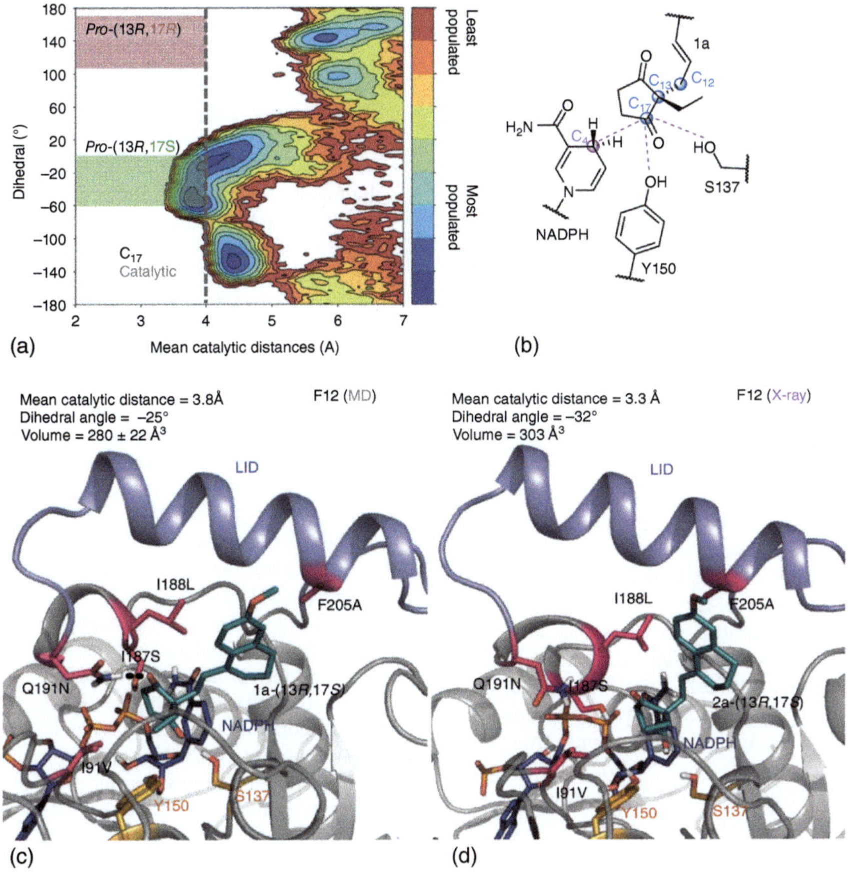

Figure 8.20 Molecular dynamics analysis of variant F12. (a) Distribution of distance and dihedral angle during five independent MD simulations, the schematic representation of distance and dihedral angle is shown in (b). Two representative snapshots from the catalytically competent minima pro-(13*R*,17*S*) for F12 (c) and an F12 X-ray structure co-crystalized with the (13*R*,17*S*)-**2a** product and NADPH co-factor (d). Source: Ref. [65]/Springer Nature.

evolution and rational enzyme design. The two approaches have merged. Since computer capacity will continue to increase with better quantum computers, it will be possible to develop more sophisticated QM/MM techniques, so that the reliability of predictions will be increased. This includes better open-shell systems for studying radical intermediates in enzyme-catalyzed reactions. Researchers are also advised to figure out in which cases machine learning (ML) and deep learning (DL) notably support protein engineering. Following the initial studies of hitherto unexplored fleeting chiral intermediates [8, 64], more investigations will certainly provide new and mechanistically unique lessons, and indeed the first reports concerning this issue have already appeared [67].

References

1. (a) Hutchison, C.A. III, Phillips, S., Edgell, M.H. et al. (1978). Mutagenesis at a specific position in a DNA sequence. *J. Biol. Chem.* **253** (18): 6551–6560. (b) Smith, M. (1994). Synthetic DNA and biology (nobel lecture). *Angew. Chem. Int. Ed.* **33** (12): 1214–1221.
2. (a) Fersht, A. (1999). *Structure and Mechanism in Protein Science*. New York: W.H. Freeman and Company. (b) Fersht, A.R., Shi, J.P., Wilkinson, A.J. et al. (1984). Analysis of enzyme structure and activity by protein engineering. *Angew. Chem. Int. Ed.* **23** (7): 467–473.
3. (a) Carter, P.J., Winter, G., Wilkinson, A.J., and Fersht, A.R. (1984). The use of double mutants to detect structural changes in the active site of the tyrosyl-tRNA synthetase (*Bacillus stearothermophilus*). *Cell* **38** (3): 835–840. (b) Wells, J.A., Powers, D.B., Bott, R.R. et al. (1987). Designing substrate specificity by protein engineering of electrostatic interactions. *Proc. Natl. Acad. Sci. U.S.A.* **84** (5): 1219–1223. (c) Wells, J.A. (1990). Additivity of mutational effects in proteins. *Biochemistry* **29** (37): 8509–8517. (d) Horovitz, A. (1996). Double-mutant cycles: a powerful tool for analyzing protein structure and function. *Fold Des.* **1** (6): R121–R126. (e) Shortle, D. (1992). Mutational studies of protein structures and their stabilities. *Q. Rev. Biophys.* **25** (2): 205–250. (f) Laskowski, M. Jr., Kato, I., Ardelt, W. et al. (1987). Ovomucoid third domains from 100 avian species: Isolation, sequences, and hypervariability of enzyme-inhibitor contact residues. *Biochemistry* **26** (1): 202–221. (g) Nelson, H.C. and Sauer, R.T. (1985). Lambda repressor mutations that increase the affinity and specificity of operator binding. *Cell* **42** (2): 549–558. (h) Mildvan, A.S., Weber, D.J., and Kuliopulos, A. (1992). Quantitative interpretations of double mutations of enzymes. *Arch. Biochem. Biophys.* **294** (2): 327–340. (i) Mildvan, A.S. (2004). Inverse thinking about double mutants of enzymes. *Biochemistry* **43** (46): 14517–14520. (j) Huang, Z., Wagner, C.R., and Benkovic, S.J. (1994). Nonadditivity of mutational effects at the folate binding site of *Escherichia coli* dihydrofolate reductase. *Biochemistry* **33** (38): 11576–11585. (k) Wagner, C.R., Huang, Z., Singleton, S.F., and Benkovic, S.J. (1995). Molecular basis for nonadditive mutational effects in *Escherichia coli* dihydrofolate reductase. *Biochemistry* **34** (48): 15671–15680.
4. (a) Skinner, M.M. and Terwilliger, T.C. (1996). Potential use of additivity of mutational effects in simplifying protein engineering. *Proc. Natl. Acad. Sci. U.S.A.* **93** (20): 10753–10757. (b) Aita, T., Uchiyama, H., Inaoka, T. et al. (2000). Analysis of a local fitness landscape with a model of the rough Mt. Fuji-type landscape: application to prolyl endopeptidase and thermolysin. *Biopolymers* **54** (1): 64–79. (c) Lehmann, M., Loch, C., Middendorf, A. et al. (2002). The consensus concept for thermostability engineering of proteins: further proof of concept. *Protein Eng.* **15** (5): 403–411. (d) Sullivan, B.J., Nguyen, T., Durani, V. et al. (2012). Stabilizing proteins from sequence statistics: the interplay of conservation and correlation in triosephosphate isomerase stability. *J. Mol. Biol.* **420** (4-5): 384–399. (e) Declerck, N., Machius, M., Joyet, P. et al. (2002). Engineering the thermostability of *Bacillus licheniformis* α-amylase. *Biologia* **57**: 203–211.
5. Reetz, M.T. (2016). *Directed Evolution of Selective Enzymes: Catalysts for Organic Chemistry and Biotechnology*. Weinheim: Wiley-VCH.
6. Li, G. and Reetz, M.T. (2016). Learning lessons from directed evolution of stereoselective enzymes. *Org. Chem. Front.* **3** (10): 1350–1358.
7. (a) Bocola, M., Otte, N., Jaeger, K.E. et al. (2004). Learning from directed evolution: theoretical investigations into cooperative mutations in lipase enantioselectivity. *ChemBioChem* **5**

(2): 214–223. (b) Reetz, M.T., Puls, M., Carballeira, J.D. et al. (2007). Learning from directed evolution: further lessons from theoretical investigations into cooperative mutations in lipase enantioselectivity. *ChemBioChem* **8**: 106–112.

8. Reetz, M.T. and Garcia-Borràs, M. (2021). The unexplored importance of fleeting chiral intermediates in enzyme-catalyzed reactions. *J. Am. Chem. Soc.* **143** (37): 14939–14950.

9. Gumulya, Y., Sanchis, J., and Reetz, M.T. (2012). Many pathways in laboratory evolution can lead to improved enzymes: how to escape from local minima. *ChemBioChem* **13** (7): 1060–1066.

10. Weinreich, D.M., Delaney, N.F., DePristo, M.A., and Hartl, D.L. (2006). Darwinian evolution can follow only very few mutational pathways to fitter proteins. *Science* **312**: 111–114.

11. Qu, G., Li, A., Acevedo-Rocha, C.G. et al. (2020). The crucial role of methodology development in directed evolution of selective enzymes. *Angew. Chem. Int. Ed.* **59** (32): 13204–13231.

12. Sun, Z., Lonsdale, R., Kong, X.-D. et al. (2015). Reshaping an enzyme binding pocket for enhanced and inverted stereoselectivity: use of smallest amino acid alphabet in directed evolution. *Angew. Chem. Int. Ed.* **54**: 12410–12415.

13. Reetz, M.T., Prasad, S., Carballeira, J.D. et al. (2010). Iterative saturation mutagenesis accelerates laboratory evolution of enzyme stereoselectivity: rigorous comparison with traditional methods. *J. Am. Chem. Soc.* **132**: 9144–9152.

14. (a) Reetz, M.T., Wilensek, S., Zha, D., and Jaeger, K.-E. (2001). Directed evolution of an enantioselective enzyme through combinatorial multiple-cassette mutagenesis. *Angew. Chem. Int. Ed.* **40** (19): 3589–3591.

15. Recent examples of cooperative non-additive mutational effects in directed evolution of stereoselective enzymes: (a) Zhang, Z.-G., Lonsdale, R., Sanchis, J., and Reetz, M.T. (2014). Extreme synergistic mutational effects in the directed evolution of a Baeyer–Villiger monooxygenase as catalyst for asymmetric sulfoxidation. *J. Am. Chem. Soc.* **136**: 17262–17272. (b) Bartsch, S., Kourist, R., and Bornscheuer, U.T. (2008). Complete inversion of enantioselectivity towards acetylated tertiary alcohols by a double mutant of a *Bacillus subtilis* esterase. *Angew. Chem. Int. Ed.* **47**: 1508–1511. (c) Reetz, M.T., Soni, P., Acevedo, J.P., and Sanchis, J. (2009). Creation of an amino acid network of structurally coupled residues in the directed evolution of a thermostable enzyme. *Angew. Chem. Int. Ed.* **48**: 8268–8272. (d) Sandström, A.G., Wikmark, Y., Engström, K. et al. (2012). Combinatorial reshaping of the *Candida antarctica* lipase a substrate pocket for enantioselectivity using an extremely condensed library. *Proc. Natl. Acad. Sci. U.S.A.* **109** (1): 78–83.

16. Reetz, M.T. (2013). The importance of additive and non-additive mutational effects in protein engineering. *Angew. Chem. Int. Ed.* **52** (10): 2658–2666.

17. Bornscheuer, U.T. and Kazlauskas, R.J. (2005). *Hydrolases in Organic Synthesis: Regio- and Stereoselective Biotransformations*, 2e. Weinheim: Wiley-VCH.

18. Nardini, M., Lang, D.A., Liebeton, K. et al. (2000). Crystal structure of *Pseudomonas aeruginosa* lipase in the open conformation: the prototype for family I.1 of bacterial lipases. *J. Biol. Chem.* **275** (40): 31219–31225.

19. Reetz, M.T., Carballeira, J.D., Peyralans, J. et al. (2006). Expanding the substrate scope of enzymes: combining mutations obtained by CASTing. *Chem. Eur. J.* **12**: 6031–6038.

20. Reetz, M.T. and Sanchis, J. (2008). Constructing and analyzing the fitness landscape of an experimental evolutionary process. *ChemBioChem* **9**: 2260–2267.

21. Stemmer, W.P.C. (1994). Rapid evolution of a protein in vitro by DNA shuffling. *Nature* **370** (6488): 389–391.

22. Ma, F., Chung, M.T., Yao, Y. et al. (2018). Efficient molecular evolution to generate enantioselective enzymes using a dual-channel microfluidic droplet screening platform. *Nat. Commun.* **9** (1): 1030.

23. Becker, S., Höbenreich, H., Vogel, A. et al. (2008). Single-cell high-throughput screening to identify enantioselective hydrolytic enzymes. *Angew. Chem. Int. Ed.* **47** (27): 5085–5088.
24. (a) Yaginuma, K., Aoki, W., Miura, N. et al. (2019). High-throughput identification of peptide agonists against GPCRs by co-culture of mammalian reporter cells and peptide-secreting yeast cells using droplet microfluidics. *Sci. Rep.* **9** (1): 10920. (b) Yanakieva, D., Elter, A., Bratsch, J. et al. (2020). FACS-based functional protein screening via microfluidic co-encapsulation of yeast secretor and mammalian reporter cells. *Sci. Rep.* **10** (1): 10182.
25. Reetz, M.T., Bocola, M., Wang, L.W. et al. (2009). Directed evolution of an enantioselective epoxide hydrolase: uncovering the source of enantioselectivity at each evolutionary stage. *J. Am. Chem. Soc.* **131** (21): 7334–7343.
26. Reetz, M.T., Wang, L.W., and Bocola, M. (2006). Directed evolution of enantioselective enzymes: iterative cycles of CASTing for probing protein-sequence space. *Angew. Chem. Int. Ed.* **45** (8): 1236–1241.
27. Zou, J., Hallberg, B.M., Bergfors, T. et al. (2000). Structure of *Aspergillus niger* epoxide hydrolase at 1.8 Å resolution: implications for the structure and function of the mammalian microsomal class of epoxide hydrolases. *Structure* **8** (2): 111–122.
28. Morisseau, C., Archelas, A., Guitton, C. et al. (1999). Purification and characterization of a highly enantioselective epoxide hydrolase from *Aspergillus niger*. *Eur. J. Biochem.* **263** (2): 386–395.
29. Bruice, T.C. (2002). A view at the millennium: the efficiency of enzymatic catalysis. *Acc. Chem. Res.* **35** (3): 139–148.
30. Lind, M.E. and Himo, F. (2013). Quantum chemistry as a tool in asymmetric biocatalysis: limonene epoxide hydrolase test case. *Angew. Chem. Int. Ed.* **52** (17): 4563–4567.
31. König, G., Reetz, M.T., and Thiel, W. (2018). 1-Butanol as a solvent for efficient extraction of polar compounds from aqueous medium: theoretical and practical aspects. *J. Phys. Chem. B* **122** (27): 6975–6988.
32. Reviews of ene-reductases (enoate reductases): (a) Gatti, F.G., Parmeggiani, F., and Sacchetti, A. (2014). Synthetic strategies based on C=C bioreductions for the preparation of biologically active molecules. In: *Synthetic Methods for Biologically Active Molecules* (ed. E. Brenna), 27–47. Weinheim: Wiley-VCH. (b) Winkler, C.K., Tasnàdi, G., Clay, D. et al. (2012). Asymmetric bioreduction of activated alkenes to industrially relevant optically active compounds. *J. Biotechnol.* **162** (4): 381–389. (c) Amato, E.D. and Stewart, J.D. (2015). Applications of protein engineering to members of the old yellow enzyme family. *Biotechnol. Adv.* **33** (5): 624–631. (d) Zheng, L., Lin, J., Zhang, B. et al. (2018). Identification of a yeast old yellow enzyme for highly enantioselective reduction of citral isomers to (*R*)-citronellal. *Bioresour. Bioprocess.* **5** (1): 9. (e) Kumar, R.T., Sreedharan, R., Ghosh, P. et al. (2022). Ene-reductase: a multifaceted biocatalyst in organic synthesis. *Chem. Eur. J.* https://doi.org/10.1002/chem.202103949.
33. Reviews and key studies of protein engineering of ene-reductases: (a) Kille, S. and Reetz, M.T. (2013). Protein engineering: development of novel enzymes for the improved reduction of C=C double bonds. In: *Synthetic Methods for Biologically Active Molecules: Exploiting the Potential of Bioreductions* (ed. E. Brenna), 139–181. Weinheim: Wiley-VCH. (b) Toogood, H.S. and Scrutton, N.S. (2019). Discovery, characterisation, engineering and applications of ene reductases for industrial biocatalysis. *ACS Catal.* **8** (4): 3532–3549. (c) Deng, J., Yao, Z., Chen, K. et al. (2016). Towards the computational design and engineering of enzyme enantioselectivity: a case study by a carbonyl reductase from *Gluconobacter oxydans*. *J. Biotechnol.* **217**: 31–40. (d) Wang, T., Wei, R., Feng, Y. et al. (2021). Engineering of yeast old yellow enzyme OYE3

enables its capability discriminating of (*E*)-citral and (*Z*)-citral. *Molecules* **26** (16): 5040. (e) Parmeggiani, F., Brenna, E., Colombo, D. et al. (2022). "A Study in Yellow": investigations in the stereoselectivity of ene-reductases. *ChemBioChem* **23** (1): e202100445.

34. Lonsdale, R. and Reetz, M.T. (2015). Reduction of α,ß-unsaturated ketones by Old Yellow Enzymes: mechanistic insights from quantum mechanics/molecular mechanics calculations. *J. Am. Chem. Soc.* **137**: 14733–14742.

35. (a) Walton, A.Z., Conerly, W.C., Pompeu, Y. et al. (2011). Biocatalytic reductions of Baylis–Hillman adducts. *ACS Catal.* **1** (9): 989–993. (b) Pompeu, Y.A., Sullivan, B., and Stewart, J.D. (2013). X-ray crystallography reveals how subtle changes control the orientation of substrate binding in an alkene reductase. *ACS Catal.* **3** (10): 2376–2390. (c) Brenna, E., Crotti, M., Gatti, F.G. et al. (2015). Opposite enantioselectivity in the bioreduction of (*Z*)-beta-aryl-beta-cyanoacrylates mediated by the tryptophan 116 mutants of Old Yellow Enzyme 1: synthetic approach to (*R*)- and (*S*)-beta-aryl-gamma-lactams. *Adv. Synth. Catal.* **357** (8): 1849–1860.

36. Fox, K.M. and Karplus, P.A. (1994). Old Yellow Enzyme at 2-angstrom resolution – overall structure, ligand-binding, and comparison with related flavoproteins. *Structure* **2** (11): 1089–1105.

37. Padhi, S.K., Bougioukou, D.J., and Stewart, J.D. (2009). Site-saturation mutagenesis of tryptophan 116 of *Saccharomyces pastorianus* Old Yellow Enzyme uncovers stereocomplementary variants. *J. Am. Chem. Soc.* **131** (9): 3271–3280.

38. Fraaije, M.W. and Mattevi, A. (2000). Flavoenzymes: diverse catalysts with recurrent features. *Trends Biochem.* **25** (3): 126–132.

39. Horita, S., Kataoka, M., Kitamura, N. et al. (2015). An engineered Old Yellow Enzyme that enables efficient synthesis of (4*R*,6*R*)-actinol in a one-pot reduction system. *ChemBioChem* **16** (3): 440–445.

40. Horita, S., Kataoka, M., Kitamura, N. et al. (2019). Structural basis of different substrate preferences of two old yellow enzymes from yeasts in the asymmetric reduction of enone compounds. *Biosci. Biotechnol., Biochem.* **83** (3): 456–462.

41. (a) Fitzpatrick, T.B., Amrhein, N., and Macheroux, P. (2003). Characterization of YqjM, an Old Yellow Enzyme homolog from *Bacillus subtilis* involved in the oxidative stress response. *J. Biol. Chem.* **278** (22): 19891–19897. (b) Stuermer, R., Hauer, B., Hall, M., and Faber, K. (2007). Asymmetric bioreduction of activated C=C bonds using enoate reductases from the Old Yellow Enzyme family. *Curr. Opin. Chem. Biol.* **11** (2): 203–213.

42. (a) Ortiz de Montellano, P.R. (2005). *Cytochrome P450: Structure, Mechanism, and Biochemistry*, 3e. Berlin: Springer. (b) Isin, E.M. and Guengerich, F.P. (2007). Complex reactions catalyzed by cytochrome P450 enzymes. *Biochim. Biophys. Acta, Gen. Subj.* **1770** (3): 314–329. (c) Munro, A.W., Girvan, H.M., and McLean, K.J. (2007). Variations on a (t)heme – novel mechanisms, redox partners and catalytic functions in the cytochrome P450 superfamily. *Nat. Prod. Rep.* **24** (3): 585–609. (d) Ortiz de Montellano, P.R. (2010). Hydrocarbon hydroxylation by cytochrome P450 enzymes. *Chem. Rev.* **110** (2): 932–948. (e) Shaik, S., Cohen, S., Wang, Y. et al. (2010). P450 enzymes: their structure, reactivity, and selectivity-modeled by QM/MM calculations. *Chem. Rev.* **110** (2): 949–1017. (f) Urlacher, V.B. and Girhard, M. (2012). *Enzyme Catalysis in Organic Synthesis*, 3e, vol. **3** (ed. K. Drauz, H. Gröger and O. May), 1227–1267. Weinheim: Wiley-VCH. (g) Li, R.J., Tian, K., Li, X. et al. (2022). Engineering P450 monooxygenases for highly regioselective and active *p*-hydroxylation of *m*-alkylphenols. *ACS Catal.* **12**: 5939–5948. (h) Permana, D., Niesel, K., Ford, M.J., and Ichinose, H. (2022). Latent functions and applications of cytochrome P450 monooxygenases from *Thamnidium elegans*: a novel biocatalyst for 14α-hydroxylation of testosterone. *ACS Omega* **7** (16): 13932–13941.

43. Sun, Z. and Reetz, M.T. (2019). Controlling the regio- and stereoselectivity of cytochrome P450 monooxygenases by protein engineering. In: *Dioxygen-Dependent Heme Enzymes* (ed. M. Ikeda-Saito and E. Raven), 274–291. The Royal Society of Chemistry.
44. Hogg, J.A. (1992). Steroids, the steroid community, and Upjohn in perspective: a profile of innovation. *Steroids* **57** (12): 593–616.
45. (a) Fasan, R., Chen, M.M., Crook, N.C., and Arnold, F.H. (2007). Engineered alkane-hydroxylating cytochrome P450(BM3) exhibiting nativelike catalytic properties. *Angew. Chem. Int. Ed.* **46** (44): 8414–8418. (b) Xu, F., Bell, S.G., Lednik, J. et al. (2005). The heme monooxygenase cytochrome P450cam can be engineered to oxidize ethane to ethanol. *Angew. Chem. Int. Ed.* **44** (26): 4029–4032.
46. (a) Roiban, G.D. and Reetz, M.T. (2015). Expanding the toolbox of organic chemists: directed evolution of P450 monooxygenases as catalysts in regio- and stereoselective oxidative hydroxylation. *Chem. Commun.* **51** (12): 2208–2224. (b) Whitehouse, C.J., Bell, S.G., and Wong, L.L. (2012). P450(BM3) (CYP102A1): connecting the dots. *Chem. Soc. Rev.* **41** (3): 1218–1260. (c) Lewis, J.C., Coelho, P.S., and Arnold, F.H. (2011). Enzymatic functionalization of carbon-hydrogen bonds. *Chem. Soc. Rev.* **40** (4): 2003–2021. (d) Fasan, R. (2012). Tuning P450 enzymes as oxidation catalysts. *ACS Catal.* **2** (4): 647–666. (e) Khatri, Y., Hannemann, F., Girhard, M. et al. (2013). Novel family members of CYP109 from *Sorangium cellulosum* So ce56 exhibit characteristic biochemical and biophysical properties. *Biotechnol. Appl. Biochem.* **60** (1): 18–29. (f) Holtmann, D., Fraaije, M.W., Arends, I.W. et al. (2014). The taming of oxygen: biocatalytic oxyfunctionalisations. *Chem. Commun.* **50** (87): 13180–13200. (g) Ariyasu, S., Stanfield, J.K., Aiba, Y., and Shoji, O. (2020). Expanding the applicability of cytochrome P450s and other haemoproteins. *Curr. Opin. Chem. Biol.* **59**: 155–163. (h) Yang, Y. and Arnold, F.H. (2021). Navigating the unnatural reaction space: directed evolution of heme proteins for selective carbene nitrene transfer. *Acc. Chem. Res.* **54** (5): 1209–1225.
47. (a) Fischer, E. (1894). Einfluss der configuration auf die wirkung der enzyme. *Ber. Dtsch. Chem. Ges.* **27**: 2984–2993. (b) Lichtenthaler, F.W. (1995). 100 years "Schlüssel-Schloss-Prinzip": what made Emil Fischer use this analogy? *Angew. Chem. Int. Ed.* **33** (23–24): 2364–2374.
48. (a) Pauling, L. (1948). Nature of forces between large molecules of biological interest. *Nature* **161** (4097): 707–709. (b) Amyes, T.L. and Richard, J.P. (2013). Specificity in transition state binding: the Pauling model revisited. *Biochemistry* **52** (12): 2021–2035.
49. (a) Lonsdale, R., Harvey, J.N., and Mulholland, A.J. (2010). Inclusion of dispersion effects significantly improves accuracy of calculated reaction barriers for cytochrome P450 catalyzed reactions. *J. Phys. Chem. Lett.* **1** (21): 3232–3237. (b) Lonsdale, R., Harvey, J.N., and Mulholland, A.J. (2010). Compound I reactivity defines alkene oxidation selectivity in cytochrome P450cam. *J. Phys. Chem. B* **114** (2): 1156–1162.
50. (a) Kille, S., Zilly, F.E., Acevedo, J.P., and Reetz, M.T. (2011). Regio- and stereoselectivity of P450-catalysed hydroxylation of steroids controlled by laboratory evolution. *Nat. Chem.* **3** (9): 738–743. (b) Agudo, R., Roiban, G.D., and Reetz, M.T. (2012). Achieving regio- and enantioselectivity of P450-catalyzed oxidative CH activation of small functionalized molecules by structure-guided directed evolution. *ChemBioChem* **13** (10): 1465–1473. (c) Agudo, R., Roiban, G.-D., Lonsdale, R. et al. (2015). Biocatalytic route to chiral acyloins: P450-catalyzed regio- and enantioselective α-hydroxylation of ketones. *J. Org. Chem.* **80** (2): 950–956. (d) Roiban, G.D., Agudo, R., Ilie, A.

et al. (2014). CH-activating oxidative hydroxylation of 1-tetralones and related compounds with high regio- and stereoselectivity. *Chem. Commun.* **50** (92): 14310–14313. (e) Roiban, G.D., Agudo, R., and Reetz, M.T. (2014). Cytochrome P450 catalyzed oxidative hydroxylation of achiral organic compounds with simultaneous creation of two chirality centers in a single C–H activation step. *Angew. Chem. Int. Ed.* **53** (33): 8659–8663. (f) Ritter, C., Nett, N., Acevedo-Rocha, C.G. et al. (2015). Bioorthogonal enzymatic activation of caged compounds. *Angew. Chem. Int. Ed.* **54**: 13440–13443.

51. (a) Narhi, L.O. and Fulco, A.J. (1986). Characterization of a catalytically self-sufficient 119,000-dalton cytochrome P-450 monooxygenase induced by barbiturates in *Bacillus megaterium*. *J. Biol. Chem.* **261** (16): 7160–7169. (b) Munro, A.W., Leys, D.G., McLean, K.J. et al. (2002). P450 BM3: the very model of a modern flavocytochrome. *Trends Biochem. Sci.* **27** (5): 250–257. (c) Dubey, K.D., Wang, B., and Shaik, S. (2016). Molecular dynamics and QM/MM calculations predict the substrate-induced gating of cytochrome P450 BM3 and the regio- and stereoselectivity of fatty acid hydroxylation. *J. Am. Chem. Soc.* **138** (3): 837–845.

52. (a) Haines, D.C., Tomchick, D.R., Machius, M., and Peterson, J.A. (2001). Pivotal role of water in the mechanism of P450BM-3. *Biochemistry* **40** (45): 13456–13465. (b) Haines, D.C., Chen, B., Tomchick, D.R. et al. (2008). Crystal structure of inhibitor-bound P450 BM-3 reveals open conformation of substrate access channel. *Biochemistry* **47** (12): 3662–3670.

53. Weber, E., Seifert, A., Antonovici, M. et al. (2011). Screening of a minimal enriched P450 BM3 mutant library for hydroxylation of cyclic and acyclic alkanes. *Chem. Commun.* **47** (3): 944–946.

54. Carmichael, A.B. and Wong, L.L. (2001). Protein engineering of *Bacillus megaterium* CYP102. The oxidation of polycyclic aromatic hydrocarbons. *Eur. J. Biochem.* **268** (10): 3117–3125.

55. Acevedo-Rocha, C.G., Li, A., D'Amore, L. et al. (2021). Pervasive cooperative mutational effects on multiple catalytic enzyme traits emerge via long-range conformational dynamics. *Nat. Commun.* **12** (1): 1621.

56. (a) van der Meer, J.Y., Biewenga, L., and Poelarends, G.J. (2016). The generation and exploitation of protein mutability landscapes for enzyme engineering. *ChemBioChem* **17** (19): 1792–1799. (b) van der Meer, J.Y., Poddar, H., Baas, B.J. et al. (2016). Using mutability landscapes of a promiscuous tautomerase to guide the engineering of enantioselective Michaelases. *Nat. Commun.* **7**: 10911. (c) Hecht, M., Bromberg, Y., and Rost, B. (2013). News from the protein mutability landscape. *J. Mol. Biol.* **425** (21): 3937–3948.

57. Acevedo-Rocha, C.G., Gamble, C., Lonsdale, R. et al. (2018). P450-catalyzed regio- and diastereoselective steroid hydroxylation: efficient directed evolution enabled by mutability landscaping. *ACS Catal.* **8**: 3395–3410.

58. Zilly, F.E., Acevedo, J.P., Augustyniak, W. et al. (2011). Tuning a P450 enzyme for methane oxidation. *Angew. Chem. Int. Ed.* **50**: 2720–2724. Corrigendum: (2013) 52: 13503.

59. (a) Renz, M. and Meunier, B. (1999). 100 years of Baeyer–Villiger oxidations. *Eur. J. Org. Chem.* **4**: 737–750. (b) ten Brink, G.J., Arends, I.W., and Sheldon, R.A. (2004). The Baeyer–Villiger reaction: new developments toward greener procedures. *Chem. Rev.* **104** (9): 4105–4124.

60. Li, G., Garcia-Borràs, M., Fürst, M.J.L.J. et al. (2018). Overriding traditional electronic effects in biocatalytic Baeyer–Villiger reactions by directed evolution. *J. Am. Chem. Soc.* **140** (33): 10464–10472.

61. (a) Grein, F., Chen, A.C., Edwards, D., and Crudden, C.M. (2006). Theoretical and experimental studies on the Baeyer–Villiger oxidation of ketones and the effect of alpha-halo substituents. *J. Org. Chem.* **71** (3): 861–872. (b) Alvarez-Idaboy, J.R.,

Reyes, L., and Cruz, J. (2006). A new specific mechanism for the acid catalysis of the addition step in the Baeyer–Villiger rearrangement. *Org. Lett.* **8** (9): 1763–1765. (c) Alvarez-Idaboy, J.R., Reyes, L., and Mora-Diez, N. (2007). The mechanism of the Baeyer–Villiger rearrangement: quantum chemistry and TST study supported by experimental kinetic data. *Org. Biomol. Chem.* **5** (22): 3682–3689. (d) Vil', V.A., Dos Passos Gomes, G., Bityukov, O.V. et al. (2018). Interrupted Baeyer–Villiger rearrangement: building A stereoelectronic trap for the Criegee intermediate. *Angew. Chem. Int. Ed.* **57** (13): 3372–3376.

62. (a) Drauz, K., Gröger, H., and May, O. (2012). *Enzyme Catalysis in Organic Synthesis*, 3e (ed. K. Drauz, H. Gröger and O. May). Weinheim: Wiley-VCH. (b) Kayser, M.M. (2009). 'Designer reagents' recombinant microorganisms: new and powerful tools for organic synthesis. *Tetrahedron* **65** (5): 947–974. (c) Pazmiño, D.E.T., Dudek, H.M., and Fraaije, M.W. (2010). Baeyer–Villiger monooxygenases: recent advances and future challenges. *Curr. Opin. Chem. Biol.* **14** (2): 138–144. (d) Leisch, H., Morley, K., and Lau, P.C. (2011). Baeyer–Villiger monooxygenases: more than just green chemistry. *Chem. Rev.* **111** (7): 4165–4222.

63. Wang, J.B., Li, G., and Reetz, M.T. (2017). Enzymatic site-selectivity enabled by structure-guided directed evolution. *Chem. Commun.* **53** (28): 3916–3928.

64. Dong, Y., Li, T., Zhang, S. et al. (2022). Biocatalytic Baeyer–Villiger reactions: uncovering the source of regioselectivity at each evolutionary stage of a mutant with scrutiny of fleeting chiral intermediates. *ACS Catal.* **12**: 3669–3680.

65. Chen, X., Zhang, H., Maria-Solano, M.A. et al. (2019). Efficient reductive desymmetrization of bulky 1,3-cyclodiketones enabled by structure-guided directed evolution of a carbonyl reductase. *Nat. Catal.* **2**: 931–941.

66. Lerchner, A., Jarasch, A., Meining, W. et al. (2013). Crystallographic analysis and structure-guided engineering of NADPH-dependent *Ralstonia* sp. alcohol dehydrogenase toward NADH cosubstrate specificity. *Biotechnol. Bioeng.* **110** (11): 2803–2814.

67. (a) Zheng, M., Li, Y., Dong, W. et al. (2022). Depolymerase-catalyzed polyethylene terephthalate hydrolysis: a unified mechanism revealed by quantum mechanics/molecular mechanics analysis. *ACS Sustainable Chem. Eng.* **10**: 7341–7348. (b) Zhuang, J., Zhang, F., Tang, X. et al. (2022). Insights into the enzymatic catalytic mechanism of bCinS: the importance of protein conformational change. *Catal. Sci. Technol.* **12** (5): 1651–1662. (c) Soler, J., Gergel, S., Hammer, S., and Garcia-Borràs, M. (2022). Enzymatic control over reactive intermediates enables direct oxidation of alkenes to carbonyls by a P450 iron-oxo species. *ChemRxiv* https://doi.org/10.26434/chemrxiv-2022-6lpif.

9
Perspectives for Future Work

9.1
Introductory Remarks

Throughout this monograph, references to applications in synthetic organic chemistry, pharmaceutical chemistry, and biotechnology have already been made, and a few of them are summarized here in Table 9.1. Enantio-, diastereo-, and regio-selectivity, as well as activity stand at the heart of these disciplines, with enzymes occupying a rapidly growing place therein. In this chapter, some perspectives for future research in protein engineering of selective enzymes are presented, with the emphasis being on further kinds of applications in organic and pharmaceutical chemistry (Section 9.2) and biotechnology (Section 9.3). Section 9.4 focuses on patent issues, and Section 9.5 offers some final comments.

9.2
Extending Applications in Organic and Pharmaceutical Chemistry

Nature has evolved a huge collection of enzymes in all living matter, which are essential for a variety of different reasons. Important for this monograph are the enzymes found in bacteria, plants, and animals that selectively catalyze millions of different transformations and can be induced to serve chemists [10]. Enzymes have been classified by the Enzyme Commission (EC), part of which is shown in Table 9.2, a list of natural catalysts that kindle the imagination of what could possibly be accomplished with their use. These are the enzymes that nature utilizes in the biosynthesis of structurally simple and complex natural products. Especially organic chemists active in the field of natural products synthesis will realize that this is all that nature "needs," i.e. it does not require any of the chiral synthetic transition metal catalysts that leaders in organometallic chemistry have invented, nor any synthetic organocatalysts. Thus, researchers could, in principle, rely solely on enzymes as catalysts when facing challenging transformations, provided the methods of directed evolution and/or rational enzyme design are not forgotten. However, this is not the message of this monograph in the year 2023. *Rather, at the current state of synthetic organic and pharmaceutical*

Enzyme Engineering: Selective Catalysts for Applications in Biotechnology, Organic Chemistry, and Life Science, First Edition. Manfred T. Reetz, Zhoutong Sun, and Ge Qu.

chemistry, the two approaches to catalysis, enzymes and chiral synthetic catalysts, are complementary.

A central theme throughout this monograph is stereoselectivity, because this catalytic property is crucial for a multitude of synthetic efforts in organic and pharmaceutical chemistry. Chemists were not interested in directed evolution of enzyme stability, an attitude that slowly changed with the advent of protein engineering of enzyme selectivity (Section 1.2). It is abundantly clear that chiral therapeutic drugs are usually active in the desired manner only when one of the enantiomers is administered to the patient, the mirror-image compound causing adverse effects or imparting no biological influence [11]. For this reason, the US Food and Drug Administration (FDA) and health agencies of most other countries in the world, require toxicity and other biological data of both enantiomers of chiral drug candidate, and subsequently allow only the active form to be marketed. It is thus desirable, if not mandatory for a drug company, to have access to both stereoisomeric forms. A rare exception is a chiral drug called bicalutamide (Casodex®), which is administered as a racemate for treating prostrate cancer [12]. Both enantiomers show positive effects in studies, but the (*R*)-form is more active.

With regard to future research, Table 9.1 suggests that extensive research should be invested both in identifying new members of a given enzyme type by genome mining and in protein engineering whenever the wild type (WT) does not fulfill the requirements of a researcher. Recent reviews of protein engineering used in the production of pharmaceuticals and pharmaceutical intermediates are not only informative with respect to past achievements, they also identify particularly difficult problems to be solved by future research [13]. A few of these challenges include [14]:

- Practical implementation of any given artificial multistep pathway that can be envisioned by pharmaceutical chemists involves only unnatural substrates in designer cells.
- Metabolic engineering of secondary metabolites in a general manner.
- Designing clever cascade processes.
- Construction of cell-free systems as chemical factories.
- Solving the problem of some suboptimal expression systems.
- Using QM/MM to predict all steps in a stereoselective FRISM process so that only the final multi-mutational variant needs to be screened.
- Solving the problem of low coupling efficiency in some P450-catalyzed transformations and uncovering its structural basis.
- Solving the problem of simultaneous (rather than sequential) protein engineering of several essential catalytic traits such as selectivity, activity, and stability.
- Identifying essential protein–protein interactions in the development of novel therapeutic concepts.
- Developing novel screening or selection systems for enzymes for which such high-throughput assays have not yet been reported.
- Inducing high activity of hyperthermally stable enzymes at room temperature.

Table 9.1 Selected examples of therapeutic drugs, their access being possible by protein engineering of the final compound or of a chiral intermediate leading to the final pharmaceutical.

Compound	Comment	Enzyme	References
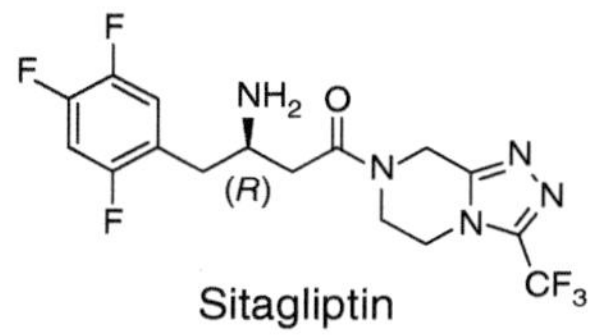 Sitagliptin	Diabetes treatment	Engineered transaminase (ATA-117)	[1]
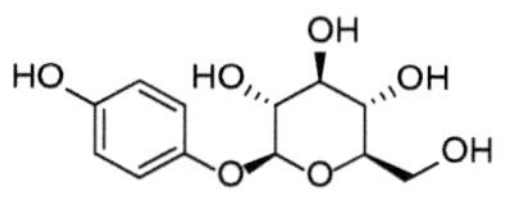 Arbutin	Treatment of skin diseases	Engineered P450-BM3	[2]
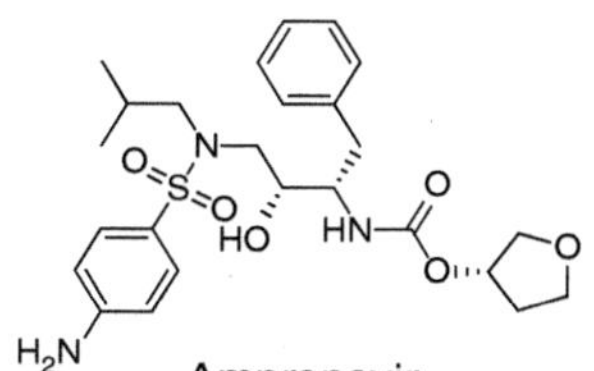 Amprenavir	Anti-HIV agent	Engineered alcohol dehydrogenase (TbSADH)	[3]
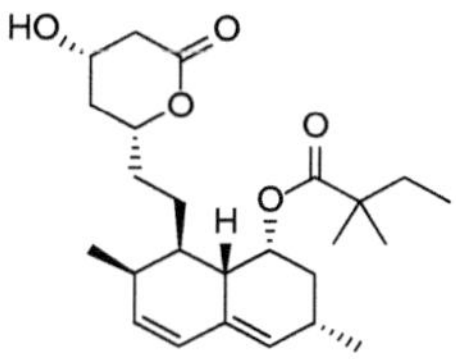 Simvastatin	Cholesterol-lowering agent	Engineered transesterase (LovD)	[4]
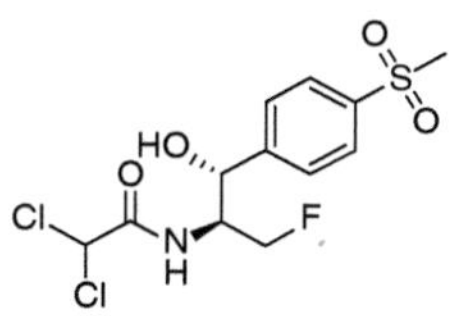 Florfenicol	Anti-bacteria	Engineered transketolase (TK) and ω-transaminase (TA)	[5]
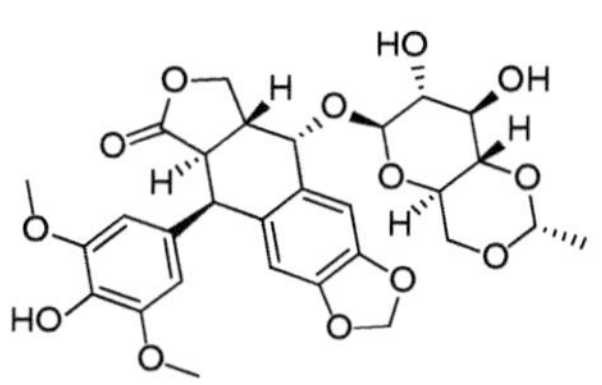 Etoposide	Treatment of malignancies	Engineered glycosyltransferase (UGT78D2)	[6]

Table 9.1 (Continued)

Compound	Comment	Enzyme	References
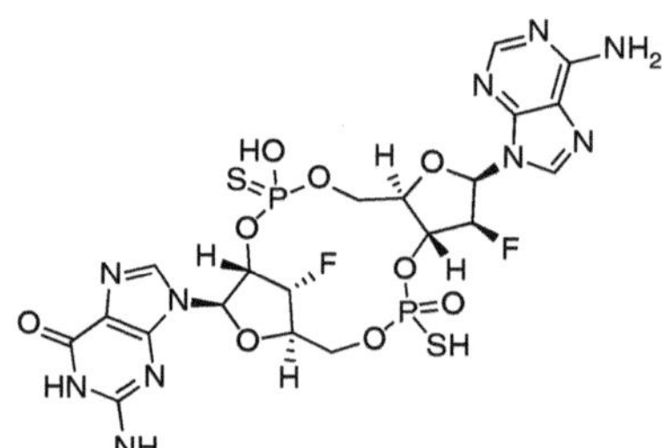 MK-1454	Immuno-oncology therapeutic	Engineered animal cyclic guanosine–adenosine synthase (cGAS)	[7]
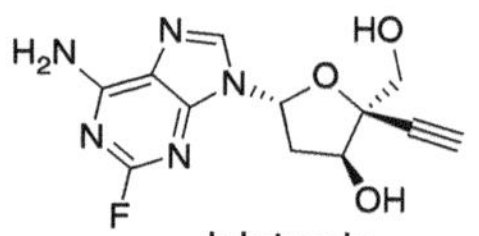 Islatravir	HIV treatment	Five engineered enzymes	[8]
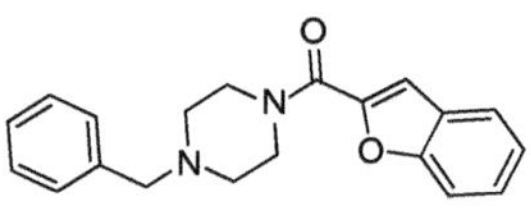 Befuraline	Antidepressant drug	Engineered carboxylic acid reductase (CAR*mm*-A)	[9]

Retrosynthetic methods and cascade construction using only enzymes or in combination with synthetic transition metal catalysts, or organocatalysts offer huge possibilities for future research [13j, 15]. The only limiting factor is the imagination of the researcher! This involves a transdisciplinary approach. Collaborating with protein process engineers [16], who consider the principles of green chemistry when constructing ecologically and economically viable production systems, is recommended.

Another point concerns the production of vaccines with the help of protein engineering, which has not been attempted very often. Rare examples concern the use of combinatorial active-site saturation test (CAST)/iterative saturation mutagenesis (ISM) or error-prone polymerase chain reaction (epPCR) in vaccine development [17]. Other studies of harnessing protein engineering for vaccine production also deserve mention [18]. The reader should be reminded that the first vaccine against a cancer disease was developed years ago by Harald zur Hausen in Germany, who obtained the Nobel Prize for Physiology and Medicine in 2008. As one of the pioneers specializing in onco-viruses [19], he discovered the papilloma virus and developed a vaccine against cancer in the cervix. Millions of women have profited from this basic research to the present day [20]. Fortunately, this type of research is continuing to the present day [21].

In the corona-pandemic, unexpected phenomena influencing the human immune system became apparent. New techniques for vaccine development

Table 9.2 Classification of enzymes by the Enzyme Commission (EC).

Class	Reaction catalyzed	Typical reaction	Enzyme example(s)
EC 1 Oxidoreductases	Oxidation/reduction reactions by transferring H and O atoms or electrons from one substance to the other	$AH_2 + B \rightleftharpoons A + BH_2$	Dehydrogenase, P450
EC 2 Transferases	Transferring a methyl-, acyl-, amino-, phosphate-, or other functional group from one substance to the other	$AX + B \rightleftharpoons A + BX$	Transaminase, kinase
EC 3 Hydrolases	Forming two products from one substrate via hydrolysis	$A - B + H_2O \rightleftharpoons AH + BOH$	Lipase, epoxide hydrolase
EC 4 Lyases	Cleaving C—C, C—O, C—N, or other bonds by non-hydrolytic addition	$A = B + X - Y \rightleftharpoons \underset{\mid \atop X}{A} - \underset{\mid \atop Y}{B}$	Decarboxylase
EC 5 Isomerases	Intramolecule rearrangement	$A \rightleftharpoons B$	Isomerase, mutase
EC 6 Ligases	Combining two substrates by synthesis of new C—O, C—S, C—N, or C—C bonds with the consumption of NTP	$A + B + NTP \rightleftharpoons A - B + NDP + P$; $A + B + NTP \rightleftharpoons A - B + NMP + PP$	Synthetase
EC 7 Translocases	reactions across membranes	$\underset{\text{(Side I)}}{AX + B} \parallel \rightleftharpoons A + X + \underset{\text{(Side II)}}{\parallel B}$	Transporter

based on RNA technology have been developed. Indeed, RNA technology can not only be applied to vaccine development but also in the treatment of diseases such as cancer [22]. In this pandemic, the continuous occurrence of new mutants constitutes a major threat to human health. More research is required to understand how they are formed. Are they formed sequentially as point mutations one by one, or does the simultaneous formation of several mutations with the formation of new multi-mutational variants cause the problem?

Relevant are also therapeutic drugs once patients have become victims of these viruses, for which several compounds have been suggested that were previously developed for other indications. An interesting further approach would be to subject these pharmaceuticals to oxidative hydroxylation using P450 mutants, in hope that such "metabolites" could be even more potent. A particularly promising recent development is the establishment of an antiviral defense strategy based on the construction of targeted siRNAs [23].

In another effort, increasing the utility of site-directed mutagenesis has been reported using the CASPR-Cas system, the aim being to reduce errors as much as possible [24]. Here as well, further innovative work could be rewarding.

Developing treatments for ailments that occur in the brain due to the malfunctioning of proteins is a particularly important venture for the future. Decades ago, Stanley B. Prusiner showed that this is involved in Jakob–Creutzfeldt syndrome, Alzheimer's and Parkinson's diseases, and others as well. Prusiner received the Nobel Prize in Physiology and Medicine in 1997 "for the discovery of prions – a new biological principle of infection" [25]. He celebrated his 80th birthday in May 2022, and is still scientifically active [26], as are hundreds of other researchers. Mutagenesis studies may help in identifying the details of the mechanism in some cases.

Finally, the often used term "biologics" needs mention, which includes vaccines, blood components, allergenics, gene therapies, somatic cells, tissues, and recombinant therapeutic proteins, comprising sugars, proteins, nucleic acids, or combinations thereof. So-called "biosimilars" offer innovative perspectives [27]. Monoclonal antibodies have also been developed, as in the treatment of rheumatoid arthritis, ulcerative colitis, Crohn's disease, and other ailments [28], and these efforts are continuing. Some aspects have already been highlighted in this book. A whole journal has been dedicated to these research areas, namely the open access journal *Biologics*. Some recent articles and reviews therein concern antimicrobial peptides for treating diabetes mellitus [29], nucleic acid vaccines [30], and peptides in COVID-19 clinical trials [31]. Here as well, mutagenesis studies could provide new insights.

9.3 Extending Applications in Biotechnology

Applications of protein engineering in biotechnology, which are not directly related to the production of pharmaceuticals or medical issues also constitute an exciting research field. Generally, the same approaches to semi-rational directed evolution and rational enzyme design are employed, with CAST/ISM and focused rational iterative site-specific mutagenesis (FRISM) often being the preferred options [14]. Producing monomers by protein engineering of appropriate enzymes for the polymer industry is one of several possibilities [32]. For example, subjecting P450 monooxygenases to rational enzyme design for obtaining mutants that catalyze the regioselective terminal oxidative hydroxylation of medium-length fatty acids is of great interest in the polyester and polyamide industries (Scheme 9.1) [33]. In future work, process engineering and upscale reactions would contribute to the significance of this concept.

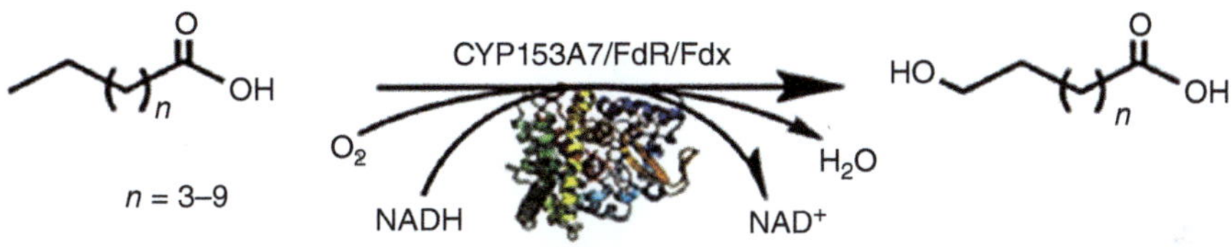

Scheme 9.1 Terminal hydroxylation of medium-chain fatty acids to ω-hydroxy fatty acids catalyzed by CYP153A7 in a regioselective fashion. Source: Ref. [33]/John Wiley & Sons.

Another important application concerns microbial degradation of plastics, not only for cleaning the oceans around the world. A bio-based circular plastic economy has been envisioned (Figure 9.1) [34].

The reader is referred to additional articles and reviews, which suggest that more intensive research in this area is necessary [35]. Only a few studies are cited here, many more contributions have actually appeared. An interesting research area concerns the use (valorization) of lignin. Globally, over 50 million tons of lignin is produced annually in paper and bioethanol factories [36]. Lignin valorization can have different forms and purposes (Figure 9.2a) [36, 37]. In some of these efforts, protein engineering was effectively employed to achieve maximum effects, as in the valorization of low-value lignin monomers with the formation of high-value monomeric compounds of interest in different areas including pharmaceutical intermediates [36]. With the aim of valorizing such low-value lignin monomers, an (*R*)-β-selective phenylalanine aminomutase from *Taxus chinensis* (TchPAM) was computationally designed to transform *trans*-*p*-hydroxycinnamic acid to (*R*)-ß-tyrosine with excellent enantiopurity (ee > 99%) (Figure 9.2b) [36]. In this endeavor, the PyRosetta program [38] was employed to rationally design the adjacent residues based on the binding energy calculation of small molecule–protein interactions (Figure 9.2c), which led to the identification of the best mutant Tyr424Asn [36].

In other future biotechnological applications, protein engineering using the appropriate mutagenesis method can be expected to play a key role.

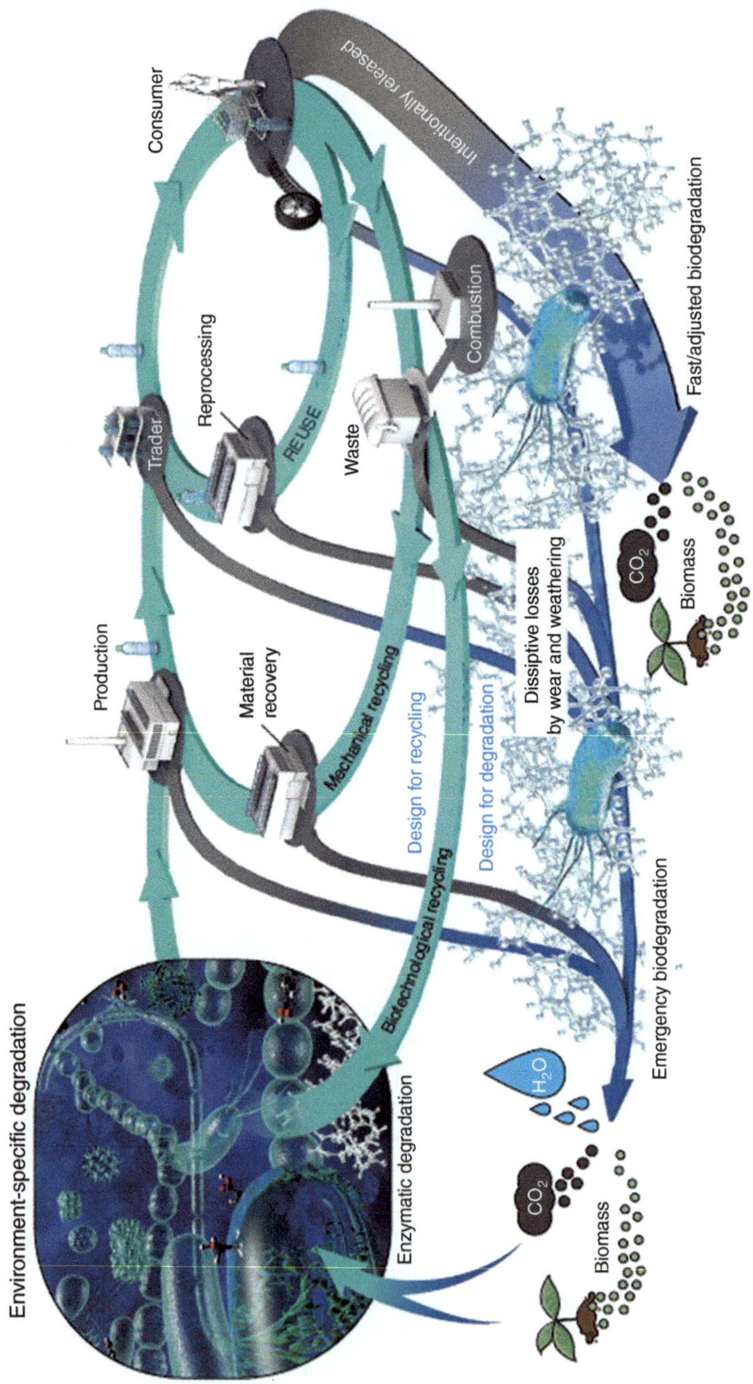

Figure 9.1 Illustration of the bio-based circular plastic economy in the future. Source: Wei et al. [34a]/with permission of Springer Nature.

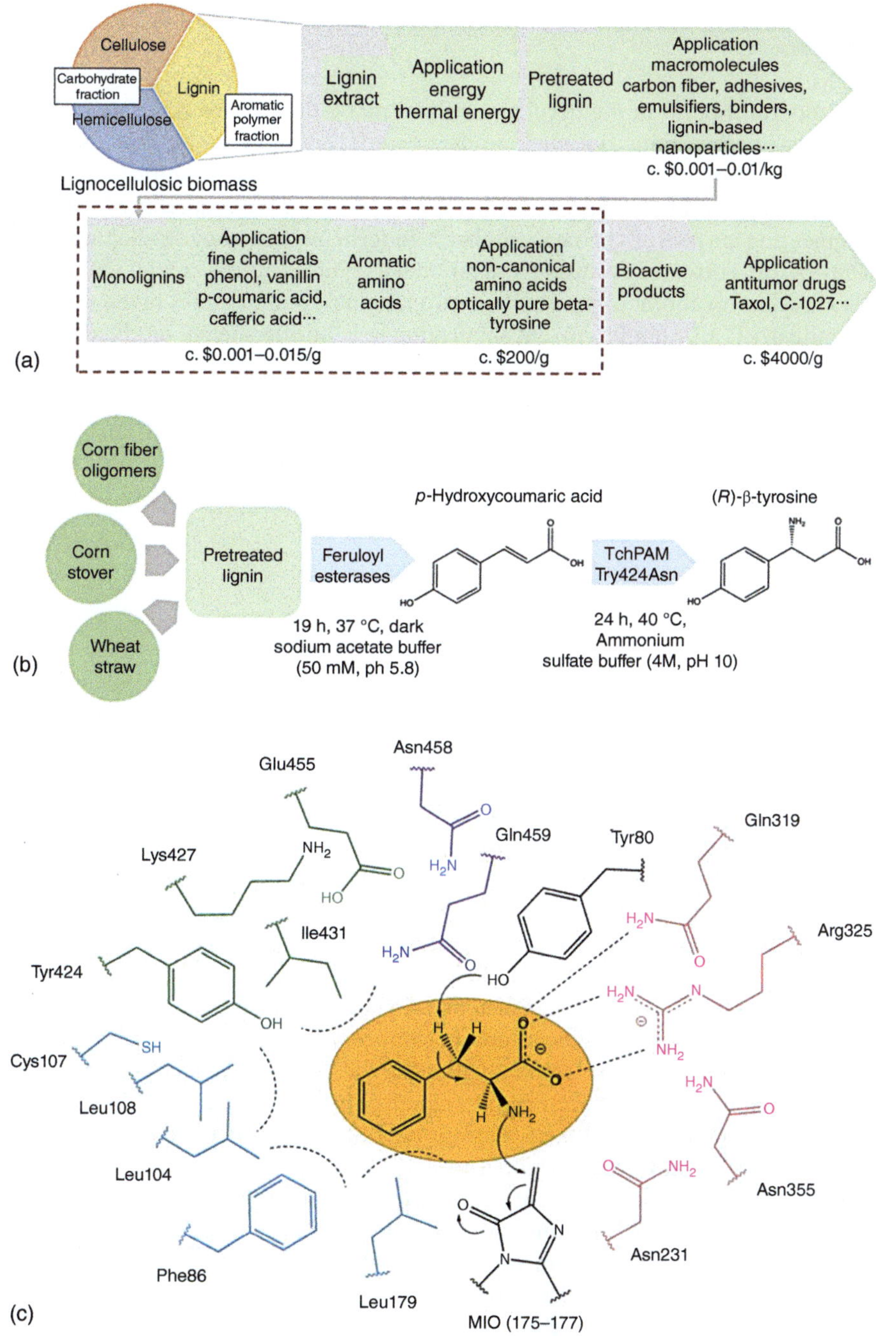

Figure 9.2 Computer-guided design of a phenylalanine aminomutase for the β-tyrosine production in lignin valorization. (a) A sustainable value chain in this process; (b) catalytic process for the biosynthesis of (*R*)-β-tyrosine from pretreated lignin; (c) interactions between (*S*)-α-phenylalanine and the neighbouring residues in TchPAM. Source: Ref. [36]/MDPI/CC BY 4.0.

9.4 Patent Issues

An important point in the development and marketing of new pharmaceutical products or other industrial compounds is speed [13i]. If marketing is delayed by as little as a few months, then the respective company suffers significant losses in collecting revenues. Thus, not only for this reason is, the reliability of protein engineering is a central theme in this book. Indeed, methodology development in semi-rational directed evolution and rational enzyme design is crucial [39]. This also touches on another important question, namely patent issues in the chemical industry [13i]. Should a patent application be submitted before production, or should a company proceed with production without a patent and without making the gene sequence of an engineered mutant available? Many companies around the world choose the latter option; academic researchers who consult with pharmaceutical and biotech companies know that this is often the case. It also means that some advances in protein engineering may go unnoticed.

9.5 Final Comments

The authors of this monograph are convinced that those readers, who have "made it through" the nine chapters have learned a lot. They are well suited to tackle the future problems in biology and chemical biology. Current techniques and strategies of protein engineering [14, 39] are quite different from the state of affairs 20–25 years ago, which at the time initiated a new era concerning the directed evolution of stereoselective enzymes as catalysts in organic chemistry [40]. Advanced directed evolution methods and rational enzyme design techniques have merged [41]. Throughout this book, the crucial role of methodology development is evident, with enantio-, diastereo-, and regio-selectivity playing dominant roles [14, 39, 41, 42]. Reliability and speed in controlling these catalytic properties have transformed protein engineering into a tool that is now routinely applied in organic and pharmaceutical chemistry and biotechnology. Further methodology development will make these tools even sharper.

The insights gained concerning genes, enzymes, and other proteins enable researchers in future efforts to address new questions that are on the horizon, or possibly not even imagined to date. One of many examples concerns the identification and treatment of ailments in the brain [25]. Unveiling the secrets of the human immune system and its manipulation is perhaps the greatest challenge, which can be expected to gain increased attention (and more financial support) in the future. Creative chemical biologists and curious organic chemists, certainly those familiar with the techniques and methods described in this monograph, are well suited to address complex questions beyond their present research interests. Ethical issues must not be ignored, e.g. when assessing the quality of scientific contributions, as in the field of directed evolution [43], and when applying gene editing using the CRISPR-technology in clinical treatments [44]. Biology and chemical biology directed toward enabling sustainable and economically viable systems for mankind are the dominating themes in the twenty-first century, an exciting time for researchers active in basic and applied science in a transdisciplinary manner.

References

1. Savile, C.K., Janey, J.M., Mundorff, E.C. et al. (2010). Biocatalytic asymmetric synthesis of chiral amines from ketones applied to sitagliptin manufacture. *Science* **329** (5989): 305–309.
2. (a) Zhou, H., Wang, B., Wang, F. et al. (2019). Chemo- and regioselective dihydroxylation of benzene to hydroquinone enabled by engineered cytochrome P450 monooxygenase. *Angew. Chem. Int. Ed.* **58** (3): 764–768. (b) Zhou, H., Zhao, J., Li, A., and Reetz, M.T. (2019). Chemical and biocatalytic routes to arbutin. *Molecules* **24** (18): 3303.
3. Sun, Z., Lonsdale, R., Ilie, A. et al. (2016). Catalytic asymmetric reduction of difficult-to-reduce ketones: triple code saturation mutagenesis of an alcohol dehydrogenase. *ACS Catal.* **6** (3): 1598–1605.
4. Jiménez-Osés, G., Osuna, S., Gao, X. et al. (2014). The role of distant mutations and allosteric regulation on LovD active site dynamics. *Nat. Chem. Biol.* **10** (6): 431–436.
5. Liu, Q., Xie, X.Y., Tang, M.C. et al. (2021). One-pot asymmetric synthesis of an aminodiol intermediate of florfenicol using engineered transketolase and transaminase. *ACS Catal.* **11**: 7477–7488.
6. Jia, K.Z., Zhu, L.W., Qu, X.D. et al. (2019). Enzymatic O-glycosylation of etoposide aglycone by exploration of the substrate promiscuity for glycosyltransferases. *ACS Synth. Biol.* **8**: 2718–2725.
7. McIntosh, J.A., Liu, Z., Andresen, B.M. et al. (2022). A kinase-cGAS cascade to synthesize a therapeutic STING activator. *Nature* **603** (7901): 439–444.
8. Huffman, M.A., Fryszkowska, A., Alvizo, O. et al. (2019). Design of an in vitro biocatalytic cascade for the manufacture of islatravir. *Science* **366** (6470): 1255–1259.
9. Lubberink, M., Schnepel, C., Citoler, J. et al. (2019). Biocatalytic monoacylation of symmetrical diamines and its application to the synthesis of pharmaceutically relevant amides. *ACS Catal.* **10** (17): 10005–10009.
10. (a) Drauz, K., Gröger, H., and May, O. (2012). *Enzyme Catalysis in Organic Synthesis*, 3e. Weinheim: Wiley-VCH. (b) Faber, K. (2011). *Biotransformations in Organic Chemistry*, 6e. Heidelberg: Springer. (c) Liese, A., Seelbach, K., and Wandrey, C. (2006). *Industrial Biotransformations*, 2e. Weinheim: Wiley-VCH. (d) Reetz, M.T. (2016). *Directed Evolution of Selective Enzymes: Catalysts for Organic Chemistry and Biotechnology.* Weinheim: Wiley-VCH.
11. Crossley, R. (1995). *Chirality and Biological Activity of Drugs.* Boca Raton/New York: CRC Press.
12. Nageswara Rao, R., Narasa Raju, A., and Nagaraju, D. (2006). An improved and validated LC method for resolution of bicalutamide enantiomers using amylose tris-(3,5-dimethylphenylcarbamate) as a chiral stationary phase. *J. Pharm. Biomed. Anal.* **42** (3): 347–353.
13. (a) Li, G., Wang, J.B., and Reetz, M.T. (2018). Biocatalysts for the pharmaceutical industry created by structure-guided directed evolution of stereoselective enzymes. *Bioorg. Med. Chem.* **26** (7): 1241–1251. (b) Sun, H., Zhang, H., Ang, E.L., and Zhao, H. (2018). Biocatalysis for the synthesis of pharmaceuticals and pharmaceutical intermediates. *Bioorg. Med. Chem.* **26** (7): 1275–1284. (c) Zetzsche, L.E., Chakrabarty, S., and Narayan, A.R.H. (2022). The transformative power of biocatalysis in convergent synthesis. *J. Am. Chem. Soc.* **144** (12): 5214–5225.

(d) Di Nardo, G. and Gilardi, G. (2020). Natural compounds as pharmaceuticals: the key role of cytochromes P450 reactivity. *Trends Biochem. Sci.* **45** (6): 511–525. (e) Tobin, P.H., Richards, D.H., Callender, R.A., and Wilson, C.J. (2014). Protein engineering: a new frontier for biological therapeutics. *Curr. Drug Metab.* **15** (7): 743–756. (f) Simić, S., Zukić, E., Schmermund, L. et al. (2022). Shortening synthetic routes to small molecule active pharmaceutical ingredients employing biocatalytic methods. *Chem. Rev.* **122** (1): 1052–1126. (g) Hauer, B. (2020). Embracing nature's catalysts: a viewpoint on the future of biocatalysis. *ACS Catal.* **10** (15): 8418–8427. (h) Hanefeld, U., Hollmann, F., and Paul, C.E. (2022). Biocatalysis making waves in organic chemistry. *Chem. Soc. Rev.* **51** (2): 594–627. (i) Truppo, M.D. (2017). Biocatalysis in the pharmaceutical industry: the need for speed. *ACS Med. Chem. Lett.* **8** (5): 476–480. (j) Bell, E.L., Finnigan, W., France, S.P. et al. (2021). Biocatalysis. *Nat. Rev. Methods Primers* **1**: 46.

14. (b) Reetz, M.T. (2013). Biocatalysis in organic chemistry and biotechnology: past, present, and future. *J. Am. Chem. Soc.* **135** (34): 12480–12496.

15. (a) Turner, N.J. and O'Reilly, E. (2013). Biocatalytic retrosynthesis. *Nat. Chem. Biol.* **9** (5): 285–288. (b) Liu, Y., Wang, W., and Zeng, A.P. (2022). Biosynthesizing structurally diverse diols via a general route combining oxidative and reductive formations of OH-groups. *Nat. Commun.* **13** (1): 1595. (c) Muschiol, J., Peters, C., Oberleitner, N. et al. (2015). Cascade catalysis – strategies and challenges en route to preparative synthetic biology. *Chem. Commun.* **51** (27): 5798–5811. (d) Liese, A. and Pesci, L. (2015). Enzyme classification and nomenclature and biocatalytic retrosynthesis. In: *Science of Synthesis, 1: Biocatalysis in Organic Synthesis* (ed. K. Faber, W.-D. Fessner and N. Turner), 41–74. Thieme. (e) Li, A., Ilie, A., Sun, Z. et al. (2016). Whole-cell catalyzed multiple regio- and stereoselective functionalization in cascade reactions enabled by directed evolution. *Angew. Chem. Int. Ed.* **55**: 12026–12029.

16. (a) Woodley, J.M. (2020). New frontiers in biocatalysis for sustainable synthesis. *Curr. Opin. Green Sustainable Chem.* **21**: 22–26. (b) Sheldon, R.A., Brady, D., and Bode, M.L. (2020). The Hitchhiker's guide to biocatalysis: recent advances in the use of enzymes in organic synthesis. *Chem. Sci.* **11** (10): 2587–2605. (c) Finnigan, W., Hepworth, L.J., Flitsch, S.L., and Turner, N.J. (2021). RetroBioCat as a computer-aided synthesis planning tool for biocatalytic reactions and cascades. *Nat. Catal.* **4** (2): 98–104.

17. (a) Ihssen, J., Haas, J., Kowarik, M. et al. (2015). Increased efficiency of *Campylobacter jejuni N*-oligosaccharyltransferase PglB by structure-guided engineering. *Open Biol.* **5** (4): 140227. (b) Ye, J., Wen, F., Xu, Y. et al. (2015). Error-prone PCR-based mutagenesis strategy for rapidly generating high yield influenza vaccine candidates. *Virology* **482**: 234–243. (c) Horiya, S., MacPherson, I.S., and Krauss, I.J. (2014). Recent strategies targeting HIV glycans in vaccine design. *Nat. Chem. Biol.* **10** (12): 990–999.

18. (a) Campeotto, I., Goldenzweig, A., Davey, J. et al. (2017). One-step design of a stable variant of the malaria invasion protein RH5 for use as a vaccine immunogen. *Proc. Natl. Acad. Sci. U.S.A.* **114** (5): 998–1002. (b) Soltan, M.A., Eldeen, M.A., Elbassiouny, N. et al. (2021). In silico designing of a multitope vaccine against *Rhizopus microspores* with potential activity against other mucormycosis causing fungi. *Cells* **10** (11): 3014. (c) Bar-Peled, Y., Huang, J., Nuñez, I.A. et al. (2019). Structural and antigenic characterization of a computationally-optimized H5 hemagglutinin influenza vaccine. *Vaccine* **37** (41): 6022–6029.

19. (a) Dürst, M., Gissmann, L., Ikenberg, H., and zur Hausen, H. (1983). A papillomavirus DNA from a cervical carcinoma and its prevalence in cancer biopsy samples from different geographic regions. *Proc. Natl. Acad. Sci. U.S.A.* **80** (12): 3812–3815. (b) Boshart, M., Gissmann, L.,

Ikenberg, H. et al. (1984). A new type of papillomavirus DNA, its presence in genital cancer biopsies and in cell lines derived from cervical cancer. *EMBO J.* **3** (5): 1151–1157.

20. de Villiers, E.-M. and zur Hausen, H. (ed.) (2009). *The Still Elusive Human Pathogens*. Berlin: Springer.
21. Chang, Y., Moore, P.S., and Weiss, R.A. (2017). Human oncogenic viruses: nature and discovery. *Philos. Trans.* **372**: 1–9.
22. Waters, M.D., Dhawan, A., Marrs, T. et al. (ed.) (2022). *The Coronavirus Pandemic and the Future*. Royal Society of Chemistry.
23. Traube, F.R., Stern, M., Tölke, A.J. et al. (2022). Suppression of SARS-CoV-2 replication with stabilized and Click-chemistry modified siRNAs. *Angew. Chem. Int. Ed.* https://doi.org/10.1002/anie.202204556.
24. She, W., Ni, J., Shui, K. et al. (2018). Rapid and error-free site-directed mutagenesis by a PCR-free in vitro CRISPR/Cas9-mediated mutagenic system. *ACS Synth. Biol.* **7** (9): 2236–2244.
25. (a) Prusiner, S.B. (1998). Prions. *Proc. Natl. Acad. Sci. U.S.A.* **95** (23): 13363–13383. (b) Prusiner, S.B. (2014). *Madness and Memory. The Discovery of Prions – A New Biological Principle of Disease*. New Haven: Yale University Press.
26. (a) Lester, E., Ooi, F.K., Bakkar, N. et al. (2021). Tau aggregates are RNA-protein assemblies that mislocalize multiple nuclear speckle components. *Neuron* **109** (10): 1675–1691.e9. (b) Ayers, J.I., Lee, J., Monteiro, O. et al. (2022). Different α-synuclein prion strains cause dementia with Lewy bodies and multiple system atrophy. *Proc. Natl. Acad. Sci. U.S.A.* **119** (6): e2113489119.
27. Niazi, S.K. (2022). The coming of age of biosimilars: a personal perspective. *Biologics* **2**: 107–127.
28. Walsh, G. (2018). Biopharmaceutical benchmarks 2018. *Nat. Biotechnol.* **36**: 1136–1145.
29. Depta, J., Małkowska, P., Wysokińska, M. et al. (2022). Therapeutic role of antimicrobial peptides in diabetes mellitus. *Biologics* **2**: 92–106.
30. Chavda, V.P., Hossain, M.K., Beladiya, J., and Apostolopoulos, V. (2021). Nucleic acid vaccines for COVID-19: a paradigm shift in the vaccine development arena. *Biologics* **1**: 337–356.
31. Hilpert, K. (2021). Peptides in COVID-19 clinical trials—a snapshot. *Biologics* **1**: 300–311.
32. (a) Parra, L.P., Acevedo, J.P., and Reetz, M.T. (2015). Directed evolution of phenylacetone monooxygenase as an active catalyst for the Baeyer–Villiger conversion of cyclohexanone to caprolactone. *Biotechnol. Bioeng.* **112** (7): 1354–1364. (b) Srinivasamurthy, V., Boettcher, D., Engel, J. et al. (2020). A whole-cell process for the production of ε-caprolactone in aqueous media. *Process Biochem.* **88** (1): 22–30.
33. Dong, Y.L., Chong, G.G., Li, C.X. et al. (2022). Carving the active site of CYP153A7 monooxygenase for improving terminal hydroxylation of medium-chain fatty acids. *ChemBioChem* **23** (9): e202200063.
34. (a) Wei, R., Tiso, T., Bertling, J. et al. (2020). Possibilities and limitations of biotechnological plastic degradation and recycling. *Nat. Catal.* **3**: 867–871. (b) Tiso, T., Winter, B., Wei, R. et al. (2022). The metabolic potential of plastics as biotechnological carbon sources – review and targets for the future. *Metab. Eng.* **71**: 77–98.
35. (a) Tournier, V., Topham, C.M., Gilles, A. et al. (2020). An engineered PET depolymerase to break down and recycle plastic bottles. *Nature* **580** (7802): 216–219. (b) Lu, H., Diaz, D.J., Czarnecki, N.J. et al. (2022). Machine learning-aided engineering of hydrolases for PET depolymerization. *Nature* **604** (7907): 662–667. (c) Chen, C.C., Dai, L., Ma, L., and Guo, R.T. (2020). Enzymatic degradation of plant biomass and synthetic polymers. *Nat. Rev. Chem.* **4**: 114–126. (d) Kim, J., Hwang, S., and Lee, S.M. (2022). Metabolic engineering for the utilization of carbohydrate portions of lignocellulosic biomass. *Metab. Eng.* **71**: 2–12. (e) Brott, S., Pfaff, L., Schuricht, J. et al. (2021). Engineering and evaluation of thermostable *Is*PETase

variants for PET degradation. *Eng. Life Sci.* **22** (3–4): 192–203.

36. Peng, F., Aliyu, H., Delavault, A. et al. (2021). Computational-designed enzyme for β-tyrosine production in lignin valorization. *Catalysts* **11**: 1310.

37. (a) Mahmood, N., Yuan, Z.S., Schmidt, J., and Xu, C. (2016). Depolymerization of lignins and their applications for the preparation of polyols and rigid polyurethane foams: a review. *Renewable Sustainable Energy Rev.* **60**: 317–329. (b) Yuliestyan, A., García-Morales, M., Moreno, E. et al. (2017). Assessment of modified lignin cationic emulsifier for bitumen emulsions used in road paving. *Mater. Des.* **131**: 242–251. (c) Xie, S., Li, Q., Karki, P. et al. (2017). Lignin as renewable and superior asphalt binder modifier. *ACS Sustainable Chem. Eng.* **5**: 2817–2823. (d) Chauhan, P.S. (2020). Lignin nanoparticles: eco-friendly and versatile tool for new era. *Bioresour. Technol. Rep.* **9**: 100374. (e) Becker, J. and Wittmann, C. (2019). A field of dreams: lignin valorization into chemicals, materials, fuels, and health-care products. *Biotechnol. Adv.* **37**: 107360.

38. Chaudhury, S., Lyskov, S., and Gray, J.J. (2010). PyRosetta: a script-based interface for implementing molecular modeling algorithms using Rosetta. *Bioinformatics* **26**: 689–691.

39. Qu, G., Li, A., Acevedo-Rocha, C.G. et al. (2020). The crucial role of methodology development in directed evolution of selective enzymes. *Angew. Chem. Int. Ed.* **59** (32): 13204–13231.

40. Reetz, M.T., Zonta, A., Schimossek, K. et al. (1997). Creation of enantioselective biocatalysts for organic chemistry by in vitro evolution. *Angew. Chem. Int. Ed.* **36** (24): 2830–2832.

41. (a) Reetz, M.T. (2022). Making enzymes suitable for organic chemistry by rational protein design. *ChemBioChem* **7**: e202200049. (b) Reetz, M.T. (2022). Witnessing the birth of directed evolution of stereoselective enzymes as catalysts in organic chemistry. *Adv. Synth. Catal.* https://doi.org/10.1002/adsc.202200466.

42. Acevedo-Rocha, C.G., Hollmann, F., Sanchis, J., and Sun, Z. (2020). A pioneering career in catalysis: Manfred T. Reetz. *ACS Catal.* **10**: 15123–15139.

43. Reetz, M.T. (2021). Effective mentoring and the problem of assessing quality in science. *Helv. Chim. Acta* **104**: e2100124.

44. Doudna, J.A. and Sternberg, S.H. (2018). *A Crack in Creation: Gene Editing and the Unthinkable Power to Corol Evolution.* Boston: Mariner Books.

Index

Enzyme Engineering: Selective Catalysts for Applications in Biotechnology, Organic Chemistry, and Life Science, First Edition. Manfred T. Reetz, Zhoutong Sun, and Ge Qu.